Sensors Applications
Volume 3
Sensors in Medicine and Health Care

Sensors Applications

H.K. Tönshoff, I. Inasaki

Sensors in Manufacturing
2001
ISBN 3-527-29558-5

O. Gassmann, H. Meixner

Sensors in Intelligent Buildings
2001
ISBN 3-527-29557-7

J. Marek, H.-P. Trah, Y. Suzuki, I. Yokomori

Sensors for Automotive Technology
2003
ISBN 3-527-29553-4

G. Tschulena, A. Lahrmann

Sensors in Household Appliances
2003
ISBN 3-527-30362-6

Related Wiley-VCH titles:

W. Göpel, J. Hesse, J.N. Zemel

Sensors Vol. 1–9

ISBN 3-527-26538-4

H. Baltes, G.K. Fedder, J.G. Korvink

Sensors Update

ISSN 1432-2404

T.C. Pearce, S.S. Schiffman, J.W. Gardner, H.T. Nagle

Handbook of Machine Olfaction
Electronic Nose Technology

ISBN 3-527-30358-8

H. Baltes, O. Brand, G.K. Fedder, C. Hierold, J.G. Korvink, O. Tabata

Advanced Micro & Nanosystems

Vol. 1

Enabling Technology for MEMS and Nanodevices
2004
ISBN 3-527-30746-X

Vol. 2
O. Brand, G.K. Fedder (volume eds.)

CMOS – MEMS
2005
ISBN 3-527-31080-0

Sensors Applications
Volume 3

Sensors in Medicine and Health Care

Edited by
P. Å. Öberg, T. Togawa, F. A. Spelman

Series Editors:
J. Hesse, J. W. Gardner, W. Göpel

WILEY-
VCH

WILEY-VCH Verlag GmbH & Co. KGaA

Series Editors

Prof. Dr. J. Hesse
formerly Carl Zeiss Jena
Bismarckallee 32c
14193 Berlin
Germany

Prof. J. W. Gardner
University of Warwick
Division of Electrical & Electronic Engineering
Coventry CV4 7AL
United Kingdom

Prof. Dr. W. Göpel †
Institut für Physikalische
und Theoretische Chemie
Universität Tübingen
Auf der Morgenstelle 8
72076 Tübingen
Germany

Volume Editors

Prof. P. Åke Öberg
Dept. of Biomedical Engineering
Linköpings Universitet
58185 Linköping
Sweden

Prof. Francis A. Spelman
Dept. of Bioengineering
University of Washington
Box 357962
Seattle, WA 98195-7962
USA
Advanced Cochlear Systems
Snoqualmie, WA, USA

Prof. Tatsuo Togawa
School of Human Sciences
Waseda University
2-579-15, Mikajima, Tokorozawa-shi
Saitama 359-1192
Japan

Library of Congress Card No.: applied for

British Library Cataloguing-in-Publication Data
A catalogue record for this book is available from the British Library.

Bibliographic information published by Die Deutsche Bibliothek
Die Deutsche Bibliothek lists this publication in the Deutsche Nationalbibliografie; detailed bibliographic data is available in the Internet at <http://dnb.ddb.de>

Printed in the Federal Republic of Germany
Printed on acid-free paper

Composition K+V Fotosatz GmbH, Beerfelden
Printing Betz-Druck GmbH, Darmstadt
Bookbinding Buchbinderei J. Schäffer GmbH & Co. KG, Grünstadt

ISBN 3-527-29556-9

Preface to the Series

As the use of microelectronics became increasingly indispensable in measurement and control technology, so there was an increasing need for suitable sensors. From the mid-Seventies onwards sensors technology developed by leaps and bounds and within ten years had reached the point where it seemed desirable to publish a survey of what had been achieved so far. At the request of publishers WILEY-VCH, the task of editing was taken on by Wolfgang Göpel of the University of Tübingen (Germany), Joachim Hesse of Carl Zeiss (Germany) and Jay Zemel of the University of Philadelphia (USA), and between 1989 and 1995 a series called *Sensors* was published in 8 volumes covering the field to date. The material was grouped and presented according to the underlying physical principles and reflected the degree of maturity of the respective methods and products. It was written primarily with researchers and design engineers in mind, and new developments have been published each year in one or two supplementary volumes called *Sensors Update.*

Both the publishers and the series editors, however, were agreed from the start that eventually sensor users would want to see publications only dealing with their own specific technical or scientific fields. Sure enough, during the Nineties we saw significant developments in applications for sensor technology, and it is now an indispensable part of many industrial processes and systems. It is timely, therefore, to launch a new series, *Sensors Applications.* WILEY-VCH again commissioned Wolfgang Göpel and Joachim Hesse to plan the series, but sadly Wolfgang Göpel suffered a fatal accident in June 1999 and did not live to see publication. We are fortunate that Julian Gardner of the University of Warwick has been able to take his place, but Wolfgang Göpel remains a co-editor posthumously and will not be forgotten.

The series of *Sensors Applications* will deal with the use of sensors in the key technical and economic sectors and systems: *Sensors in Manufacturing, Intelligent Buildings, Medicine and Health Care, Automotive Technology, Aerospace Technology, Environmental Technology* and *Household Appliances.* Each volume will be edited by specialists in the field. Individual volumes may differ in certain respects as dictated by the topic, but the emphasis in each case will be on the process or system in question: which sensor is used, where, how and why, and exactly what the benefits are to the user. The process or system itself will of course be outlined and the volume will close with a look ahead to likely developments and applications in the future. Actual sensor functions will only be described where it seems necessary for an understanding of how they relate to the process or system. The basic principles can always be found in the earlier series of *Sensors* and *Sensors Update*.

The series editors would like to express their warm appreciation in the colleagues who have contributed their expertise as volume editors or authors. We are deeply indebted to the publisher and would like to thank in particular Dr. Peter Gregory, Dr. Jörn Ritterbusch and Dr. Claudia Barzen for their constructive assistance both with the editorial detail and the publishing venture in general. We trust that our endeavors will meet with the reader's approval.

Oberkochen and Coventry, November 2000

JOACHIM HESSE
JULIAN W. GARDNER

Preface to Volume 3 of "Sensors Applications"

Diagnosis of disease and its therapy are problems in sensing and control. The patient who seeks information or care wants to learn the state of her organism. The same person, having been diagnosed with a problem seeks to monitor her condition and ensure that the solution offered by the physician is appropriate. In that sense, there are similarities between medical, aerospace, automotive and atmospheric sensing.

The authors of this volume of the Wiley Sensors Applications Series cover biomedical sensing in breadth: ranging from fundamental modalities like optics and imaging, ranging to applications such as hemodynamics, neonatal monitoring and prostheses for the deaf. Each topic is reviewed in depth, so that a practicing biomedical engineer or a bioengineering graduate student could gain insight into a specific topic and learn to apply the principles that are given.

The co-editors P. Å. Öberg, F. A. Spelman and T. Togawa give an introductory review of the history of medical sensing, and use historical examples to point to the future. What will the bioengineers of the future provide to aid diagnosis? Will the dreams of completely non-invasive sensing be reached in the future? Will they be realized in the near future?

Dr. Öberg follows the introduction with an in-depth exposition of optical sensing in medical care. The chapter leads the reader through the fundamental principles of optics and uses those principles to base a discussion of applications of biomedical optics.

Drs. J.D. Newman and A.P.F. Turner review glucose sensing, both invasive and non-invasive approaches. The chapter is tantalizing: if glucose can be sensed, can its level be controlled as well? That is the grail of glucose sensing. Drs. Newman and Turner ask not only if a sensor be built, but whether it can be manufactured.

Dr. S. Ueno presents the principles and application of magnetic resonance imaging (MRI). He describes the techniques employed to acquire detailed images, and delves into the ways that anatomy and physiology can be joined with a sensitive imaging modality.

Dr. K. Yamakoshi describes non-invasive measurement of hemodynamic variables. He provides the principles that have lain behind the sphygmomanometer for more than 100 years, techniques that were limited to measurements of sedentary subjects. He expands his work to review methods by which hemodynamics can be monitored non-invasively in ambulatory subjects. The work is exciting because it invites the possibility of giving information about patients during normal activities as contrasted to measurements that are made while people are sedentary and possibly anxious in a medical setting.

Drs. A. Johansson and B. Hök introduce the respiratory system and methods by which it can be monitored in the clinic and at rest. They present sensors to observe respiratory flows and pressures directly, as well as some devices that measure respiratory rate to indicate the condition of the respiratory system. They provide a clear exposition of the benefits and limits of each of the sensing modalities that they describe.

Dr. P. Rolfe addresses fetal and neonatal monitoring. He covers measurements of both the mother and the infant, and describes the information gained by each measurement. He leads the reader through sensing techniques and describes sampling and processing issues as well.

Dr. T. Tamura brings the reader into the realm of motion and energy analysis. His chapter on body motion analysis develops both direct and indirect methods of sensing, and then shows the applications of those methods. He leads the reader to conclusions about the energy that is consumed by humans in motion.

Drs. B. Hensel, G. Czgan, I. Weiss, and T. Nappholz present information about cardiac pacemaking. They take the reader from an understanding of electrodes that are used both as stimulus sources and as sensors: bidirectional devices. They write about the processing necessary to achieve control of the rhythm of the heart, and offer information about continuous, long-term control of the heart, and the success of the work done in the area.

Dr. F.A. Spelman presents information about cochlear implants. The implant is presented as a substitute for a physiological sensing modality. The principles of design are given, descriptions of the success of the device are offered, and questions whose answers will lead to future designs are presented as well.

Drs. P.J. French, D. Tanase, and J.F.L. Goosen provide an enlightening chapter on the design and application of catheter-based sensors. They give a broad spectrum of applications ranging from blood flow to urology, describing the need for and application of navigational techniques to ensure that the locations of catheter tips are known and controlled. Their practical approach to sensor development and application gives the reader a view of both the process and application by which biomedical engineers approach problems.

Dr. T. Togawa closes this volume with an exposition of home health care and telecare. He describes several sensing applications and the ways by which they can provide information from patients to physicians at a distance. This provocative chapter can lead the reader to think about ways by which health care can be provided efficiently, at low cost and to people who, because of separation from medical centers, would otherwise not be served.

Each author or group of authors has provided an extensive bibliography, so that the readers of this volume can go to the original sources behind the chapters presented here. While the bibliography is not exhaustive, it will lead the inquisitive reader to a rich trove of information.

We thank every author for assembling a comprehensive and interesting chapter. The work done by each is substantial, and, we hope will benefit you as a reader and user of this volume. Special thanks are due to Dr. Martin Ottmar of Wiley-VCH as well as to his staff. Dr. Ottmar was incisive in his comments, helpful at all levels of production and patient to a fault. This volume has benefited greatly from his contributions. Finally, our families deserve thanks for patience and support during the production of this work. The book couldn't be complete without them.

P. Åke Öberg — Linköping, Sweden
Tatsuo Togawa — Saitama, Japan
Francis A. Spelman — Seattle, WA, USA

Contents

List of Contributors

G. Czygan
Zentralinstitut für Biomedizinische Technik
Friedrich-Alexander-Universität Erlangen-Nürnberg
Turnstrasse 5
91054 Erlangen
Germany

P. French
Electronic Instrumentation Laboratory
Department of Microelectronics
Faculty of Information Technology and Systems
Delft University of Technology
Mekelweg 4, room 13.260
2628 CD Delft
The Netherlands

J. F. L. Goosen
Electronic Instrumentation Laboratory
Department of Microelectronics
Faculty of Information Technology and Systems
Delft University of Technology
Mekelweg 4, room 13.260
2628 CD Delft
The Netherlands

B. Hensel
Max Schaldach-Stiftungsprofessur für Biomedizinische Technik
Friedrich-Alexander-Universität Erlangen-Nürnberg
Turnstrasse 5
91054 Erlangen
Germany

B. Hök
Hök Instruments
Flottiljgattan 49
72131 Västerås
Sweden

N. Iriguchi
Siemens-Asahi Medical Technologies
Tokyo 141-8644
Japan

A. Johansson
Department of Biomedical Engineering
Linköping University
University Hospital
58185 Linköping
Sweden

T. Nappholz
Zentralinstitut für Biomedizinische Technik
Friedrich-Alexander-Universität Erlangen-Nürnberg
Turnstrasse 5
91054 Erlangen
Germany

J. D. Newman
Cranfield University at Silsoe
Barton Road
Silsoe MK45 4DT
UK

P. Åke Öberg
Department of Biomedical Engineering
Linköping University
58185 Linköping
Sweden

P. Rolfe
Daisy Lake
Oakley
Market Drayton TF9 2QW
UK

F. A. Spelman
Dept. of Bioengineering
University of Washington
Box 357962
Seattle, WA 98195-7962
USA

T. Tamura
Department for Gerontotechnology
National Institute for Longevity Sciences
36-3 Gengo, Morioka, Ohbu
Aich 474-8511
Japan

D. Tanase
Electronic Instrumentation Laboratory
Department of Microelectronics
Faculty of Information Technology and Systems
Delft University of Technology
Mekelweg 4, room 13.260
2628 CD Delft
The Netherlands

T. Togawa
School of Human Sciences
Waseda University
2-579-15, Mikajima, Tokorozawa-shi
Saitama 359-1192
Japan

P. F. Turner
Cranfield University at Silsoe
Barton Road
Silsoe MK45 4DT
UK

S. Ueno
Department of Biomedical Engineering
Graduate School of Medicine
University of Tokyo
7-3-1 Hongo, Bunkyo-ku
Tokyo 113-0033
Japan

I. Weiss
Zentralinstitut für Biomedizinische Technik
Friedrich-Alexander-Universität Erlangen-Nürnberg
Turnstrasse 5
91054 Erlangen
Germany

K. Yamakoshi
Department of Human and Medical Systems
Graduate School of Natural Science and Technology
Kanazawa University
2-40-20 Kodatsuno
Kanzawa 920
Japan

List of Abbreviations

2D	two dimensional
3D	three dimensional
3D-RA	3D rotational angiography
ABPM	ambulatory BP monitoring
ACOM	ambulatory CO monitoring
ADC	apparent diffusion coefficient
ARTMA	Advanced Real-Time Motion Analysis
AV	anteversion
BBB	blood-brain barrier
BioMEMS	MEMS applied to medicine and biology
BOLD	blood oxygenation level development
BPD	bi-parietal diameter
CBF	cerebral blood flow
CBV	cerebral blood volume
CCD	charge-coupled device
CFNN	compensatory fuzzy neural networks
CMOS	complementary metal oxide semiconductor
CNEP	continuous negative extra-thoracic pressure
CNS	central nervous system
CO	cardiac output
CSF	cerebrospinal fluid
CSI	catheter-system interface
CSI	chemical shift images
CT	computed tomography
DBP	diastolic blood pressure
DLT	direct linear transformation
DPF	differential pathlength factor
DTG	dynamically tuned gyroscope
DURS	direct ultrasound ranging system
DWI	diffusion-weighted imaging
ECG	electrocardiograph
EEG	electroencephalogram
EFM	electronic fetal monitoring

EHG	electrohystography
EM	electromagnetic
EMF	electromotive force
EMG	electromyogram
EOG	electrooculography
EPI	echo-planar imaging
EPS	electrode positioning system
FD	frequency domain
FE	field echo imaging
FES	functional electrical stimulation
FHR	fetal heart rate
FID	free induction decay
FLASH	fast low-angle shot
fMRI	functional MRI
FSR	force-sensitive resistor
FT	Fourier transform
GDH	glucose dehydrogenases
Gox	glucose oxidase
HFP	high-frequency power
HR	heart rate
HSP	hydrostatic pressure difference
IC	integrated circuit
ICP	intracranial pressure
IPPV	intermittent positive pressure ventilation
IR	infrared
IR-SE	inversion-recovery spin echo imaging
ISF	interstitial fluid
ISFET	ion-sensitive field effect transistor
IUGR	intra-uterine growth retardation
IUP	intra-uterine pressure
IVUS	intravascular ultrasound
LDF	laser Doppler flowmetry
LED	light emitting diode
LF	low frequency
LFP	low-frequency power
LITE	low-intensity treadmill exercise
MABP	mean arterial blood pressure
MAP	maximum amplitude of pulsation
MBP	mean blood pressure
MEMS	microelectromechanical systems
MFP	medium-frequency power
MIP	maximum intensity projection
MPG	motion-probing gradient
MRA	magnetic resonance angiography
MRI	magnetic resonance imaging

MRS	magnetic resonance spectroscopy
MSA	muscle sympathetic activity
MSS	Magnetic Surgery System
MTI	magnetic resonance imaging
MV	minute ventilation
NAD	Ch3
NADH	Ch3
NHE	normal hydrogen electrode
NIRS	near infrared spectrophotometry
NMR	nuclear magnetic resonance
Ox	oxidized species
PC	phased-contrast
PGp	photoplethysmography
PPG	photoelectric plethysmography
PPM	pacing pulses per minute
PQQ	pyrroloquinoline quinone
PTA	percutaneous transluminal angioplasty
PTCA	percutaneous coronary angioplasty
PU	polyurethane
PVC	polyvinylchloride
PVD	physical vapor deposition
PVDF	polyvinylidene fluoride
PWM	pulse-width modulated
QD	quadrature demodulation
QT	cardiac repolarization
R	reduced species
RIP	respiratory inductance plethysmograph
RPP	rate pressure product
RRF	rate response factor
RSA	respiratory sinus arrhythmia
SBP	systolic blood pressure
SE	spin echo imaging
SEP	systolic end point
SNR	signal-to-noise ratio
SQUID	superconducting quantum interference device
SV	stroke volume
TD	time domain
THI	tissue hemoglobin index
TOF	time-of-flight
TOI	tissue oxygenation index
TPR	total peripheral resistance
TPSF	temporal point-spread function
TRS	time-resolved spectroscopy
TTI	transthoracic impedance plethysmograph
US	ultrasound

UV	ultraviolet
VER	ventricular evoked response
VIP	ventricular inotropic parameter
WAMAS	Wearable Accelerometric Motion Analysis System

1
Introduction

P. Å. Öberg, F. A. Spelman, and T. Togawa

Measurement is the key to understanding biology and for the diagnosis of pathology. Measurement requires precise sensing to be successful, to allow scientists to advance their knowledge and physicians to control and cure abnormal conditions. In this book, we discuss both measurement and control of physiological variables.

Sensing can focus on the whole human organism as it does in the case of whole-body sensing systems; it can stress the measurement of critical variables such as glucose concentration, temperature, or pressure; it can define the behaviors of particular cells such as single neurons in the brain.

Control can be holistic, as it is in the case of home health care of the elderly or when the condition of a baby is monitored and controlled during the process of birth. Control can be specific when glucose is monitored and held within normal physiological limits, or it can be focused on a particular somatic sense as it is in the cochlear implant that provides hearing to the deaf.

Sensing is critical to all the cases above. It demonstrates a clear need for appropriate transducers that detect one form of energy and convert it to another, either at the input or the output of a specific instrument. Further, sensing implies signal processing. In the chapters that follow, the reader will be exposed to input sensors, output transducers, and the processing that connects them. While sensors are the focus of this volume, output transducers come into play, as does signal processing. As you read the chapters that follow, you will be impressed with the breadth and imagination that are the hallmarks of bioengineering.

1.1
Historical Breakthroughs in Medical Sensing Science

1.1.1
Plethysmography

Plethysmography is one of the earliest methods developed to make non-invasive blood flow measurements in the extremities. It is still one of the most frequently used and accurate methods used to assess peripheral blood flow. A great deal of

Sensors in Medicine and Health Care
Edited by P. Å. Öberg, T. Togawa, and F. A. Spelman

ISBN: 3-527-29556-9

our knowledge of vascular physiology in health and disease has been derived from data obtained plethysmographically.

Glisson [1] and Swammerdam [2] first employed plethysmography to study isolated muscle contraction. It was not until the latter half of the nineteenth century that this measurement principle was applied to blood flow measurements. Francois-Frank [3] published the first blood flow results in 1876 utilizing the venous occlusion technique. Brodie and Russel [4] studied renal blood flow by enclosing the kidney in a closed chamber in which the volume changes were studied while venous outflow was occluded. Hewlett and Zwaluwenburg [5] investigated blood flow in human limbs with plethysmography, thereby introducing the venous occlusion plethysmograph. Hyman and Winson have reviewed the early development of plethysmography [6].

The word plethysmography is derived from the Greek word for increase (in volume) *plethysmos* and the word to record, to write, *grafein*. Thus, the term plethysmography describes the basic principle of the technique, that is to record changes in the volume of parts of the body. The principle can be applied to the heart, liver, kidney, and to vascular measurements of the limbs and parts thereof. Most transient changes in the volume of organs are related to changes in blood content. Thus, plethysmography can serve to record blood volume as well as changes in blood volume. Since blood flow can be expressed as blood volume change per unit time, we can use plethysmography to measure blood flow. Indeed, plethysmography has been used primarily to assess blood flow to organs.

The most common application is venous occlusion plethysmography, usually used to diagnose obstructions in limb blood vessels. In this method the early influx of arterial blood to a limb is recorded when the venous drainage is stopped temporarily. The volume changes can be assessed in a number of ways, including water-filled cuffs, air-filled cuffs, strain gauges, electrical impedance measurements, and photoelectric probes.

1.1.2
Blood Pressure Measurements

The first blood pressure measurement was made with a very simple sensor. In 1733, the Reverend Stephen Hales [7] introduced glass tubes into the carotid arteries of horses and measured the height of the blood column that arose in the tube. He found it to be 8 feet and 3 inches. The first blood pressure measurement was made. About a century later, Poisieulle [8] studied blood pressure using a mercury manometer, a method for which he received a Gold medal from the Royal Academy of Medicine in Paris. The manometer of Poisieulle was connected to the artery with a leaden cannula.

Ludwig [9], a German professor in Comparative Anatomy at Marburg, improved Poisieulle's device with a recording technique. He recorded the motion of the mercury column on a revolving smoked drum (the kymograph), and dynamic blood pressure measurements were made. Favre [10] used Ludwig's manometer clinically, recording human systolic blood pressure for the first time.

During the 150 years that have elapsed since Ludwig's first invasive human measurement, invasive blood pressure sensors have undergone remarkable development. The most recent devices are based on catheters with microelectronic components, having an outer diameter of 0.35 mm (0.014 inches), which can be introduced into the coronary vessels of the heart. These catheters [11] are used to diagnose coronary obstructive diseases before and after treatment with balloon angioplasty or with intravascular stents. It soon became evident that the early devices were not particularly well-suited for clinical work.

Indirect (non-invasive) blood pressure devices were developed to satisfy the clinical need for systolic blood pressure measurements in humans. The first sphygmomanometer is attributed to Vierordt [12]. Herisson [13] performed the first indirect measurement of blood pressure in humans. The sphygmomanometer used a mercury reservoir covered with a rubber membrane from which a glass column arose. The mercury container was pressed against the radial artery until the oscillations of the mercury column stopped. This point was taken as the systolic pressure level. Inspired by Vierordt and Herisson, many physiologists designed blood-pressure sensing devices in the middle and end of the nineteenth century.

A breakthrough in blood pressure sensors came 1896 when Riva-Rocci [14] presented a new device utilizing an inflatable rubber bladder enclosed in leather, surrounding the upper arm: the modern sphygmomanometer. The pressure in the cuff was increased until the palpated pulse in the radial artery disappeared.

A second milestone in the development of indirect blood pressure measurement was the discovery by a Russian surgeon Korotkov in 1905 [15]. He discovered that if one increased the pressure in the Riva-Rocci cuff to above systolic pressure levels and then slowly decreased the cuff pressure while listening to the sounds from the artery one could associate the sound characteristics with the corresponding pressure levels in the vessel. As long as the cuff pressure exceeds the arterial pressures no sounds are generated. However, when cuff pressure is decreased below systolic pressure and remains above diastolic pressure, the pressure levels correspond to unique changes in the sound from the vessel.

Korotkov's discovery had, and still has, an enormous influence on the quality of non-invasive blood pressure recordings. His method is still frequently used in routine health care, with only small technical improvements. The discoveries of Riva-Rocci and Korotkov are the basis of today's computerized automatic blood-pressure sensors and monitors.

1.1.3
Electrophysiology and Einthoven's Galvanometer

Animals are electrically activated. Animal cells generate electrical activity when they process information or contract. They can be stimulated electrically as well. The history of electrophysiology is long, and guided by distinguished physicians, physiologists, physicists, chemists, and engineers. One interesting story of many is that Volta, having known of Galvani's experiments with the stimulation of nerve, placed an electrode in each ear and detected 'sound...like the bubbling of a viscid fluid...' [16]. The recognition that cellular activity was coupled to electrical signals and that electrical stimulation could excite cells launched the study of electrophysiology.

Electrophysiology has spawned the diagnostic measurement of biopotentials. The most common potential measured is that produced by the heart, the electrocardiogram (ECG), and followed in the clinic by the measurement of the signals that are produced by the brain, the electroencephalogram (EEG), and those produced by somatic muscles, the electromyogram (EMG). The three biopotentials will be discussed in reverse order, since the ECG will be covered in more detail than will the EEG and EMG.

The EMG can be recorded either from the surface or from within muscle tissue. The magnitude of the signal is a nonlinear function of the force exerted by the muscle, but is a linear function of the number of fibers that are recruited to exert that force. Surface recordings produce signals from large volumes of tissue. Indeed, the signals are produced by volumes that are too large for clear interpretation. Intramuscular recordings can be focused on small groups of muscle units and trade patient discomfort for desired detail [17]. The interpretation of the EMG is still an active topic of research [18].

Hans Berger, a German psychiatrist, announced publicly that he could record tiny electric signals from the brain, using external monitoring techniques. His recordings were remarkable considering that they were made in the late 1920s, before the advent of modern electronic instrumentation [19]. While Berger's discovery showed that the general state of the brain could be assessed, it was left to British scientist, W. G. Walter, to demonstrate the diagnostic value of the measurement of the electroencephalogram (EEG). Walter's contribution was one of instrument development, in which he employed a larger number of smaller electrodes than did Berger, and was able to use them to focus on activity in specific regions of the brain [1, 19].

Walter's instrument deserves brief description: he employed 22 cathode ray tubes each connected to a pair of electrodes, to record the activities of different locations within the brain. The display was photographed to develop a snapshot of activity at a particular instant of time [19]. The story shows that innovative instrumentation comes in several eras and from physicians, such as Walter, who have backgrounds in engineering and a keen interest in detailed diagnosis and specific therapy.

The ECG has a long pedigree. In 1842, Carlo Matteucci demonstrated that electric current accompanies each heart beat [20]. The phenomenon remained a laboratory curiosity until the development of the capillary electrometer in 1872. Using a similar technique, Augustus Waller published the first human electrocardiogram in 1887 [20]. In 1889, Willem Einthoven defined the term 'electrocardiogram' for the surface potential field that is produced by the heart. Despite being credited for that definition, Einthoven attributed it to Waller [20]. Work continued with the capillary galvanometer, a device that required tedious mathematical correction to produce a faithful recording of the ECG, until 1902, when Einthoven employed the string galvanometer, a device initially developed by Clement Ader to assess the signals carried by undersea telegraph lines [20].

Einthoven's modification of the string galvanometer made possible the clinical use of the ECG. The string galvanometer is a display transducer that employs the optical magnification of the mechanical deflection of a current-carrying quartz string that is suspended in a DC magnetic field. When electric current flows through the string, it deflects. If the string is coated with an opaque material, such as silver, then it casts a shadow across a slit that is illuminated. As the string deflects, its shadow moves with greater deflection than does the string, and the motion of the shadow can be displayed on film [20]. The string galvanometer responds to small DC currents that flow through it.

Einthoven's application of Ader's transducer is a classic and fascinating illustration of bioengineering. The capillary galvanometer that preceded the string galvanometer required elaborate and tedious correction to obtain a faithful recording of the ECG. The string galvanometer produced faithful recordings. Unfortunately, the electrode potentials drifted and drift was observed in the recordings. Later investigators AC coupled the ECG signal, using a time constant of 3.2 s for clinical recordings [21].

Einthoven proposed the use of the *Einthoven Triangle* to record the ECG. Lead I measures the potential difference between the left arm (positive) and the right arm (negative); lead II measures the potential difference between the right arm (negative) and the left leg (positive); lead III measures the potential difference between the left arm (negative) and the left leg (positive). The triangle is used to determine the electrical axis of the heart [21].

The application of the Ader instrument requires the development of an input transducer, the correction of the output transducer, and the use of an available recording technique. These principles will be illustrated throughout this book in many different applications. The input transducer for the ECG will be discussed briefly below.

1.1.3.1 **Electrocardiogram**

Einthoven's original design of electrodes, the input transducers for the ECG, is straightforward. He filled buckets with salt water and had the subjects immerse their hands and left legs in the saline-filled buckets. Clearly, the system was awkward in a hospital setting. Today's electrodes are much more convenient and made of materials that produce stable potential measurements. The surfaces of the electrodes are Ag:AgCl [21], their areas are about 1 cm^2, and the electrodes are disposable and made of flexible polymers [22]. The electrodes have adhesive backing and can be applied directly to the skin. They are easily and painlessly removed. ECG monitoring can be completed in less than 15 min, and is routine in cardiology.

1.1.3.2 **Electroencephalogram**

The electroencephalogram is recorded with a large number of surface electrodes [23], which must be placed comfortably on the scalp for long periods of time. The electrodes have to be small and cause minimal cosmetic disturbance, but they must record reliably for hours. The electrodes are made of Ag:AgCl, with areas of less than 1 cm^2 [17]. They are fixed to the scalp with conductive glue, called colloidon, and further stabilized with rubber straps that hold them to the head of the patient [17]. At times, subsurface electrodes can be used, but they are not used routinely [21].

1.1.3.3 Electromyogram

The electromyogram can be measured either with external or with internal electrodes. In the former case, the electrodes are similar to those used for the electrocardiogram. Subsurface electrodes are needle electrodes and can have a variety of configurations. The configurations of intramuscular electrodes vary from single contacts to closely-spaced contacts and coaxial contacts, often in an 18 gauge or smaller needle [21]. The configuration of the intramuscular contacts determines their receptive fields: a single contact has a broader field than does either a closely spaced pair (a dipole) or a coaxial design. Thus, the user can choose the field of activity over which he or she measures muscular function.

1.1.3.4 Microelectrodes and Intracellular Measurements

The determination of the potentials produced by complex brain structures leads to the assessment of brain organization and function. Neurophysiologists seek to measure the action potentials that are produced by single neurons or groups of cells. The advantage of intracellular measurements is that the data that they acquire comes from a single cell, whose location is well defined. The advantage of extracellular measurements that are made with microelectrodes is that the data that they produce comes from small groups of cells, cells that may act in concert with each other. Each recording produces unique engineering challenges.

The electrodes that are used for intracellular measurements are glass micropipettes. To enter the intracellular space requires a tip diameter of 1 µm, and a glass pipette that is filled with a conductive solution, usually 3 M KCl [23]. Several engineering issues are relevant to the use of the micropipette. First, the recording amplifier must have an extremely small bias current, less than 0.1 pA. Second, the input impedance of the recording amplifier must be very high, of the order of gigaohms. Third, the geometry of the electrode defines a distributed, parasitic capacitance that shunts the high resistance of the tip impedance to the reference potential. The time constant of the circuit distorts the signal produced by an active neuron. Recording amplifiers employ positive feedback to neutralize the capacitance (negative capacitance amplifiers) and to record the action potential faithfully [21].

Microelectrodes that are used for extracellular recording use tips 10–50 µm in diameter. In the larger case, the tips are often planar electrode surfaces on silicon [24]. The impedances of microelectrodes range from several hundred kiloohms to a few megaohms. The input impedances of the amplifiers that are used to record from extracellular microelectrodes are usually greater than 100 MΩ. To reduce the effects of parasitic capacitance, the amplifiers are placed close to the microelectrodes. Bioengineers have developed multi-channel electrode arrays; those arrays are now being used routinely to record from deep and shallow structures in the brain [25, 26]. The development of arrays is exciting, since it offers the opportunity for the simultaneous recording of several events within the central nervous system. While the data produced by such systems are enlightening, they require sophisticated signal processing techniques to isolate individual cells and to interpret the meaning of the recorded data.

1.1.4
Pulse Oximetry

Hemoglobin has an unusual feature in its optical property so that its absorption spectrum varies greatly depending on oxygen saturation. Thereby the non-invasive measurement of blood oxygenation can be achieved. After preceding early works by Matthes (1935) and Millikan (1942), Wood and Geraci (1949) established ear oximetry in which arterial blood oxygen saturation could be determined from tissue absorption in the ear pinna at two wavelengths [27–29]. However, this method required a calibration for bloodless tissue under application of a cuff pressure about 200 mmHg.

The idea of using pulsatile light variation to measure arterial oxygen saturation was conceived by Aoyagi [30]. Takuo Aoyagi, working at the Nihon Kohden Corporation, noted the pulsatile variation in the dye densitometer signal for cardiac output measurement, and concluded that the pulsatile changes were due to pulsatile alteration of the intervening arterial blood volume. He made a prototype pulse oximeter in 1973 and presented a paper at the Annual Conference of the Japanese Society of Medical Electronics and Biological Engineering in 1974. The significant feature of this idea is that the absorption of the arterial blood can be extracted selectively owing to the pulsatile nature of blood flow, and that arterial oxygen saturation can be determined without calibration for the bloodless tissue. Thereafter, the fingertip probe was developed by Minolta. Its clinical evaluation was made by Yoshiya et al. [31]. However, early models were quite sensitive to motion, and the accuracy was not acceptable for clinical monitoring. The technique was abandoned for several years. Then an anesthesiologist, William New, at Stanford University Medical School, and an engineer, Jack Lloyd, recognized the potential importance and market for a convenient oximeter in many clinical situations. They founded the Nelcor Company. After the evaluation of the Nelcor pulse oximeter was published by Yederman and New in 1983, its use and development spread rapidly [32]: instruments were released by many manufacturers and became widely used in anesthesia and critical-care situations.

1.1.5
Body Temperature Measurement

Body temperature measurement has a long history. The oldest description of body temperature measurement is found in the text of Hippocrates in the 4th century

BC [33]. According to Potter's text, Hippocrates wrote the procedure of finding a region in the chest where the pus causes pleurisy, '... soak a piece of fine linen in warm moist finely triturated Eretrian earth; then wrap this all the way around his thorax, and wherever it first dries, that is where you must cauterize... [34]'. It was shown that this procedure is quite effective [35]. However, this is not an absolute temperature measurement but a relative measurement of temperature distribution. Therefore, this is more precisely described as the first thermography.

The first clinical thermometer was the Santorio's air thermometer in 16th century [33]. It had a bulb filled with air and the temperature could be read by the shift of meniscus due to the thermal expansion of the air at a capillary partially filled with water. When the air thermometer measures body temperature, the bulb is placed in the mouth of the subject. At the end of 19th century, the mercury-in-glass clinical thermometer that measures and holds the maximum temperature was developed by Ehrle [33]. The mercury-in-glass clinical thermometer is appreciated for its remarkable reliability, as well as its convenience of handling and low cost. Even today, no other thermometer can fully replace it. However, many situations arise where the mercury-in-glass thermometer is unacceptable because of its slow response, large heat capacity, and inconvenience during continuous monitoring. Also, mercury contamination has become a problem in hospitals. Electronic clinical thermometers with digital display for oral, axillary, or rectal measurement are becoming popular. They make more rapid measurements than do mercury thermometers, and require no sterilization, since they employ sterile probe covers.

Body temperature measurement at the tympanum has made rapid progress. It was first introduced in thermal physiology where variable body temperature has to be measured rapidly and faithfully. For that purpose, a fine thermocouple wire having contact with the tympanum was used [36]. Although this method has been used widely in physiological studies, clinical applications were limited because the contact probe is difficult to handle and the risk of perforation of the tympanum must be considered. Intelligent Medical Systems developed a non-contacting tympanic thermometer. The thermometer employed a thermopile infrared detector according to the patent applied for by O'Hare and Phillips in 1986. Shinozaki et al. evaluated the thermometer clinically [37]. The salient feature of the non-contacting tympanic thermometer is its fast response. Temperature can be determined precisely within 2 s, a more than 60-fold decrease in delay from oral thermometry that takes 3 min.

1.2 The Future

The future of medical and biological sensing is likely to be based on small biological sensors, as well as on non-invasive techniques. We offer below a few examples as brief summaries of MEMS and BioMEMS sensors, including cell-based sensors and optical biopsy sensors. Both techniques may provide a revolutionary approach to sensing in general and to biological sensing in particular.

1.2.1
MEMS and BioMEMS Sensors

Many sensor problems in clinical medicine can probably be solved by applying new knowledge from the rapidly growing MEMS, BioMEMS, and microfluidics fields. The progress in microfabrication and micromachining is transforming the field of solid-state sensors into what today is called microelectromechanical systems (MEMS) [38]. The advantage of MEMS in sensor applications lies in the fact that components can be batch fabricated, enabling miniaturization, low-cost, and integrated solutions to advanced sensor problems. MEMS has so far been mainly utilized in the automotive and aircraft industries. Typical examples are pressure sensors and accelerometers for air-bag control.

MEMS is mostly associated with fabrication in silicon. Photolithography in combination with etching processes (wet and dry etching) makes it possible to manufacture advanced tools and sensors with very small dimensions.

Silicon probes for neural activity monitoring, microcutters for eye surgery, and probes for atomic force microscopes are all examples of medical applications of MEMS technology. MEMS applied to medicine and biology is often called BioMEMS [39]. BioMEMS differs from traditional MEMS in a number of ways. BioMEMS is often implemented as a group of analytical techniques, used in many research areas such as biochemistry, biology, and medicine. The number of materials used in BioMEMS is much higher than in traditional MEMS. Glass slides, polypropylene, and polymeric materials as well as rigid plastic materials are often used in BioMEMS in order to achieve biocompatibility. The plastic materials also allow mass fabrication at low prices.

Typical applications of BioMEMS are a variety of lab-on-a-chip methods for analyzing body fluids. The range of applications for BioMEMS technology has expanded from the molecular scale near single cells to larger scales applied to complex biological tissues. Closely related to BioMEMS is the field of microfluidics. Microfluidic devices and systems have created an enormous interest in the life sciences and medicine as they offer many advantages over existing macroscale systems. Microfluidic systems are often compact in size, disposable, and admit analyte analysis at higher speed using decreased sample volumes. Typical application areas are fluid delivery, microvalves and micropumps, mixers and reactors with very small volumes as well as flow sensors. We have probably seen only the beginning of a new era in sensor technology in medicine, demonstrated by the fascinating progress in MEMS and BioMEMS sensing.

1.2.2
Cell-Based Biosensors

The term ‘biosensors’ can have two different definitions.

- Biosensors include all kinds of technical sensors that are used to study biological and medical variables. This definition is indeed very wide. For instance, an

ECG electrode used to pick up an electrocardiogram from the human body can be a biosensor. Similarly, a silicon substrate that holds chemically sensitive compounds in wells to produce fluorescence in the presence of a drug such as epinephrine is also a biosensor.
- In a stricter sense, a biosensor is a transducer in which the sensing mechanism or principle is of biological origin. Very often the 'sensors' of the system are biomolecules, single cells, or groups of cells ('tissue'). In these sensors the measurement process is very often electronic, that is electrodes and amplifiers, optical fibers, and optical detectors or processors of ionizing or non-ionizing radiation.

An early and well-known example of 'biosensors' in the latter sense were the canaries used by coal miners to detect high levels of methane gas. By observing the behavior of the birds, miners could avoid lethal concentrations of the gas.

There is a growing interest in the use of cell-based biosensors in a variety of scientific fields, for instance in environmental monitoring, defense applications, and medical diagnosis.

The reasons that the development of cell-based biosensors is considered a promising future field in sensor science are the following.

- Biosensor elements have a naturally evolved selectivity to biological analytes.
- Biosensors have the capacity to respond to stimuli from an analyte in a physiological way.
- Biosensors have high sensitivity and broad dynamic range.
- Biosensors could provide solutions to sensor problems with rapid, sensitive, low-cost technology to monitor analyte concentrations.

The field of cell-based biosensors is traditionally divided in a number of subgroups.

- Molecular biosensors
  - antibody sensors
  - nucleic acid sensors
  - enzyme sensors
  - sensors of specific ion channels
- Cellular biosensors
  - sensors of cellular microorganisms
  - sensors of intracellular potentials
  - sensors of extracellular potentials
  - impedance sensors
  - metabolic sensors
  - cell fluorescence sensors

Although many problems remain to be solved to take advantage of biological mechanisms in sensor solutions the field is both dynamic and promising because of the many potential clinical applications, ranging from pharmacological screening to environmental monitoring [40–42].

1.2.3
Optical Biopsies

Over the last decade a number of impressive advances have been made in the field of biomedical optics. The establishment of new technology such as optical fibers, photodetectors, and new light sources has facilitated the design of innovative new diagnostic techniques [43, 44]. A typical example is what is generally called an optical biopsy. A biopsy is usually a tissue sample taken from a patient and analyzed in the cytology laboratory. Very often biopsies are used to diagnose cancer. Traditional biopsy techniques normally require complex sample handling and long analysis times to answer a diagnostic question.

Bio-optical techniques illuminate new ways of obtaining biopsy information. Their use may shorten the diagnosis time and make it less invasive. An optical biopsy can be analyzed without taking any tissue sample from the organ under study. An analysis of absorbed, emitted, or Raman-scattered light from an optically excited tissue provides signatures of morphological, chemical, or physiological changes that may help the physician to determine if the tissue is normal, benign, or malignant.

There are many examples of successful attempts to utilize this new technique in diagnostic work. Optical mammography [45] research develops optical sensors to detect cancerous lesions of human breasts at an early stage. The sensors use non-invasive and non-ionizing near-infrared light. The number of possible applications is not limited to mammography. Potential diagnoses are numerous: esophageal cancers and dysplasia, bladder tumors, and head and neck carcinogenesis. Optical markers of a disease are important in this field: increased vascularity or changes in presence of metabolic cofactors are typical examples. The future advantages expected are minimally invasive diagnostic procedures accompanied by rapid and correct answers to diagnostic questions. Several applications can result in technical solutions that make it possible to diagnose diseases at the bedside, diseases that today require laboratory assistance to define.

1.3
References

1 Glisson, F., History of Plethysmography: Tractatus de ventriculo intestines (1622), reprinted in *J. Cardiovasc. Surg.* (Eds.: C. E. Hyman and T. Winsor) **1961**, p. 506.

2 Swammerdam, J., Biblia Naturae, Vol. 3 (Ed.: Boerhaave) 1737, reprinted in *Theory and Practice of Blood Flow Measurements* (Ed.: J. P. Woodcock). Butterworths, London **1975**.

3 Francois-Franck, C. E., Du volume des organs dans ses rapports avec la circulation du sang. *Physiol. Exp. (Paris)* **1876**, *2*, 1.

4 Brodie, T. G. and A. E. Russel, On the determination of the rate of blood flow through an organ. *J. Physiol. (Lond.)* **1905**, *32*, 47P.

5 Hewlett, A. W. and J. Zwaluwenburg, The rate of blood flow in the arm. *Heart* **1909**, *1*, 87.

6 Hyman, C. and T. Winson, History of Plethysmography. *J. Cardiovasc. Surg.* **1961**, *2*, 506.

7 Hales, S., Statical essays: containing haemostatics. Vol. ii. W. Innys & R. Manby, London **1733**.

8 Poisieulle, J. L. M., Recherches sur la force du Coeur aortique. Thesis no. 166. Didot, Paris **1828**.

9 Ludwig, C., Beiträge zur Kenntniss des Einflusses der respirationsbewegungen auf den Blutlauf im Aortensysteme. *Mullers Arch. Anat. Physiol. Wissensch. Med.* **1847**, 242–302.

10 Faivre, J., Etudes experimentales sur les lesions organiques du Coeur. *Gaz. Med. Paris* **1856**, 727.

11 Anonymous, Pressure Wire Sensor. RADI Medical, Uppsala, Sweden **2003**.

12 Vierordt, K., Die Lehre vom Arterienpuls. F. Vieweg, Braunschweig **1855**.

13 Herisson, J., The Sphygmomanometer, an Instrument that makes the Arteries Apparent to the Eye. Longman, Rees, Orme, and Brown, London **1855**.

14 Riva-Rocci, S., Un nuovo sfigmomanometro. *Gazz. Med. Torino* **1896**, *50*, 981–996.

15 Korothov, N. C., To the question of methods of determining the blood pressure (from the clinic of Professor C. P. Federoff). *Rep. Imp. Mil. Acad.* **1905**, *11*, 365–367.

16 Simmons, F. B., Electrical Stimulation of the Auditory Nerve in Man. *Arch. Otolaryngol.* **1966**, *84* (July), 24–76.

17 Clark, J. W., The Origin of Biopotentials, in Medical Instrumentation: Application and Design (Ed.: J. G. Webster), Wiley, New York **1998**, 138–139.

18 McGill, K. C., Automatic Decomposition of the Electromyogram. Stanford University **2002**.

19 Sabbatini, R. M. E., The History of the Electroencephalogram. www.epub.org **1997**.

20 Jenkens, D., A (not so) brief history of electrocardiography, **1996**, *http://www.ecglibrary.com/ecghist.html*, accessed January, **2004**.

21 Geddes, L. A. and L. E. Baker, Principles of Applied Biomedical Instrumentation, 3rd edn. Wiley, New York **1989**, 961.

22 Harrell Medical Inc., Cardiology Products, Burdick, Inc. Cardio Sens/Ultra II, *http://www.harrellmedical.com*, accessed January, **2004**.

23 Neuman, M. R., Biopotential Electrodes, in Medical Instrumentation: Application and Design (Ed.: J. G. Webster), Wiley, New York **1998**, 183–232.

24 Middlebrooks, J. C. and J. A. Bierer, Auditory cortical images of cochlear-implant stimuli: coding of stimulus channel and current level. *J. Neurophysiol.* **2002**, *87*, 493–507.

25 Bierer, J. A. and J. C. Middlebrooks, Auditory cortical images of cochlear-implant stimuli: dependence on electrode configuration. *J. Neurophysiol.* **2002**, *87*, 478–492.

26 Middlebrooks, J. C. and J. A. Bierer, Cortical images of cochlear implant stimuli, in *Proc. 2001 Conf. Implantable Cochlear Prostheses*, Pacific Grove, CA, USA **2001**.

27 Matthes, K., Untersuchungen über die Sauerstoffsättigung des menschlichen Arterienblutes. *Arch. Exp. Pathol. Pharmacol.* **1935**, *176*, 693–696.

28 Millikan, G.A., Oximeter, an instrument for measuring continuously oxygen saturation of arterial blood in man. *Rev. Sci. Instrum.* **1942**, *13*, 434–444.

29 Wood, E. and J. E. Geraci, Photoelectric determination of arterial oxygen saturation of arterial blood in man. *J. Lab. Clin. Med.* **1949**, *34*, 387–401.

30 Severinghaus, J. W. and P. Astrup, History of blood gas analysis. VI. Oximetry. *J. Clin. Monitoring* **1986**, *2*, 270–288.

31 Yoshiya, I., Y. Shimada, and K. Tanaka, Spectrophotometric monitoring of arterial oxygen saturation in the fingertip. *Med. Biol. Eng. Comput.* **1980**, *18*, 27–32.

32 Yederman, M. and W. New, Evaluation of pulse oximetry. *Anesthesiology* **1983**, *59*, 349–352.

33 Epstein, E., Die Entwicklung der klinischen Thermometry. *Eng. Inn. Med.* **1928**, *33*, 407–503.

34 Potter, P., Hippocrates, Vol. 6. Harvard Univ. Press, Cambridge, MA **1988**.

35 Otsuka, K. and T. Togawa, Hippocratic thermography. *Physiol. Meas.* **1997**, *18*, 227–232.

36 Benzinger, T. H. and G. W. Taylor, Cranial measurement of internal temperature in man, in Temperature: Its Measurement and Control in Science and Industry (Ed.: J. D. Hardy). Reinhold, New York **1963**.

37 Shinozaki, T., R. Deane, and F. M. Perkins, Infrared tympanic thermometer: Evaluation of a new clinical thermometer. *Crit. Care Med.* **1988**, *16*, 148–150.

38 Maluf, N., An Introduction to Microelectromechanical Systems Engineering. Artech House, Norwood, MA **2000**.

39 Tay, F. E. H., Microfluidics and BioMEMS Applications. Kluwer, New York **2001**.

40 Pancrazio, J. J., J. P. Whelan, and D. A. Borkholder, Development and application of cell-based sensors. *Ann. Biomed. Eng.* **1999**, *27*, 697–711.

41 Rogers, K. R., Biosensors for environmental applications. *Biosens. Bioelectron.* **1995**, *10*, 533–541.

42 Wilkins, E. and P. Atanasov, Glucose monitoring: state of the art and future possibilities. *Med. Eng. Phys.* **1996**, *18*, 273–288.

43 Katzir, A., Laser and Optical Fibers in Medicine. Academic, London **1999**.

44 Tuchin, V., Tissue Optics. Light Scattering Methods and Instruments for Medical Diagnosis. *SPIE* **2000**, TT38.

45 Alfano, R. R. (Ed.) Advances in Optical Biopsy and Optical Mammography. *Ann. N.Y. Acad. Sci.* **1998**, 838.

2
Optical Sensors in Medical Care

P. Åke Öberg

2.1
Optics in Medicine

Optical assessment of physiological parameters constitutes a group of measurement principles, which are very attractive because they often provide simple, noninvasive, continuous physiological monitoring conditions.

Optical principles have been used in medical diagnosis since the dawn of the medical sciences. Well-known examples are the typical pallor of an anemic patient or the yellowish skin of a baby with jaundice. Skin colors, and changes in skin colors, are typical of a number of infectious diseases. The human body often responds with skin rashes, typical for each infection, such as in chickenpox or measles. Infected wounds develop a red-purple-blue color. The color of urine and feces and how the optical properties of these human waste products change with various pathological conditions of the body have been used for diagnosis since medieval times. Bleeding in the gastrointestinal canal change the color of the feces from brown to black. Urine is, for instance, discolored by bacterial infections in the urinary tract or by bleeding in the kidneys. Thus, spectral information is very important for the correct diagnosis of disease: knowledge that has been utilized by physicians for hundreds of years.

Today, spectral changes can be accurately recorded and analyzed by means of optical instruments and computers, which supply the doctor with important information as a basis for diagnosis and treatment. Reflection spectroscopy, which is an important methodology in this connection, is based on the absorbency of applied light in a sample under test. Every tissue type spontaneously emits fluorescent radiation, so called autofluorescence, which can be used for diagnostic purposes. Studies of autofluorescence have revealed that tumors and pathological tissues have a changed autofluorescence pattern, which opens possibilities for new diagnostic principles, by which neoplastic tissues can be distinguished from healthy tissue. Radiation from the sun has been used for treatment of skin disorders for a long time. In both Roman and Greek cultures a daily sunbath was regarded as important treatment for staying healthy. The interest in optical diagnosis has increased in recent years because of the availability of new instruments and methods.

Sensors in Medicine and Health Care
Edited by P. Å. Öberg, T. Togawa, and F. A. Spelman

ISBN: 3-527-29556-9

For a long time optical instruments have been used for analysis of samples taken from the patient in the laboratory. In such instruments an optical beam typically passes a cuvette containing the sample and properties of the sample can be extracted by analyzing absorption or scattering of light from the cuvette.

A newer trend is that data are collected directly from a patient eliminating the need of sample collection. For instance, tiny optical fibers can be inserted in muscles, introduced into blood vessels, and into most cavities of the human body. This means that we can extract diagnostic information from most part of the human body with little discomfort for the patient. Thus, 'optical biopsies' can be taken by studying reflected or absorbed light from the interior of an organ without the painful or risky removal of tissue samples for later laboratory investigations. The risk for sample substitution is also markedly decreased by this approach. Optical biopsies give promises of new diagnostic approaches. Absorbency, emission, or Raman-scattered light can include a signature of normal, malignant, or benign tissue. Very often, the techniques used for optical biopsies are minimally invasive and can be used with little discomfort for the patient. Many of the important optical measurement principles have been developed in order to facilitate assessments of the circulatory conditions in the human body.

Since the days of Stephen Hales [1], who was the first to perform arterial pressure measurements in 1733, we have experienced an extensive and innovative development of principles for pressure and flow measurements in the major vessels of the human body. The lodestar of this development has been the need to quantify the magnitude and time course of blood flow and pressure in all parts of the vascular system. Methodological development was necessary as a tool for increased understanding of the basic circulatory mechanisms, and also for the diagnosis of conditions of health and disease of the circulatory system.

The more important steps in methodological development are related to plethysmography, electromagnetic (EM) flow measurements, dilution measurements, and radioisotopes. The majority of flow measurement methods have been developed for use in cardiac flow assessments for the mapping of flow conditions in the major vessels and in the vessels located in the periphery of the vascular tree. The reason for the concentration of development efforts in this part of the circulatory systems is easy to understand.

- Blood flow is a well-defined variable in the cavities of the heart or the major arteries and veins.
- Blood flow in the supplying and draining vessels of an organ is of central importance when the basic physiology of an organ is studied.
- In industrialized countries, approximately 50% of deaths are related to circulatory disorders.

There exists a considerable clinical interest in problems such as arteriosclerosis, obstructive and/or occlusive diseases in the heart, and of the major arteries. The end stages of these diseases are often dramatic and attract a great deal of interest from many researchers with theoretical as well as clinical interests. In spite of the early discoveries of the existence of a microcirculatory network [2], the methods we have to study this part the circulatory system are less developed in comparison with those available for flow measurements in the major vessels. Also here the reasons are obvious.

- The complexity of the 3D microcirculatory network is one contributing factor.
- The flow of blood in the microcirculation has many different functions such as temperature regulation and oxygen, hormone, and enzyme transport as well as the washout of metabolites.

Most flow-measuring probes available today cause trauma that may seriously disturb the blood flow of the microcirculation in the part of the organ into which the probe is introduced. Flow measurements in the microcirculation require non-invasive procedures or approaches causing little or no trauma. It is also important to find measurement principles (current, heat transport, laser light) that do not disturb the circulation.

Later in this chapter three methods based on optical principles, all mirroring microcirculatory conditions, will be presented. The scientific field in which optical principles are applied to solve diagnostic or therapeutic problems is often called *bio-optics* or *biomedical optics* [3].

2.1.1 The Diagnostic/Therapeutic Window

Biological tissues are strongly inhomogeneous from the optical point of view and have a refractive index greater than one. This means that a light beam directed towards a tissue volume is partly reflected at the air–tissue interface and partly penetrates the tissue. Absorption and scattering will broaden the beam and reduce the intensity of the penetrating part. The scattering events mainly take place in collagen fibers and cellular organelles. Absorbed light is converted into heat or will be re-radiated as fluorescent radiation.

Absorption spectra are strongly dependent on tissue composition and especially on water content. At ultraviolet (UV) and infrared (IR) wavelengths most of the light is absorbed in water and an incident beam only penetrates a few cell diameters into the tissue. Visible light with shorter wavelengths (green, yellow) penetrates 0.5–2.5 mm and undergoes an exponential decay in intensity. Both scattering and absorption occur in this spectral region. In the near IR and IR parts of the spectrum scattering is the most pronounced optical phenomenon to consider. In these parts of the spectrum, photons penetrate approximately 10 mm into the tissue.

The transmitted light undergoes scattering in the tissue in a very complicated way. Light is refracted in the microscopic inhomogeneities, formed by cell membranes, collagen fibers, and sub-cellular structures. Pigments such as oxyhemoglobin, reduced hemoglobin, and bilirubin absorb the diffuse light. Anderson and Parrish [4] have given the wavelength-dependent absorption characteristics of these pigments. In human skin, the spectral range 600–1600 nm has very good penetration for light. This part of the spectrum is often called the 'diagnostic window' (or sometimes the 'therapeutic window') because it makes deeper tissue layers accessible.

Thus, optical sensors make it possible to extract important information about a patient's vital health conditions without taking blood samples or using an indwelling catheter.

2.1.2
Propagation of Light in Tissue

Measurement of tissue properties to extract physiological information is often very complex in practice. Mathematical tools have been developed as an aid to model photon transport and scattering in tissue. Problems of this kind can be solved with analytical and/or transport theory. The latter is the most rigorous but has the disadvantage that the complexity of solving the problem often limits the applicability.

However, transport theory has been found to be more easily adoptable.

2.1.3
Transport Theory

Transport theory has been mainly applied to solve problems concerning propagation of electromagnetic waves in scattering and absorbing media. Ishimaru [5] has extensively studied scattering and transport of photons in turbid media. Henyey and Greenstein [6] have studied diffuse scattering of light in galaxies. Their results have, interestingly enough, been of major importance for understanding photon transport in tissue.

The radiative transport equation forms the basis of any treatment of transport of light in tissue

$$\hat{s} \cdot \nabla L(r, \hat{s}) = -\mu_t L(r, \hat{s}) + \mu_s \int_{4\pi} P(\hat{s}, \hat{s}') L(r, \hat{s}) d\varpi' + S(r, \hat{s}) \tag{2.1}$$

where

$L(r, \hat{s})$ is the radiance (W/m$^2 \cdot sr$) at position r in the direction $\hat{s}$

$\mu_t = \mu_a + \mu_s$ is the attenuation coefficient (cm^{-1})

μ_a and μ_s are the absorption and scattering coefficients, respectively

$P(\hat{s}, \hat{s}')$ is the scattering phase function giving the scattering contribution from direction $\hat{s}'$ into $\hat{s}$ where $\hat{s}'$ and $\hat{s}$ are unit vectors

$S(r, \hat{s})$ is a source term.

The scattering-phase function is often described by the Henyey-Greenstein [6] phase function

$$f(\theta) = \frac{1}{4\pi} \frac{1 - g^2}{(1 + g^2 - 2g \cos \theta)^{3/2}} \tag{2.2}$$

g is the anisotropy coefficient, which is equal to the average cosine of the scattering angle

$$g = \int_{-1}^{+1} \cos\theta f(\cos\theta) d(\cos\theta) \tag{2.3}$$

The transport equation can be solved analytically but exact solutions are possible only in a limited number of cases.

Diffusion theory and Monte Carlo models are more widely used for understanding photon migration in tissue.

Both models are based on the assumption that photons travelling in tissue can be regarded as neutral particles and that wave phenomena can be disregarded.

2.1.4
Diffusion Theory and Monte Carlo Models

The difficulties in finding exact solutions of the transport equation has generated approximations based on diffusion theory, which can be derived from the full Boltzmann transport equation. The most common approximation is the diffusion equation

$$\frac{1}{c}\frac{\partial}{\partial t}\phi(r,t) - D\nabla^2\phi(r,t) + \mu_a\phi(r,t) = S(r,t) \tag{2.4}$$

where $\phi(r,t)$ is the diffuse radiant fluence rate, $S(r,t)$ the photon source, and D is the diffusion coefficient, which can be expressed as

$$D = \frac{1}{3\{\mu_a + (1-g)\mu_{tr}\}} \tag{2.5}$$

where μ_{tr} is the transport coefficient.

In Monte Carlo modeling the paths of individual photons are calculated. A random number generator is used for sampling discrete events from interaction coefficients and phase functions. One advantage is that complex geometries can be modeled in simplified ways. The disadvantage of the Monte Carlo method is that long calculations time is often required to get accurate results. Graaff et al. [7] improved the Monte Carlo simulation technique. These authors give a detailed presentation of this simulation method.

There are numerous applications of optical sensor principles that have reached clinical utilization. Most of them are used in the hospital laboratory disciplines in blood and biochemistry measurements. In this chapter we are focusing on three patient monitoring applications:

- near IR spectroscopy
- pulse oximetry
- laser Doppler flowmetry/imaging

These three methods have been chosen as examples of bio-optical progress and because they represent technologies at various stages of introduction into use in health care:

- near IR spectroscopy is offering a great potential for non-invasive brain blood-flow measurements
- pulse oximetry is a clinically well-established method for non-invasive monitoring of oxygen saturation
- laser Doppler flowmetry is the preferred method for non-invasive microvascular blood flow assessments

2.2 Near IR Spectroscopy

Near IR spectroscopy (NIRS) is based upon two important principles.

- Biological tissue is relatively transparent to light in the wavelength interval 600–1000 nm.
- There are substances in the tissue, the absorption of which is dependent on the prevailing oxygen levels.

This means that NIRS wavelengths can be used investigate deeper tissue volumes and the principle can be used to monitor changes in the concentration of compounds that are present in the tissue volume under study. Of particular interest are absorbing compounds, the concentrations of which vary with the oxygenation state of the tissue. These are usually oxyhemoglobin (HbO_2), deoxyhemoglobin (HbO), and oxidized cytochrome oxidase (C_tO_x).

Photon–tissue interaction in the near IR range involves both absorption and scattering, both of which are wavelength dependent and must be taken into account at measurements in the NIRS wavelengths interval.

The NIRS principle can be explained as an application of the Beer–Lambert law, which can be used to calculate absorption in media that do not scatter photons.

Fig. 2.1 illustrates the Beer-Lambert law. A laboratory cuvette contains a solution with a compound having the concentration C. Light travels a distance d through

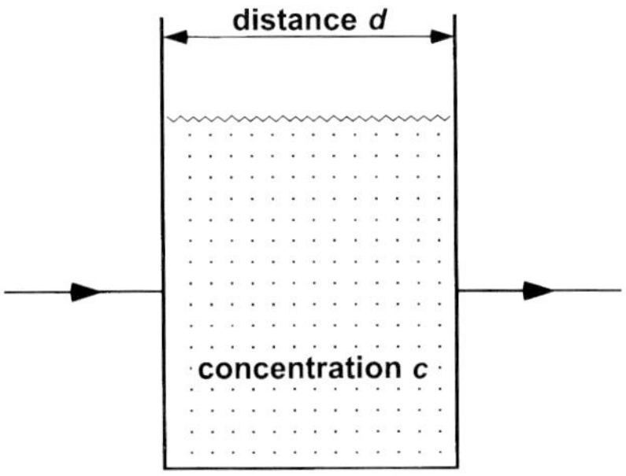

Fig. 2.1 The Beer-Lambert law for a absorbing, non-scattering compound in a cuvette.

the solution. If the solution is non-scattering, light travels in straight paths through the medium. The incident light intensity is I_0 and the intensity leaving the cuvette is I_t. Some light is absorbed in the medium, so $I_0 > I_t$. The loss A (attenuation) is usually measured in 'optical density' (OD).

Beer-Lambert's law can be expressed as

$$A = \log_{10} \frac{I_0}{I_t} = a \cdot c \cdot l \qquad (2.6)$$

where
A is the attenuation 'OD'
I_0 is the incident light intensity
I_t is the transmitted light intensity
a is the specific extension coefficient
c is the concentration of absorbing compound
l is the distance light travels in the absorbing compound

The product $a \cdot c$ is the absorption coefficient μ_a.

$$\mu_a = a \cdot c \ (\mathrm{cm}^{-1}) \qquad (2.7)$$

In biological tissue there exist many types of absorbers and only those that are oxygenation-related are of NIRS importance. Other types of absorbers are always present and contribute to the total attenuation of the tissue under study. The latter type of absorbers can be sorted under 'fixed scattering and absorbing media' (Table 2.1).

Let us briefly review the properties of some of them. Water is present in all types of biological tissue to an extent of 70–90% (80% in brain tissue) of the total tissue volume. Absorption of water is relatively low in the interval 300–900 nm. An absorption peak is located at 970 nm and above 1700 nm there is a substantial increase in the absorption. Water is also absorbing strongly in the UV range (<300 nm).

Skin, bone, and surface tissues, such as muscle tissue, contribute to the total absorption pattern in most NIRS measurements. Skin contains melanin that absorbs, especially in the UV range. Dermis, a part of the skin structure, contains the absorbers hemoglobin and cytochrome oxidase. Concentration changes in these two components in superficial tissue affect the interpretation of the data from deeper lying tissue, which may be of primary interest.

Tab. 2.1 Fixed and oxygenation-dependent absorption and scattering media in tissue.

Fixed scattering and absorbing media	Oxygenization-dependent absorption
Melanin	Oxygenized hemoglobin (HbO_2)
Bilirubin	Deoxygenized hemoglobin (HbO)
Bone	Oxygenized cytochrome oxidase (C_tO_x)

Hemoglobin and the hemoglobin groups are the most important absorbers in the 'diagnostic window'. The absorption spectrum of oxygenated hemoglobin (HbO_2) and deoxyhemoglobin (HbO) differ substantially from each other, which forms the basis of blood oxygen saturation measurements, discussed later in this chapter.

This spectral difference explains the well-known color difference between arterial (oxygenated) and venous (reduced) blood. Arterial blood is bright red whereas venous blood is purple or blue [8].

The two oxidation components can be separated at many wavelengths in the spectrum. At the so called isobestic point (805 nm) the absorption spectra of the two hemoglobins states coincide [8]. This wavelength is often used as a reference point in instruments for blood oxygen saturation measurements.

Cytochrome oxidase (C_tO_x) is the terminal enzyme in the cell respiratory pathway. It is located in the mitochondrial membrane of the cell. Cytochromes are electron transporting, heme-containing proteins and are required in the process of oxidative phosphorylation to convert metabolites and oxygen into CO_2 and high-energy phosphates. In this process cytochromes are oxidized and reduced. Consequently, the red-ox states of cytochromes are probes for the respiratory state of cells. By monitoring spectral changes the cytochrome oxidation state in tissue can be determined [9, 10].

These mechanisms make it possible to quantify the oxygenation of tissue continuously. Brain oxygenation, in particular, has been successfully studied with techniques based on cytochrome changes.

2.2.1 Scattering

When photons are travelling in a tissue matrix some photons are absorbed and energy is dissipated as thermal energy in the medium. When near IR light is propagating in tissue most of the collisions are elastic, which means that energy is lost. Instead, photons are setting up secondary 'light oscillators' (dipoles) that will change the directions of light into scattered light.

The direction of scattering is very much dependent of the wavelength in relation to the size of the scattering particles. In biological tissue a wide distribution of scatterer sizes occur, from sub-cellular structures to collagen fibers or even larger structures. Thus, the resulting scattering pattern is complex and includes directed scattering (forward, or Mie scattering) as well as diffuse scattering (Rayleigh scattering).

The scattering properties of a cell or a particle is expressed as the total scattering cross section, σ_S, and measured in mm^2. The concentration of scattering cells or 'particles' in a tissue volume is called the density ρ and is measured in mm^{-3}. The scattering coefficient μ_s is expressed as

$$\mu_s = \rho \cdot \sigma_S \ (mm^{-1}) \tag{2.8}$$

In most biological tissue scattering is the dominating optical phenomenon. Attenuation of light due to scattering is given by the equation

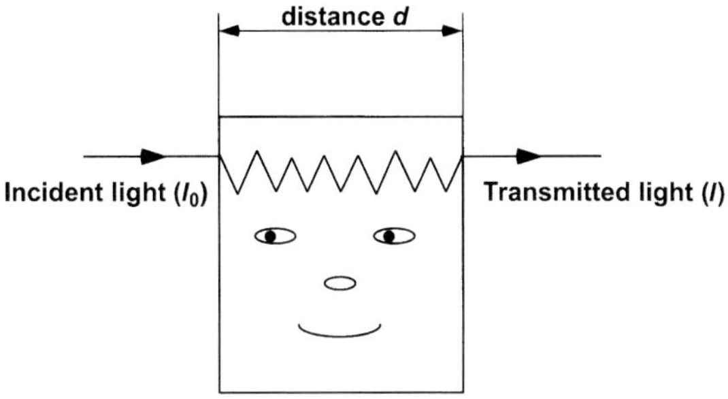

Fig. 2.2 Scattering in tissue increases the path-length of photons compared with the geometrical distance.

$$A = \ln\left(\frac{I_0}{I_t}\right) = \sigma_S \cdot \rho \cdot d \tag{2.9}$$

where d is the distance in meters (m).

In this simple model we assume that scattering only occurs once. In tissue, however, photons are scattered many times before leaving the tissue volume or reaching a photodetector in a measurement device (Fig. 2.2).

The theoretical description of light transport in turbid media is complicated. Various mathematical models have been developed for this purpose. Monte Carlo modeling [11] and transport theory [12] are two often used approaches to light transport calculations as described earlier in this chapter.

In spite of the theoretical complexity, a modified version of the Beer-Lambert law can be quite helpful in understanding the basic scattering processes.

The modified Beer-Lambert law can be written

$$A = \log_{10}\frac{I_0}{I_t} = a \cdot c \cdot d \cdot B + G \tag{2.10}$$

where
$B = \beta/d$ is the differential path-length factor

Generally, G is an unknown factor, which means that this equation cannot be solved. However, if the differential attenuation is measured, G can be eliminated and the attenuation calculated. B can be determined with different techniques, including time- and phase-resolved methods.

2.2.2 Brain Spectroscopy

The major application of NIRS so far has been spectroscopic monitoring of tissue oxygenation of the infant brain as well as adult brain. A schematic set-up for NIRS measurements is shown in Fig. 2.3.

Optical fibers carry photons from the laser diodes at four different wavelengths in the near IR domain. The optical fibers end in so-called optodes (compare 'elec-

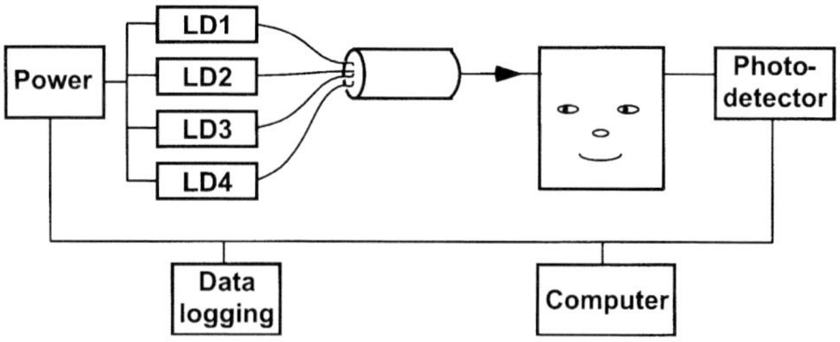

Fig. 2.3 The NIRO-300 (Hamamatsu Photonics KK) instrument for brain blood flow measurement at four different wavelengths.

trodes'), which often contain prisms giving an incident beam perpendicular to the skin surface. Light is received in a similar device. The distance in between the optodes is an important parameter [13]. After a few millimeters of travelling in tissue the photons are diffusely scattered and the actual distance the photons travel before reaching the receiving optode is often 4–6 times longer than the geometrical distance. The factor is often called the differential path-length factor (B).

If d, a, and B are known we can now calculate the change in attenuation

$$\Delta A = \Delta c \cdot a \cdot d \cdot B \tag{2.11}$$

where

a, d, and B are measured in units per μmol of chromophore per liter of tissue per cm

Δc = concentration change

ΔA = change in attenuation

A number of chromophores, each with a unique extinction coefficient will contribute to the total attenuation. Therefore, the change in concentration in a number of chromophores in several wavelengths have to be calculated in order to estimate ΔA.

Changes in concentration of HbO and HbO_2 can be used to calculate hemodynamic parameters such as cerebral blood flow (CBF) and cerebral blood volume (CBV).

2.2.3 Fick's Law Applied to Brain Blood Flow

NIRS estimations of CBF is based on the Fick principle [14], which forms the basis of a variety of minimal invasive cardiovascular clinical measurements. This method was originally described by Fick in 1870. The principle states that blood flow through an organ can be estimated by the organ uptake or release of a substance, divided by the arteriovenous concentration difference of the same substance (Fig. 2.4).

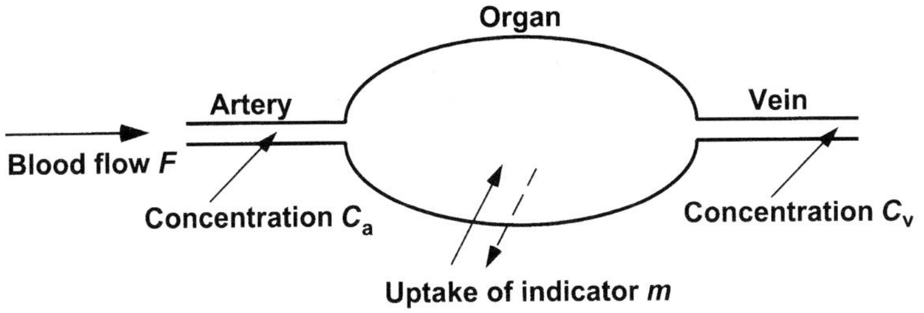

Fig. 2.4 Fick's principle for blood flow measurements.

$$F = \frac{m}{C_a - C_v} \tag{2.12}$$

Kety and Schmidt [15] developed methods to determine m, a parameter that is not easily obtained through direct measurements. m can be estimated as the product of the concentration of the tracer substance in tissue C_T and the weight of the tissue volume under study. The blood perfusion can be estimated in milliliters of blood per 100 grams of tissue and minute. So, brain blood flow can be expressed as

$$\frac{F}{W} = \frac{C_T}{\int_0^t (C_a - C_v)dt} \tag{2.13}$$

where
F = blood flow through the brain
W = unit weight of brain tissue
C_a = arterial tracer substance concentration
C_v = venous tracer substance concentration
C_T = tracer substance concentration in tissue
t = time

In the NIRS brain blood flow application C_v is zero because the measurements are arranged so that the sampling is finished before the tracer substance has occurred at the venous side of the circulatory system. Thus, Equation (13) can be simplified

$$\frac{F}{W} = \frac{C_T}{\int_0^t C_a dt} \tag{2.14}$$

In NIRS brain blood flow measurements the 'tracer' substance is HbO_2 and the changes in HbO_2 concentration forms the basis for flow measurements.

A sudden increase in arterial oxygen saturation is induced through oxygen inhalation. During stationary conditions the small amount of oxygen dissolved in plasma can be disregarded and the actual amount of tracer substance can be calculated as the integral over the change in S_aO_2 with time, multiplied with the total concentration of hemoglobin in whole blood. The advantage of the NIRS application of Fick's principle is that the concentration of the tracer can be measured directly and easily.

The total hemoglobin concentration is constant during the measurement, so it can be assumed that the changes in HbO_2 and Hb are equal and of opposite sign.

The method is based on an accurate determination of the arterial HbO_2 concentration and a simultaneous measurement of arterial oxygen saturation by pulse oximetry. Elwell [16] has given a more extensive theoretical treatment.

2.2.4 Practical Details

It is important that the increase in HbO_2 is rapid. This can be obtained via an external breathing circuit to which the supplies of hypoxic and oxygenated gas mixtures are connected. However, in ventilated patients the dead space of the ventilation circuit impedes a rapid induction of the tracer. Manual ventilation with 100% oxygen for a short period can overcome this problem. A good lung function is important to be able to create a quick rise in HbO_2. Impaired lung function results in a less pronounced step function in HbO_2. In this application it is important to be able to measure saturation in a beat-by-beat mode. Motion artifacts should be avoided by fixing the sensor to the skin surface. It is also important to reduce stray light from the surroundings, which can disturb the measurements.

In the measurement procedure the oxygen fraction of the inhaled gas is lowered, resulting in a S_aO_2 of 92–94%. When the tracer is formed the oxygen concentration should be elevated to 10% for one or two breaths after which the concentration should return to pre-baseline levels. To avoid autoregulatory changes in brain blood flow the changes in S_aO_2 should be kept as small as possible: $<8\%$.

A venous blood sample must be withdrawn to determine the total hemoglobin concentration C_T. The NIRS method is based on the assumption that the tracer concentration is negligible before the measurements start. A second assumption is when the tracer reaches the brain the small changes in concentration that occurs does not disturb the circulatory conditions of the brain.

2.2.5 NIRS Instrumentation

The Hammamatsu Company in Japan is a leader in the field of NIRS. The monitor NIRO 300 (near IR oxygenation monitor) has its main application areas in brain blood flow, muscle oxygenation in sports medicine, and high altitude and aviation medicine. The NIRO 300 monitor presents TOI (tissue oxygenation index), THI (tissue hemoglobin index), concentration changes in HbO_2, Hb, and cytochrome levels. NIRO 300 works at four different wavelengths from laser diodes. The detectors are semiconductor photo diodes. Detectors and light sources are arranged in optodes.

In adults reflection mode measurements are preferred. This means that the photons only travel in a shorter path through the brain and not transcranially, as is often the case in neonatal measurements. The laser diodes are located in the main unit of the instrument and light is carried to the skin surface via optical fibers. In the patient end each optode has a small prism allowing the light to be transmitted perpendicular to the skin surface, which makes it possible to attach the fiber optic bundles along the skin surface.

2.3
Pulse Oximetry

The development of pulse oximetry for oxygen saturation monitoring is very much based on the early findings in photoelectric plethysmography in the 1930s. Hertzman [17, 18] was the first to use the descriptive term photoelectric plethysmography (PPG). The first reports on the successful use of this principle were published in the mid-1930s by Molitor and Kniazuk [19] and Hanzlik [20]. The PPG method is simple, although the underlying detailed optical mechanisms remain unknown. A beam of light is directed toward the part of the tissue in which blood flow (or volume) is going to be measured. Reflected, transmitted and scattered light leaving this volume is collected and brought to impinge on a photodetector surface. A signal modulated by the attenuation or scattering of light in the blood volume can be recorded. Two different signal components can be derived from the photodetector, the a.c. and d.c. components. The a.c. component is pulsatile and synchronous with the heartbeat and the other (d.c. part) is constant. The physiological significance of the two components is still under debate.

PPG has used clinically mainly in the pulse oximetry application, discussed later in this chapter, but also for monitoring of skin perfusion, venous reflux conditions, and skin flap blood flow during plastic surgery. Challoner [21], Roberts [22], and Bernstein [23] have reviewed the PPG literature.

Oxygen levels of blood and tissue are maintained by the respiratory and circulatory systems of the human body. The average partial pressure of oxygen (pO_2) is about 5.3 kPa (40 mmHg) in tissue. At this level about 75% of the hemoglobin is saturated with oxygen. When pO_2 is increased to 8.0–13.3 kPa (60–100 mmHg) the saturation level is 90% and above.

Low oxygen supply to tissues is resulting in hypoxia, which, within minutes, can reduce the work capacity of muscles and lead to cell death.

The partial pressure of oxygen in tissues can be reduced to about 0.4 kPa (3 mmHg) and still support the basic metabolic requirements. To maintain this tissue level the capillary pO_2 must be close to 5.3 kPa (40 mmHg), which corresponds to an arterial level of 12.4 kPa (95 mmHg).

In whole blood oxygen is transported in two ways. Approximately 2% of the total amount of O_2 is physically dissolved in blood plasma and thus transported via the plasma flow.

The rest, 98% is loosely bound to hemoglobin in the form of oxyhemoglobin, HbO_2, in the erythrocytes, the red blood cells. Both the amount of oxygen dissolved in plasma and bound to hemoglobin are dependent of the partial pressure of oxygen pO_2. There is a linear relation between pO_2 and dissolved oxygen amount but a sigmoid shaped hemoglobin oxygen dissociation curve describing the relation between pO_2 and the corresponding saturation level (Fig. 2.5).

Variations in pH, temperature, and pCO_2 change the affinity of Hb for O_2 and shift the location of the oxyhemoglobin dissociation curves. Under normal physiological conditions arterial blood has a pO_2 of 13.3 kPa (100 mmHg). The corresponding oxygen saturation S_aO_2 level is 97–98%. On the venous side of the circulatory system the corresponding pO_2 values are considerably lower.

Historically, the O_2 and CO_2 levels of human blood has been analyzed by taking a blood sample from an indwelling catheter or from an arterial puncture. The sample is then analyzed in the laboratory using established instrumentation principles. This procedure has a number of disadvantages. Infection risks, discomfort to the patient, and difficulties with sample handling makes a non-invasive, direct method attractive. Today, pulse oximetry offers an interesting alternative for oxygen saturation measurements.

Oxygenated whole blood absorbs strongly in the blue region of the spectrum and has a bright red color. Reduced blood absorbs in the whole visible region and has a dark or even bluish color. The optical absorption spectra for HbO_2 and HbO differ substantially and this difference forms the basis for spectroscopic determination of the two types of hemoglobin in whole blood. Absorption spectra for the hemoglobin states are seen in Fig. 2.6.

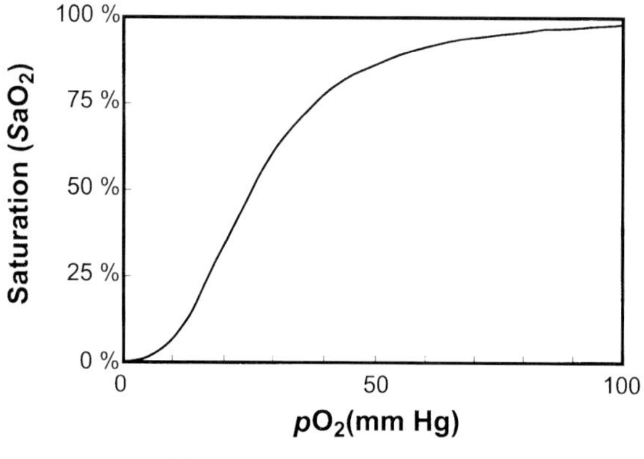

Fig. 2.5 Hemoglobin–oxygen dissociation curve. A sigmoid relation exists between the partial pressure of oxygen and the oxygen saturation.

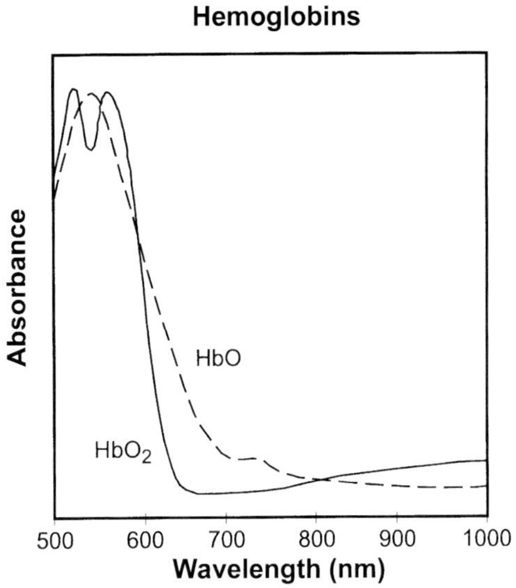

Fig. 2.6 Absorption spectra for HbO_2 and HbO. Maximum difference occurs at about 660 nm. The isobestic point is at 805 nm.

In whole blood there are also other types of hemoglobins present: carbo-oxyhemoglobin and met-hemoglobin. These are the dysfunctional hemoglobins that do not bind oxygen reversibly. Because of the two kinds of hemoglobins (functional and dysfunctional) functional oxygen saturation is defined as

$$SO_2 = \frac{HbO_2}{HbO + HbO_2} \cdot 100\% \tag{2.15}$$

Non-invasive oximetry is based on several important discoveries through the years. In 1943 Horecker [8] presented the absorption spectra for HbO_2 and HbO. The Beer-Lambert law for light absorption was applied to measurements of solutions of hemoglobin derivatives.

Hüfner [24] published a measurement principle, which later was generally applied to oximetry. He analyzed two compartment systems by measuring the absorption of light in hemoglobin derivatives at two wavelengths. He chose one wavelength in which the difference in absorption between HbO_2 and HbO was significant (660 nm) and another in which the absorption coefficients are equal for the two states, the isobestic point (805 nm). This two-wavelength method has been extensively used in S_aO_2 determinations.

Kramer and collaborators [25] contributed with the important finding that Beer-Lamberts law could be applied to non-hemolysed whole blood or suspensions of erythrocytes. The same group demonstrated nonlinearities when this law was applied to scattering solutions. The nonlinearity is an effect of scattering, reflection, and refraction in the solution.

Wood and Gerachi [26] developed a non-invasive method for arterial S_aO_2 measurements *in vivo* from the ear pinna. By compressing the tissue under study a baseline from bloodless tissue could be obtained.

In 1974 Aoyagi [27] suggested a method to eliminate the effect of many interfering absorbers by only regarding the optical differences occurring at every heart beat, every 'pulse'. The pulse oximeter was thereby invented. Nilsson [28] reviewed the development of early oximetry development in 1960. Severinghaus and Astrup [29]) presented a classical review of pulse oximetry development in 1986.

2.3.1 Theory

As seen in Fig. 2.6 the spectra for HbO_2 and HbO differ from each other. S_aO_2, the ratio of HbO to total hemoglobin can be analyzed by absorption measurements at two or more wavelengths in the visible and near-IR part of the spectrum. The absorption of light in hemoglobin solutions follows the Beer-Lambert law

$$\mathrm{OD} = \log_{10}\frac{I_\mathrm{t}}{I_0} = -a \cdot c \cdot l \tag{2.16}$$

As indicated earlier the Beer-Lambert law is only approximately correct for scattering media (see the NIR section).

The hemoglobin molecule is one of the strongest absorbing molecules in the tissue. However, scattering in turbid media causes most of the attenuation of light. The effective hemoglobin concentration is about 150 μM. With the absorption coefficients for HbO_2 and HbO in the order of 100–1000 $\mathrm{M^{-1}cm^{-1}}$ the hemoglobin content is approximately 0.2 OD of the total optical density. In addition, only 10% is pulsatile, so the pulse signal is only a few percent of the total amplitude.

For a two-wavelength system (λ_1 and λ_2) the attenuation can be written

$$\begin{aligned}\mathrm{OD}_{\lambda_1} &= A_{\lambda_1} = a_{0_1} \cdot c_0 \cdot l_0 + a_{\mathrm{r}_1} \cdot c_\mathrm{r} \cdot l_\mathrm{r} + a_{x_1} \cdot c_x \cdot l_x + A_{y_1}\\ \mathrm{OD}_{\lambda_2} &= A_{\lambda_2} = a_{0_2} \cdot c_0 \cdot l_0 + a_{\mathrm{r}_2} \cdot c_\mathrm{r} \cdot l_\mathrm{r} + a_{x_2} \cdot c_x \cdot l_x + A_{y_2}\end{aligned} \tag{2.17}$$

The blood volume changes during each cardiac cycle cause a modulation of the measured absorbencies. If each of the equations is differentiated with the respect to time the two last terms in each equation will disappear because the product $c_x l_x$ does not change during the cardiac cycle (outside the circulatory system), so $\mathrm{d}(c_x l_x)/\mathrm{d}t = 0$.

Non-specific effects of light attenuation is assumed to be constant during the cardiac cycle, so $\mathrm{d}(A_y)/\mathrm{d}t = 0$. By analyzing the remaining part of the equations it is reasonable to assume that the change in absorbency occurring is due to changes in the optical path lengths at the two wavelengths ($\mathrm{d}l_0/\mathrm{d}t$, $\mathrm{d}A_y/\mathrm{d}t$). We can then form the quotient R between the rates of changes in absorbency of the two wavelengths

$$R = \frac{\mathrm{d}(\mathrm{OD}_{\lambda_1})/\mathrm{d}t}{\mathrm{d}(\mathrm{OD}_{\lambda_2})/\mathrm{d}t} = \frac{-\mathrm{d}\left(\log\frac{I_1}{I_0}\right)/\mathrm{d}t}{-\mathrm{d}\left(\log\frac{I_2}{I_0}\right)/\mathrm{d}t} = \frac{\frac{\Delta I_1}{I_1}}{\frac{\Delta I_2}{I_2}} = \frac{a_{0_1}\cdot c_0 + a_{r_1}\cdot c_r}{a_{0_2}\cdot c_0 \cdot a_{r_2}\cdot c_r} \tag{2.18}$$

Functional oxygen saturation S can be expressed as

$$S = \frac{c_0}{c_0 + c_r} \tag{2.19}$$

By combining Equations (18) and (19) we obtain

$$S = \frac{a_{r_1} - a_{r_2}\cdot R}{(a_{r_1} - a_{0_1}) - (a_{r_2} - a_{0_2})R} \tag{2.20}$$

This equation gives the relation between R, which we can measure, and S, which is the blood oxygen saturation we want to monitor.

2.3.2 Empirical Calibration

The mathematical model presented above and technical variability in the optical probe parameters requires that we calibrate the pulse oximeter instruments empirically. In this calibration procedure we estimate the relation between S and R in the equation

$$S = \frac{a - b\cdot R}{c - d\cdot R} \tag{2.21}$$

From the empirical calibration procedure we can determine a, b, c, and d. With this knowledge we can determine the oxygen saturation non-invasively and continuously with an accuracy of within a few per cent.

2.3.3 Clinical Use

Pulse oximetry is today used routinely in operating theaters and in critical care units at hospitals around the world. The advantages of this method is that it provides accurate, continuous, and reliable information about an important circulatory parameter, the oxygen saturation level, sometimes called the 'fifth vital sign'.

Pulse oximetry is simple to use in the clinical environment. It can be easily attached to the patient. Usually, a finger is used in adults. Reflection mode sensors are available, which opens additional locations for sensor positioning. However, this type of sensor is considered less accurate and more exposed to interference.

The disadvantages of pulse oximetry are the sensitivity to interferences from external light sources and other electromagnetic fields. False readings can also occur if the patient has high levels of dysfunctional hemoglobins, which can be the case in smokers. Circulatory disorders resulting in weak peripheral pulse or poorly perfused tissue can also limit the applicability of pulse oximetry.

2.4
Laser Doppler Flowmetry

Laser Doppler Flowmetry (LDF) is a continuous and non-invasive method for tissue blood flow monitoring, utilizing the Doppler shift of laser light as the information carrier. The method has shown its potential usefulness in clinical assessment of blood flow within disciplines such as dermatology, plastic surgery, and gastrointestinal surgery. In experimental medicine, laser Doppler flowmetry has been used in the study of a large number of problem areas related to skin as well as in various internal organs and skeletal muscles. Temporal as well as spatial variations can easily be monitored under continuous conditions. Several commercial instruments are available today, some of which present color-coded images of microcirculatory flow patterns, giving spatial and temporal information.

2.4.1
Light Scattering and Doppler Shift of Laser Light

Laser Doppler measurements are based on light scattering in tissue, aspects of which have been presented earlier in this chapter. Of particular interest are the differences between scattering from fixed and moving cells/structures. Photons which interacts with moving red blood cells (erythrocytes) will undergo a Doppler (frequency) shift whereas photons hitting fixed structures will be scattered with unchanged frequency.

2.4.1.1 Elastic and Quasi-Elastic Scattering

When incident light with a frequency of ω_{i} interacts with stationary molecules, the secondary (scattered) radiation ω_{sc} has the same frequency as the incident radiation.

$$\omega_{\mathrm{i}} = \omega_{\mathrm{sc}} \tag{2.22}$$

We call this scattering type elastic scattering. However, when the incident light hits a cell or structure that is moving, a red blood cell in a capillary loop, the frequency of scattered radiation from the moving cell will differ from the frequency of incident light with a factor $\Delta\omega_{\mathrm{sc}}$. The shift in frequency is the well-known Doppler shift caused by the motion of the particle relative to the observer. Because of the low velocity of the erythrocytes in the tissue, the Doppler shifts are usually quite small.

$$\omega_i \gg \Delta\omega_{sc} \tag{2.23}$$

In blood flow studies, usually

$$\frac{\Delta\omega_{sc}}{\omega_i} \approx 10^{-11} \tag{2.24}$$

This is called quasielastic scattering and prevails in most microvascular blood flow applications.

2.4.1.2 Doppler Shift

The frequency shift resulting from relative motion of the source and observer of a propagating wave is known as the Doppler effect. The magnitude of the frequency shift is given by

$$\Delta f = \frac{v}{c} V_0 \tag{2.25}$$

where
v is the velocity of the source with respect to the observer
c is the velocity of the carrier wave
V_0 is the unshifted frequency

The velocity of the source can be determined if c and V_0 are known. Christian Doppler [30] used this principle for the measurement of the speed of galaxies. In laboratory measurements it has been used for velocity measurements in a variety of applications from wind tunnel experiments to the flow of protoplasm in individual cells. LDF also has an excellent spatial resolution. Generally, the volume resolution is of the order of 1 mm^3. In some types of tissue, such as highly absorbing liver tissue, this resolution becomes less. On the other hand, neural tissues, which have excellent scattering properties, can modify the spatial resolution to values much greater than 1 mm^3.

Velocities can be measured in the range 10^{-5} to 10 m/s. In capillary blood flow applications, the erythrocyte velocity is around 10^{-3} m/s. In such an application, the Doppler shift of laser light is quite small. Such small frequency shifts are hard to measure directly. Instead, the frequency-shifted light is mixed with unshifted light on the surface of a nonlinear photodetector to extract the information of interest, that is, the frequency difference of the Doppler frequency (heterodyne detection). In blood flow studies in the microcirculation the frequency difference is in the order of 10^1–10^4 Hz.

2.4.2 Instrumentation

In a typical flow measurement experiment, light from a laser passes through a polarizing element. The incident beam impinges on the medium in which flow shall be determined (Fig. 2.7). The position of the detector defines the scattering angle ϕ. The scattered light is collected on a photodetector surface.

The angle between the incident light and the scattered light falling on the detector surface is called the scattering angle ϕ. There are two Doppler shifts to consider; motion with respect to source changes the frequency of scattered light and motion relative to the detector shifts the frequency observed. It can be shown that the result and total Doppler shifts are

$$\Delta f = (\mathbf{K}_s - \mathbf{K}_0) \cdot \frac{\mathbf{V}}{2\pi} = \mathbf{K} \cdot \frac{\mathbf{V}}{2\pi} \tag{2.26}$$

where
V is the velocity vector, $\mathbf{K}_s$ and $\mathbf{K}_0$, respectively

$$|\mathbf{K}| = |\mathbf{K}_s - \mathbf{K}_0| = \frac{4\pi}{\lambda} \sin\frac{\theta}{2} \tag{2.27}$$

The unshifted light is usually brought to the detector in the form of reflected light from stationary surfaces. In other applications, though, a portion of the incident light must be directed towards the photodetector where it is mixed with the portion of light that has passed the scattering volume under study.

The signal from the photodetector can be processed as discrete photo pulses (photon counting) or as an analogue signal. Signal processing gives either the autocorrelation function or the frequency spectrum of the signal. The power spectral density can be obtained by a Fourier transformation of the autocorrelation function. The choice of autocorrelation function vs. power spectral density for presentation of the measurement depends on the application and the traditions within the different fields in which the LD method is applied.

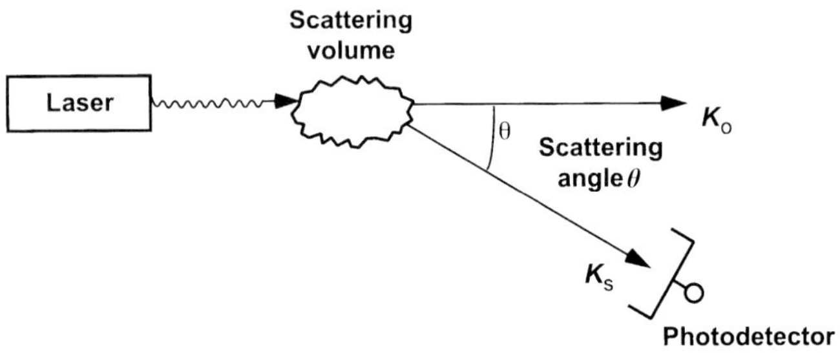

Fig. 2.7 Schematic representation of the light scattering experiment.

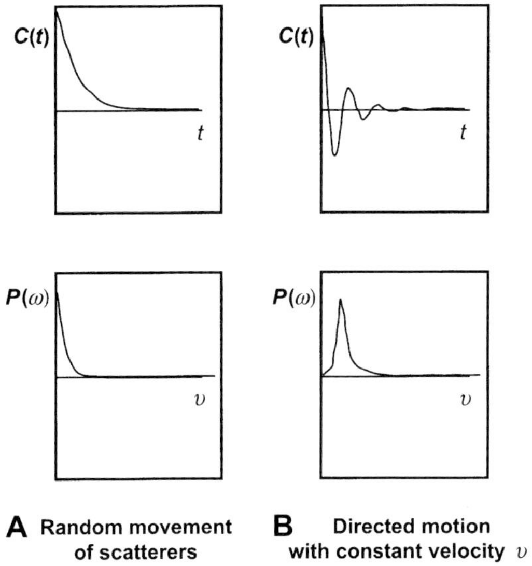

Fig. 2.8 The autocorrelation function and power spectral density: A) at random moving particles; B) constant directed velocity. Observe the peak in $P(\omega)$ corresponding to the Doppler frequency.

As a rule, correlation is used for higher frequency shifts, whereas the power spectral density representation is more appropriate for low frequency applications. Fig. 2.8 shows the typical shape of correlation curves as well as the spectral density curves for two types of LDF recordings.

A more detailed presentation of the physics of LDF methodology (light scattering in tissue, photon diffusion, models for interaction between light and tissue) can be studied in numerous books [31–34] and articles [35–45]. In instruments for microvascular blood flow measurements, the laser light is launched and the scattered light is picked up by optical fibers, or, in the case of imaging devices, an airborne scanning laser beam [46].

2.4.3
Fiber Optics Geometry and Fiber Types

The most common way to arrange the optical fibers in LDF is illustrated in Fig. 2.9.

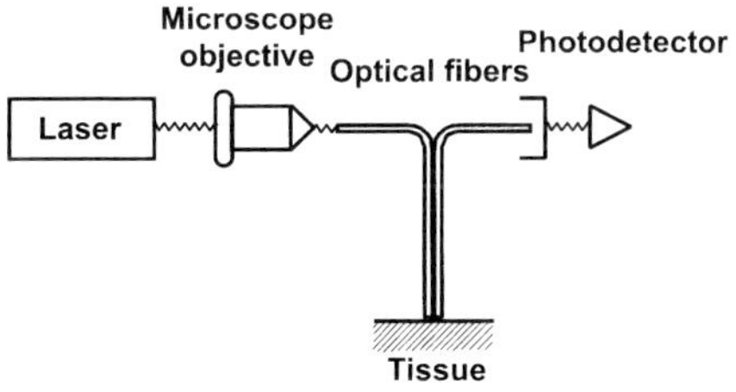

Fig. 2.9 Typical laser Doppler set-up.

The laser light is coupled into the optical fibers by means of a microscope objective (×5 to ×40). A variety of optical fiber types have been used with diameters from 50 to 2000 μm. Plastic as well as silica fibers are commonly used.

Nilsson et al. [43] presented a variant in which one central fiber is leading light to the tissue and a number of fibers surrounding the central fiber were used to pick up the scattered light. Later this design was simplified to a three-fiber probe with the fibers arranged close to each other in a triangular configuration. The distance to the tissue surface under study was standardized to 0.8 mm [43, 47]. Bonner et al. [48] used two 125 μm graded index fibers. In their design the fibers were separated 0.5 mm, which was regarded to be an optimum.

Salerud and Öberg [49] suggested the use of a single optical fiber technique for use in deep tissue perfusion measurements. By using small fiber diameters and by introducing the fiber into the tissue volume in an atraumatic way, blood flow from deeply located tissue volumes can be recorded. With this technique Larsson et al. [50–53] recorded intramuscular blood flow in human trapezoid muscles under different workloads. Vasomotion, a local periodic flow variation (4–9 per minute) has often been seen in these recordings, which is an indication that the introduction of the fiber into the muscle tissue does not seriously traumatize the tissue volume under study.

2.4.4
Signal Processing Principles

An important problem in Laser Doppler Flowmetry is to find algorithms that from the Doppler shifts can assess complicated blood flow patterns. Several approaches to signal processing and interpretation have been presented. The ideal algorithm calculates the magnitudes of flow in all tissue types and under various hematocrit and vascular geometries, taking into account the effects of multiple scattering. So far, no such algorithm exists. The various approaches that have been suggested are the following. Stern et al. [54] used a 'heuristic flow parameter'

$$F = \sqrt{\int \omega^2 \cdot P(\omega) \mathrm{d}\omega} \tag{2.28}$$

where $P(\omega)$ is the power spectral density of the Doppler signal. This algorithm is based on empirical findings, but the authors have shown that in a number of applications it was found to give an accurate estimate of flow. In practice, the flow value can be calculated as

$$F = \frac{\sqrt{R^2 - S \cdot I}}{I} \tag{2.29}$$

where R is the output of an RMS-detector that receives the signal, I is the mean photocurrent, and S is a gain constant.

Normalization with respect to photocurrent serves to make the flow parameter independent of laser intensity and reflectivity of the tissue.

Bonner and Nossal [39] developed a mathematical model for LDF measurements of tissue blood flow. Their model is based on the following assumptions. The scattering cross section for particles of the size of a red blood cell is sharply peaked in the forward direction (Mie scattering). The radiation in a tissue matrix mostly contains photons that have been scattered at least once by surrounding tissue elements. A small percentage of the backscattered light comes from photons scattered only by moving erythrocytes. Such a complex tissue matrix can generally not be characterized in detail. As long as the tissue scattering cross section is much greater than that of the moving cells, a model can be designed, which can be of guidance in instrument design. In a tissue matrix, photons generally undergo several collisions before they interact with a blood cell. This means that the direction of light will be randomized before the collision between a photon and moving blood cells occurs. Bonner and Nossal showed that the normalized first moment of the spectrum $\langle\omega\rangle$ is proportional to the RMS speed of moving particles $\langle V\rangle^{1/2}$.

$$\langle\omega\rangle = \frac{\langle V^2\rangle^{1/2}}{\sqrt{12\xi}\cdot a}\cdot\beta f(\overline{m}) \tag{2.30}$$

where a is the radius of an average spherical scatterer, ξ is an empirical factor related to the shape of a cell, and β is an instrumental factor, namely, the optical coherence of the signal at the detector surface and $\overline{m}$ is the average number of collisions that a detected photon makes with a moving cell, $f(\overline{m})$ is a function of the hematocrit of tissue and is linear if $\overline{m}<1$: low hematocrit and low perfusion conditions. In experiments with a flow model system based on hollow silicon fibers surrounded by a scattering medium, these authors could demonstrate that the Doppler shift varied linearly with the mean velocity of the flowing cells. Nilsson et al. [47], independently, came to a similar conclusion by using a more experimental approach. They verified that for low and moderate red cell volume fractions, the first moment of the power spectral density is approximately proportional to the flux

$$\text{Flux} = \int_0^{\omega} \omega\cdot P(\omega)\mathrm{d}\omega \tag{2.31}$$

However, for higher local hematocrits and higher flow velocities, this algorithm underestimates the flux value. Ahn et al. [55, 69] evaluated the first moment algorithm in experiments on isolated parts of the small intestine. They found that the first moment algorithm seriously underestimated the flow at higher flow rates (>100 mL min^{-1} 100 g tissue).

2.4.5 Calibration and Standardization of LDF Flow Meters

Most LDF devices need to be calibrated at regular intervals by comparing them with a standard to maintain accuracy over longer periods of time. The standard must have a well-specified accuracy and long-term stability. An instrument under calibration is adjusted to give readings coinciding with the setting of the standard. An alternative is to plot calibration curves: diagrams describing the difference between the 'true' value (standard) and the reading of instruments under calibration. The calibration curve is then used for correcting the actual measurements obtained with the instrument.

Such calibrations are not possible with laser Doppler instruments because there is no 'gold standard' for blood flow in tissue. Nevertheless, the readings of the LDF method can be compared with a variety of methods for blood flow measurements. Such comparisons cannot generally be regarded as 'calibrations' in the true sense of this concept because other methods have usually unspecified accuracies or even measure a different physiological variable related to the flux of erythrocytes. Nevertheless, comparisons of this type can give useful information in a specific experiment or preparation.

Examples of comparisons with the Xenon isotope technique are available in the literature [56–60]. Comparisons with radioactive microspheres are reported elsewhere [61–64].

The conclusion that can be drawn from experimental calibration of laser Doppler instruments is that a flow meter can be calibrated in absolute units ($mL\ min^{-1}\ 100\ g^{-1}$) for a particular measurement site in a specific preparation. However, the calibration factor obtained in this way is unique for the particular site and tissue preparation. It should not be used in other preparations and in other tissue models. In the light of these results, the attempts by commercial producers to market factory-calibrated LDF instruments must be seriously questioned. Fig. 2.10 shows a calibration experiment in a rat liver model. Two calibrations are performed in two liver preparations.

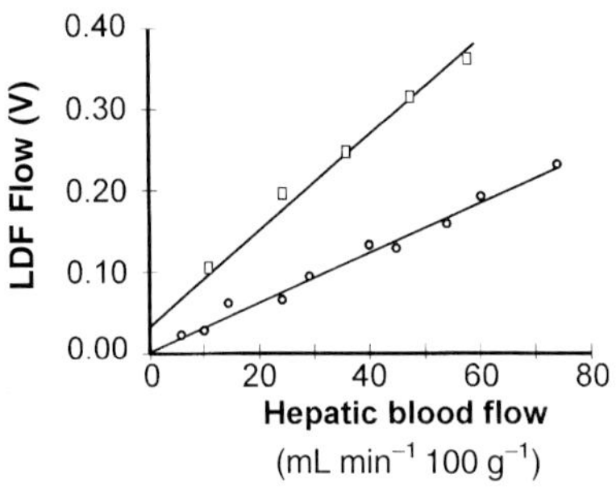

Fig. 2.10 A calibration experiment in laser Doppler flowmetry. Hepatic blood flow is compared with the LDF signal in a two-rat liver experiment (from Shepherd et al. [68] with permission).

2.4.6
Standardization

To standardize an instrument or a method is to make sure it maintains stability, linearity, and reproducibility. Standardization does not mean that an absolute value of flow is achievable. Several methods have been suggested for the standardization of LDF instruments.

Several authors have designed rotating discs for the standardization of LDF instruments. A slowly revolving disc with scatterers is used as scattering model, the rotational speed of which can be varied. A fixed block with stationary scatterers (Delrin) between the rotating disc and optical part of the instrument under calibration simulates the static tissue, which scatters unshifted photons back to the photodetector. Several authors have presented devices of this kind [65–68].

Another way of standardizing a laser Doppler instrument is to use the Brownian motion of particles in a suspension. The Brownian motion is temperature dependent but under well-controlled temperature conditions the particle motion can be used for standardizations. For short-term use (1–2 h) regular milk can be used but for longer periods (2–3 months) it is important to use a special suspension that is not affected by sedimentation or particle aggregation.

2.4.7
Applications of the Laser Doppler Principle

The applications in which the laser Doppler method has been utilized so far are numerous and about 2000 papers have been published so far in which LDF has been used to study microvascular blood flow. It is beyond the scope of this chapter to cover all the areas in which Laser Doppler Flowmetry has been used. The reader is referred to Shepherd and Öberg [70] and Belcaro et al. [71] for more detailed presentations of applications. The following summary gives a general idea of which medical fields the methods have been used in so far. The laser Doppler method has been used mostly in experimental studies. Some areas of major interest for LDF studies are the following.

- Cochlear blood flow
- Neurological applications
  - peripheral nerves
  - CNS blood flow
- Kidney and liver blood flow
- Skeletal muscle studies
- Bone blood flow studies
- Skin blood flow studies
  - effect of local anesthesia
  - skin physiology
  - provocation studies
  - skin pharmacology studies

- Plastic surgery
- Obstructive and occlusive diseases
- Odontological studies
  - gingival studies
  - pulp blood flow
- Retinal blood flow
- Gastrointestinal blood flow

2.5
Conclusions

The pros and cons of laser Doppler flowmetry can be concluded the following way.

2.5.1
Advantages

- LDF is a continuous, totally non-invasive method for microvascular blood flow measurements with a large dynamic range.
- It is convenient to use in clinical studies.
- In most applications there exists a linear relationship between the flux of red cells and the output of an LDF instrument.
- LD flowmeters measure a well-defined physical quantity: the Doppler shift of monochromatic light.
- A well-defined and stable zero level can be identified. The basic principle is technically adaptable to a variety of medical and biological applications.

2.5.2
Disadvantages

- LDF is a relative method and can be calibrated in absolute units only in exceptional cases.
- There exist no universal calibration factor for all kinds of tissues.
- Only superficial tissue layers can be studied because of the limited penetration depth of laser light in tissue.
- The method cannot differentiate between thermoregulatory and nutritive blood flow.
- Optical fibers are sensitive to motion, giving rise to motion artifacts.
- Laser instabilities and ambient light can cause measurement errors.
- The size of the probed volume varies between tissue types.

2.6 References

1 Hales, S. Statistical essays: containing haemostatics or an account of some hydraulic experiments made on the blood and blood vessels of animals (1733). In Classics of Cardiology Vol. I (Willius, F. A. and Keys, T. E. eds). Dover Publications, New York **1961**, 125.

2 Leuwenhoek, A. V. On the circulation of blood. 65th Letter of the Royal Society 1688. Facsimile with introduction by A. Schierbeck. Graaf, London **1962**, 165.

3 Soares, O. D. D. and Scheggi, A. M. (eds). Bioptics: Optics in Biomedicine and Environmental Sciences. *SPIE*, **1992**, *1524*.

4 Anderson, R. R. and Parrish, J. A. Optical Properties of Human Skin. In The Science of Photomedicine (Regan, J. D. and Parrish, J. A. eds). Plenum Press, New York **1987**, 147–194.

5 Ishimaru, A. Wave Propagation and Scattering in Random Media. Academic, New York **1978**.

6 Henyey, L. G. and Greenstein, J. L. Diffuse radiation in the galaxy. *J. Astrophys.* **1941**, *93*, 70–83.

7 Graaff, R., Koelink, M. H., de Mul, F. F. M., Zijlstra, W. G., Dassel, A. C. M., and Aarnoudse, J. G. Condensed Monte Carlo simulations for the description of light transport. *Appl. Optics* **1991**, *32*, 426–434.

8 Horecker, B. L. The absorption spectra of haemoglobin and its derivative in the visible and near infra-red regions. *J. Biol. Chem.* **1943**, *148*, 173–183.

9 Jöbsis, F. F. Non-invasive infrared monitoring of cerebral and myocardial oxygen sufficiency and circulatory parameters. *Science* **1977**, *198*, 1264–1267.

10 Jöbsis, F. F., Keizer, L. H., La Manna, J. C. and Rosenthal, M. Reflectance spectroscopy of cytochrome aa3 in vivo. *J. Appl. Physiol.* **1977**, *43*, 858–872.

11 Morin, R. L. Monte Carlo Simulation in the Radiological Sciences. CRC Press, Boca Raton, FL **1988**.

12 Bohren, C. F. and Hoffman, D. R. Absorption and scattering of light by small particles. Wiley, New York **1983**.

13 Duncan H., Meek J. H., Clemence M., Elwell C. E., Tyszczuk L., Cope M., and Delpy D. T. Optical pathlength measurements on adult head, calf and forearm and the head of newborn infants using phase-resolved optical spectroscopy. *Phys. Med. Biol.* **1991**, *40*, 295–304.

14 Fick, A. Über die Messung des Blutquantums in den Herzventrikeln. Verh. Deutsch. Phys. Med. Ges. 2, XVI (1870), cited by H. E. Hoff and H. J. Scott, *N. Engl. J. Med.* **1948**, *239*, 122.

15 Kety, S. S. and Schmidt, C. F. The nitrous oxide method for the quantitative determination of cerebral blood flow in man: theory, procedure and normal values. *J. Clin. Invest.* **1948**, *27*, 476–483.

16 Elwell, C. E. A practical users guide to near infrared spectroscopy. Hamamatsu Photonics KK **1995**.

17 Hertzman, A. B. Photoelectric plethysmography of the nasal septum in man. *Proc. Soc. Exp. Med.* **1937**, *124*, 328.

18 Hertzman, A. B. The blood supply of various skin areas as estimated by the photoelectric plethysmograph. *Am J. Physiol.* **1938**, *124*, 328–340.

19 Molitor, H. and Kniazuk, M. A new bloodless method for continuous recording of peripheral circulatory changes. *J. Pharmacol. Exp. Ther.* **1936**, *57*, 6.

20 Hanzlik, P. J., Deeds, F. and Terada, B. A simple method for demonstrating changes in blood supply to the ear and effects of some measures. *J. Pharm. Exp. Therap.* **1936**, *56*, 194–204.

21 Challoner, A. V. J. Photoelectric plethysmography for estimating cutaneous blood flow. In Non-invasive Physiological Measurements, Vol. 1 (Rolfe, P. ed). Academic Press, London **1979**, 127.

22 Roberts, V. C., Photoplethysmography – fundamental aspects of the optical properties of blood in motion. *Trans. Inst. MC* **1979**, *4*, 1982.

23 Bernstein, E. F. Non-invasive diagnosis. Techniques in Vascular Disease, 2nd edn. Mosby, St Louis **1982**.

24 Hüfner, G. Über die gleichzeitige quantitative Bestimmung zweier Farbstoffe im

Blute mit Hilfe des Spectrophotometers. *Arch. Anat. Physiol. Abt.* **1990**, 39–48.
25 KRAMER, K., ELAM, Z. O., SAXTON, G. A. and ELAM, W. N. Jr. Influence of oxygen saturation, erythrocyte concentration and optical depth upon the red and near-infrared light transmittance of whole blood. *Am. J. Physiol.* **1951**, *165*, 229–246.
26 WOOD, E. H. and GERACHI, J. E. Photoelectric determination of arterial oxygen saturation in the finger tip. *J. Lab. Clin. Med.* **1948**, *34*, 387–401.
27 AOYAGI, T., KISHI, T., YAMAGUCHI, K. and WATANABE, S. Improvement of earpiece oximeter. In Proc. 13th Jap. Med. Electronics and Biological Engineering Conference, Osaka **1974**, 90.
28 NILSSON, N. J. Oximetry. *Physiol. Rev.* **1960**, *40*, 1–26.
29 SEVERINGHAUS, J. W. and ASTRUP, P. B. History of blood gas analysis VI, Oximetry. *J. Clin. Monit.* **1986**, *2*, 135.
30 DOPPLER, C. Über das farbige Licht der Doppelsterne und einiger anderer Gestirne des Himmels. *Abh. königl. böhm. Ges. Wiss.* **1843**, *2*, 465–482.
31 CHU, B. Laser Light Scattering. Academic Press, New York **1974**.
32 BERNE, B. J. and PECORA, R. Dynamic Light Scattering. Wiley, New York **1979**.
33 DRAIN, L. E. The Laser Doppler Technique. Wiley, New York **1980**.
34 CUMMINS, H. Z. and SWINNEY, H. L. Light beating spectroscopy. in Progress in Optics, Vol. 8 (WOLF, E. ed). North-Holland, Amsterdam **1970**.
35 DUTEIL, L., BERNENGO, J. C. and SCHALLA, W. A double wavelength laser Doppler system to investigate skin microcirculation. *IEEE Trans. Biomed. Eng.* **1985**, *BME-32*, 439.
36 NOSSAL, R. Laser light scattering. *Meth. Exp. Phys.* **1982**, *20*, 299.
37 JOHNSON, C. Optical diffusion in blood. *IEEE Trans. Biomed. Eng.* **1970**, *BME-17*, 129.
38 KHAN, A., SCHALL, L. M., TUR, E., MAYBACH, H. I. and GUY, R. H. Blood flow in psoriatic skin lesions: the effect of treatment. *Br. J. Dermatol.* **1987**, *117*, 193.
39 BONNER, R. F. and NOSSAL, R. Model for laser Doppler measurements of blood flow in tissue. *Appl. Opt.* **1981**, *20*, 2097.
40 BONNER, R. F., NOSSAL, R., HAVLIN, S. and WEISS, G. H. Model for photon migration in turbid media. *J. Opt. Soc. Am.* **1987**, *A4*, 423.
41 MARET, G. and WOLF, P. E. Multiple light scattering from distorted media. The effect of Brownian motion of scatterers. *Z. Phys. B* **1987**, *65*, 309.
42 STEINKE, J. M. and SHEPHERD, A. P. Diffusion model of the optical absorbance of whole blood. *J. Opt. Soc. Am.* **1988**, *5*, 813.
43 NILSSON, G. E., TENLAND, T. and ÖBERG, P. Å. A new instrument for continuous measurement of tissue blood flow by light beating spectroscopy. *IEEE Trans. Biomed. Eng.* **1980**, *BME-27*, 12.
44 STERN, M. D. Laser Doppler velocimetry in blood and multiple scattering fluids: theory. *Appl. Opt.* **1985**, *24*, 1968.
45 TWERSKY, V. Absorption and multiple scattering by biological suspensions. *J. Opt. Soc. Am.* **1970**, *60*, 1084.
46 WÅRDELL, K., JACOBSON, A. and NILSSON, G. E. Laser Doppler perfusion imaging by dynamic light scattering. *IEEE Trans. Biomed. Eng.* **1993**, *40*, 309–316.
47 NILSSON, G. E., TENLAND, T. and ÖBERG, P. Å. Evaluation of laser Doppler flowmeter for measurements of tissue blood flow. *IEEE Trans. Biomed. Eng.* **1980**, *BME-27*, 597.
48 BONNER, R. F., CLEM, T. R., BOWEN, P. D. and BOWMAN, R. L. Laser-Doppler real-time monitor of pulsatile and mean blood flow in tissue microcirculation. In Scattering Techniques Applied to Supra-Molecular and Nonequilibrium Systems, Vol. 73, **1981**. CHEN, S.-H., CHU, B., NOSSAL, R. (eds), Plenum Press, New York, 685–701.
49 SALERUD, E. G. and ÖBERG, P. Å. Single-fibre laser Doppler flowmetry. A method for deep tissue perfusion measurements. *Med. Biol. Eng. Comput.* **1987**, *25*, 329.
50 LARSSON, S. E., BODEGÅRD, L., HENRIKSSON, K. G. and ÖBERG, P. Å. Chronic trapezius myalgia. Morphology and blood flow studied in 17 patients. *Acta Orthop. Scand.* **1990**, *61*, 394–398.
51 LARSSON, S. E., HONGMING CAI and ÖBERG, P. Å. Microcirculation in the Upper Trapezius Muscle during Varying Levels of Static Contraction. *Eur. J. Exp. Musculoskel. Res.* **1993**, *2*, 143–149.

52 Larsson, S. E., Ålund, M., HongMing Cai and Öberg, P. Å. Chronic pain after soft-tissue injury of the cervical spine: trapezius muscle blood flow and electromyography at static loads and fatigue. *PAIN* **1994**, *57*, 173–180.

53 Larsson, S. E., HongMing Cai and Öberg, P. Å. Blood flow in the trapezius and supraspinatus muscles during varying levels of static contraction and fatigue. *Eur. J. Musculoskel. Res.* **1994**, *3*, 99–106.

54 Stern, M. D., Lappe, D. L., Bowen, P. D., Chimosky, J. E., Holloway, G. A., Keiser, H. R. and Bowman, R. L. Continuous measurements of tissue blood flow by laser-Doppler spectroscopy. *Am. J. Physiol.* **1977**, *232*, H441.

55 Ahn, H., Johansson, K., Lundgren, O. and Nilsson, G. E. In-vivo evaluation of signal processors for laser Doppler tissue flowmeters. *Med. Biol. Eng. Comput.* **1987**, *25*, 207.

56 Holloway, G. A. Cutaneous blood flow responses to injection trauma measured by laser Doppler. *J. Invest. Dermatol.* **1980**, *74*, 1.

57 Kastrup, J., Bulow, J. and Lassen, N. A. A comparison between ^{133}Xenon washout technique and laser Doppler flowmetry in the measurement of local vasoconstrictor effect on the microcirculation in subcutaneous tissue and skin. *Clin. Physiol.* **1987**, *7*, 403.

58 Neufeld, G. R., Galante, S. R., Whang, J. M., de Vries, D., Baumgardner, J. E., Graves, D. J. and Quinn, J. A. Skin blood flow from gas transport: helium, xenon and laser Doppler compared. *Microvasc. Res.* **1988**, *35*, 143.

59 Engelhart, M. and Kristensen, J. K. Evaluation of cutaneous blood flow responses by ^{133}Xenon washout and a laser-Doppler flowmeter. *J. Invest. Dermatol.* **1983**, *80*, 12.

60 Nicholson, C. D., Schmitt, R. M. and Wilke, R. The effect of acute and chronic femoral artery ligation on the blood flow through the gastrocnemius muscle of the rat examined using laser Doppler flowmetry and xenon-133 washout. *Int. J. Microcirc. Clin. Exp.* **1985**, *4*, 57.

61 Engelhart, M., Petersen, L. J. and Kristensen, J. K. The local regulation of blood flow evaluated simultaneously by 133-Xenon washout and laser Doppler flowmetry. *J. Invest. Dermatol.* **1988**, *91*, 451.

62 Eyre, J. A., Essex, T. J. H., Flecknell, P. A., Bartholomew, P. H. and Sinclair, J. I. A comparison between measurements of cerebral blood flow in the rabbit using laser Doppler spectroscopy and radionuclide labelled microspheres. *Clin. Phys. Physiol. Meas.* **1988**, *9*, 65.

63 Matsen, F. A., Wyss, C. R., Robertson, C. L., Öberg, P. Å. and Holloway, G. A. The relationship of transcutaneous pO_2 and laser Doppler measurement in a human model of local arterial insufficiency. *Surg. Gynecol. Obstet.* **1984**, *1599*, 418.

64 Kvietys, P. R., Shepherd, A. P. and Granger, D. N. Laser-Doppler H_2 clearance and microsphere estimates of mucosal blood flow. *Am. J. Physiol.* **1985**, *249*, G221.

65 Öberg, P. Å. A method for frequency stabilization of multimode He-Ne lasers in laser-Doppler flowmetry. Proc. World Congr. Med. Phys. Biomed. Eng. BE18-E.1 San Antonio, **1988**.

66 de Mul, F. F. M., van Spijker, J., van der Plas, D., Greve, J., Aarnoudse, J. G. and Smits, T. M. Mini laser-Doppler (blood) flow monitor with the laser source and detection integrated in the probe. *Appl. Opt.* **1984**, *23*, 2970.

67 Jentink, H. W., Hermsen, R. G. A. M., de Mul, F. F. M., Suiches, H. E., Aarnoudse, J. G. and Greve, J. Tissue perfusion measurements using a mini diode laser Doppler perfusion sensor. In Microsensors and Catheter-Based Imaging Technology. *SPIE* **1988**, *904*.

68 Shepherd, A. P., Riedel, G. L., Kiel, J. W., Haumschild, D. J. and Maxwell, L. C. Evaluation of an infrared laser Doppler blood flowmeter. *Am. J. Physiol.* **1987**, *252*, G832.

69 Ahn, H., Lindhagen, J., Nilsson, G. E., Öberg, P. Å. and Lundgren, O. Assessment of flow in the small intestine with laser Doppler flowmetry. *Scand. J. Gastroenterol.* **1986**, *21*, 863.

70 Shepherd, A. P. and Öberg, P. Å. Laser-Doppler Blood Flowmetry. Kluwer **1990**.

71 Belcaro, G., Hoffman, U., Bollinger, A. and Nicolaides, A. Laser Doppler. Med-Orion, London **1994**.

3 Biosensors for Monitoring Glucose

J. D. Newman and A. P. F. Turner

3.1 Introduction

The concept of the biosensor evolved from the work of Leland C. Clark Jr. on the measurement of oxygen and, in particular, the development of the oxygen electrode (Fig. 3.1), which commonly carries his name [1]. Based on this experience and addressing his desire to expand the range of analytes that could be measured in the body, he made a landmark address in 1962, at a New York Academy of Sciences symposium, in which he described how 'to make electrochemical sensors (pH, polarographic, potentiometric, or conductometric) more intelligent' by adding 'enzyme transducers as membrane-enclosed sandwiches'. The concept was illustrated by an experiment in which glucose oxidase was entrapped at a Clark oxygen electrode using dialysis membrane. The decrease in measured oxygen concentration was proportional to glucose concentration. In the published paper, the term *enzyme electrode* was coined [2].

Fig. 3.1 Clark oxygen electrodes.

Sensors in Medicine and Health Care
Edited by P. Å. Öberg, T. Togawa, and F. A. Spelman

ISBN: 3-527-29556-9

Whereas it is clear what is meant by the phrase *enzyme electrode*, the same is not true of the term *biosensor*. Over the years, this term has been variously applied to a number of devices either used to monitor living systems or incorporating biotic elements. A recent IUPAC committee has been trying to unravel a literature that, at one time or another, has used the term to describe a thermometer, a mass spectrometer, *Daphnia* in pond water, electrophysiology equipment, chemical labels for imaging, and ion-selective electrodes. The consensus, however, is that the term should be reserved for use in its modern context, that of a sensor incorporating a biological element such as an enzyme, antibody, nucleic acid, microorganism, or cell. For our purposes, a biosensor will be defined as

> 'a compact analytical device incorporating a biological or biologically-derived sensing element either integrated within or intimately associated with a physicochemical transducer. The usual aim of a biosensor is to produce either discrete or continuous digital electronic signals that are proportional to a single analyte or a related group of analytes.'

The basic concept is to convert a biologically-induced recognition event into a usable signal. This event is typically a catalytic or binding reaction, which is selective, due to the nature of the biological components used. In order to convert this into a readable signal, a transducer is used to convert the (bio)chemical signal into an electronic one, which can be processed in some way, usually with a microprocessor. The literature now contains descriptions of a huge variety of devices exploiting enzymes, nucleic acids, cell receptors, antibodies, and intact cells, in combination with electrochemical, optical, piezoelectric, and thermometric transducers [3–6].

Clark's ideas became commercial reality in 1975 with the successful re-launch (first launch 1973) of the Yellow Springs Instrument Company (Ohio) glucose analyzer based on the amperometric detection of hydrogen peroxide. This was the first of many biosensor-based laboratory analyzers to be built by companies around the world.

It should be mentioned here that the first successful blood glucose meters were not biosensors according to the above definition and that, for many years, biosensors accounted for only a small proportion of commercial sales. Initially, the dominant technology was based on a reflectometer invented by Anton H. (Tom) Clemens, whose work led to the launch of the Ames Reflectance Meter. At the time, Ames was a division of Miles Laboratories in Elkhart, IN, USA and is now part of Bayer Corporation. Interestingly, work on the meter was started in 1966, 4 years after Clark's description of the glucose biosensor, but development of the reflectometer was much faster. Several prototypes were built in 1968 for field trials and a patent application was filed in April of that year. A US patent (no. 3604815) for the meter was granted on 14 September 1971, about 2 years after it went on the market. The original Ames Reflectance Meter was expensive and required a prescription. It was also rather large and heavy, weighing approximately 1 kg. Despite this, the instrument was a success and eventually led to the Eyetone, then to the Ames Glucometer, and eventually to a great variety of other products.

3.2
Diabetes and the Need for Glucose Monitoring

Diabetes is a serious world problem. It is currently estimated that there are over 150 million diabetics worldwide. Worse still, incidence of the disease has risen by an alarming 11% over the last 5 years, and a further doubling of new cases is predicted in the next 25 years. There are three types of diabetes.

- *Type 1 diabetes* usually affects the young and occurs when the pancreas no longer produces any (or very little) insulin. The body needs insulin to use sugar for energy. Approximately 10% of diabetics have Type 1 diabetes.
- *Type 2 diabetes* commonly affects middle-aged or older patients and occurs when the pancreas does not produce enough insulin or when the body does not use the insulin that is produced effectively; 90% of people with diabetes have Type 2 disease.
- *Gestational diabetes* is a temporary condition that occurs during pregnancy. It affects 2–4% of all pregnancies with an increased risk of developing diabetes for both mother and child.

Diabetes is one of the leading causes of death by disease. When left untreated or improperly managed, the high levels of blood sugar associated with diabetes slowly damage both the small and large blood vessels in the body, resulting in a variety of complications. For example:

- incidence of heart disease is raised between two- and four-fold on average,
- the condition is a leading cause of adult blindness,
- diabetes is responsible for up to 30% of all new cases of serious kidney disease,
- over 50% of all non-traumatic limb amputations are due to diabetes,
- diabetes is a major cause of erectile dysfunction.

With careful management, these complications can be delayed and even prevented. For many patients, this involves the regular measurement of blood glucose levels. The American Diabetes Association's Consensus Statement on Self-Monitoring of Blood Glucose recommends that people with Type 1 diabetes in intensive therapy should test at least four times a day. Even people with Type 2 in good control are encouraged to check twice a day.

3.3
Monitoring Principles: Transducers

As mentioned previously, a transducer is a device that converts the biological recognition event into a signal that can be further processed. This signal is then usually converted into a digital electronic format. Transducers can take many forms, relying on electrochemical, optical, piezoelectric, or thermal principles. In principle, any transducer could be used in a biosensor for the measurement of glucose, but in practice, electrochemical methods have dominated. Devices of this type offer suitable sensitivity and reproducibility and, importantly, can be manufactured inexpensively.

The electrochemical method that is most often employed is amperometry. This relies on the application of a potential between two electrodes and measurement of the resultant current flow. This is usually achieved in a three electrode cell. The potential is applied between a reference electrode (typically Ag:AgCl or calomel) and a working electrode, which is usually constructed of a noble metal or carbon. A third electrode is included in the cell. Termed the counter (or auxiliary) electrode, its function is to carry current, which would otherwise have to pass through the reference electrode, upsetting its half-cell potential. This is the principle by which the Clark oxygen electrode operates. In order to understand how biosensors based on this electrode function, we must first consider how oxygen measurements are carried out with such a device.

If an electrode of noble metal such as platinum or gold is made 0.6–0.8 V negative with respect to a suitable reference electrode, the oxygen dissolved in the solution between the electrodes is reduced at the surface of the noble metal. This phenomenon can be observed from a current–voltage diagram, termed a voltammogram (Fig. 3.2). In this instance, the working electrode becomes the cathode. As the negative voltage is increased, the current increases initially but soon it becomes saturated. In this plateau region of the voltammogram, the reduction of oxygen at the cathode is so fast that the rate of reaction is limited by the diffusion of oxygen to the cathode surface.

When the negative bias voltage is further increased, the current increases rapidly due to other reactions, mainly the reduction of water to hydrogen. If a fixed voltage in the plateau region is applied to the cathode, the current output of the electrode can be linearly calibrated to the dissolved oxygen. It has to be noted that the current is proportional not to the actual concentration but to the activity or equivalent partial pressure of dissolved oxygen, which is often referred to as *oxygen tension*. A fixed potential of between –0.6 and –0.8 V is usually selected as the polarization voltage when using Ag:AgCl as the reference electrode.

The cathodic reaction proceeds as follows:

$$O_2 + 2H_2O + 4e^- \rightarrow 4OH^-$$

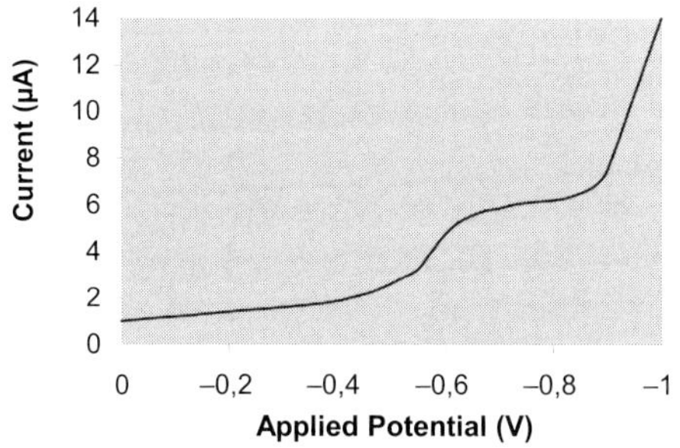

Fig. 3.2 Linear sweep voltammogram featuring oxygen reduction plateau.

The current, I, measured during the electrolysis is a direct measure of the rate of the electrochemical reaction at the electrode as described by Faraday's law

$$I = nF(\mathrm{d}N/\mathrm{d}t)$$

where $\mathrm{d}N/\mathrm{d}t$ is the oxidation or reduction rate (mol s^{-1}), N is the number of moles of material present, n is the number of electrons taking part in the electrochemical reaction, t is time, and F is the Faraday constant (96 485 C mol^{-1}).

The rate of reaction ($\mathrm{d}N/\mathrm{d}t$) is given by

$$\mathrm{d}N/\mathrm{d}t = JA$$

where J is the molar flux of the electroactive species (mol s^{-1} m^{-2}) and A is the electrode area (m^2). Combining the above equations leads to the following relationship, which describes the current at the electrode

$$I = nFAJ$$

The electrochemical reaction depends upon both the rate of electron transfer at the interface and on the mass transport of analyte to the electrode from the bulk solution. The most important mass transport processes in an electrochemical measurement are diffusion and forced convection.

Taking a generalized reaction, it can be seen that concentration gradients arise at stationary electrodes as an electrochemical reaction proceeds. Before electrolysis is started, the concentration of the oxidized species (Ox) is uniform throughout the solution (assuming that at the start of the experiment, the solution has been well mixed and is completely homogeneous). As a current flows, the concentration of Ox at the electrode surface becomes less than in the bulk solution, due to the electrochemical conversion of Ox into R. In an unstirred solution, relying solely on diffusion, the flux of Ox to the electrode surface at time t, at a distance x from the electrode surface, is proportional to the concentration gradient and, therefore, obeys Fick's first law of diffusion

$$J(x,t) = D_O \delta(C_O(x,t))/\delta t$$

where C_O is the concentration of Ox in bulk solution, D_O is the diffusion coefficient of Ox in the solution, and x represents distance from the electrode surface.

Since the current flowing in the electrochemical cell is dependent upon the flux of material at the electrode surface (where $x=0$), the current for an electrode with area A can be defined as

$$I = nFAJ(0,t)\ .$$

Furthermore, it is possible to substitute the previous equation (for J) into this, to give

$$I = nFAD_0\delta(C_0(0,t))/\delta x\ .$$

When a potential is applied to the cell, the concentration of the species that is either being oxidized or reduced at the working electrode falls to zero. Diffusion of the species will then occur from bulk solution towards the electrode surface. If the solution is stirred (or the electrode agitated), and the material at the surface of the electrode is instantaneously oxidized or reduced (thereby maintaining a zero concentration at the electrode surface), then a profile of concentration with respect to the distance from the electrode surface can be represented as in Fig. 3.3.

Since the solution is stirred (and, therefore, well mixed), the concentration in the bulk region is constant throughout ($C = C_0$). This is referred to as the convective region. Closer to the electrode, the concentration drops, falling to zero at the electrode surface. This is referred to as the Nernst diffusion layer and has a thickness *d*, which depends upon the nature of the solution into which it extends. In most cases, in aqueous, well-stirred solutions the thickness of the diffusion layer is between 1 and 10 μm.

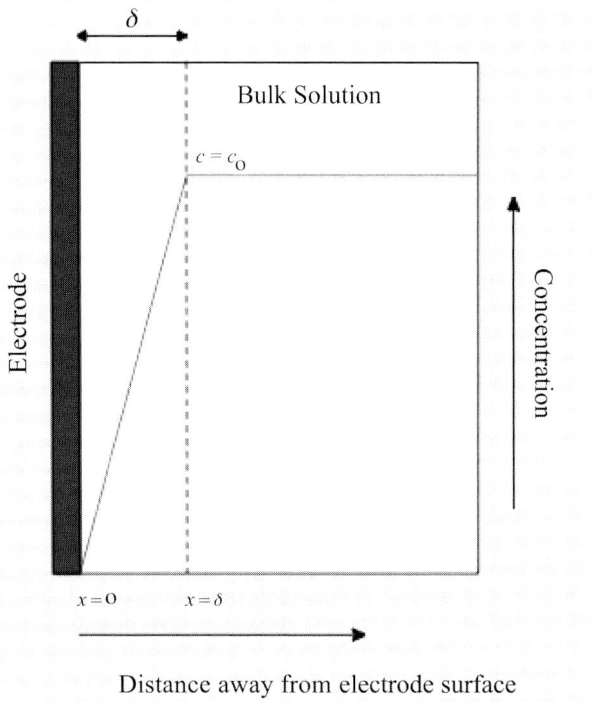

Fig. 3.3 Concentration profile of electroactive species close to electrode surface.

In practice, this has a number of implications. A thin layer of solution immediately adjacent to the electrode surface is stationary, regardless of the conditions in the bulk solution. Within this layer, diffusion controls the transfer of the electroactive species from the bulk layer to the electrode surface, which occurs as a result of a concentration gradient. Beyond this thin layer, the concentration of the material remains effectively constant (at C_0) as a result of convective mass transfer.

As can be seen in Fig. 3.3, the concentration gradient is represented as being linear. This is not strictly true, but the approximation is close enough in most practical applications. Thus, the gradient can be given by

$$dC/dx = (C_0 - C_e)/\delta \, .$$

However, since the concentration at the electrode surface (C_e) is zero, this equation can be simplified to

$$dC/dx = C_0/\delta \, .$$

If we use this relationship in the earlier equation, describing the limiting current for a stirred solution, we can obtain

$$I_L = nFAD_0C_0/\delta \, .$$

This is an important relationship, since it states that the limiting current obtained from the oxidation or reduction of a species in a stirred solution is directly proportional to the concentration and the diffusion coefficient, and inversely proportional to the thickness of the Nernst diffusion layer.

The above case, however, does not apply in the majority of blood glucose biosensor applications, where the solution (blood) is applied to an electrode and is not stirred. In the former case, we had a stable diffusion profile at the electrode surface, in which the concentration gradient was constant with respect to time. It, therefore, resulted in a steady, unchanging, current.

In the latter case, where a (typically) planar electrode is placed in an unstirred solution, or where a static sample is applied to the electrode, a similar situation occurs when a potential is first applied. At $t=0$, the concentration at the electrode surface ($x=0$) is reduced to zero, establishing a concentration gradient, as before. However, in this case the solution is quiescent and the diffusion layer does not remain at a constant thickness, getting thicker and moving further into the bulk solution over time. This results in a change in the slope of the diffusion gradient, which manifests itself as a non-steady-state current. In this case, the behavior is governed by Fick's second law of diffusion

$$\delta(C_0(x,t))/t = D_0\delta^2(C_0(x,t))\delta x^2 \, .$$

Using the Laplace transform gives a solution for concentration in terms of the parameters x and t:

$$C_{(x,t)} = C\mathrm{erf}\left(x/\sqrt{(\Pi Dt)}\right)$$

where erf is the error function.

Taking the differential of this equation and using the boundary condition $x=0$, for the electrode surface, we can obtain an expression for the concentration gradient at the electrode surface

$$\mathrm{d}C/\mathrm{d}x(x=0) = C/(\sqrt{\Pi Dt}) \; .$$

This can be used, as before, to obtain the current, which in this case is governed by semi-infinite linear diffusion to a horizontal plane

$$I = nFAC_0\sqrt{(D/\Pi t)} \; .$$

According to this expression, which is known as the Cottrell Equation, the product of current and square root of time is constant, provided the area and diffusion coefficient are constant. Furthermore, it confirms that no stable time-dependent current is ever achieved for diffusion to a planar electrode in quiescent conditions.

When the electrodes and the electrolyte are separated from the measurement medium by a polymer membrane that is permeable to the dissolved gas but excludes most ions and other species, and when most of the mass-transfer resistance is confined in the membrane, the device can measure oxygen tension in various liquids. This is the basic operating principle of Clark oxygen electrode, shown in Fig. 3.1. Other electrode configurations, featuring carbon and other working electrodes, can be constructed and operated via amperometry in a similar manner [7].

Optical biosensors for monitoring glucose have been less successful commercially, despite the fact that it is relatively simple to construct optical glucose biosensors. A common format is to use optical fibers, which have a chemically-modified tip containing either a pH-sensitive material [8] or a fluorophore [9]. Thermal transducers have also been described, but are less common still [10].

3.4 Monitoring Principles: Enzymes

The most commonly used enzymes in the design of glucose biosensors contain redox groups that change redox state during the biochemical reaction. Enzymes of this type are glucose oxidase (GOx) and glucose dehydrogenases (GDH). In nature, oxidase enzymes such as GOx act by oxidizing their substrates, accepting electrons in the process and thereby changing to an inactivated reduced state. These enzymes are normally returned to their active oxidized state by transferring these electrons to molecular oxygen, resulting in the production of hydrogen peroxide (H_2O_2)

$$\text{glucose} + O_2 \rightarrow \text{gluconolactone} + H_2O_2 \ .$$

Glucose may also be oxidized by GDH. One mechanism relies on NAD^+ acting as a cofactor, rather than oxygen as a cosubstrate. In this case NADH is produced

$$\text{glucose} + NAD^+ \rightarrow \text{gluconolactone} + NADH$$

A third mechanism involves the use of quinoprotein glucose dehydrogenase, which requires neither oxygen nor NAD^+. Quinoproteins form a class of enzymes, which require ortho-quinone cofactors to oxidize a wide variety of alcohols and amines to their corresponding aldehydes and ketones. The soluble quinoprotein GDH uses pyrroloquinoline quinone (PQQ) as a cofactor:

$$\text{glucose} + PQQ(ox) \rightarrow \text{gluconolactone} + PQQ(red) \ .$$

Each of the above enzyme mechanisms can be utilized in glucose biosensors but feature different advantages and drawbacks.

- The oxidase enzyme is inexpensive but requires oxygen as a cosubstrate. Consequently, as oxygen is depleted in the sample, performance decreases, whether one is monitoring oxygen depletion, or hydrogen peroxide production.
- NAD^+ dependent GDH, on the other hand, is oxygen independent, and has the added attraction of being a well-established probe for monitoring biochemical reactions. The drawback is that the cofactors are relatively expensive.
- PQQ-GDH is a particularly efficient enzyme system, with a rapid electron transfer rate, but it, too, is comparatively costly.

So, combining these enzyme-based reactions with a suitable transducer enables selective sensor devices to be fabricated. Electrochemical systems are the most popular for this purpose.

Furthermore, we can use the relationships derived earlier, and extend them to explain the behavior of enzyme electrodes using a simple model. Assuming we have a thin, immobilized enzyme layer on the surface of an amperometric electrode system, we can consider the space adjacent to the working electrode (Fig. 3.4) and divide it into three broad regions:

- a convective region ($x > L$), where the analyte concentration is effectively constant ($C = S_{bulk}$),
- a diffusion-limiting region ($0 < x < L$) where a pure diffusion process takes place, and
- the region where the enzyme reaction takes place ($x < 0$).

We have assumed that the system is independent of the rate of supply of any cosubstrate for the enzyme, that the enzyme is distributed uniformly, and that the sensor signal depends on product generation, which is in turn related to the flow of substrate into the sensor. Boundary conditions are established at the two interfaces ($c = S_1$) and very deep within the enzyme layer ($c = 0$). Using S_1 and the ap-

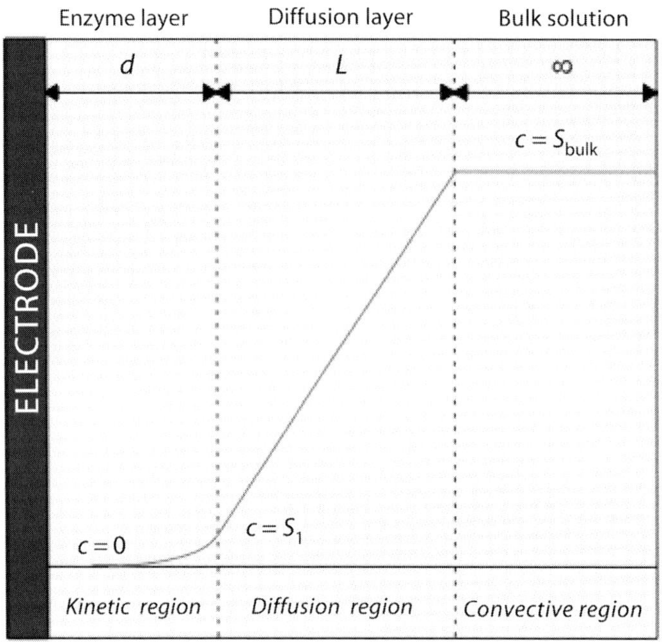

Fig. 3.4 Diagram of analyte concentration profiles adjacent to an enzyme electrode surface.

propriate boundary conditions it is possible to express the concentration profile within the enzyme region by a simple exponential

$$C(x) = S_1 \exp\left(\sqrt{(a/D_e)}x\right)$$

where $x>0$ the term a is V_{max}/K_m and D_e is the diffusion coefficient of the substrate within the enzyme layer.

If the enzyme reaction is slower than the overall diffusion process, then the sensor response is kinetically controlled. In this case, assuming a boundary condition where only half of the enzyme-generated product diffuses to the electrode, the current can be expressed as

$$I_k = nFALK_2[E_\Sigma][S]/2(K_m + S) .$$

This gives a maximum current response I_{max} (when $[S] \gg K_m$), which is directly proportional to the enzyme loading and the thickness L

$$I_{max} = nFALK_2[E_\Sigma]/2 .$$

Combining the above relationships, we can obtain a general expression for the kinetic current

$$I_k = I_{max}[S]/(K_m + S) \ .$$

The above equation reveals that the expression for the kinetic current is similar to the Michaelis-Menten equation describing enzyme kinetics, which is not too surprising since the expression was derived assuming that the response of the electrode was controlled only by enzyme kinetics. The only difference between the two equations is that the velocity terms in the Michaelis-Menten equation have been replaced by expressions of current. Consequently, the calibration curves obtained from these devices are the classic rectangular hyperbolas.

On the other hand, if the enzymatic reaction in the enzyme layer of thickness d is faster than the transport process (i.e., the concentration S_1 approaches 0), then the sensor response is under mass transport control. The concentration S_1 is negligible if the substrate concentration reaches zero inside the layer. This is possible when the dimensionless parameter σ^2 is greater than unity where

$$\sigma^2 = aD^2/D_e = V_{max}d^2/D_e K_m \ .$$

This parameter essentially compares the rate of enzyme reaction (V_{max}/K_m) with diffusion through the enzyme layer (d^2/D_e). If $\sigma^2 < 1$, then enzyme kinetics predominate. If on the other hand $\sigma^2 > 1$, the response is under diffusion control. In an analogous manner to the kinetic situation, one can derive a general expression for a sensor operating in the diffusion-limiting regime, such that

$$I_d = 2I_{max} D_e[S]/K_2[E_\Sigma]d^2 \ .$$

Comparing the two limiting situations reveals that I_d is dependent on both d and D, whereas I_k is independent of D and increases with L. Since S_1 represents the maximum possible concentration in the enzyme layer, the linear range of the sensor will be determined by S_1 (i.e., when $S_1 \ll K_m$). It is possible to determine an expression for S_1 as a function of S_{bulk} such that

$$S_1 = S_{bulk}/1 + (L/D_s)\sqrt{(aD_e)} \ .$$

The above expression implies that S_1 can be optimized by careful selection of the parameters: D_s, $\sqrt{(aD_e)}$, and L. In operational terms, however, increasing the thickness of the diffusion layer or decreasing the diffusivity may be of limited value because of the relationship between the characteristic diffusion time across the membrane (τ) and D_s and L

$$\tau = L^2/D_s \ .$$

This section has dealt only with a specific model that may not apply to all enzyme electrode configurations. It does, however, illustrate the theoretical problems, and their implications, which should be considered in the design of an enzyme electrode.

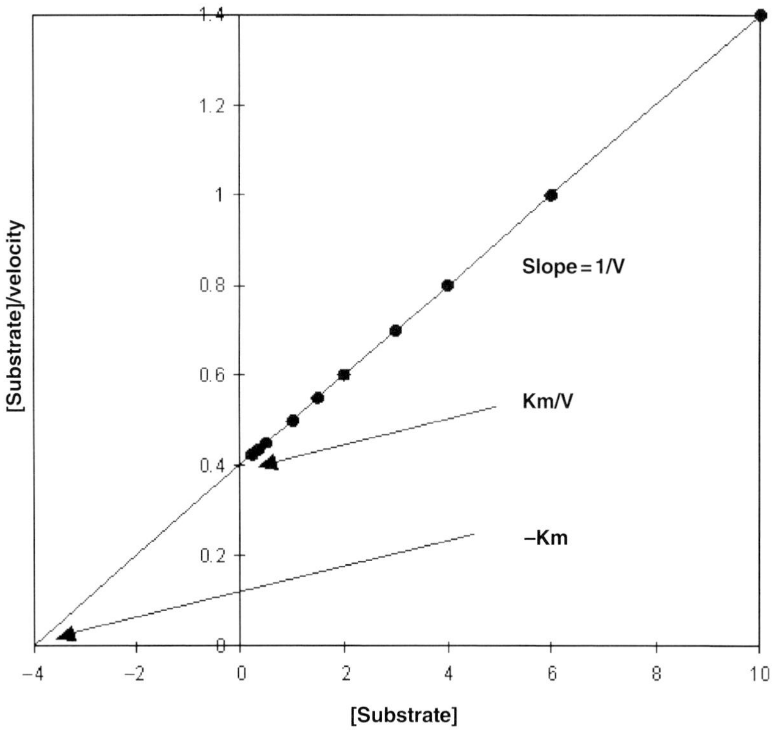

Fig. 3.5 Hanes plot.

One last problem remains: how do we tell if the electrode is mass transport limited or limited by enzyme kinetics? The answer to this problem can be solved, quite simply, by using a Hanes plot from classical enzyme kinetics. This is another conversion of the Michaelis-Menten equation to give the equation of a straight line

$$a/v = 1a/V + K_m/V \ .$$

Thus, a plot (Fig. 3.5) of substrate concentration divided by velocity (a/v) versus substrate concentration (a), yielding a straight line with an intercept on the horizontal axis of $-K_m$ and on the vertical axis of K_m/V. The slope is $1/V$. As can be seen, it is simple to read K_m directly from the plot and then trivial to obtain V_{max} from the K_m/V intercept.

3.5 Manufacturing Issues

Screen-printing is a thick-film process that has been used for many years in artistic applications and, more recently, for the production of miniature, robust, and cheap electronic circuits. During the 1980s, the process was adapted for the production of amperometric biosensors and has had a huge impact on their commercialization. Many of the most successful electrochemically based devices to date have used the technique.

The process involves squeezing a suspension (usually referred to as an ink, due to the origins of the technique) through a patterned screen onto a solid support (substrate) held on the reverse of the screen. Hence, the process is very simple and rapid. Multiple repeat patterns are usually designed onto a single screen, further enhancing production speed. A small pilot-scale machine, such as that shown in Fig. 3.6, is able to produce many thousands of sensors per day. Larger machines are available, and can greatly exceed this production figure.

The process has been one of the major reasons for the commercial success of many biosensors and is the process by which MediSense (now Abbott) produce over 1 billion biosensor strips annually.

The main developments, from a biosensor viewpoint, have involved ink formulation. The full commercial impact of the technology was initially limited by the range of inks available for the construction of inexpensive devices. The main rea-

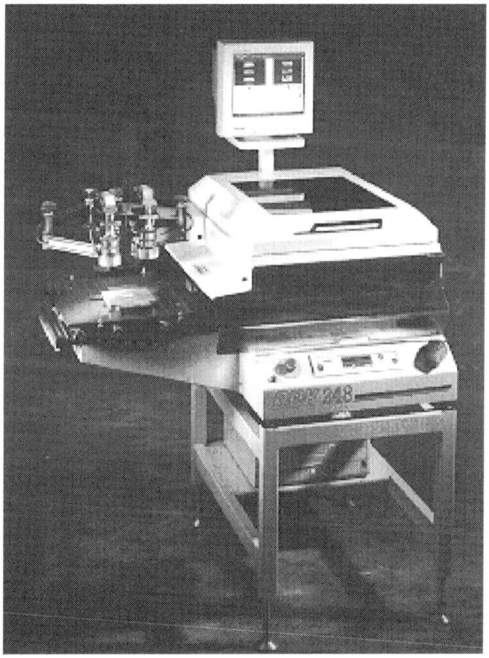

Fig. 3.6 Laboratory-scale screen-printer.

son for this was that the initial applications of the process involved the use of inks that required high temperature-curing processes. Ceramic substrates were often used, mainly for their temperature resistance and rigidity. However, this raised production costs in a very cost-sensitive marketplace.

Today carbon inks are widely available for low temperature applications and, increasingly, other metal-based systems, those for many metals are not. Temperatures of up to 850 °C are often required to cure inks employing gold and many other useful electrode materials and, hence, costly ceramic substrates are often required. It is not impossible to produce relatively inexpensive devices using ceramic bases, but polymeric materials are cheaper and easier to process.

Amongst others, Gwent Electronic Materials (Pontypool, Gwent, UK) are developing a new range of low-fire materials. Using a new technology, the firing temperature of metallic-based pastes has been reduced from the normal temperatures of 400–450 °C to 150–350 °C. This type of system means that it is now possible to fire onto polymeric rather than ceramic materials. Additionally, they have found that it is possible to produce a precious metal coating, which is able to connect directly to base metal systems at a temperature below which oxidation would normally be a problem. These modifications to the metallic film give the ability to render the surface easily solderable. It is expected that these developments will open up new opportunities for inexpensive, precious metal-based chemical sensors and biosensors.

There is now a wide array of inks suitable for producing biosensors. Most of the applications to date have involved electrochemical devices, but the technique is applicable to the production of any planar device.

3.6
First Generation Amperometric Glucose Biosensors

The earliest approaches to the construction of amperometric glucose biosensors have been alluded to in the preceding sections. GOx can be immobilized close to an electrode and the depletion of oxygen can be monitored, as we have seen, using a Clark oxygen electrode (Fig. 3.7).

From the reaction scheme in Fig. 3.7, it is apparent that it is also possible to measure glucose via the oxidation of the hydrogen peroxide produced by the enzymatic reaction, since this is directly proportional to the concentration of glucose. This can also be measured amperometrically at a potential of approximately +0.7 V vs. Ag:AgCl, when a platinum working electrode is used

$$H_2O_2 \rightarrow 2H^+ + O_2 + 2e^- \ .$$

The first commercially successful glucose biosensors were based on the above principle of measurement of hydrogen peroxide. The YSI (OH, USA) model 23A instrument (Fig. 3.8), which was launched in 1975, was designed to measure glucose in whole blood. Since then, this product and its descendents have become a

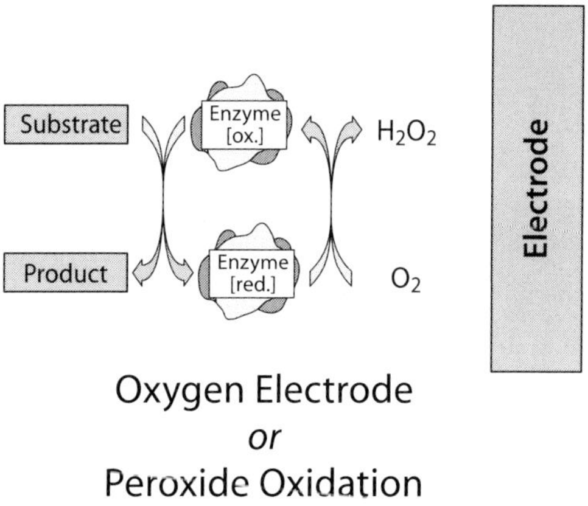

Fig. 3.7 Diagram of first generation glucose biosensor.

standard for clinical diagnostic work at many sites in the hospital, including: coronary care, emergency room, operating room, stat laboratory, chemistry laboratory, endocrinology/diabetes, pediatrics/neonatal, central laboratory, pulmonary laboratory, cardiac rehabilitation, and clinical research.

The original model 23A analyzer has been superseded by more advanced machines, most recently by the model 2300 STAT Plus, which also contains a lactate analyzer. This instrument allows glucose measurements to be made in whole blood, plasma or serum and lactate measurements in whole blood, plasma or cerebrospinal fluid. The analyzer's hematocrit correction option also allows glucose results expressed as plasma even when analyzing whole blood.

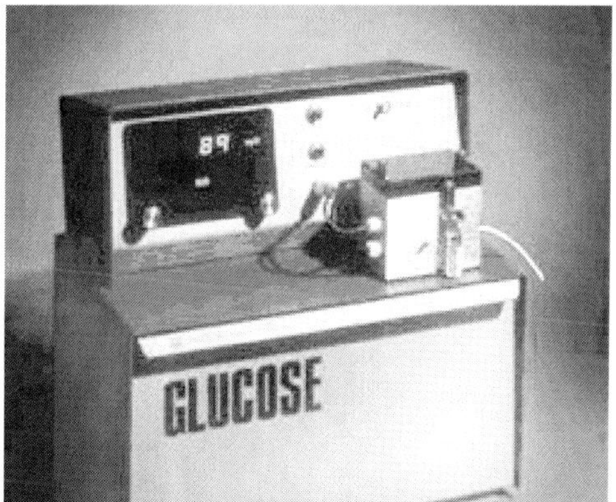

Fig. 3.8 YSI 23A glucose biosensor.

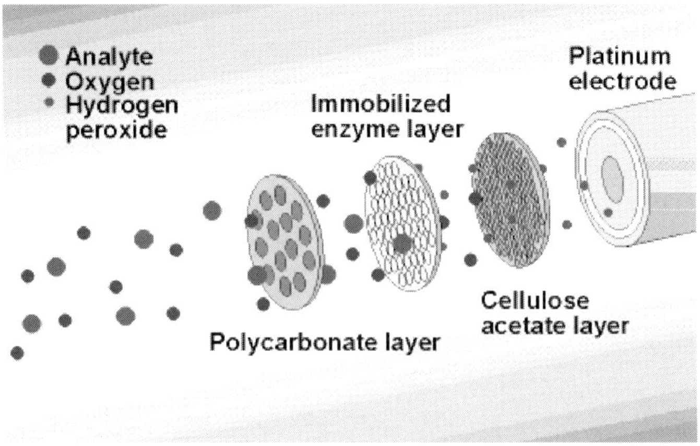

Fig. 3.9 Diagram of YSI biosensor construction principle.

The construction of the biosensor has changed little over the years. For glucose analysis, GOx is immobilized between two membrane layers. The outer polycarbonate membrane retains the enzyme, allows glucose to pass, but prevents many larger molecules from entering, thus reducing interference. The glucose enters the enzyme layer, where it is oxidized, producing hydrogen peroxide. This passes through the cellulose acetate membrane to a platinum electrode where it is measured amperometrically. The second membrane acts as a further size exclusion barrier preventing many other potentially interfering (electroactive) compounds from reaching the electrode surface. The construction is illustrated in Fig. 3.9.

3.7 Catalytic Transducers

There are several problems associated with the use of the electrode configuration shown in Fig. 3.9. The construction is relatively expensive, since it uses platinum electrodes. It is robust, but it is not straightforward to miniaturize and simplify its method of production. In addition, the rather high detection voltage required makes the system prone to interference in the absence of the membrane structures utilized.

One method of simplification involves the use of electrode materials that lower the required applied voltage for hydrogen peroxide detection. The rates of many electrochemical reactions can be tremendously enhanced by the deposition of very small particles of metals or metal oxides on the surface of the electrode [11]. The highly catalytic nature of these surfaces is strongly dependent on the particle size of the deposited metals, which should be comparable to that of the electrical double layer. When deposited on carbon electrodes, for example, the difference in the electronic work functions of the metal and the carbon result in an increase in the electron density on the metal [7], resulting in an electrocatalytic surface.

The most commonly used metals for deposition onto electrodes and used in biosensors are platinum [12] and rhodium [13]. Incorporation of the metal into the carbon matrix can be achieved by

- electrochemical deposition [14],
- sputtering [15], or
- straightforward mixing of the metal into a carbon paste [16].

The ability of these materials to reduce the oxidation potential of hydrogen peroxide is extremely attractive. In addition, the amperometric signal is greatly enhanced, producing a high signal-to-noise ratio. There is, however, a drawback. The electrocatalytic behavior is not very selective, resulting in lowered oxidation potentials for many other, potentially interfering compounds, including reducing sugars, paracetamol (acetaminofen), and uric acid, which may be present in significant concentrations in blood.

To overcome this problem, selectivity may be enhanced in much the same way as for the earlier, YSI-type, biosensors. Membranes, typically constructed from polymers such as cellulose acetate, polyurethane, Nafion and many others, can be applied to the electrode surface to exclude interferences by size or charge exclusion [7]. On first sight, this may appear to offer no advantage over the previous type of biosensor, but the material costs are much lower, and the resultant device is amenable to mass production, using techniques such as screen-printing [13]. Such devices can be produced in large numbers and can be inexpensive enough to be disposable, if required. Re-use of the devices, however, is possible and biosensors based on such principles have been used in applications where they have made over 1000 measurements over a 7 day period [17].

3.8 Mediated Devices

Direct electron transfer from the enzyme to the electrode would greatly enhance performance, potentially greatly reducing interference and the need for co-factors. However, for the vast majority of enzymes, direct electron transfer is not straightforward. This difficulty may be due to the location of the electroactive center of the enzyme, which is often deep within its structure, or it may be due to electrode surface effects connected with immobilization and orientation of the macromolecule.

Although very specific with regards to the reducing substrate, enzymes such as glucose oxidase show a high degree of flexibility with regard to the second substrate: the electron acceptor. Thus, many inorganic redox couples and organic dyes have been successfully utilized as electron sinks for the glucose oxidase-catalyzed oxidation of glucose. In the early 1980s, it was realized that this method of

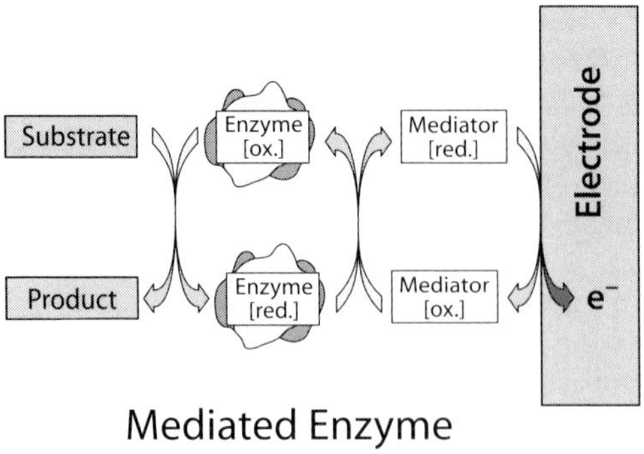

Fig. 3.10 Diagram of mediated biosensor.

operation facilitated the transfer of electrons in enzyme electrodes, which was independent of the local oxygen concentration and allowed operation at much lower potentials, eliminating many of the problems associated with interference noted with the previously discussed devices. A group of workers based at Cranfield and Oxford Universities, whilst working on a fuel cell project, realized that the electron transfer compounds that they were using to increase the efficiency of their fuel cells, held a great deal of promise in biosensor applications [18]. These redox couples, known as mediators, are able to shuttle electrons between the redox center of the enzyme and the electrode. Depending on the compound used, they can also be regenerated at potentials where interference from species such as ascorbate, urate, and paracetamol is minimal. The principle of operation is shown in Fig. 3.10.

A vast number of compounds are capable of acting as enzyme mediators, and the groups that are most frequently used in the construction of enzyme electrodes, are detailed below (Tab. 3.1). Of these, mediators based on metal complexes are the most popular. Arguably the most important examples of this class are mediators based on ferrocene and its derivatives. This can be attributed to three main factors: they have a wide range of redox potentials; the redox poten-

Tab. 3.1 Electrochemical mediators commonly used in enzyme electrodes.

Mediator	E^0 *(V)*
Ferrocene 1,1-dicarboxylic acid	0.64
$Ferrocene^{+/0}$	0.44
Potassium ferricyanide/ferrocyanide	0.36
1,4-Benzoquinone/hydroquinone	0.28
Phenazine methosulphate	0.08

Formel potentials E^0, are quoted versus normal hydrogen electrode (NHE) at pH 7.0 and standard conditions tials are independent of pH; and the synthetic schemes involved in making derivatives are usually straightforward.

In order to ensure the electron transfer, the mediator must be present in both its oxidized and reduced forms, which must remain in the vicinity of the electrode. In some formats, this means that the mediator must be insoluble, but in certain designs, it is possible to use one that is soluble, as we shall see later, when some commercially-successful designs are discussed. The mediator should also have a rate constant for the reaction with glucose oxidase that is sufficiently competitive with the natural mediator, oxygen.

In many cases, especially for the measurement of blood glucose, it is acceptable to use a biosensor design that is suitable only for a single measurement. This is convenient from the point of view of using a mediated biosensor format, since one of the major drawbacks of most mediators is that they are relatively soluble, leading to short operational lifetimes and irreproducible results. This is even true of one of the most popular mediators, ferrocene, which, despite being relatively insoluble in its oxidized form, becomes quite soluble when reduced to the ferricinium ion.

3.9 Currently-Available Home Blood Glucose Monitors

In order to adjust insulin dosage, food intake, and exercise, the diabetic patient needs to monitor blood glucose levels. The required monitoring frequency depends on the type and severity of the diabetes and the lifestyle of the individual, but typically, measurements are made 2–4 times daily, although a higher frequency may be required for certain patients.

Most glucose monitors operate by measuring glucose in blood samples obtained from the finger, using a small lancet. This has several implications, which have influenced the design of monitors and the development work being undertaken by the companies involved in their sale.

- Since blood is being handled, a disposable sensor design is preferred.
- Minimizing required sample size is a priority.
- Finger-pricking is uncomfortable, so other sampling sites have been investigated.
- Instruments have become more automated.

The first really successful blood glucose biosensor for home-use was a mediated device based around a disposable, screen-printed sensor design. It was developed by a company that was originally called Genetics International, in conjunction with the Universities of Cranfield and Oxford. The company was renamed Medisense and the product that was launched was the ExacTech device (Fig. 3.11). The Company has since been bought by Abbott Laboratories, one of the world's largest healthcare companies.

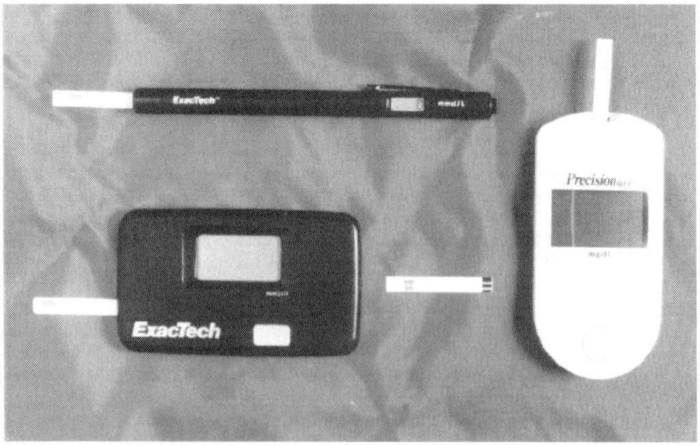

Fig. 3.11 ExacTech and precision QID biosensors.

Abbott still uses the Medisense name and the Company now manufactures and distributes glucose monitoring equipment for both home and hospital use. Largely through Medisense, Abbott offers a wide range of biosensors, including the following products: ExacTech, ExacTech R · S · G, Precision PCx, Precision Q · I · D Pen, Precision Q · I · D, and Precision G.

All current Medisense products utilize similar technology. The basic concept involves an amperometric biosensor, employing glucose oxidase and electrochemical mediators on a disposable strip format. Various improvements have been incorporated since the launch of the original devices in 1987.

Improvements in ergonomics have led to an easier-to-read display and a simpler-to-use instrument. The clinically-relevant range can be measured in 30 s in whole blood. The R · S · G device eliminates the need for calibration. The PCx instrument requires just 3.5 μL of whole blood to provide accurate results within 20 s and an audible signal that tells the operator that an adequate sample has been provided. Automatic initiation of the test procedure occurs once a sample has been detected; and laser barcode scanning for all data entry eliminates manual entry errors. Results for up to 4000 tests can be stored in the monitor itself, and can be easily retrieved for review. Precision PCx test strips provide glucose results from 20 to 600 mg per dL across a 20–70% hematocrit range.

The latest product is the Precision Xtra, the first personal blood glucose meter that also allows people with diabetes to test their blood for ketones. This is an important addition to the range, since over one million diabetics, in the US alone, are at risk of developing diabetic ketoacidosis (DKA), a life-threatening condition that can result in coma or death.

Precision Xtra tests ketone levels in a small drop of blood, resulting in a digital reading and removing the difficult interpretation of color changes, commonly associated with many rival technologies. More importantly, Precision Xtra gives patients and their doctors an accurate measure of their ketones in the blood in real time, which was not possible with existing devices that measure ketones in urine.

Bayer Diagnostics (Leverkusen, Germany) designs, manufactures and markets clinical diagnostics systems for the major industry markets of self-testing, point-of-care testing, laboratory testing, critical care and nucleic acid diagnostics. The Company's primary biosensor products are for rapid blood glucose monitoring, an area it first became involved in when it introduced Dextrostix, the first dry reagent blood sugar test, in 1964.

Bayer market a range of instruments and accessories (Fig. 3.12) for home use. The Glucometer DEX contains a 100-test memory and an automated 10-test cartridge, which eliminates the need to open and handle individual test strips. The DEX cartridge automatically calibrates the meter for all 10 tests. Each biosensor is constructed using a capillary gap that automatically draws the blood sample (3–4 μL) into the test sensor. The electrochemical system also uses mediated technology, but in this case, the mediator is soluble, since the capillary-fill design provides a contained electrochemical cell. Other glucose biosensor products from Bayer utilize similar sensing principles, but contain a variety of different features to aid in diabetes management. These include the Glucometer Esprit, Elite, and Elite XL instruments. The most advanced instrument is the Glucometer Encore instrument, which includes quality assurance (QA) testing, automatic data storage and recall. Effectively the same device is sold by the originators of the technology, Kyoto Daiichi Kagaku (Kyoto, Japan) and Menarini (Firenze, Italy).

Lifescan (Milpitas, CA, USA) produces numerous glucose biosensor products within its One Touch product range (Fig. 3.13). Its current meters are the following:

- The One Touch Ultra Blood Glucose Monitoring System. The system features a five second test time, advanced electrochemical biosensor test strips that require a 1 μL blood sample, and the user option of finger or arm testing. It has a data port and uses the LifeScan In Touch software. This meter is plasma-blood calibrated. Regular cleaning is not required.
- SureStep Meter (which is marketed as GlucoTouch in Europe and as NovoAssist in Japan). The current version of this meter adds a data port. It requires a 10–

Fig. 3.12 Bayer Diagnostics Vaculance device.

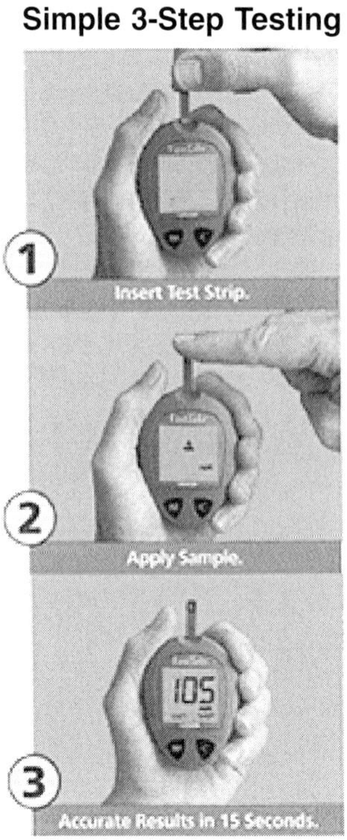

Fig. 3.13 Lifescan instrument.

30 μL drop of blood and tests in 15–30 s. This meter is plasma-blood calibrated. It uses color reflectance technology, which means that regular cleaning is required.

- One Touch Profile. Its data port allows it to work with diabetes management software. This is a whole blood-calibrated meter. This is the only meter now on the market that uses event markers help you flag test results with activities that affect blood glucose levels.
- One Touch Basic. The current version of this meter adds a data port. It requires a 10 μL drop of blood and tests in 45 s. This is a whole blood-calibrated meter, which uses color reflectance technology.

Roche Diagnostics (Basle, Switzerland) produces and markets the Accu-chek® range of blood glucose biosensors (Fig. 3.14), primarily aimed at patient self testing, which is replacing the company's range of photometer-based test strip technologies for this application. Both types of product are sold under the Accu-chek name in the USA, but under different names elsewhere. As with the other market leaders, Roche Diagnostics have based their devices on disposable biosensor elements, using an amperometric approach. The Accu-chek Advantage uses glucose oxidase in common with other devices, but the Accu-chek Advantage H is based on the enzyme PQQ glucose dehydrogenase allowing it to exhibit complete

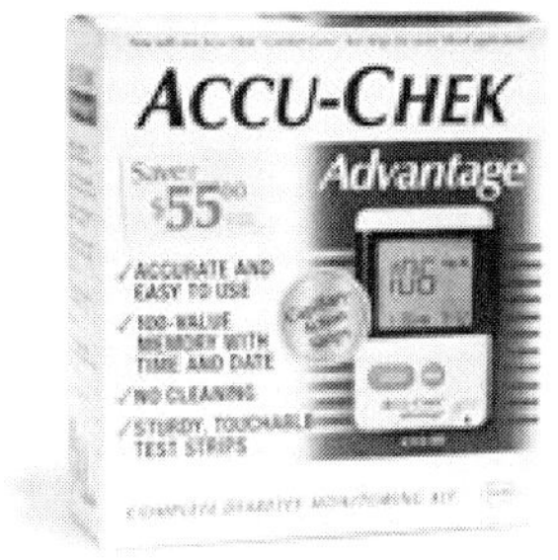

Fig. 3.14 Accu-chek advantage glucose monitor.

independence of oxygen concentration and to be used on both capillary blood and venous blood.

A recent product launched by Therasense (Alameda, CA, USA), the FreeStyle instrument, allows diabetic patients to take a blood sample from multiple sites including their fingertips, forearm, upper arm, thigh, calf and fleshy part of the hand. With a blood sample requirement of only 0.3 μL, this minimizes the pain of testing. This has some interesting implications since blood glucose levels may not always be constant throughout the body.

Generally, blood in the arteries has the highest glucose levels, followed by capillary blood (finger), and then venous blood. In a fasting state, the difference between venous and capillary glucose is very small. Following a meal, however, a capillary sample obtained from a finger may be up to 70 mg/dL higher than a laboratory blood glucose level using a sample taken from a vein. Furthermore, the rate of blood flow is faster in the fingers than it is in the thigh, calf, forearm, upper arm or hand, which can result in a difference in glucose measurements in these sites after eating or during a rapid drop in blood glucose.

Broadly, studies have shown the following:

- When glucose is changing rapidly, the arm value can lag behind the finger value, meaning that a change in blood glucose will be observed in finger blood sooner than from the forearm or other alternate sites.
- Vigorously rubbing the test site (until it feels warm) just before lancing will stimulate blood flow at the skin surface and will minimize the differences in blood glucose.
- If the glucose level is suspected to be low (due to having symptoms of low blood glucose, when a meal is delayed, after taking insulin, or during/after exercise), it is preferable to test on the finger, since a rapidly falling glucose will be detected sooner at the finger than at other sites.

3.10
Currently-Available Laboratory Analyzers for Monitoring Glucose

Several manufacturers currently produce larger analyzers for use in the laboratory. Many of these are aimed at use in hospital laboratories, where constraints are quite different from those of the previously described sensors for home-use. High throughputs are usual, so sensors are contained within large automated instruments. Sensor lifetime and lack of maintenance are much more important than low cost in this case, so robust designs are utilized. Accuracy can be enhanced by automated calibration, so reproducibility between devices is less of an issue than with one-shot biosensors.

AVL (Schaffhausen, Switzerland) produces sensors for the measurement of parameters from a single drop of blood, serum, or other biological fluid. The OMNI benchtop instrument concept is modular, which means that sensors can be chosen for specific parameter configurations. As well as containing glucose biosensors, the instrument contains sensors for pH, blood gases, hemoglobin and hemoglobin derivatives, various electrolytes, hematocrit, lactate, and urea. The biosensors within the OMNI instrument are zero-maintenance, flow-through, thick-film devices enabling simultaneous measurement of glucose, lactate, and urea.

Eppendorf (Hamburg, Germany) manufactures and sells numerous diagnostic instruments including photometric and flame photometric analyzers, as well as bioanalytical devices. The company produces two glucose analyzers, the EBIO Plus and the EBIO Compact, both of which are automated instruments for measuring glucose in whole blood, serum or plasma. The EBIO Plus also measures blood lactate. All measurements are made using amperometric enzyme electrodes (glucose oxidase and lactate oxidase are used), which feature a membrane system. The sensors are able to make up to 3000 measurements over a 15 day period before it is necessary to replace the membrane. Up to 80 analyses per hour are possible with the Compact instrument and 160 analyses per hour can be carried out using the EBIO Plus. Both instruments feature sample trays, automatic repeat testing, quality control samples and the ability to measure uncentrifuged samples.

i-STAT Corporation (NJ, USA) develops, manufactures and markets point-of-care medical diagnostic products for blood analysis. Through the use of advanced semiconductor manufacturing technology and established principles of electrochemistry, the i-STAT instrument (Fig. 3.15) is a hand-held automated blood analyzer. It is capable of performing a panel of commonly ordered blood tests on 2–3 drops of blood in approximately 2 min at the patient's side. Disposable test cartridges contain an array of biosensors microfabricated onto silicon chips. The current menu of tests includes: blood gases (pH, PCO_2, PO_2), sodium, potassium, chloride, glucose, creatinine, urea nitrogen (BUN), ionized calcium, and hematocrit. These cartridges are also used in Hewlett-Packard's Omnicare system.

Nova Biomedical (Waltham, MA, USA) produces whole blood analyzers to support the care of critically ill patients. Instruments are available for testing in the central laboratory, emergency department, surgery, critical care area, and in respiratory care. Several blood electrolyte/chemistry analyzers are available, many of which feature biosensors. The instruments are designed with simple one-button

Fig. 3.15 i-STAT analyzer.

operation. They can accept whole blood, plasma, serum, CSF, and urine samples. Routine testing throughput of up to 475 tests per hour is possible. Amperometric biosensors (Fig. 3.16) are used, which are based on the Clark oxygen electrode principle, with hydrogen peroxide measurement at a platinum anode.

Radiometer (Brønshøj, Denmark) have been involved in the blood gas business for over 40 years. The company produces numerous blood analyzers, including the ABL 700 series. The biosensor components are similar to those used in other machines of this type. They are in the form of cartridges (Fig. 3.17) that clip into the instrument.

Smaller, but otherwise similar to the ABL700 range, is the ABL70 analyzer that has been designed to meet special needs for near patient applications such as intensive care, emergency, operating rooms, and cardiovascular surgery. The ABL70 combines the cost savings and accuracy known from benchtop analyzers with the speed and convenience of a bedside device. Finally, the most basic machine is the ABL555, which provides the most essential parameters usually needed: pH, pCO_2, pO_2, cK^+, cNa^+, Hct, plus a choice of cCa^{2+}, cCl^-, cGlu, or cLac.

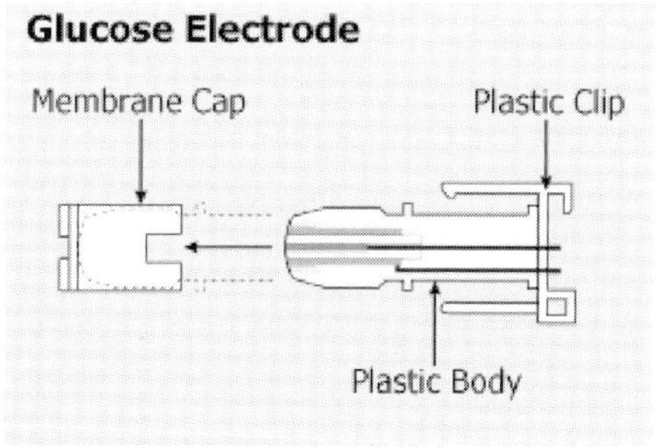

Fig. 3.16 Nova biosensor construction principle.

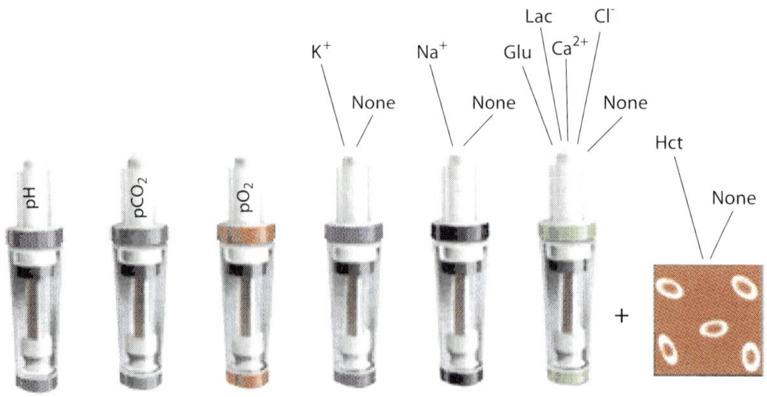

Fig. 3.17 Radiometer sensor and biosensor cartridges.

3.11
Direct Electron Transfer Systems

An elegant solution to elimination of interference in biosensors is to seek connection of the redox center of the enzyme to the electrode via a molecular wire. This can be achieved by modifying the enzyme, the transducer, or both. There is a considerable body of work in this area, including the following examples.

Glucose oxidase was wired with {poly[1-vinylimidazole osmium (4,4′-dimethyl bipyridine)$_2$ Cl)]}$^{(+/2+)}$ was used to 'wire' the enzyme. An enzyme electrode was then constructed by mixing the wired glucose oxidase into carbon paste. The electrodes that were uncoated, but cross-linked with poly(ethylene glycol) diglycidyl ether, responded linearly to glucose concentrations up to 60 mM. The device effectively rejected common interferences such as ascorbic acid and paracetamol when present at their physiological concentrations.

Alternatively, enzymes can be functionalized using artificial cofactor derivatives [19]. Glucose oxidase, for example, has been deposited on an electrode surface functionalized with a relay (PQQ)-FAD monolayer. The resulting sensor exhibited bioelectrocatalytic properties with unusually high current densities (300 mA cm^{-2} at 80 mM of glucose). Furthermore, the efficient wiring of the enzyme makes glucose sensing unaffected by oxygen fluctuations and minimizes the effect of other interferents.

Another approach is effected via a single molecule with delocalized electrons. Heteroarene oligomers, consisting of two pyridinium groups, linked by thiophene units of variable length (thienoviologens) are promising candidates for such conducting molecular wires and may be used in conjunction with self-assembly techniques to produce an insulated electrode that transfers electrons specifically along predetermined molecular paths [20]. This design should produce enzyme electrodes free from electrochemical interference.

3.12
Implantable Glucose Sensors

The 'holy grail' in terms of diabetes treatment is the successful introduction of an artificial pancreas. In order to facilitate continuous subcutaneous insulin infusion, with a portable and stable delivery system, it is necessary to undertake frequent, or preferably continuous glucose measurements. An implantable glucose measurement system is seen as a key component of such a closed-loop glycemic control system. Many of the approaches described in previous sections have been applied in the quest for reliable *in vivo* sensing devices. Such sensors can be broadly classified into intravenous and subcutaneous categories.

Unfortunately, the development of *in vivo* measurement systems is not straightforward. There are numerous problems that have been difficult to overcome.

- Sensor stability, which is particularly problematic when a biosensor approach is employed.
- Calibration: it is not a trivial process to relate the sensor reading to blood glucose levels.
- Biocompatibility: rejection of the sensor by the body is difficult to suppress.

The biocompatibility issue is, arguably, the greatest hurdle to be overcome. There are numerous effects, but from a sensing point of view, the major problem is build-up of tissue around the sensor, as a result of its implantation, affects diffusion to and from the sensor, resulting in readings that do not accurately reflect current blood glucose levels.

A major advance in the *in vivo* application of glucose biosensors was reported by Shichiri [21] who described the first needle-type enzyme electrode for subcutaneous implantation in 1982. Companies are still pursuing this possibility, but no device for general use is available yet. However, several sensors are in development and are thought to be close to market. Therasense (Alameda, CA, USA), for example, is developing a miniaturized subcutaneous sensor designed for insertion into the skin by the user to continuously monitor glucose levels and provide immediate results while also storing the results for future analysis. It is a wireless system, intended for use with a 'pager unit' that displays glucose levels. It can be conveniently carried or kept close to the bed at night.

Another company, MiniMed (Sylmar, CA, USA) has been developing an artificial, implantable insulin pump, which is implanted under the skin of the abdomen through a minor surgical procedure. The pump is designed to be refilled in a clinic every 3 months. For a patient, this can mean going from four injections per day to four injections per year. The implantable pump delivers basal rates of insulin and larger bolus doses, both controlled today by the patient through the use of a hand-held communicator that uses radio frequency telemetry. The MiniMed implantable pump has been in use since 1990, since when approximately 700 patients have received MiniMed implantable pumps through clinical trials. The Company's current implantable pump, the Model 2007, received European approval in 2000 and has a projected 10-year battery life. In early 2000, human testing of an implantable blood glucose sensor (Fig. 3.18) was started. The sensor is located in a central vein leading to the heart through a minor surgical procedure. It records glucose levels once per minute and is designed to be replaced after 1–2 years. The sensor is designed to communicate directly with the implantable pump, with the eventual goal of 'closed-loop' control using continuous glucose information to automatically regulate insulin delivery.

Minimed has recently received limited FDA clearance for its product, which currently restricts access of the information obtained, to the physician. No access by the patient is currently accepted.

A different approach is being taken by Animas (Frazer, PA, USA), with the aim of developing a long-term (>5 years), implantable sensor, which will provide continuous and accurate monitoring of blood glucose levels. This device will be equipped with alarms to give warnings of impending hypoglycemia and hyperglycemia. Ultimately the aim is to link the sensor to an insulin infusion pump to provide closed-loop control of blood glucose levels.

The Animas sensor measures the near-infrared absorption of blood, and it will be implanted across a vein with readings transmitted via radio wave telemetry to a small display unit worn on the wrist. Hence, there will be no percutaneous wires. The display unit will be about the size of a wrist-watch and the implanted sensor about the size of a pager.

Spectroscopic determination of glucose in blood has been difficult because glucose is in relatively low concentration and its spectra overlaps that of other blood constituents, such as proteins, urea, uric acid, hemoglobin, and even water. Many attempts to measure glucose through the skin or mucous membranes have failed, due to loss of light energy in the intervening tissues and the problem of extracting the blood spectra from that of intervening tissues. The Animas sensor side-

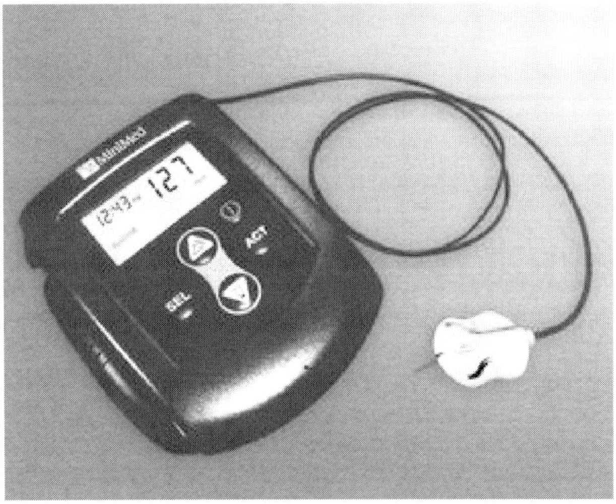

Fig. 3.18 Minimed implantable glucose sensor.

tracks these limitations because the sensor is implanted and, therefore, has direct access to blood. A further advantage is that the system offers the potential to overcome many of the limitations of other chemical sensors and biosensors associated with biocompatibility and the encapsulation tissue that forms around implanted sensor devices. At the selected wavelengths used, this tissue is transparent.

At present, the device has been demonstrated over a course of at least two weeks on an *in vivo* basis in dogs. The company is in the progress of designing a miniaturized sensor, which is planned for preliminary human trials in late 2003.

3.13 Minimally-Invasive Systems

In an attempt to minimize the discomfort caused by regular finger-pricking and to overcome the significant problems inherent in the introduction of *in vivo* devices, some workers have looked at ways of measurement that are minimally invasive. The general aim of this work is to avoid, where possible, the need for traditional blood sampling. Most efforts have targeted methods of drawing fluid through the skin without conventional puncture.

Early work involved abrasion of the skin, but a more recent approach is based on electro-osmosis and electrochemical measurement of interstitial fluid:

The GlucoWatch produced by Cygnus Inc. (Redwood City, CA, USA). The GlucoWatch is a wrist-worn device intended for detecting trends and tracking patterns in glucose levels in adults with diabetes. The device is intended for use at home and in healthcare facilities to supplement, not replace, information obtained from standard home blood glucose-monitoring devices. Due to the frequency of the automatic and non-invasive measurements, the GlucoWatch biographer has the potential to provide previously unavailable information about blood glucose, including: automatic and frequent measurements and alerts for high and low glucose levels.

Following a three-hour warm-up period, the device is capable of providing up to three glucose readings per hour for 12 hours after a single point calibration with the results from a standard finger-stick meter. The GlucoWatch biographer uses reverse iontophoresis to collect glucose samples through intact skin (Fig. 3.19).

The glucose molecules are collected in gel collection discs that are part of a single-use AutoSensor. The gel collection discs contain the enzyme glucose oxidase. As glucose enters the discs, it reacts with the glucose oxidase in the gel to form hydrogen peroxide. A biosensor in contact with each gel collection disc detects the hydrogen peroxide, generating an electronic signal. The biographer uses the calibration value previously entered by the patient to convert the signal into a glucose measurement. The glucose measurement is then displayed on the biographer and stored in memory. In 2001, the GlucoWatch biographer received approval from the US FDA. At the time of writing, the launch of the product had not yet taken place, but was expected in the near future.

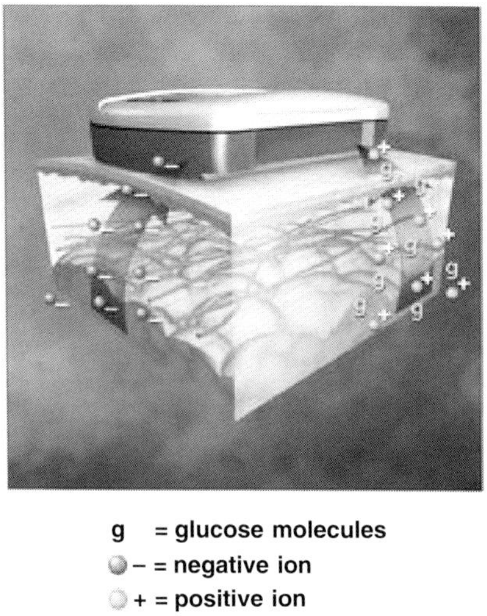

Fig. 3.19 Glucowatch operating principle.

Integ (St. Paul, MN, USA) has been developing the LifeGuide hand-held instrument that, when pressed against the skin on the forearm, uses a small needle to collect interstitial fluid, which is then analyzed by infrared spectroscopy. A 1 μL sample of interstitial fluid is collected through a 1.4 mm hole in the dermis. No blood is drawn, and the method is less painful than with conventional lancets. A recent research study demonstrated that the dermal interstitial fluid results closely tracked blood results. However, initial tests using infrared sensors have not been sufficiently accurate and it is believed that a more conventional approach using glucose oxidase biosensors is being pursued.

The approach taken by SpectRx (Norcross, GA, USA) is to measure glucose levels in interstitial fluid (ISF). The ISF is collected through an array of microscopic holes, created with a laser in the stratum corneum, the dead outer layer of skin, and measured in a patch containing a glucose sensor. ISF is the clear fluid under the skin through which glucose and other nutrients travel from the blood stream to the cells. Interstitial fluid and blood generally contain very similar glucose concentrations, as long as the blood glucose levels are steady and not rising or falling. The problem with this approach may be when glucose levels are rapidly falling or rising, in which case the interstitial fluid glucose concentration may trail or possibly precede that found in serum or blood. There is therefore a slightly increased risk of hyper- or hypoglycemia with insufficient warning. The company is currently assessing these issues and has attracted commercial interest and backing from Abbott Laboratories.

Bioject Inc. (Portland, OR, USA) is best known for it's innovative drug delivery systems, based on needle-free injection. The company has been adapting this technology and using it to develop a glucose monitoring system that will continuously measure glucose levels on a 24 h basis. The system is a patch-like sensor coupled

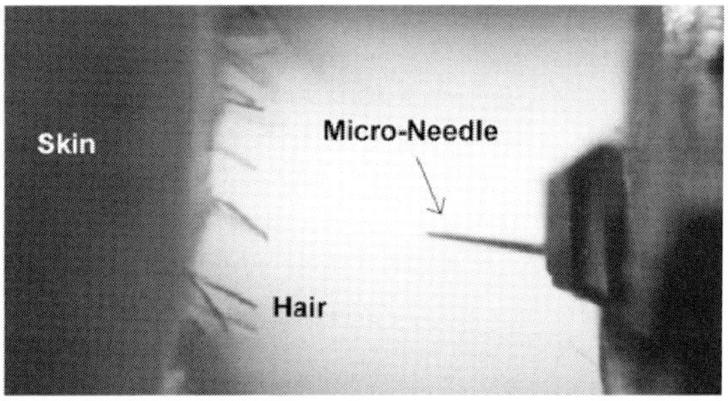

Fig. 3.20 Silicon micro-needle.

with a smart card similar in size to a credit card that can be kept in the pocket. The sensor and the card are connected by a telemetric system.

An interesting sampling system is being developed by Kumetrix (Union City, CA, USA), based on silicon micro-needles similar in size to a human hair. The aim is to develop a hand-held, battery-powered, electronic monitor, which accepts a cartridge loaded with disposable sampling devices. Each disposable consists of the micro-needle and a receptacle into which the blood sample is drawn. To take a measurement, the patient will load the cartridge into the electronic monitor and simply press the monitor against the skin. This action will cause the micro-needle (Fig. 3.20) to penetrate the skin and draw a very small volume of blood (less than 100 nL) into the disposable element, where the blood-glucose concentration will be measured.

3.14 Non-Invasive Systems

There are numerous potential methods for the measurement of glucose non-invasively. These include the following:

- Direct measurement of interstitial fluid or sweat. This has the potential advantage in the use of conventional analytical measurement techniques.
- Transcutaneous infrared spectroscopy. Near IR spectroscopy has been used successfully to measure cerebral hemodynamics as the spectral features of hemoglobin and oxyhemoglobin are strong and different. The same is not the case for glucose, where variations are subtle and not unique.
- Fluorescence lifetime or r.f. spectroscopy.
- Pulse laser photoacoustic spectroscopy. Instead of measuring transmitted or absorbed light, an optically generated pressure wave is created within the medium and detected at any other point. The pressure and photoacoustic spectra are dependent on laser energy, optical absorption coefficient, thermal expansion coefficient, velocity of sound, and specific heat. One can potentially use variations in these parameters with concentration to attempt measurement. This has been demonstrated in preliminary work for glucose.

- Far infrared (FIR) spectroscopy. This technique monitors FIR emitted by the body and thus requires no external energy source. Glucose, for example, absorbs in a band around 9.4 m, but the signal is very small and current detectors require cryogenic cooling.
- Breath analysis could be carried out via gas chromatography-mass spectrometry (GC-MS), of the components in exhaled air.
- Optical analysis through the eye. For example, spectral analysis of the chemical components in vitreous humor.
- Polarimetric techniques. The rotation of linearly polarized light is used to quantify concentrations of a chiral solution. Such an approach has been suggested for glucose determination, but the key problem *in vivo* is that the skin is highly scattering, which depolarizes most of the light. It could be applied through the eye, but problems remain with corneal birefringence, eye motion artifacts, and the time lag between blood and vitreal concentrations.

Despite the difficulties involved in using NIR spectroscopy for the measurement of glucose *in vivo,* the approach has been pursued and it is now thought that many of these problems have been overcome.

Diasense (Pittsburgh, USA) has been developing the Diasensor (Fig. 3.21) non-invasive glucose monitor. The Diasensor illuminates a small area of skin on the patient's arm with near-infrared light (700–2500 nm wavelength). The light penetrates into the skin tissue and the subpapillary vasculature (the capillary bed just below the epidermis), where some of it is absorbed by the various tissue and blood components, of which glucose is one such component. The Diasensor then measures the infrared spectrum of back-scattered light. This spectrum is a complex mixture of chemical components, such as that represented by the blood and tissues of the skin, and is the sum of the individual absorption spectra of all components. The contribution of each component's spectrum to the sum is weighted by its concentration and its ability to absorb infrared light (i.e., its absorption coefficient).

Glucose concentration is elucidated via a multivariate mathematical analysis. This is accomplished by first measuring a series of infrared spectra obtained from samples with known concentrations of the component of interest. This is referred to as the calibration sample set (some have referred to it as a 'learning' set). A multivariate analytical method is then used to model the mathematical relationship between a measured spectrum and the concentration of the component. This relationship can be expressed as a vector. The result obtained by multiplying a measured spectrum of a sample containing an unknown concentration of the component by this vector yields the concentration of the component. In this way, the concentration of glucose can be determined by measuring the infrared spectrum of the skin.

Because of the unique physiological and spectral characteristics of skin, it is not possible to calibrate such a sensor using artificial samples that contain glucose. Instead, spectra measured from the patient must be used for calibration. Also, because of significant person-to-person differences in skin morphology and physiol-

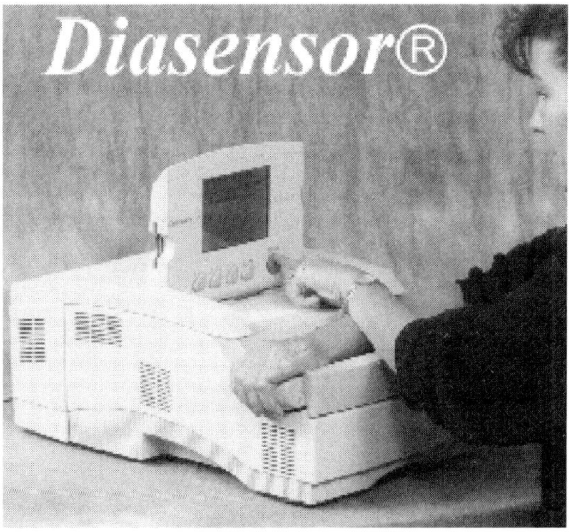

Fig. 3.21 Diasensor non-invasive glucose monitor.

ogy, the sensor must be individually calibrated to each patient. The calibration must also be performed over some reasonable period of time in order to break up any coincidental correlation that extraneous variables that are known to affect the measurement of an infrared spectrum, may have with glucose concentration. Such variables could include environmental variables such as temperature and humidity or patient variables such as skin temperature and skin hydration. In this way, the effect of such variables on the measurement of blood glucose is eliminated or at least reduced.

The current design of the Diasensor incorporates a modem that allows diabetics to transmit glucose readings via the Internet. This permits both a central monitoring department and the diabetic's own physician to monitor glucose levels and assist in glucose control.

Currently marketed in the EU, the Diasensor is the first commercially-available painless non-invasive monitor for glucose measurements in people with diabetes. At this time, the company is working with Joslin Diabetes Center, an affiliate of Harvard Medical School, to conduct FDA-approved clinical trials on the Diasensor in the USA. At the conclusion of the trials, the clinical data will be submitted to the FDA for further consideration of a modular Pre-Market Approval (PMA) application.

3.15
References

1 Clark, L.C. Jr. Monitor and control of blood and tissue oxygen tensions. *Trans. Am. Soc. Artif. Intern. Organs* **1956**, *2*, 41–48.
2 Clark, L.C. Jr., Lyons, C. Electrode systems for continuous monitoring in cardiovascular surgery. *Ann. NY Acad. Sci.* **1962**, *102*, 29–45.
3 Turner, A.P.F. *Advances in Biosensors I.* JAI Press, London **1991**.
4 Turner, A.P.F. *Advances in Biosensors II.* JAI Press, London **1992**.
5 Turner, A.P.F. *Advances in Biosensors Suppl I.* JAI Press, London **1993**.
6 Turner, A.P.F. *Advances in Biosensors III.*, JAI Press, London **1995**.
7 Newman, J.D., White, S.F., Tothill, I.E., Turner, A.P.F. Catalytic materials, membranes and fabrication technologies suitable for the construction of amperometric biosensors. *Anal. Chem.* **1995**, *67*, 4594–4599.
8 Koncki, R., Lenarczuk, T., Radomska, A., Glab, S. Optical biosensors based on Prussian Blue films. *Analyst* **2001**, *126*, 1080–1085.
9 de Marcos, S., Galindo, J., Sierra, J.F., Galban, J., Castillo, J.R. An optical glucose biosensor based on derived glucose oxidase immobilised onto a sol-gel matrix. *Sens. Actuators B* **1999**, *57*, 227–232.
10 Xie, B., Ramanathan, K., Danielsson, B. Mini/micro thermal biosensors and other related devices for biochemical/clinical analysis and monitoring. *TRAC (Trends in Anal. Chem.)* **2000**, *19*, 340–349.
11 Gorton, L. A carbon electrode sputtered with palladium and gold for the amperometric detection of hydrogen peroxide. *Anal. Chim. Acta* **1985**, *178*, 247–253.
12 White, S.F., Turner, A.P.F., Schmid, R.D., Bilitewski, U., Bradley, J. Investigations of platinized and rhodinized carbon electrodes for use in glucose sensors. *Electroanalysis* **1994**, *6*, 625–632.
13 White, S.F., Tothill, I.E., Newman, J.D., Turner, A.P.F. Development of a mass-producible glucose biosensor and flow-injection analysis system suitable for on-line monitoring during fermentations. *Anal. Chim. Acta.* **1996**, *321*, 165–172.
14 Gunasingham, H., Tan, C.B. Glucose enzyme electrode preparation by the codeposition of platinum and glucose oxidase onto glassy carbon. *Electroanal.*, **1989**, *1(3)*, 223–227.
15 Gorton, L., Svensson, T. An investigation of the influences of the background material and layer thickness of sputtered palladium gold on carbon electrodes for the amperometric determination of hydrogen peroxide. *J. Mol. Catal.*, **1986**, *38(1/2)*, 49–60.
16 Wang, J., Naser, N., Angnes, L., Wu, H., Chen, L. Metal dispersed carbon paste electrodes. *Anal. Chem.*, **1992**, *64(11)*, 1285–1288.
17 Tothill, I.E., Newman, J.D., White, S.F., Turner, A.P.F. Monitoring of the glucose concentration during microbial fermentation using a novel mass-producible biosensor suitable for on-line use. *Enz. Microb. Technol.* **1997**, *20*, 590–596.
18 Cass, A.E.G., Davis, G., Francis, G.D., Hill, H.A.O., Aston, W.J., Higgins, I.J., Plotkin, E.V., Scott, L.D.L., Turner, A.P.F. Ferrocene-mediated enzyme electrode for amperometric determination of glucose. *Anal. Chem.* **1984**, *56*, 667–671.
19 Katz, E., Willner, I. Biomolecular Architecture: Routes to Biosensors, Bioelectronics and Biofuel Cells, in *Proc. 1st Int. Symp. Macro- and Supramolecular Architectures and Materials (MAM-01): Biological and Synthetic Systems.* Kwangju, South Korea, 10–14 April **2001**.
20 Albers, M.W., Lekkala, J.O., Jeuken, L., Canters, G.W., Turner, A.P.F. Design of novel molecular wires for realizing long distance electron transfer. *Bioelectrochem. Bioenerg.* **1997**, *42*, 25–33.
21 Shichiri, M., Kawamori, R., Yamaski, R., Hakai, Y., Abe, H. Wearable artificial endocrine pancreas with needle-type glucose sensor. *Lancet* **1982**, *ii*, 1129–1131.

4
Biomagnetic Imaging: Principles of Magnetic Resonance Imaging and Emerging Techniques in Progress

S. Ueno and N. Iriguchi

4.1
Introduction

Magnetic resonance imaging (MRI) is a powerful way of producing medical images of the human body non-invasively. The first successful nuclear magnetic resonance (NMR) experiments were performed by two independent groups in the USA (Edward Purcell et al. at Harvard and Felix Bloch et al. at Stanford) in late December 1945 [1, 2]. Edward Purcell used a 0.7 T magnet and an 830 mL sample of paraffin wax, and obtained a proton signal of SNR (signal-to-noise ratio) 20. He submitted a paper to the *Physical Review* one day before Christmas, stating 'We have observed the absorption of radio-frequency energy due to transitions in a solid material'. Felix Bloch had moved from Switzerland to Stanford. He observed a nuclear induction of 0.1 mL water at 7.8 MHz. In 1952 they received the Nobel Prize for their world first observations.

Magnetic resonance signals are high-frequency signals, generated by nuclei that spin and make steady precessions in a static magnetic field. The nuclei resonate to an external RF field so as to cohere in precession phase to the RF field. A signal can be detected with a resonating electrical coil that is positioned close to the nuclei. The signal is an induced, macroscopic electromotive force (EMF), which has the same frequency as that of the spin precessions, and so small in amplitude as typically in microvolts.

NMR techniques were developed in the early days as important tools mainly for chemical analysis. The early experiments were limited in scope by the relatively poor instrumentation available then. In the late 1960s, superconducting magnets were introduced to NMR experiments and revolutionized the scope of NMR together with the emergence of Fourier-transform (FT) NMR [3].

Raymond Damadian noted in the early 1970s that the spin-lattice relaxation time T_1 and spin-spin relaxation time T_2 of living tissues vary significantly even with normal tissues, and relaxation times of tumor tissues are generally much longer than those of normal tissues [4]. Since then, relaxation times T_1 and T_2 have been recognized as important parameters in developing and designing magnetic resonance imaging machines, and obtaining and evaluating magnetic resonance images. Paul Lauterbur published the first NMR pictorial map of two

Sensors in Medicine and Health Care
Edited by P. Å. Öberg, T. Togawa, and F. A. Spelman

ISBN: 3-527-29556-9

1 mm tubes of water in 1973 [5]. He called MR imaging 'zeugmatography.' There have been continuing developments in instrumentation and applications that have led to the present situation.

MRI is now established as an important modality in clinical practice. The application techniques of MRI have diversified in recent years. These are, for example, magnetic resonance angiography (MRA), diffusion-weighted imaging (DWI), magnetic resonance spectroscopy (MRS), and functional MRI (fMRI). Scanning time has been dramatically shortened by the development of echo-planar imaging (EPI). Since there are clear attractions in using non-invasive methods for the study of living systems, MRI techniques appear to have a very bright future. Therefore, in this chapter, we overview MRI techniques [6], then discuss promising applications, and introduce emerging techniques to investigate tissue characteristics non-invasively.

4.2 Magnetic Resonance Signal

4.2.1 Electromotive Force (EMF)

When proton spins are placed in a static magnetic field B_0, the nuclear spins begin to make precessions. The phases of precessions are at random because they are primarily governed by uncertainty. Therefore, the direction of the net macroscopic vector of magnetization $\boldsymbol{M}_0$ is expected to be that of the z axis, the axis of the static magnetic field B_0.

When the radio-frequency (RF) field of the magnetic flux density B_1^* and the Larmor angular frequency ω_0 which is equal to γB_0 is applied to the nuclear spins for a duration Δt and in the direction of the x axis perpendicular to the z axis, the nuclear spins make precessions around the x axis by the angle $\theta = \gamma B_1^* \Delta t$, where γ is the gyromagnetic ratio proper to the nuclei. This occurs because the B_1^* field is the only apparent field experienced by the nuclear spin in the rotating frame. Thus, the vector of magnetization $\boldsymbol{M}_0$ makes a precession around the x axis, and the flip angle θ is the phase of the macroscopic precession of the magnetization $\boldsymbol{M}_0$ around the x axis. Numerous spins making precessions individually and directed parallel to the static magnetic field are flipped as a whole by resonance to the RF field in the direction perpendicular to the static magnetic field.

When the RF coil and the sample are supplied with the RF power P, the current i^* of the coil is $i^* = (P/R)^{1/2}$, where R is the equivalent, gross resistance of the RF coil and the sample. If B_1 is defined as the magnetic flux density produced by a coil carrying unit current, the magnetic flux density B_1^* which is actually applied to the sample is $B_1^* = B_1(P/R)^{1/2}$. The B_1 value of the RF coil can be determined also by the geometrical configuration of the RF coil.

When the flip angle θ of the magnetization $\boldsymbol{M}_0^*$ of a volume of a sample has been made perpendicular to the static magnetic field by a RF field ($B_1 \cos \omega_0 t$) gen-

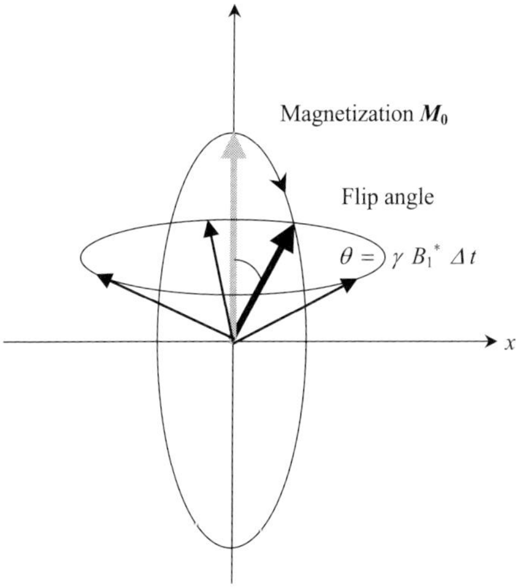

Fig. 4.1 The precession of the magnetization $\boldsymbol{M}_0$ around the x axis.

erated by a coil carrying unit current $i = \cos\omega_0 t$, magnetic resonance signals can be acquired with the same coil. The electromotive force (EMF) ξ_S induced by the magnetization $\boldsymbol{M}_0^*$ is then given by

$$\xi_S = -d(B_1 \boldsymbol{M}_0^*)/\mathrm{d}t\,.$$

Since the RF field is $B_1 \cos\omega_0 t$, the EMF can be given by

$$\xi_S = \omega_0 B_1 M_0^* \sin\omega_0 t$$

The number N^* of spins in the volume is $N^* = 10^3 cvA$, where A is Avogadro's number (6.02×10^{23}), v is the volume of the sample in m^3, c is the concentration of spins in M. Then

$$M_0^* = 10^3 cvA\gamma^2 (h/2\pi)^2 B_0 I(I+1)/3kT$$

where h is Planck's constant, I is the spin quantum number, k is Boltzmann's constant, and T is the temperature in Kelvin. If the concentration c, the RF field B_1, and the volume v are known, then the EMF ξ_S can be numerically calculated, since all other constants are known.

4.2.2 Relaxation Times

Tissues have proper characteristics in nuclear relaxation. Relaxation is the term given to a process of transition from an unstable, high-energy state to a stable, low-energy state.

When proton spins are exposed to a static magnetic field, the individual magnetic moments align parallel or anti-parallel with the static magnetic field. Spins whose expected axes of precessions are parallel to the external magnetic field are called α spins and in a low-energy, stable state. On the other hand, spins whose precession axes are anti-parallel to the external magnetic field are called β spins and in a high-energy, unstable state. In the state of thermal equilibrium, spins are in one of those two states, and the number of α spins is slightly greater than that of β spins. As small as this slight excess is, it accounts for the small macroscopic net magnetization M_0 directing parallel to the static magnetic field, and it is this differential that accounts for the NMR signal on which MRI is based.

When a 90° RF pulse is applied, α spins are excited and elevated to the high-energy state, and the number of α spins becomes the same as that of β spins. When a 180° RF pulse is applied, the number of β spins becomes greater than that of α spins. If excited spins are left in the thermal equilibrium, they will be allowed to lose energy to become low-energy, stable α state. This is the process of T_1 or longitudinal relaxation, or spin-lattice relaxation.

T_1 relaxation is an exponential process, and T_1 is the time constant of the process. T_1 relaxation is also the macroscopic phenomenon where the magnetization is allowed to return to the initial state of the thermal equilibrium. Therefore, it takes a long time to obtain an MR image of a sample with a long T_1 by repeatedly acquiring signals from those long-T_1 spins. With a clinical MRI machine, a lengthy time is inconvenient for a patient who is not allowed to move. Therefore, in general, T_1 should be preferably short. For example, the proton T_1 value of the white matter of the human brain is typically 1000 ms at 1 T.

When molecules are restricted in movement like those in solids, T_1 relaxation rarely occurs, and T_1 value are even as long as several hours. Nuclei are often the elements of large molecules, and the term lattice of spin-lattice relaxation usually means the whole environment of the molecular complex.

The main cause of relaxation phenomena of excited spins in water is supposed to be the dipole–dipole interactions. One water molecule has two protons. Each proton rotates at a high speed with an electrical positive charge, and consequently, each proton generates a magnetic field around itself like a magnetic dipole. In turn, each proton is exposed to magnetic fields generated by other close protons. Simultaneously, water molecules make thermodynamic motions. A correlation time is the time constant in that molecules lose as a whole a certain state by thermodynamic motions, and the more rapid the thermodynamic motions are, the sooner the molecules lose the certain state. When a correlation time of thermodynamic motions is similar to that of the period of spin precessions, many protons can be exposed to RF fields of a frequency similar to that of precessions, and therefore, similar to the resonant frequency but with various frequencies like a white noise. In consequence, spins excited by an external RF field with a monotone, resonant frequency are relaxed, since the excited state can no longer be maintained for a long time. That is the T_1 relaxation, caused by the dipole–dipole interactions.

T_2 relaxation is the macroscopic phenomenon where the magnetization decays in amplitude in the xy plane perpendicular to the main static magnetic field.

Spins have magnetic features affecting each other. Therefore, the magnetic flux density of the external static field actually experienced by a spin is varied by other spins. For the reason, spins placed in the same static field make precessions with slightly different frequencies, and phases of spin precessions become incoherent. Thus, T_2 or spin–spin relaxation takes place. When correlation times of thermodynamic molecular motions are very long, each proton is exposed to slowly-varying magnetic fields generated by other protons, and consequently, spin precessions lose the resonant coherency in phase. This is the T_2 relaxation, caused by the dipole–dipole interactions.

T_2 relaxation is also an exponential process, and T_2 is the time constant of the process, in that numerous spins, once resonating to an external RF field and cohering in phase of precessions, lose the coherency and become undetectable with a resonating electrical coil. Therefore, T_2 should be preferably long. By T_2 relaxation, the amplitude of EMF ξ_S becomes smaller and exhibits free induction decay (FID). The FID signal ξ_S can be expressed as

$$\xi_S = \xi_{S_0} \exp(i\omega_0 t) \exp(-t/T_2)$$

where the initial value of the amplitude ξ_{S_0} is

$$\xi_{S_0} = \omega_0 B_1 \boldsymbol{M}_0^*$$

In contrast to T_1 relaxation, the exchange of energy in T_2 relaxation does not take place between spins and lattices but only among spins. For example, the proton T_2 value of the white matter of the human brain is typically 100 ms. T_2 relaxation is accelerated in large molecules, and, therefore, T_2 values of the nuclei of large molecules are generally short. T_2 values of nuclei of small molecules are also short when small molecules are combined to large molecules.

When the external static magnetic field is inhomogeneous, phases of spin precessions become incoherent. This process appears like T_2 relaxation and is called T_2^* relaxation. By shimming or improving the homogeneity of the main static magnetic field, the T_2^* value can be made long. The FID signal ξ_S from spins in an inhomogeneous static magnetic field can be expressed as

$$\xi_S = \xi_{S_0} \exp(i\omega_0 t) \exp(-t/T_2^*)$$

After spins are excited by a 90° pulse, the vector of magnetization $\boldsymbol{M}_0$ on the xy plane begins to decay in amplitude by T_2^* relaxation. However, if a 180° pulse follows immediately, the phases are refocused to be coherent and the spins generate an echo [7]. This is the spin echo (SE) signal, and the 180° pulse is often called a time-reversing pulse, because it refocuses the phases of spin precessions as if time were reversed or the Second Law of Thermodynamics were violated.

The spin echo (SE) peak signal amplitude ξ_{SE} is affected by T_1 relaxation during the repetition time T_r and the T_2 relaxation during the echo time T_e. The SE signal ξ_{SE} can be hence expressed as

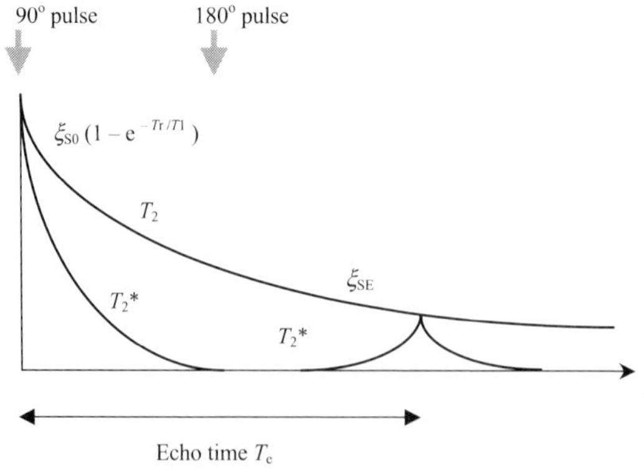

Fig. 4.2 Spin echo (SE) signal.

$$\xi_{SE} = \xi_{SO}\{1 - \exp(-T_r/T_1)\}\exp(-T_e/T_2)\exp(i\omega_0 t)\exp(-|t - T_e|/T_2^*)\ .$$

If the excitation pulse and refocusing pulse are θ_1 and $2\theta_2$ instead of 90° and 180° respectively, then the signal ξ_{SE} can be more generally expressed as

$$\xi_{SE} = \xi_{SO}\{1 - \exp(-T_r/T_1)\}\exp(-Te/T_2)\exp(i\omega_0 t)\exp(-|t - T_e|/T_2^*)\sin\theta_1\sin^2\theta_2 .$$

4.3 Overview of the Spin-Warp Imaging Method

In MRI machines, the scanning space inside the magnet is divided by three intersecting planes corresponding to three anatomical planes of the body (i.e., axial, coronal, and sagittal). The three planes are indicated by three intersecting axes, with the z axis indicating the alignment of the main static magnetic field. A gradient magnetic field can be used to assign nuclei to a position in space within the body. The gradient field is generated by a gradient coil [8]. A field from one coil opposing to a field from the other creates a net field which serves as a gradient between the two coils. Gradients can be applied in each of the three orthogonal directions (x, y, and z).

Slice selection in an MRI subject is the process of recognition of spins making precessions in a specific plane in the space. The slice selection is performed by applying a gradient magnetic field and an excitation RF pulse simultaneously. Also, the recognition of spin distribution in the 2nd and 3rd directions are performed by employing gradient coils.

The spin-warp imaging method is the traditional term for popular methods of MRI, such as field echo (FE) imaging, spin echo (SE) imaging and echo-planar imaging (EPI). The main feature of spin-warp imaging method is the peculiar process of detecting the location of a spin by varying a linear gradient magnetic field [9].

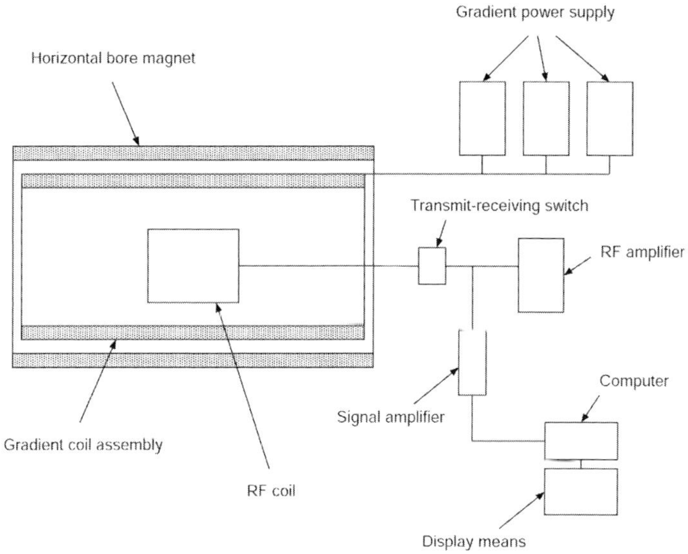

Fig. 4.3 Exemplified block diagram of an MRI scanning system.

A sample is positioned in a static magnetic field of, for example, a horizontal magnet. The magnet is equipped with a gradient field coil assembly and a RF coil. The gradient magnetic field coil assembly produces gradient magnetic fields in the same space of the homogeneous, static magnetic field, by transmitting power from a set of gradient magnetic field power suppliers. The gradient magnetic field power suppliers work by commands sent via a man–machine interface. RF power is transmitted to the RF coil from a RF power amplifier via a transmit-receiving switch, and excites proton spins in the subject. Contrarily, a magnetic resonance signal generated by proton spins is detected by the RF coil as an electromotive force and sent to a computer via the transmit-receiving switch and an amplifier equipped with a preamplifier and intermediate amplifier. Signals are sent to a display means after a data processing like a Fourier transform. The RF amplifier also works by commands sent via a man–machine interface.

4.3.1 Recognition of Spins Distributed in the First Direction

When a sample is placed in a static magnetic field B_0 and a linear, gradient magnetic field G_z is superimposed on the static field B_0, spins in the sample are excited by a RF field only in the region where the net flux density is specific to the spin resonance frequency ω_0. The RF pulses are applied in the direction perpendicular to the static magnetic field. This process is the slice-selective excitation of spins, and the selection is in the z axis direction of the applied gradient field G_z. By modulating the excitation power, the RF field is often given a square envelope function for a square excitation, a $(1/t)\sin t$ envelope function for a sinc excitation, or a Gaussian envelope function for a Gaussian excitation. Since a Fourier transformed sinc function in time is a square function in frequency, the slice profile resulted from a sinc excitation is square.

The effective frequency bandwidth of an excitation is the reciprocal of the effective duration of the applied RF pulse. When an excitation pulse has a sinc envelope function with a main lobe of 1 ms in duration at half maximum, the resulting slice has a square profile of about 1 kHz in bandwidth. The smaller the excitation bandwidth is, the thinner the resulting slice thickness is. If the frequency bandwidth is fixed and unchanged, the slice thickness in the z axis direction becomes thinner with a larger gradient field G_z.

When the gradient field G_z is 10 mT/m, for example, the proton resonance frequency is made distributed in 426 Hz/mm, because the resonance frequency of protons is approximately 42 600 000 Hz/T. When a pulse is applied with a sinc envelope with a main lobe of 2.35 ms in duration at half maximum, the Fourier transformed bandwidth is 426 Hz. Hence, a slice is selectively excited with a thickness of 1 mm.

4.3.2
Recognition of Spins Distributed in the Second Direction

The recognition of spins distributed in the 2nd, or the x axis direction can be performed by acquiring signals simultaneously applying a linear, gradient magnetic field G_x in the x axis direction. With the gradient field G_x, the resonance frequency of spins distributed in the x axis direction is linearly distributed in the direction of x axis from lower frequencies to higher frequencies. For example, if the gradient field G_x is 10 mT/m, then in the region of –0.1 m to +0.1 m inclusively in the x axis direction, the frequency deviation of resonance will be linearly distributed from –4200 Hz to +4200 Hz inclusively. This frequency bandwidth is small enough to be processed by an ordinary computer.

The signal acquired by a RF coil has various frequencies specific to the locations of spins distributed in the x axis direction. The spatial distributions of spins are restored by Fourier transformation. The process of encoding spin locations is often called reading-out of the signal. By the Fourier transformation, the spin distribution in the x axis direction can be obtained. The data are often exhibited in an absolute value $(Re^2 + Im^2)^{1/2}$ with the real amplitude Re and imaginary amplitude Im, both detected by quadrature demodulation (QD). For example, if the number of acquired data points is 512, and the half of those points are real and the rest are imaginary, then the spin distribution can be exhibited in an absolute 256-point data.

With the application of a reading-out gradient field G_x, the signal intensity rapidly diminishes. This is because the phases of spin precessions rapidly become incoherent. This decay can be altered by applying a negative gradient field $-G_x$ and then reversing the gradient field to a positive $+G_x$. By the reversing, a field echo (FE) signal is generated.

4.3.3
Recognition of Spins Distributed in the Third Direction

The recognition of spins distributed in the 3rd direction is performed by a technique called phase-encoding. It begins after the slice-selective excitation. A small gradient field G_y in the 3rd, y axis direction is applied for a certain period Δt. By the application of the gradient field G_y, a spin located far in the $+y$ direction from the original point is exposed to a higher field than that of the original point and makes a precession at a higher frequency than that of the spin located at the original point. In the period Δt, the phase of spin precession far in the $+y$ direction becomes larger than that of the spin at the original point. On the other hand, a spin located far in the $-y$ direction is exposed to a lower field than that of the original point and makes a precession at a lower frequency.

Secondly, slightly larger gradient field is applied in the same y direction for the same period Δt and also after the second slice-selective excitation. Thus, the gradient field is successively changed stepwise in amplitude and applied in 256 steps for example. By this, the farther spins are located from the original point, the greater changes of gradient field in amplitude are experienced by the spins. By Fourier transformation of the acquired 256-data set, it is recognized how far the spins are located from the original point.

A typical field echo (FE) imaging sequence is a combination of a slice-selective excitation in the z axis direction, a frequency-encoding in the x axis direction and a phase-encoding in the y axis direction. By 2D Fourier transform of a set of acquired signal data, the 2D distribution of spins is restored. Hence, the resulting image is on the plane of one axis in frequency f in Hz and another axis in phase θ in radian/step.

A spin echo (SE) imaging sequence employs a 180° pulse. When excited spins undergo the 180° pulse, an echo signal is generated at the moment in a same

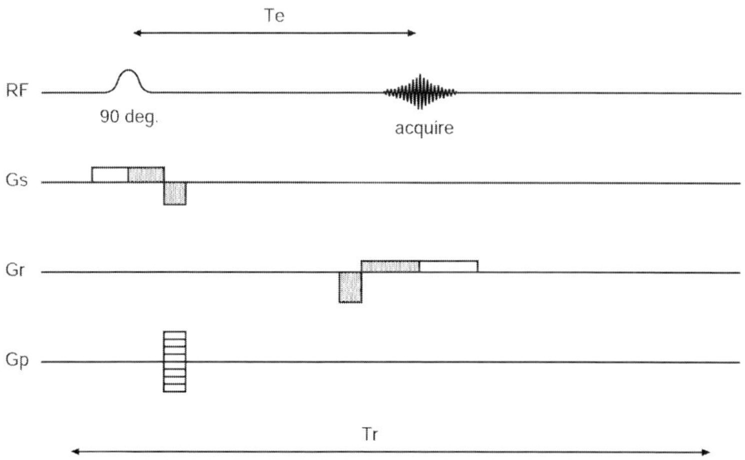

Fig. 4.4 Field echo (FE) imaging sequence.

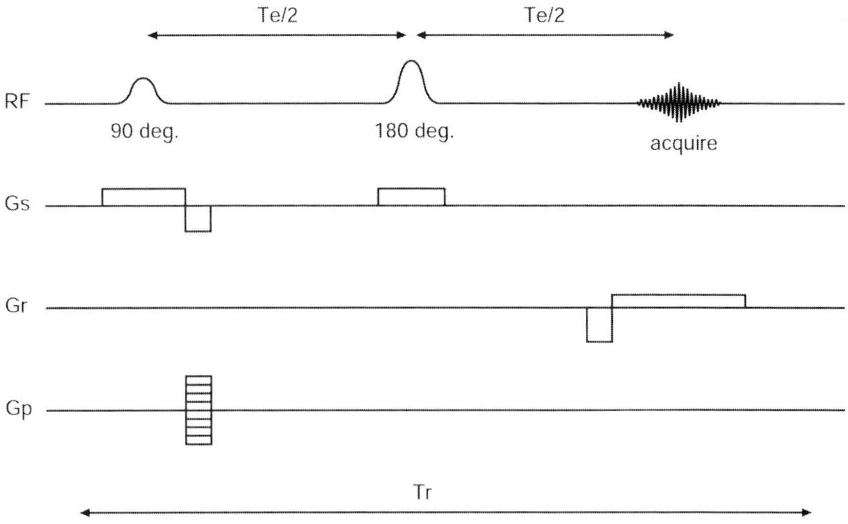

Fig. 4.5 Spin echo (SE) imaging sequence.

duration, $T_e/2$ from the time of the application of the first excitation pulse (i.e., the 90° pulse) through the moment of the 180° pulse. That is also a time, T_e, an echo time, from the application of the first excitation pulse. The 180° pulse is applied in the period $T_e/2$ after the 90° slice-selective excitation. The dephased spin phases refocus to generate an echo signal in the echo time T_e.

4.3.4
k-Space

By varying the amplitude of the gradient field in the phase-encoding direction, spins are labeled according to their phases of precessions. The successively acquired data are then put in successive order on a plane traditionally called *k*-space [10]. The echo signals put on the *k*-space are those demodulated and subtracted with the primary frequency $\omega_0 = 2\pi\nu_0$.

When a quadrature demodulation (QD) is employed, each data has a real and imaginary parts. For example, when the number of data points is 512, then 256 are real data points and the remaining 256 are imaginary data points. Hence, when the number of phase-encoding steps is 256, the vertical, phase-encoded data have 256 real data points and 256 imaginary data points. Therefore, on the *k*-space, there are a real data set of 256×256 points and an imaginary data set of 256×256 points. A 2D spin distribution can be restored by a 2D Fourier transform of the horizontal and vertical waves on the *k*-space. The spin distribution is usually displayed in absolute values, and the matrix size is typically 256×256. If an intensity diagram is gray-scaled and displayed on a CRT, then we can view an ordinary MR image.

4.3.5 Image Contrast

By applying an excitation RF pulse, the magnetization vector is given a component perpendicular to the z axis and the component decays in amplitude by T_2^* relaxation. When a field echo (FE) image sequence is used, and if both the repetition time T_r and echo time T_e are long, then the resulting FE image is a T_2^*-weighted image affected by T_2^* relaxation. The T_2^* relaxation of the brain tissue often reflects the brain activity, and T_2^*-weighted images are important in the field of functional imaging (fMRI).

By a spin echo (SE) imaging sequence, the magnetization vector is given a component perpendicular to the z axis and the component decays in amplitude by T_2 relaxation instead of T_2^* relaxation. If the repetition time T_r is much longer than T_1 and the echo time T_e is much shorter than T_2, then the resulting SE image reflects the spin density distribution.

If the echo time T_e remains short and the repetition time T_r is shortened, the sample of long T_1 undergoes successive excitations before the magnetization recovers by T_1 relaxation and generates small signal. The resulting image is a T_1-weighted image and has a constant reflecting the T_1 distribution. If both the

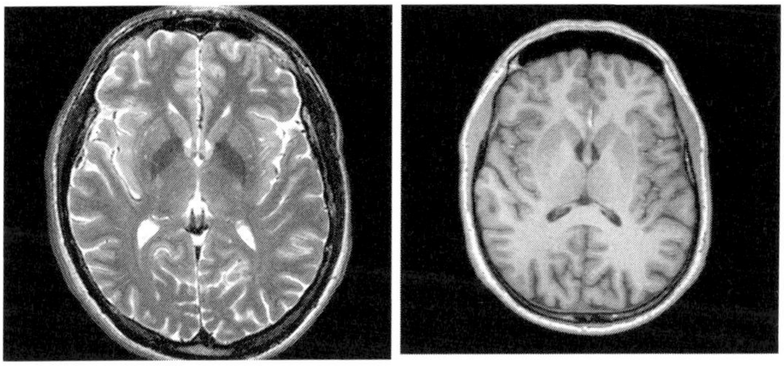

Fig. 4.6 Clinical brain image: 3T images, Siemens Magnetom.

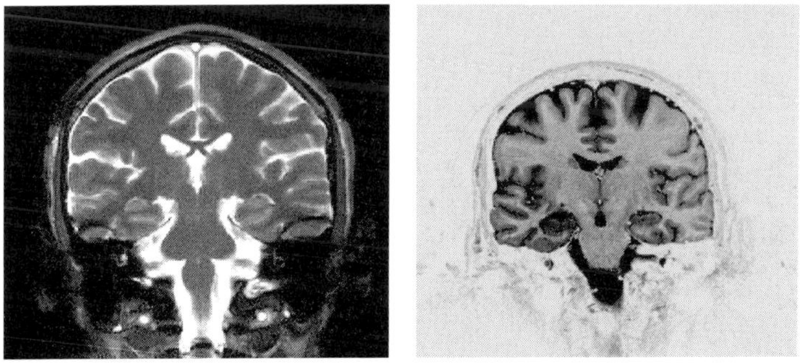

Fig. 4.7 Clinical brain image: 3T images, Siemens Magnetom.

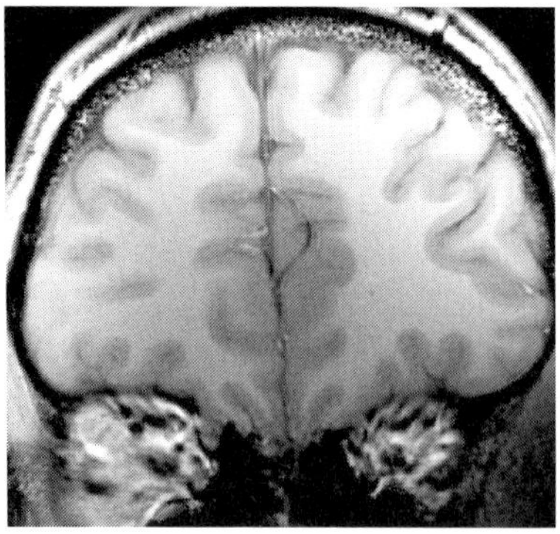

Fig. 4.8 Clinical brain image: 3T image, Siemens Magnetom, FLASH FOV 165 mm, 0.46 mm×0.43 mm×3.0 mm voxels. By permission of Larry Wald, MGH.

repetition time T_r and echo time T_e are long, then the resulting SE image is a T_2-weighted image affected by T_2 relaxation. Compared with field echo (FE) images, the influence of local inhomogeneity of the main static field is reduced in SE images.

In inversion-recovery spin echo (IR-SE) imaging, an inversion 180° prepulse is applied before SE signal acquisition. The duration from the moment of inversion 180° prepulse application through the SE excitation is called an inversion time T_i. With a long T_i, an image often becomes strongly T_1-weighted. However, the image can be also T_2-contrasted with a long T_e. In an IR-SE data, field echo (FE) data can be inserted into the *k*-space, and the image quality can be improved in anatomical details.

T_1-weighted images give higher intensity signals from the white matter than the gray matter in the brain. T_2-weighted images give higher intensity signals from the gray matter than the white matter. The contrast arises because the water protons of white matter have shorter T_1 and T_2 values than those of gray matter.

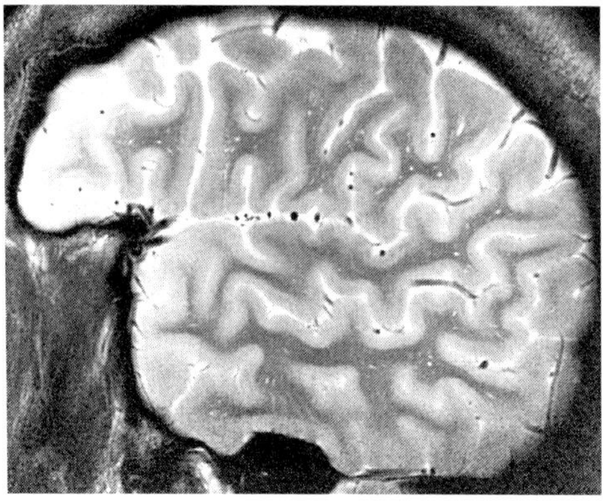

Fig. 4.9 Clinical brain image: 3T image, Siemens Magnetom, TSE FOV 140 mm, 0.33 mm×0.27 mm×2.0 mm voxels. By permission of Larry Wald, MGH.

4.4 Diversification of MRI Application Techniques

4.4.1 Magnetic Resonance Angiography (MRA)

Magnetic resonance angiography (MRA) is a powerful technique to obtain flow images. It does not require the injection of a paramagnetic contrast agent and permits the measurement of flow velocities. It is used for the assessment of abnormalities in the cerebrovascular system and peripheral vasculature.

There are two important methods for MRA. Those are the time-of-flight (TOF) method and the phased-contrast (PC) method. In a TOF method, the water protons in a slice of a certain thickness are first labeled. The labeling is performed by applying an excitation RF pulse. Since saturation makes the numbers of α and β spins equal, few signals are generated from the saturated region. In a time t after the saturation, fresh and unsaturated spins flow into the saturated region. As a result, contrast is generated between stationary and flowing water protons. If the consecutive excitation pulses are applied at intervals that are much shorter than T_1, there is an increase in signal intensity, because the inflow spins generate more signals than the stationary spins, which undergo partial saturation. Gradient-echo, FLASH (fast low-angle shot) sequences are commonly used for the 3D, time-of-flight MRA. Vascular structure can be extracted out of the full data set by virtue of their increased intensity via a post-processing routine called maximum intensity projection (MIP).

Flow can induce phase changes in MR signals. As water protons move along a gradient magnetic field, they accumulate a phase shift that is proportional to their velocity. In the PC method, a gradient field is applied in a certain direction. The faster the spins flow in the gradient direction, the more the transverse magnetization changes phase. By subtraction of data sets with and without the gradient field, a flow image is obtained. Compared with the TOF method, the PC method takes a longer time for signal acquisition, but the background tissue signals are small.

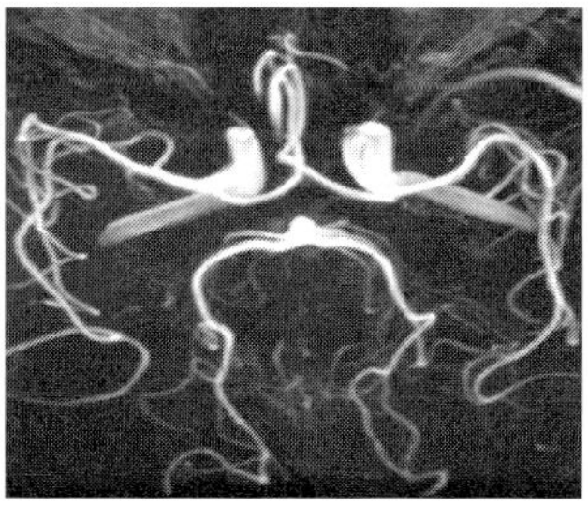

Fig. 4.10 Clinical brain image: 3T images, Siemens Magnetom.

4.4.2
Perfusion and Diffusion Imaging

Magnetic materials like transition metals and lanthanide ions have unpaired electric spins with hundreds times larger magnetic momentum than protons, and yield strong relaxation effects. As an application of those materials, intravenous administration of gadolinium compounds is widely used in the clinical field. Once gadolinium compounds are incorporated into the brain tissues through the blood-brain barrier (BBB), the compounds shorten T_1 of those tissues selectively, and by a T_1-weighted imaging sequence, those lesions can be selectively imaged.

A method called arterial spin labeling is also used in perfusion MRI. Once spins in the neck region are excited, arterial flow will carry the labeled spins into the brain tissue and influence the observed signal from brain water. Subtractions of images with and without the spin labeling permit measurements of regional perfusion. A technique of signal targeting with alternating radio-frequency (EPISTAR) is often used [11–13]. This is an echo-planar imaging (EPI) combined with time-of flight (TOF) techniques.

Diffusion measurements by NMR were introduced in the 1960s. E.O. Stejskal and J.E. Tanner introduced the method that uses motion-probing gradient (MPG) pulses [14, 15]. That is for measuring spin movements by making use of a phenomenon in which spins making precessions at a certain still positions are affected very little by a set of two gradient field pulses, or MPG pulses, with the same amplitude and opposite directions, while spins moving by diffusion are affected with those MPG pulses, and in consequence, the affected spins become incoherent in phase. The MPG two pulses can be applied with the same amplitude and direction, but before and after a 180°RF pulse.

In water, all molecules move randomly. It is often referred to as Brownian motion, and involves displacement distances that are small and comparable to cell

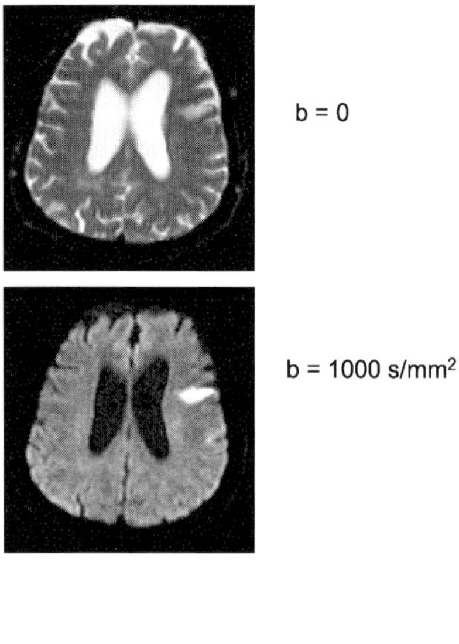

Fig. 4.11 Clinical brain image: diffusion imaging in acute stroke. Acquisition time: 1 min 8 s (21 slices, 3 b-values). Echo time: 108 ms; matrix 128×128; FOV 220 mm. By permission of Klinikum Grosshadern, Siemens Magnetom.

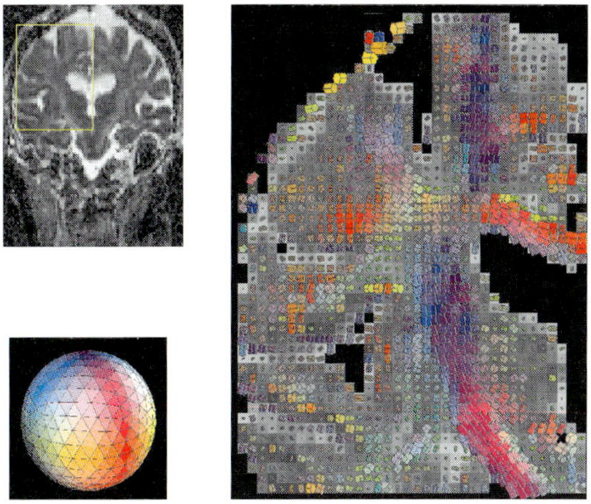

Fig. 4.12 Clinical brain image: diffusion tensor tractography of the *in vivo* human brain. The multicolor area is an enlargement of a coronal section through the brain. The eigenvalues and eigenvectors were then computed, weighted by the fractional anisotropy in the image, and overlaid on to an anatomic scan. Note the clear directionality of the fiber bundles including the corpus callosum, corticospinal tracts, and subcortical U fibers. By permission of Mette Wiegell, David Tuch, and Van Wedeen, MGH, Siemens Magnetom.

sizes. Hence, the water diffusion can be affected by the cell structure and diffusion-weighted images can reflect the change of the cell structure. In the tissue of cerebral infarction, diffusion of water is restricted and the apparent diffusion coefficient (ADC) of water becomes small. This is allegedly caused by the water shift from the extracellular spaces into the intracellular spaces [16].

Like diffusion of water molecules in water, molecular diffusion in the same molecular matrix is called self-diffusion. Self-diffusion is fundamentally isotropic. However, movements of spins in a substance like that of a living system are affected by blood flow to a certain extent beside diffusion. Therefore, a measured diffusion coefficient of water molecules in a living system is often called ADC.

Spins moving randomly under a gradient field lose signals. Hence, the diffusion can be detected by MRI. The group of D. LeBihan developed in mid-1980s an MRI technique that incorporates MPG pulses into an imaging sequence [17]. The gradient factor attenuation value, which is often referred to *b*-factor, is determined according to the MPG pulse amplitude and width δ, and the duration Δ between MPG pulses. When the *b*-factor is very small, then it is difficult to detect movements of spins only due to diffusion, but movements due to other origins like blood flow are also detected. But MPG pulses in use nowadays in the field of clinical MRI are capable to make ADC values almost equal to those of actual diffusion coefficients, by providing a large *b*-factor.

Diffusion-weighted imaging (DWI) has been widely used as a very important technique, because by a diffusion-weighted imaging, serious lesions like those of an acute cerebral infarction can be imaged in 20–30 min after the first onset, while such lesions can be imaged only in a few days after the onset by a conven-

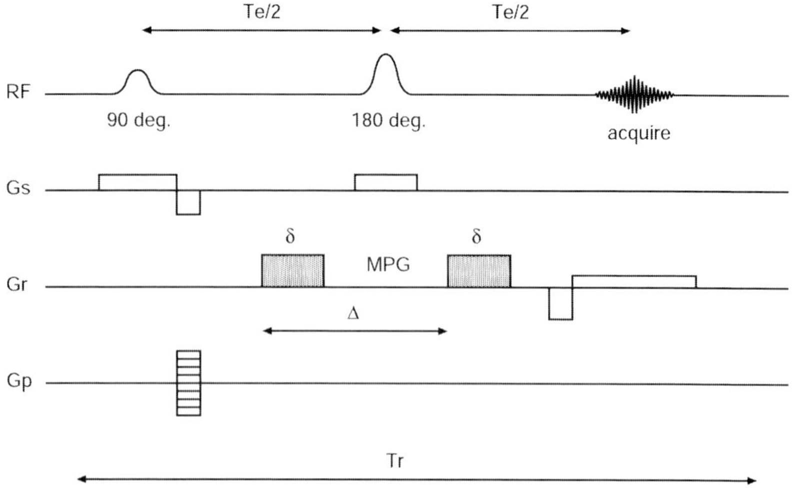

Fig. 4.13 Exemplified sequence chart of a diffusion-weighted imaging (DWI).

tional T_1-weighted or T_2-weighted imaging [18–20]. DWI is often linked to the anisotropy of the tissue structure. The anisotropy can be detected by the relations between the direction of the gradient field and the extent of signal loss by the water movements [21]. Associated with a cerebral infarction, the swelling of cells and shrinkage of extracellular spaces can lead to the reduction of electric conductivity in the early stage of infarction [22]. This suggests the importance of a non-invasive means to measure the tissue impedance of the brain.

4.4.3 Functional Imaging (fMRI)

Functional imaging (fMRI) is one of the most important MRI techniques developed in the last decade and visualizes the active regions of the brain non-invasively. fMRI uses field echo (FE) imaging sequences of long echo time T_e as 50 ms and reflects T_2^* characteristics of the tissue.

Electron spins of Fe^{2+} of deoxy-hemoglobin (Deoxy-Hb) have much greater magnetic momentum than those of proton spins, and T_2^*-weighted images of water around Deoxy-Hb generate few signals. Deoxy-Hb is paramagnetic, thus, when a brain is placed in the large static magnetic field, deoxygenated red blood cells experience a slightly larger magnetic field than surrounding tissue or plasma.

In the living brain, the gradient fields caused by the Fe^{2+} electron spins strongly influence not only the water in the blood vessels but also, to much greater extent, the outer neural tissues of the water volume of the blood vessels [23–25], and the localized magnetic field gradients can be large enough to affect MR image intensity.

The relationship among increases in neuronal electrical activity, increases in cellular metabolism, increases in oxygen extraction and blood flow are still not fully understood. However, recent research in several laboratories has clarified some important issues and there is now wide agreement on the following causal chain of events [26].

1. The brain is called upon to perform some task.
2. Neuronal activity increases in specific areas of brain gray matter.
3. Metabolic activity is increased in these areas.
4. The rates of oxygen and glucose usage in these areas are increased. (Blood oxygenation decreases in the capillary bed supplying the neuronally active tissue.)
5. Vasodilatory compounds are released at increased rate within the electrically active tissues, for as long as the extra electrical activity persists.
6. These compounds move to the capillaries and resistance arterioles supplying the electrically active tissue.
7. The resistance arterioles dilate, and this dilation is propagated along the arterioles by transmitted endotheliar depolarization.
8. Blood flow increases in the resistance arterioles.
9. Blood flow increases in the capillary bed supplied by these arterioles.
10. Capillaries experiencing increased blood flow dilate by 5–10%.
11. Oxygen supply to electrically active tissue begins to exceed demand. (Blood oxygenation increases in capillaries and the venules that drain them.)
12. Venules draining activated tissue start to increase in blood volume.
13. If enhanced neuronal activity continues, vascular and metabolic changes reach equilibrium in 1–3 min.
14. If neuronal activity returns to baseline, release of vasodilatory substances also returns to baseline.

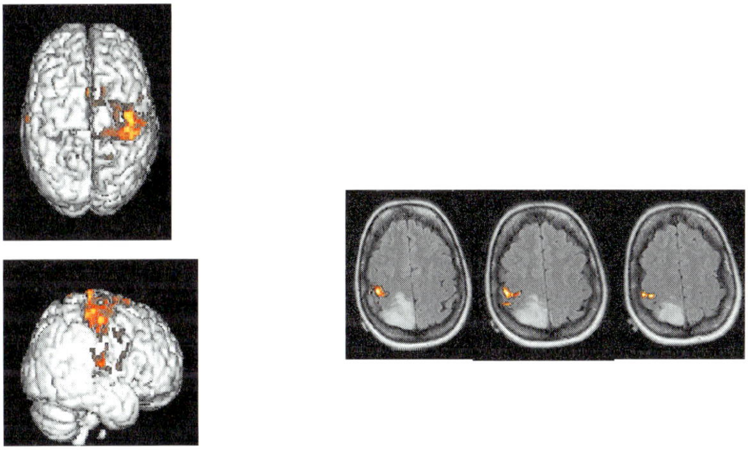

Fig. 4.14 (Left) Localization of the motor cortex in an astrocytoma case. By permission of Neuroradiology Tübingen, Siemens Magnetom. (Right) Functional data corresponding to motor cortex representation of the hand. By permission of DKFZ and Neurosurgery Heidelberg, Siemens Magnetom.

15. Blood flow returns to baseline.
16. Blood volume in draining venules remains elevated for 30–60 s after blood flow has re-equilibrated.

Associated with brain activity, the blood flow increases in the region, oxy-hemoglobin (Oxy-Hb) increases, and Deoxy-Hb decreases in relative quantity. Oxy-Hb, in which the magnetism of the iron atom is canceled by the magnetism of the attached oxygen, has very similar diamagnetic susceptibility to that of pure water and cellular tissue. From the blood oxygenation level development (BOLD) effects, the signal intensities from the active region in the obtained T_2^*-weighted images increases [27, 28]. fMRI reflects internal verbal generations and recalling of sensory or motor stimulation [29–31].

4.4.4
Magnetic Resonance Spectroscopy (MRS)

The ability to obtain localized spectra offers great potential for detecting human disease at the early stage of biochemical alteration. The proton signal from water is much greater than signals from metabolites. However, the suppression of water is expanding the use of MRI in metabolite evaluation. In the field of magnetic resonance spectroscopy (MRS), metabolite maps of specific chemicals or chemical shift images (CSI) are also obtained.

4.5
Imaging of Impedance Distribution of the Brain

Since electrical properties are important characteristics of living organisms, techniques for impedance tomography used to visualize impedance distribution have

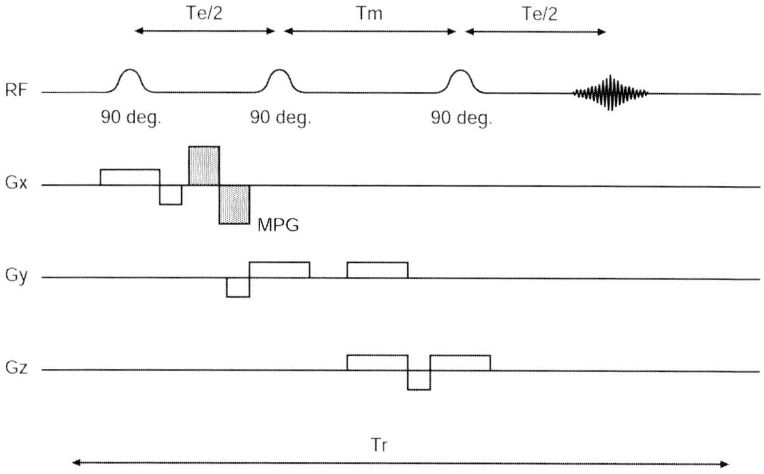

Fig. 4.15 Exemplified sequence chart of a diffusion-weighted localized MRS.

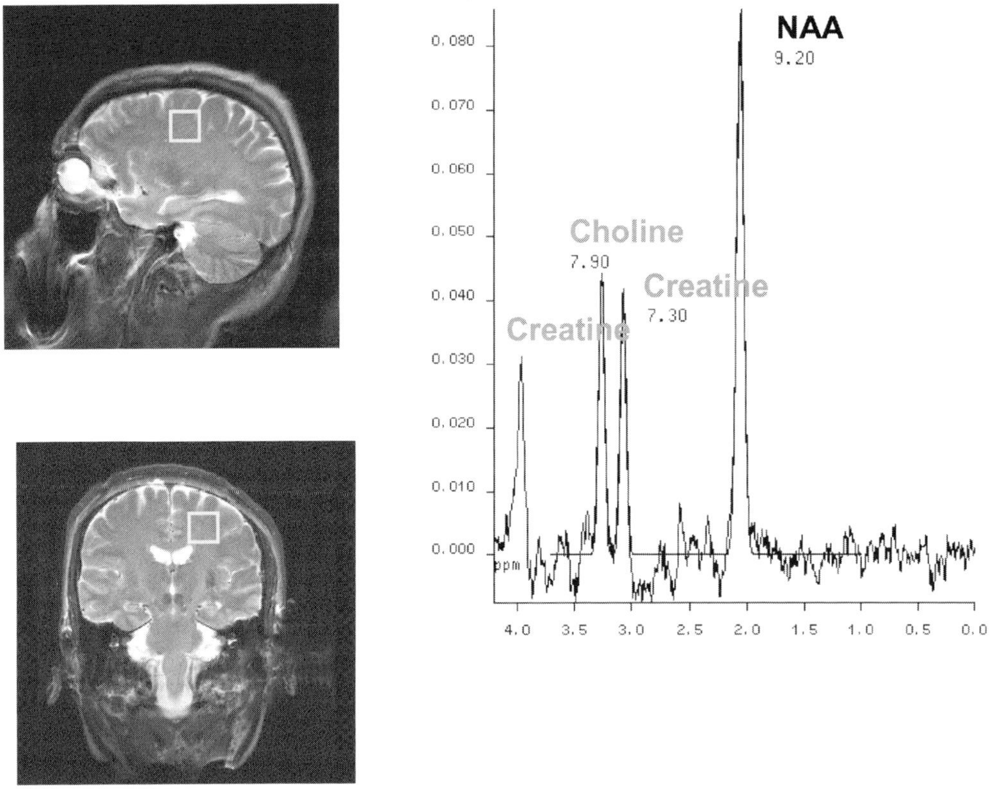

Fig. 4.16 Clinical brain image: T_e 135 ms, T_r 1.5 s, 256 averages, water suppression, 8 mL voxel. By permission of Washington University, St Louis, MO. 3T, Siemens Magnetom.

been developed with great interest [32]. This section addresses the feasibility of visualization of electrical impedance distributions of the brain, based on MRI techniques.

In MRI, when conductive tissues are subjected to an excitation RF field, eddy currents are induced in the tissues. The eddy currents generate a reactive field to reduce the applied RF field. The shielding effects can be used to visualize electrical properties of the body non-invasively. Two types of methods are described [33, 34]. One method is to use very large flip angles. The method is used to obtain conductivity-enhanced MR images at the given Larmor frequency. Another method is to apply an additional time-varying magnetic field parallel to the main static field B_0. The magnetic field is produced by the third coil, called the B_c coil. The method is used to enhance the conductivity in MR images at an arbitrary frequency. Experiments have been carried out to verify these concepts using a 7.05 T, 18.3 cm system [33, 34].

4.5.1
Principles

When conductive tissues are exposed to RF magnetic field, eddy currents are induced in the conductive tissues, resulting in the reduction of the net RF fields into the tissues. The main characters of eddy currents in the tissues are in principle movements of charged particles and rotations of electrical dipoles.

The shielding effect affects the spin dynamics of nuclei in the tissues. The flip angles are reduced in varied degrees, depending on the electrical characteristics of the tissues. Due to the absence of the transversal components of magnetization, when a precise 180, 360, or 540° excitation pulse is applied to conductive tissues, the tissues yield few signals. Meanwhile, the resistive tissues yield signals as they are less shielded than conducting tissues and undergo different flip angles simultaneously. The difference in signal, therefore, reflects the conductivity of tissues. By applying very large flip angles, conductivity-enhanced MR images can be obtained, at only the given Larmor frequency, and in the direction perpendicular to the applied RF field.

To obtain conductivity-enhanced images at an arbitrary frequency, an additional time-varying field parallel to the main static field B_0 is introduced. The purpose of the time-varying field is to perturb the main static field B_0. The perturbing field is produced by the third coil, the B_c coil. By the perturbing field or B_c field, the slice positioning of the image is affected, and the slice selection fluctuates according to the function of the B_c field. Also, the spatial information in the reading-out and phase-encoding directions is affected. Conducting tissues are less affected by the B_c field, because of the shielding effects. Since the frequency of the B_c field is independent on the given Larmor frequency, the conductivity-enhanced images are obtained at any frequency but in the direction perpendicular to the B_c field.

By triggering, the B_c field can be added to the main static field only during the period of slice selection and the artifacts in the reading-out and phase-encoding directions can be eliminated. The series of prepulses dephases nuclear spins of the sample in the neighboring region. When the sliced position is shifted by the B_c field into the neighboring regions, less nuclei are excited and the resulting image intensity weakens.

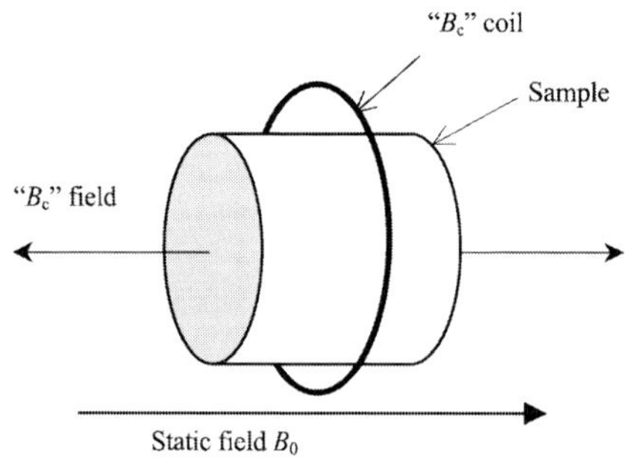

Fig. 4.17 The B_c coil produces the B_c field parallel to the main static field B_0.

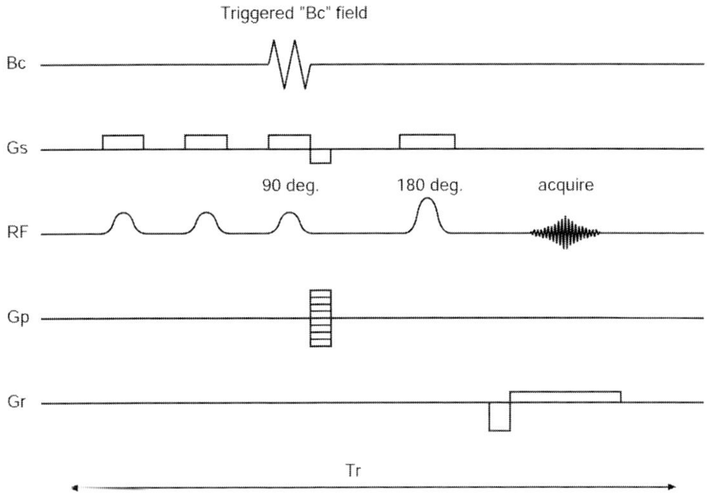

Fig. 4.18 B_c spin echo (SE) sequence.

4.5.2 Materials and Methods

A 7.05 T, 18.3 cm machine was used. The proton Larmor frequency of magnetic resonance at 7.05 T is 300 MHz. A 180° SE image of the head a 4-week old mouse was obtained by applying excitation pulses, which were selective to the cerebrospinal fluid (CSF).

A 66-turn, solenoidal coil of 6 cm in diameter was fabricated as the B_c coil to produce a low-frequency sinusoidal magnetic field. The amplitude of the field was comparable to the amplitude of the reading-out gradient field and the maximum phase-encoding field. A phantom and the mouse were subjected to imaging with a triggered B_c field of sinusoidal 100 Hz. The phantom consisted of an inner saturated saline compartment and an outer pure water compartment.

4.5.3 Results and Discussion

The 180° image showed a slight signal from the brain and muscle tissues because there were few transversal components of magnetization. On the other hand, in the same image, the resistive tissues (e.g., fatty tissues, spinal discs, and ligaments), transparent to the RF field, yielded specific signals. By applying 180° pulses to the conducting CSF and muscle tissues, resistive fatty tissues were simultaneously given different flip angles from 180° and produced image signals.

In the images of a phantom with a B_c field, the signal from pure resistive water fluctuated and diminished. One approach to evaluate an impedance-enhanced image is to compare images with and without the B_c field. An image of the mouse obtained with a 100 Hz B_c field was divided by an image obtained without the B_c field.

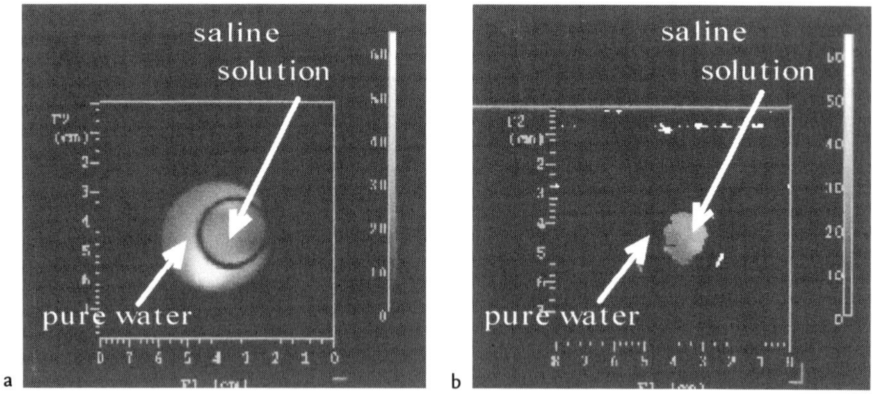

Fig. 4.19 a) Spin echo control image of a phantom. b) Impedance-enhanced image of a phantom obtained with a sinusoidal 100 Hz B_c field.

In the divided image of the mouse, the conductive brain and muscle tissues exhibited high signal intensities, whereas resistive fatty tissues exhibited low intensities.

It is known that MRI of the human brain obtained at high frequencies often exhibit unexpectedly strong signals from the geometrical center of the brain. This in contrast to the general principles that the penetration of high frequency power into a conductive sample is limited to shallow regions of the sample because of skin-depth effects and the central region of the sample must be shielded against the high frequency power. The phenomenon is known as dielectric resonance and a sort of artifact. In the 1970s, MR experiments were often associated with dielectric loss [35]. In those days, the losses were supposedly eliminated by the Faraday shield [36]. In the 1980–1990s, image artifacts related to dielectric resonance were often encountered in high frequencies [37, 38]. But those artifacts could not be removed with the use of a Faraday shield. Such peculiar phenomena associated with high frequency MRI are supposedly due to the very peculiar nature of water.

Water is composed of one oxygen atom and two hydrogen atoms, forming two 0.96 Å long O–H covalent bonds and a bond angle of 104.5° between them. Due to the asymmetric distribution of electrons in the triatomic molecules, with the electrons attracted to the oxygen nucleus, the water molecule exhibits a molecular dipole moment of 1.84 D. Water is a polar solvent and the dielectric constant is about 80 at 25 °C. The dominant electrostatic interaction in water is the 1.8 Å long hydrogen-bond (H-bond). The H-bond is stronger than an induced dipole-dipole interaction, which is known as Van der Waals interaction and 2.6 Å in length. Based on the tetrahedral bond architecture and the orientation of two unpaired electron pairs on the oxygen atom, a water molecule can form as many as four hydrogen bonds with each other in snow crystals. Liquid water has an average of 2.3 H-bonds per molecule, but the lifetime of an H-bond is so short that there is no discernible structure in liquid water.

Pure water is often conductive at high frequencies, even though it contains few ionic particles and reveals strong shielding effects against external magnetic fields

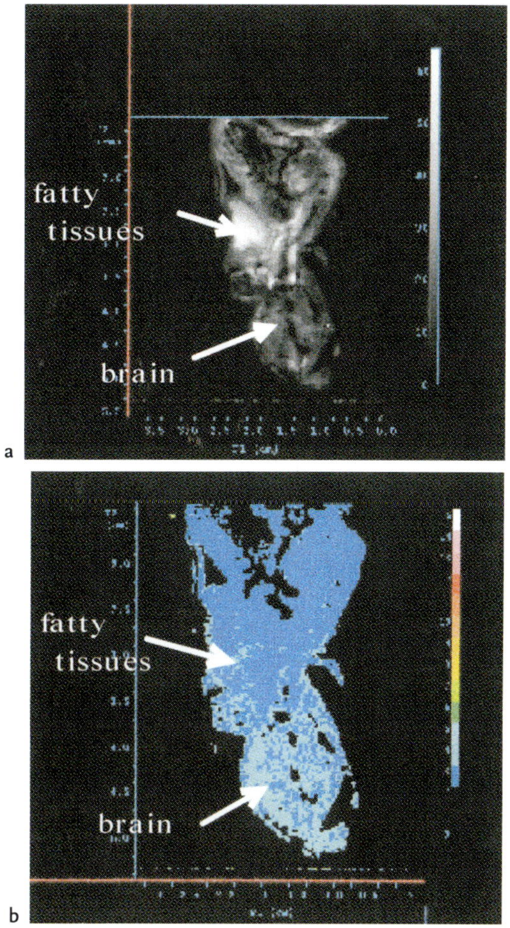

Fig. 4.20 a) Spin echo control image of a mouse. b) Impedance-enhanced, divided image of a mouse obtained with a sinusoidal 100 Hz B_c field.

[39]. Coherent rotations or resonant oscillations of water dipoles are supposedly responsible for the conductivity of pure water. Such frequencies are 100 MHz–1 GHz or higher for bound water molecules, which are partially within the range of high-frequency MRI.

Dielectric resonance of water is related to the high dielectric constant of water, however, the dielectric constant itself is supposedly not the sole cause of dielectric resonance. Glycerin, which has a relatively high dielectric constant of 43, reveals almost no dielectric resonance at 300 MHz. Rotations of water dipoles generate macroscopic eddy currents. The rotations of dipoles take place either in forced oscillations or in resonant oscillations, depending on the geometrical molecular radii with H-bonds and viscosity coefficients governed by Debye theory. As frequency becomes high, the real part of the dielectric constant of water decreases due to dielectric dispersion, and the imaginary part of the dielectric constant increases due to dielectric absorption. Rotations of dipoles tend to take place in the peripheral region of the brain more remarkably than in the central region. Meanwhile, proton spins in the central region of the brain are rather free from the molecular oscillations and the applied power is effectively used for spin excitations. This is supposedly the origin of the phenomenon that the human brain obtained at high frequencies often exhibits strong signals from the geometrical center of the brain.

When dielectric resonance is not present, the signal intensity $I(\boldsymbol{r})$ of a spin echo (SE) image ($\theta_1 - 2\theta_2$) obtained with a transmit-receiving coil is expressed as

$$I(\boldsymbol{r}) = \rho\theta(\boldsymbol{r}) \sin \theta_1(\boldsymbol{r}) \sin^2 \theta_2(\boldsymbol{r})\{1 - \exp(-T_r/T_1)\} \exp(-T_e/T_2)$$

where $\boldsymbol{r}$ denotes the vector of position and the flip angle $q(\boldsymbol{r})$ works as the transmit-receiving function. When dielectric resonance is present, the signal intensity $I(\boldsymbol{r})$ is expressed as

$$I(\boldsymbol{r}) = \rho\theta(\boldsymbol{r}) \sin\{\theta_1(\boldsymbol{r})d(\boldsymbol{r})\} \sin^2\{\theta_2(\boldsymbol{r})d(\boldsymbol{r})\}\{1 - \exp(-T_r/T_1)\} \exp(-T_e/T_2)$$

where $d(\boldsymbol{r})$ is a function of the extent of the dielectric resonance in the effective flip angle reduction. The effective flip angle $q_1(\boldsymbol{r})d(\boldsymbol{r})$ is yielded by the ratio of the signal intensities (I_2, I_1) of two SE images obtained with two flip angles ($2\theta_2$, θ_1) respectively: $\theta_1(\boldsymbol{r})d(\boldsymbol{r}) = \arccos\ \{I_2(\boldsymbol{r})/2I_1(\boldsymbol{r})\}$. Here, $\cos\{\theta_1(\boldsymbol{r})d(\boldsymbol{r})\}$ is given by $I_2(\boldsymbol{r})/2I_1(\boldsymbol{r})$, and $\sin^2\{\theta_1(\boldsymbol{r})d(\boldsymbol{r})\} = 1 - \cos^2\{\theta_1(\boldsymbol{r})d(\boldsymbol{r})\}$. If two SE image data $I_1(\boldsymbol{r})$ and $I_0(\boldsymbol{r})$ are obtained with $\pi/4$ and a very small flip angle excitations respectively, the extent $d(\boldsymbol{r})$ of dielectric resonance in flip angle reduction can be expressed as

$$d(\boldsymbol{r}) = 4\rho I_0(\boldsymbol{r}) \sin^4\{q_1(\boldsymbol{r})d(\boldsymbol{r})\} \cos^2\{\theta_1(\boldsymbol{r})d(\boldsymbol{r})\}\{1 - \exp(-T_r/T_1)\} \exp(-T_e/T_2)/I_1(\boldsymbol{r})^2.$$

Therefore, with long-T_r and short-T_e scan parameters, a map of the $d(\boldsymbol{r})$ of dielectric resonance can be produced. Also, the term $d(\boldsymbol{r})$ can be eliminated, employing two images with different flip angles [40].

Living tissues have ionized particles to produce conduction currents and water electrical dipoles those produce dielectric currents. Both give rise to the shielding effects.

Investigation of imaging of the impedance distributions of the brain based on non-invasive MRI techniques is important in brain pathology. Diffusion involves displacement distances of molecules comparable to cellular dimensions, raising the possibility that the measurement of water diffusion might provide a means of probing cellular integrity and pathology. Particularly, an investigation with electrodes on a model of the rat brain injury shows that the time course of the intensity changes in the DWI parallel the progressive shrinkage of the extracellular space as measured by the electrical impedance [19].

Both the large flip angle method and the B_c field method are sensitive to RF inhomogeneity. In addition, both the Larmor frequency B_1 field and the low frequency B_c field are easily transmitted, absorbed, and reflected by biological tissue boundaries in varying degrees. The variance depends on the geometry of the subject and the direction of the B_1 or B_c field. However, in spite of these difficulties, the methods of this study are promising to obtain tomographic images of electrical impedance distributions of living organisms.

It has been very recently reported that electrical currents compel T_1 and T_2 relaxation [41]. Further, it has been also reported that electrical currents promote movements of water molecules, resulting in a remarkable increase of apparent diffusion coefficient (ADC) [42]. The current density can be much less than 1 mA/cm^2. T_1 and T_2 can be remarkably shortened thereby without any use of paramagnetic materials. The phenomena are supposed to provide means to obtain distributions of internal electrical currents in living systems. Further, the phenomena are supposed to provide spatial distributions of electrical properties. Images, reflecting distributions of electrical properties, are supposed to be obtained by comparing T_1-weighted or T_2-weighted or diffusion-weighted images obtained with and without electrical currents. Further, images reflecting electrical currents evoked internally in the living brain are supposed to be obtained.

4.6 Concluding Remarks

By the continued progress in technologies of hardware and with the increase of utilization and deeper integration into clinical process, MRI scanners are developed in a more sophisticated way than ever before, integrating productivity in scanning reality and real-time post-processing procedures and friendliness in design concepts for clinical use. In the field of basic technology, MRI is providing a great deal of physiological information of physical, chemical and physiological properties such as temperature, mechanical elasticity, pressure, electrical conductivity, permittivity, and metabolite concentration of living systems. Therefore, in this chapter we have reviewed the MRI basic techniques and introduced emerging techniques in progress.

4.7 References

1 E. M. Purcell, H. C. Torroy, and R. V. Pound, Resonance absorption by nuclear magnetic moments in a solid. *Phys. Rev.* **1946**, *69*, 37–38.

2 F. Bloch, W. W. Hansen, and M. Packard, The nuclear induction experiment. *Phys. Rev.* **1946**, *70*, 474–485.

3 R. R. Ernst and W. A. Anderson, Application of Fourier transform spectroscopy to magnetic resonance. *Rev. Sci. Instrum.* **1966**, *37*, 93–102.

4 R. V. Damadian, Tissue detection by nuclear magnetic resonance. *Science* **1971**, *171*, 1151–1153.

5 P. C. Lauterbur, Image formation by induced local interactions: examples employing nuclear magnetic resonance. *Nature* **1973**, *242*, 190–191.

6 S. Ueno and N. Iriguchi, Principles and horizons of magnetic resonance imaging. Advances in Electromagnetic Fields in Living Systems, Vol. 3, edited by J. C. Lin, Kluwer Academic/Plenum Publishers **2000**, 39–71.

7 E. L. Hahn, Spin echoes. *Phys. Rev.* **1950**, *80*, 580–594.

8 R. Turner, Gradient coil design: a review of method. *Magn. Reson. Imag.* **1993**, *11*, 903–920.

9 W.A. Edelstein, J.M.S. Hutchison, G. Johnson, and T. Redpath, Spin warp NMR imaging and applications to whole body imaging. *Phys. Med. Biol.* **1980**, *25*, 751–756.

10 S. Ljunggren, A simple graphical representation of Fourier-based imaging method. *J. Magn. Reson.* **1983**, *54*, 338–343.

11 D.S. Williams, J.A. Detre, J.S. Leigh, and A.P. Koretsky, Magnetic resonance imaging of perfusion using spin inversion of arterial water. *Proc. Natl. Acad. Sci. USA* **1992**, *89*, 212–216.

12 W. Zhang, D.S. Williams, and A.P. Koretsky, Measurement of rat brain perfusion by NMR using spin labeling of arterial water: *in vivo* determination of the degree of spin labeling. *Magn. Reson. Med.* **1993**, *29*, 416–421.

13 R.R. Edelman, B. Siewert, D.G. Darby, V. Thangaraj, A.C. Nobre, M.M. Mesulam, and S. Warach, Qualitative mapping of cerebral blood flow and functional localization with echo-planar MR imaging and signal targeting with alternating radiofrequency. *Radiology* **1994**, *192*, 513–520.

14 E.O. Stejskal and J.E. Tanner, Spin diffusion measurements: spin-echoes in the presence of time-dependent field gradient. *J. Chem. Phys.* **1965**, *42*, 288–292.

15 E.O. Stejskal, Use of spin echo in pulsed magnetic-field gradient to study anisotropic, restricted diffusion and flow. *J. Chem. Phys.* **1965**, *43*, 3597–3603.

16 M.E. Mosley, Y. Cohen, J. Mintorovich, L. Chileuitt, H. Shimizu, J. Kucharczyk, F.M. Wendland, and P.R. Weinstein, Early detection of regional cerebral ischemia in cats: comparison of diffusion and T_2-MRI and spectroscopy. *Magn. Reson. Med.* **1990**, *14*, 330–346.

17 D. LeBihan, E. Breton, D. Lallemand, P. Grainer, E. Cabanis and M. Laval-Jeantet, MR imaging of intra voxel incoherent motions: application to diffusion and perfusion in neurologic disorders. *Radiology* **1986**, *161*, 401–407.

18 J. Mintrovich, M.E. Mosley, L. Chileuitt, H. Shimizu, Y. Cohen, and P.R. Weinstein, Comparison of diffusion- and T_2-weighted MRI for the early detection of cerebral ischemia and reperfusion in rats. *Magn. Reson. Med.* **1991**, *18*, 39–50.

19 J.M. Pennock, F.W. Cowan, J.E. Schweiso, A. Oatridge, M.A. Rutherford, L.M.S. Pubowitz, and G.M. Bydder, Clinical role of diffusion-weighted imaging: neonatal studies. *MAGMA* **1994**, *2*, 273–278.

20 S. Warach, D. Chien, W. Li, M. Ronthal, and R.R. Edelman, Fast magnetic resonance diffusion-weighted imaging of acute human stroke. *Neurology* **1992**, *42*, 1717–1723.

21 D. LeBihan and R. Turner, Diffusion and perfusion. In: *Magnetic Resonance Imaging.* Mosby Year Books, St. Louis **1992**, 353–371.

22 H.B. Verheul, R. Balazs, J.W. Berkelbach van der Sprenkel, C.A.F. Tulleken, K. Nicolay, K.S. Tamminga, and M. van Lookeren Campagne, Comparison of diffusion weighted MRI with changes in cell volume in a rat model of brain injury. *NMR Biomed.* **1994**, *7*, 96–100.

23 S. Ogawa, T.-M. Lee, A.S. Nayak, and P. Glynn, Oxygenation-sensitive contrast in magnetic resonance image of rodent brain at high magnetic field. *Magn. Reson. Med.* **1990**, *14*, 68–78.

24 S. Ogawa and T.-M. Lee, Magnetic resonance imaging of blood vessels at high field: *in vivo* and *in vitro* measurements and image simulation. *Magn. Reson. Med.* **1990**, *16*, 9–18.

25 R. Turner, D. LeBihan, C.T.W. Moonen, D. Despres, and J. Frank, Echo-planar time course MRI of cat brain by oxygenation changes. *Magn. Reson. Med.* **1991**, *22*, 159–166.

26 R. Turner and R.J. Ordidge, Technical challenges of functional magnetic resonance imaging. *IEEE Eng. Med. Biol.* **2000**, *19(5)*, 42–54.

27 J. Frahm, H. Bruhn, K.D. Merboldt, and W. Haenicke, Dynamic MR imaging of human brain oxygenation during rest and photic stimulation. *J. Magn. Reson. Imag.* **1992**, *2*, 501–505.

28 P.A. Bandettini, E.C. Wong, R.S. Hinks, R.S. Tikofsky, and J.S. Hyde, Time course EPI of human brain function during task activation. *Magn. Reson. Med.* **1992**, *25*, 390–397.

29 G. McCauthy, A.M. Blamire, D.L. Rothman, R. Gruetter, and R.G. Schulman, Echo-planar MRI studies of frontal cortex activation during word generation in humans. *Proc. Nat. Acad. Sci. USA* **1993**, *90*, 4952–4956.

30 L. Rueckert, I. Appollonio, J. Grafmaii, P. Jezzard, R. Johnson, Jr., D. LeBihan, and R. Turner, MRI functional activation of the left frontal cortex during covert word production. *J. Neuroimaging* **1994**, *4*, 67–70.

31 D. LeBihan, R. Turner, T.A. Zeffiro, C.A. Cuenod, P. Jezzard, and V. Bon nerot, Activation of human primary visual cortex during visual recall: an MRI study. *Proc. Natl. Acad. Sci. USA* **1993**, *90*, 11802–11805.

32 P. Metheall, D.C. Barger, R.H. Smallwood, and B.H. Brown, Three-dimensional electrical impedance tomography. *Nature* **1996**, *380*, 509–512.

33 S. Ueno and N. Iriguchi, Impedance magnetic resonance imaging: a method for imaging of impedance distribution based on magnetic resonance imaging. *J. Appl. Phys.* **1998**, *83*, 6450–6452.

34 Y. Yukawa, N. Iriguchi, and S. Ueno, Impedance magnetic resonance imaging with external AC field added to main static field. *IEEE Trans. Magn.* **1999**, *35*, 4121–4123.

35 D.G. Gadian and F.N.H. Robinson, Radiofrequency losses in NMR experiments on electrically conducting samples. *J. Magn. Reson.* **1979**, *34*, 449–455.

36 P. Mansfield and P.G. Morris, NMR Imaging in Biomedicine. Academic Press, NY **1982**.

37 H. Bomsdorf, T. Helzel, D. Kunz, P. Roeschmann, O. Tschendel, and J. Wieland, Spectroscopy and imaging with a 4 Tesla whole-body MR system. *NMR Biomed.* **1988**, *1*, 151–158.

38 H. Burfuss, H. Fischer, D. Hentschel, R. Ladebeck, A. Oppelt, R. Witting, W. Duerr and R. Oppelt, *In vivo* magnetic resonance imaging and spectroscopy of human with a 4-T whole-body magnet. *NMR Biomed.* **1990**, *3*, 31–45.

39 K.C. Ong, H. Wen, A.S. Chesnick, S. Duewell, F.A. Jaffer, and R.S. Balaban, Radiofrequency shielding of surface coils at 4.0 T. *J. Magn. Reson. Imag.* **1995**, *5*, 773–777.

40 H. Mihara, N. Iriguchi and S. Ueno, A method of inhomogeneity correction in MR imaging. *MAGMA* **1998**, 7, 115–120.

41 M. Sekino, K. Yamaguchi, N. Iriguchi and S. Ueno, T_1 reduction by applying a small amount of electrical currents. *Abstract Book of Ninth Scientific Meeting of ISMRM*, Glasgow **2001**.

42 K. Yamaguchi, M. Sekino, N. Iriguchi and S. Ueno, A method to enhance apparent diffusion by electrical currents, *Abstract Book of Ninth Scientific Meeting of ISMRM*, Glasgow **2001**.

5
Non-Invasive Cardiovascular Hemodynamic Measurements

K. Yamakoshi

5.1
Introduction

Blood pressure (BP) and cardiac output (CO) are physiologically fundamental variables for the assessment of cardiovascular hemodynamic function. Non-invasive measurement of such variables is desirable for medical and bio-engineering fields such as clinical diagnosis and treatment, basic and sports medicine, psycho-physiological and bio-feedback research, health science, and so on. However, the ideal technique for each measurement has yet to be developed. A number of methods have been designed, all of which have their own advantages and disadvantages [1, 2]. It is therefore important to develop simple, convenient, and reliable instrumentation capable of allowing non-invasive BP and CO measurement.

It is also well-known that cardiovascular hemodynamic variables show considerable changes in their momentary levels, which are due to physiological conditions, physical activities, psychological and mental stresses, and environmental conditions [3–5]. The changes in BP, for example, may be as much as 70 mmHg for systolic and 40 mmHg for diastolic pressure; a single measurement at a discrete point in time may therefore be of little meaning [4, 5]. This is particularly relevant to behaviorally oriented studies as well as modern screening clinics. With the latter, the emphasis is commonly on transient changes in BP and/or CO occurring in response to various stress testing procedures and clinical treatment, and on the quantification of cyclical fluctuations in these variables during daily life. Ambulatory and long-term monitoring of cardiovascular variables is one of the best ways to address these problems.

This chapter describes our current work on non-invasive BP and CO measurement techniques based on the volume-oscillometric [6] and volume-compensation methods [7–8] for the former and on electrical admittance (or impedance) cardiography [9] for the latter. For BP measurement, a local pressurization method is described that is expected to reduce the inconvenience and discomfort due to cuff occlusion of the biological segment. For CO measurement, to give a less troublesome measurement for the subject, replacement of a conventional band-electrode by a spot-electrode array for impedance cardiography is discussed, based on the measurement of thoracic impedance mapping.

Sensors in Medicine and Health Care
Edited by P. Å. Öberg, T. Togawa, and F. A. Spelman

ISBN: 3-527-29556-9

In addition to these promising approaches to BP and CO measurement, the development of two non-invasive ambulatory cardiovascular monitoring systems combining these two measurement techniques is introduced, presenting some monitoring examples including clinical applications together with circulatory autonomic analyses obtained under laboratory conditions and outside.

5.2 Blood-Pressure Measurement

5.2.1 Overview of the Measurement Methods

Several methods for non-invasive or indirect blood-pressure (BP) measurement have been developed based on the sphygmomanometric cuff method (arm occlusion) originally proposed by Scipione Riva-Rocci, such as palpatory, flush, cuff-oscillometric, auscultatory, volume-oscillometric method, and so on [1, 5]. These are classified by the methods of detecting the physical phenomena of arterial volume or blood flow changes, peculiar sounds from the artery (Korotkoff sounds) during gradual decrease (or increase) of applied counter-pressure to a biological segment. In general, it takes more than 10–20 s to determine systolic, mean, or diastolic BP values by these methods: Naturally, these methods do not allow a continuous measurement of pressure wave-forms as well as BP values on a beat-by-beat basis.

Conversely, there are three methods of measuring non-invasively pressure waveforms as well as beat-by-beat BP values: arterial tonometry [10, 11] based on coplanar measurement [12], Peňáz's method [13–16], and the volume-compensation method [7, 8]: the latter two are based on the indirect unloading or vascular unloading technique proposed by Marey [1, 5] and more recently by Shirer [17].

Arterial tonometry is a simple method of providing a continuous read-out of pulse pressure waveform without the use of an occluding cuff, being based on the principle that displacement of a force-sensitive transducer over a superficial artery like the radial artery can be made linearly proportional to the intra-arterial blood pressure [11, 12]. Calibration of BP values, if necessary, is usually made by the direct measurement, auscultatory, or cuff-oscillometric method [11, 18]. However, its application is considerably limited by the need for critical positioning of the transducer, since any slight displacement leads to marked fluctuations in the pressure record necessitating repositioning and recalibration. A new approach of tonometry using multiple pressure sensors within a single diaphragm has been made to address these practical problems [19] and such an instrument is now commercially available, although the shortcomings to be solved still remain.

Both Peňáz's and the volume-compensation method utilize the vascular unloading technique [17], requiring a volume servocontrol system to control an applied counter-pressure using an occluding cuff. Although each method is basically similar, it may be different with respect to how to set a servoreference for the servocontrol [20]. On the whole, these two methods may provide an acceptable alternative to direct BP measurement for continuous, non-invasive BP recordings [5].

In general, an occluding cuff encircling a biological segment makes the subject uncomfortable. Frequent and/or long-term measurements are therefore limited because the pain caused by blood pooling (or venous congestion) distal to the measuring site.

Here we describe a new local pressurization method intended to overcome these difficulties by using a partially occluding cuff on the finger, the radial artery, or the perforation artery in the back of the thigh. Our technique uses the volume-oscillometric or volume-compensation method. First, the principles of these two methods are briefly described, and then a local pressurization method at each measuring site is presented.

5.2.2
Brief Description of Measurement Principle and Summary of Accuracy

5.2.2.1 Volume-Oscillometric Method

This method is based on the nonlinear nature of the pressure-volume characteristics in the artery (arterial tube law) [6]. When applying a gradual change in counter-pressure (cuff pressure, P_c) externally to the biological segment, characteristic changes in the amplitude of arterial volume pulsation, ΔV, produced by the pulse pressure are observed due to the arterial tube law. When a finger is used as a measuring site, ΔV can be easily detected indirectly by a transmittance- (or reflectance-) photoplethysmography (PGp; ΔV PGp) using a near-infrared LED as a light source and a sensitive photodiode as a photodetector (Fig. 5.1). The determination of indirect systolic (SBP) and mean pressure (MBP) by this method is given as the P_c values corresponding respectively to the systolic end point (SEP) and the point of maximum amplitude of volume pulsation (MAP).

Diastolic pressure, DBP, cannot be obtained with this method directly from the characteristic changes in the amplitude of the volume pulsations, but it can be calculated, if necessary, from SBP and MBP determined above, along with a volume determinant, a_v, measured from the waveform of the volume pulsation [21, 22] as

$$\mathrm{DBP} = \mathrm{MBP} - a_v/(1 - a_v)(\mathrm{SBP} - \mathrm{MBP}) \,. \qquad (5.1)$$

This method has been used for indirect BP measurement in small animals such as rats and rabbits with reasonably high accuracy [23]. When the root of the human finger is used as a measuring site, BP values determined by this method agree well with those from direct brachial BP measurement, although they are 5–10 mmHg lower, probably due to a physiological pressure drop between the measuring sites [6]. This has also been revealed by 24 h ambulatory BP monitoring in clinical tests, compared with direct radial BP monitoring [24–26]. As a whole, this method is more accurate than a conventional cuff-oscillometric method, which has some basic and practical drawbacks, as pointed out by Mauck et al. [27] and by us [6, 23, 28]. The drawbacks are briefly summarized as follows.

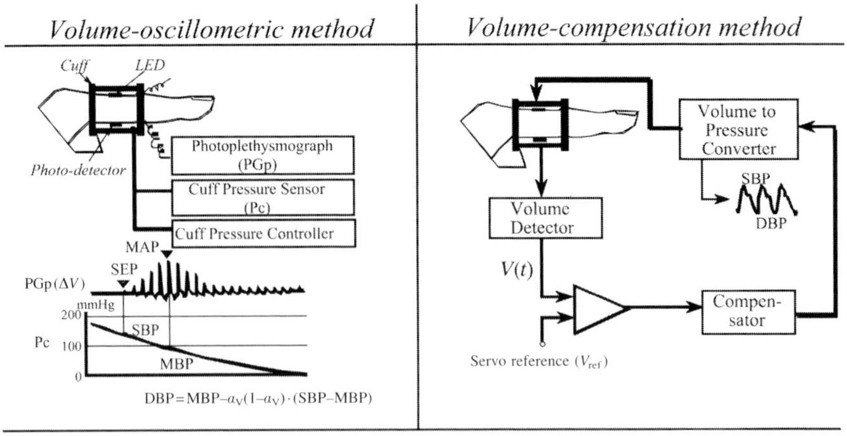

Fig. 5.1 Principles of non-invasive measurement of blood pressure: left) volume-oscillometric method; right) volume-compensation method.

1. It is inaccurate in principle to determine both SBP and DBP values. The cuff-oscillometric method gives only an estimation of MBP [1]. During a gradual change in P_c, small oscillations in cuff pressure occur, produced by transmitted arterial movements via the tissue, and the P_c value corresponding to the point of maximum oscillation gives MBP. The oscillations are above SBP and continue below DBP, so that a definite point showing both SBP and DBP cannot be exactly discriminated [6, 23, 28]. Therefore, almost all of the instruments based on this method give only estimated values of SBP and DBP according to some empirically derived algorithm.
2. There are influences on MBP from the true intra-arterial mean pressure, pulse pressure, heart rate, and arterial elastic properties [27].
3. This method gives erroneous results in BP because of poor choice of cuff size. The point of maximum oscillation of pressure shifts largely according to the air capacity of the cuff selected only when intra-arterial BP is kept constant [27, 28].

Thus, careful consideration of these matters is required when analyzing BP data obtained by the cuff-oscillometric method.

5.2.2.2 **Volume-Compensation Method**

This method is based on 'indirect unloading' of the vascular wall [17]. If the mean cuff pressure, $\bar{P}_c$, is equal to the mean intra-arterial pressure, $\bar{P}_a$, so the transmural pressure ($P_{tr}=\bar{P}_a-\bar{P}_c$) is zero, the artery is in the unloaded state. When a pressure change ($\pm\Delta P$) is imposed on $\bar{P}_a$, the vascular volume will change accordingly ($\pm\Delta V$). This volume change can be compensated for by applying the same amount of change in P_c to maintain a constant volume in the unloaded state (V_0). In this state, the cuff pressure ($P_c=\bar{P}_c\pm\Delta P$) instantaneously follows the intra-arterial pressure (P_a). Therefore, the instantaneous blood pressure can be indirectly obtained by measuring P_c.

A volume-servo system is needed for the application of this principle (Fig. 5.1). The vascular volume change is subtracted from a reference value (V_{ref}) corresponding to V_0 using a comparator to produce a servocontrol error, which controls P_c. In this way, any variation in the vascular volume due to a change in blood pressure is instantaneously compensated for by this servosystem so that the vascular volume is clamped at the reference value.

In order to determine the reference value for the servocontrol, the volume-oscillometric method can be used [7, 8, 29]. This is implemented as the mean DC level of the volume signal detected at the point of maximum volume pulsation during a gradual change in cuff pressure.

As for the volume-oscillometric method, this method has been used for indirect BP measurement on a beat-by-beat basis in a rat's tail and a rabbit foreleg with very close agreement with direct measurement [7, 20]. Also, a highly linear correlation between this method in the root of the finger and the direct measurement in a brachial artery has been determined both in the beat-by-beat systolic and diastolic pressure values as well as in their pressure wave-forms, though the indirect pressure (finger pressure) indicates pressure values approximately 10 mmHg lower than the direct arterial pressure, which may also be caused by a physiological pressure drop and/or an obstructive hydraulic pressure drop from the catheter placed in the brachial artery.

5.2.3 Blood-Pressure Measurement at the Wrist by Local Pressurization

In Japan, a wrist-sphygmomanometer based on the cuff-oscillometric method is now in widespread use for domicillary monitoring, as an alternative to the conventional arm sphygmomanometer. The wrist device is more convenient for the subject, and it is easier to make such a device small.

In this measurement, a cuff is wrapped around the wrist to occlude both the radial and the ulnar artery. The subject feels compression and discomfort due to the occlusion. Also, frequent measurements for long-term monitoring are limited by the pain caused by blood pooling (or venous congestion) distal to the measuring site, as for the arm-occluding method. Miniaturization is limited by the requirement for a band cuff and it is difficult to reduce power consumption because of the large capacity of the cuff.

To address these difficulties, we tried a local pressurization technique using a partial compression cuff to occlude only the radial artery.

5.2.3.1 Location

For BP measurement at the radial artery, an appropriate location for a partially occluding cuff at the wrist needs to be determined, in order to optimize pressure

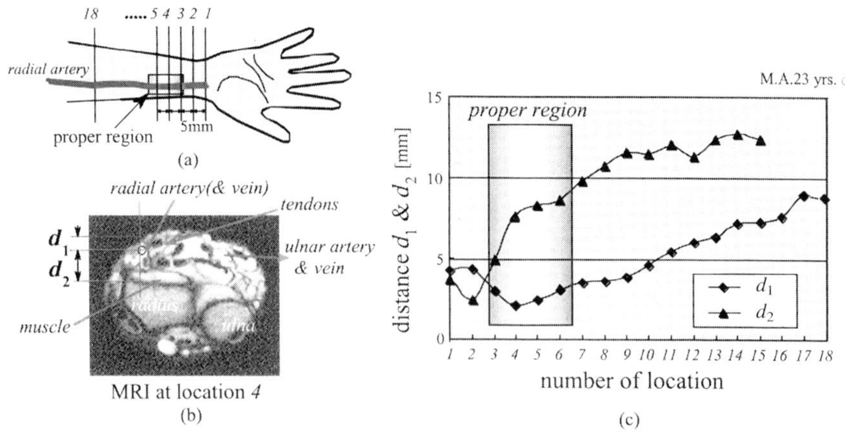

Fig. 5.2 a) Appropriate location for BP measurement at the wrist denoted by shading where a wrist-watch is usually worn. b) Example of imaging at location 4 obtained by MRI. c) Distance from the surface of the skin to the center of the radial artery (d_1) and from the center of the artery to the upper surface of the radius (d_2) against location number at the wrist.

transmission to the artery. From a practical viewpoint, the best location at the wrist is where a wrist-watch is usually worn (Fig. 5.2a). The radial artery will be displaced toward the radius when applying partial compression to the artery through the skin, so it is necessary to know the position of the artery between the skin and the surface of the radius. Accordingly, cross-sectional images of the wrist at 5 mm spacings were measured by MRI (1.5 Sigma Horizon, GEC, USA) as indicated by the numbers 1 to 18 shown in Fig. 5.1a.

Fig. 5.2b shows an example of the cross-sectional image at location 4 obtained in a young healthy man. From this image, the distance from the surface of the skin to the center of the radial artery (d_1) and that from the center of the artery to the upper surface of radius (d_2) were measured.

The result of these two distances from the distal (location 1) to the proximal portion (location 18) is shown in Fig. 5.2c. The distance d_2 is in the range 5–8 mm in the feasible region shown by the shaded rectangle. This means that less than 5–8 mm of displacement of the artery toward the bone would be required from the partial compression through the skin: from this information the cuff was designed to depress the tissue to give complete arterial occlusion.

5.2.3.2 **Pad Cuff and its Fixation**

The cuff is a square pad of length d, made of silicone rubber membrane with excellent elasticity (no hysteresis), which protrudes on the side contacting the skin sufficiently to depress the tissue and occlude the artery. The dimension d was determined from anatomical information about the wrist. There are some obstacles to effective pressure transmission like tendons around the radial artery. So the

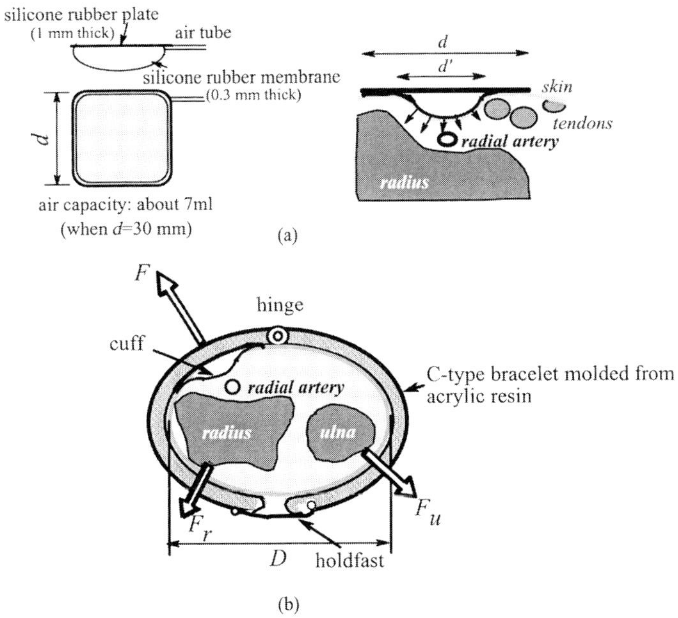

Fig. 5.3 a) Configuration of a pad cuff for partial compression of the radial artery (left) and application of the cuff to a radial portion of the wrist (right). b) Fixation of the cuff using a hinged C-bracelet. Since a force, F, acting on the bracelet due to an increase in cuff pressure reacts on both the radius, F_r, and the ulna, F_u, no circumferential (or radial) elongation occurs, and therefore the pad membrane depresses the skin to apply pressure to the artery.

width, d', capable of pressure transmission to the artery avoiding the obstacles is a limited portion of the distance between the radius styloid process and the tendon of flexor carpi radialis muscle, as shown in the right side of Fig. 5.3 a. Therefore, d' is 1/3.5 to 1/4 times the width of the wrist, D. The cuff length d should be sufficient to cover this dimension: we used 1.5–2 times d' for d.

The values of D and d' were measured by a Vernier caliper on 20 men and women (22–54 years old). They were found to be in the ranges 55–65 mm (average about 60 mm) and 12–18 mm (average about 15 mm), respectively. Thus, the length d was tentatively determined as 30 mm for the present experiment.

The cuff thus designed is fixed inside a C-bracelet (about 5 mm thick and 35 mm wide) made of acrylic resin. The center of this bracelet is hinged so as to be easily put on or taken off. A fastener locks the bracelet. Because of the bracelet being rigid, even if a force, F, acting on the bracelet due to an increase in cuff pressure reacts on both the radius, F_r, and the ulna, F_u, as shown in Fig. 5.3 b, there is no circumferential (or radial) elongation with excellent firm fixation of the cuff (the bracelet is rigid to prevent circumferential elongation even when the cuff pressure is increased). Therefore, the membrane of the cuff can depress the skin sufficiently to apply the pressure effectively to the radial artery through the subcutaneous tissue.

5.2.3.3 **Arterial Deformation Analysis**

To evaluate the performance of arterial deformation by the unit thus designed, the shape and area of the arterial cross section (A) and its pulsatile change (ΔA) were measured by ultrasonic tomography (UX-02, Rion, Japan; 30 MHz, 20 frames/s) with a specially designed image analyzer modified by commercially available software (OPTIMAS-6, Optimas, USA). For this experiment, a hole of 12 mm diameter was drilled in the part of the bracelet where the cuff was fixed, and the head of an ultrasonic probe was placed on the hole. To reduce the attenuation of ultrasound, the cavity of the hole was filled with ultrasound gel and the cuff was also filled with distilled water to control the cuff pressure hydraulically with a roller pump. Five men and two women (22–26 years old) participated in this experiment with informed consent, and the indirect blood pressure at rest in each subject was measured for reference on the contralateral side of the wrist by a commercial device (HEM-630, Omron, Japan) just before the experiment.

Fig. 5.4 shows examples of deformation and pulsatile changes of the cross-sectional area of the radial artery at the cuff pressure P_c of zero ($P_c=0$), nearly mean blood pressure ($P_c=$MBP), just below systolic blood pressure ($P_c<$SBP), and supra-systolic pressure ($P_c>$SBP), obtained in the subject. In the left part of this figure are shown the ultrasonic images at the end of diastole and at the end of systole. While the right part shows the profile traces from these images of the radial

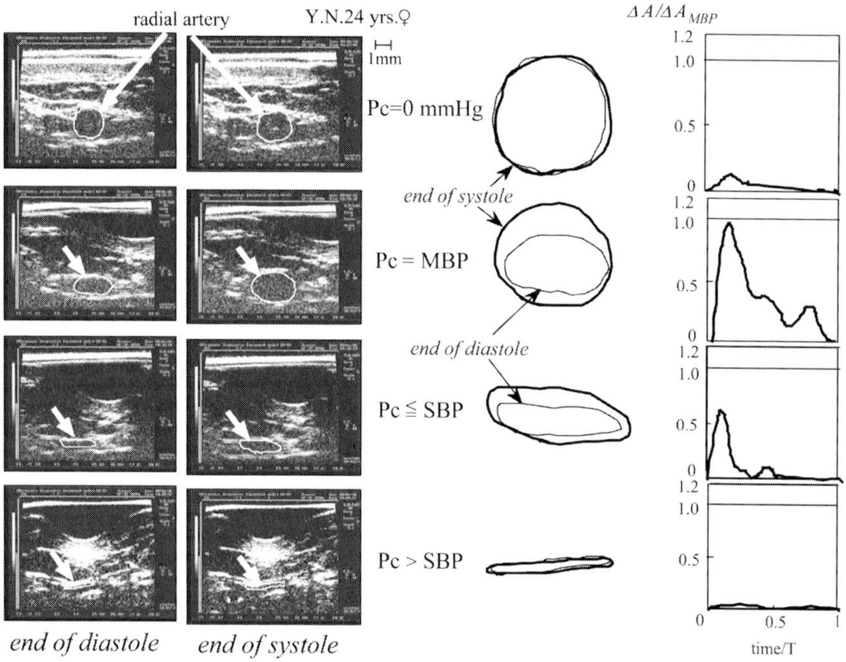

Fig. 5.4 Examples of deformation aspects and pulsatile changes of the cross-sectional area of the radial artery.

artery, together with the time course of the change in the arterial cross-sectional area (ΔA) within a cardiac cycle (T) normalized by that of the peak amplitude at P_c=MBP (ΔA_{MBP})($\Delta A/\Delta A_{MBP}$).

It is clearly shown that the radial artery is deformed in an elliptical fashion following the increase in cuff pressure and completely occluded at the supra-systolic pressure. It is also observed that the pulsatile change of the cross-sectional area is maximal at P_c=MBP and almost disappears at just above SBP. This is evidence to support the validity of the volume-oscillometric method using ultrasonic tomography for the direct measurement of the change in arterial cross-sectional area. To avoid overlapping the data obtained in each subject tested, the results are not all shown here. However, almost the same results as Fig. 5.4 were obtained in all of the subjects.

These findings strongly suggest that the cuff pressure was sufficient to deform the artery by local pressurization using the pad cuff and its fixation method.

5.2.3.4 Measurement Examples

Indirect BP measurements at the wrist using the pad cuff unit have been successfully carried out by the volume-oscillometric and/or volume-compensation method in great numbers of subjects, including healthy volunteers, in-patients, and out-patients. A reflectance photoplethysmograph was fixed inside the cuff, as shown in Fig. 5.5. It consists of a near-infrared LED (EL-302; 2.8 mm×3.2 mm×1.05 mm, Kodenshi, Japan) as a light source and a photodiode (HPI-2465R5; 4 mm×5 mm ×1.8 mm, Kodenshi, Japan) as a photodetector. For BP measurement by the volume-compensation method, a conventional band cuff for the finger was used for comparison.

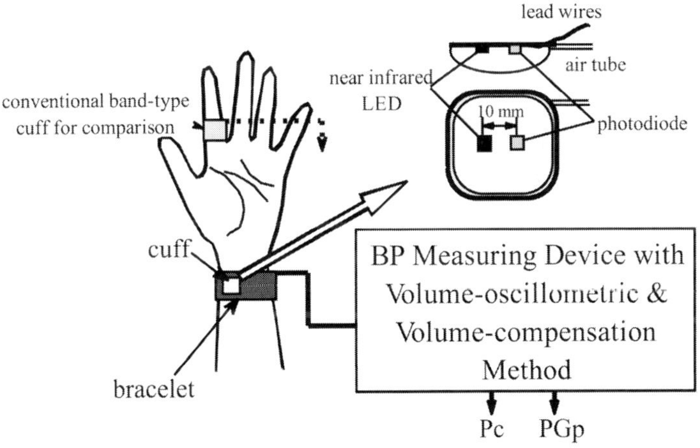

Fig. 5.5 Overview of indirect BP measurement at the wrist by volume-oscillometric and volume-compensation methods.

between two fingers, an index and a middle finger, every 30 min. This may help reduce discomfort, but it does not essentially remove the venous congestion. There has been a proposal to use a superficial temporal artery as a measuring site for BP monitoring, hoping to prevent venous congestion and let the subject move hands freely [31]. Although this approach is successful in overcoming these problems, great care is needed to attach the disk cuff firmly and accurately over the temporal artery [31].

For easy attachment, stable measurement, and applicability for long-term ambulatory BP monitoring in the finger, we describe a new design of finger BP measurement with local pressurization [33].

5.2.4.1 Description of Measurement System and Disk Cuff

The BP measurement system is shown in Fig. 5.7. The main functional modules of this system are a disk cuff, a cuff pressure controller, a vascular volume detector, and a servocontrol system. The servocontrol system consists of an 8-bit microcontroller unit (MCU; PC68HC711EQZ, Motorola, USA), which includes a CPU, an 8-channel A/D converter, pulse generators, ROM (3 kb) and RAM (512 b), and signal conditioning analogue circuits. The MCU is used for data acquisition and processing and also for automating measurement procedures. The bandwidth of the servocontrol system mainly depends on that of the cuff pressure controller, given below.

A reflectance photoplethysmograph is also used as the volume detector, which consists of an EL-302 LED and a PIN photodiode (SP-26S; 2 mm×7 mm×1 mm, Kodenshi, Japan). For reduction of external light invasion to the plethysmograph, the LED is driven by an 800 Hz pulse current, and the reflectance light signal is detected and converted to an electrical signal by the photodiode and the analogue circuits. This electrical signal goes through a high-pass filter with a cut-off fre-

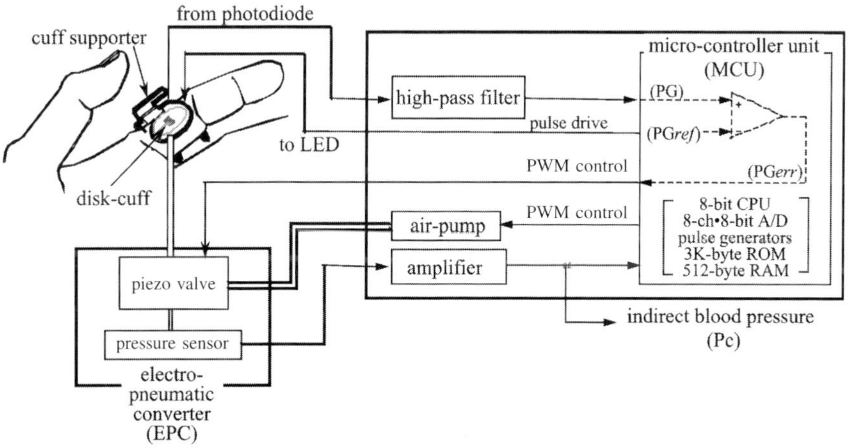

Fig. 5.7 Block diagram of finger BP measurement system with local pressurization.

quency of 160 Hz, and then is digitized by the A/D converter, synchronized to the 800 Hz pulse current driving the LED, obtaining a photoplethysmographic signal (PG).

A nozzle-flapper electropneumatic converter (EPC, 10 mm×12 mm×44 mm) is used as the cuff pressure controller. This converter adjusts leakage of the air flow from an air-pump (P36B01, Oken Seiko, Japan) to control the cuff pressure. A piezoelectric ceramic bimorph actuator (PM2740.6T, Fuji Ceramics, Japan) is used as a flapper, and a stainless tube with 2 mm inner diameter as a nozzle. The piezoelectric actuator is driven by a 125 V, 800 Hz, pulse-width modulated (PWM) signal with a duty ratio of 0–100% in 0.04% steps. The frequency response of the EPC thus designed is flat up to more than 20 Hz with pulse pressure of less than 80 mmHg. The air-pump is also driven by an 800 Hz PWM signal. The cuff pressure is measured by a pressure sensor (ADP3121, Matsushita Electric, Japan) attached to the EPC.

The volume servocontrol mentioned above is digital. The difference between the photoplethysmographic signal (PG) and a reference value (PG_{ref}), which corresponds to the unloaded vascular volume determined by the volume-oscillometric method, is given as a servo-error signal (PG_{err}). PG_{err} is modulated into the PWM signal, which drives the EPC. The EPC controls the cuff pressure (P_c) so as to diminish the PG_{err}. This feedback operation is repeated at a rate of 800 Hz so that the P_c instantaneously tracks the intra-arterial pressure. The loop gain of the servomechanism is set such that the pulsating component of the PG is reduced to less than 18% of the MAP based on our previous studies [7, 8].

As shown in Fig. 5.8, the cuff unit is composed of a disk cuff and its supporter for fixing the cuff and the finger together. The cuff is a hollow disk, 20 mm in diameter, made of clear urethane rubber membrane (0.1 mm thick) with excellent flexibility. The photoplethysmograph is placed inside the cuff membrane (Fig. 5.8c).

The cuff-supporter made of aluminum consists of two holding plates linked by a hinge like a paper clip (Fig. 5.8c). The cuff is fixed inside one of the plates. The other plate has two supporting legs forked near the hinge, providing two points of support on the finger effectively on the opposite side of the cuff. The hinge enables adjustment of the holding plate spacing to the finger size. The spacing can be fixed with a set-screw on the hinge.

The cuff fixed to one of the holding plates of the cuff-supporter is placed on the finger artery at the proximal portion of the finger (Fig. 5.8a). The spacing of the holding plates is adjusted to the finger size and then fastened with the set-screw at the hinge so as not to allow displacement between the finger and the support. The cuff-unit therefore contacts the finger at three points, the proximal portion, the back of the portion of the caput proximal phalanx (medial joint), and the portion of the basal proximal phalanx (proximal joint).

Since the cuff-supporter with the hinge fixed by the screw is rigid, the applied cuff pressure is transmitted to the finger artery through the subcutaneous tissue. The finger veins run with complicated crossings, and the finger arteries located opposite to the measuring site are not occluded. Thus the blood circulation is not

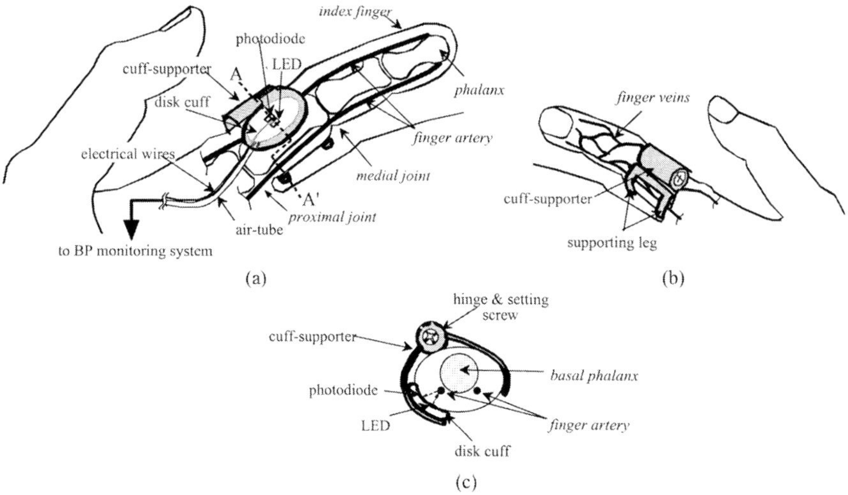

Fig. 5.8 The attachment of the cuff unit to the finger: a) palm view; b) back-of-hand view; c) cross-section view.

impeded except in the supporting portion. Therefore, this design of cuff can be expected to relieve blood pooling (venous congestion) distal to the measuring site. This results in considerable reduction of discomfort, particularly during prolonged BP monitoring.

5.2.4.2 **Evaluation of Blood Pooling in the Distal Portion of the Finger**

To assess the venous congestion due to a conventional band cuff and the present cuff-unit, blood volume measurement was made at the fingertip distal to the BP measuring site; 10 healthy subjects (2 women and 8 men, 20–22 years old) participated in this experiment. Each was asked to sit on a chair with his/her hand held horizontally at heart level. The blood volume was indirectly evaluated by another photoplethysmograph (FPG). The pertinent components of the FPG specifically designed for this experiment are a reflectance photosensor of 9 mm diameter, an EL-302 LED as a light source, and a HPI-2464R5 photodiode as a photodetector (Kodenshi, Japan). The photosensor of the FPG was attached on the palm side of the fingertip (Fig. 5.9).

Fig. 5.9 shows an example of simultaneous recordings of the cuff pressure signal (P_c) and the FPG photoplethysmographic signal (PG_{distal}) during indirect BP measurement with the band cuff and the present cuff-unit obtained in a healthy subject. The degree of venous congestion could be estimated from the PG_{distal} recording. The PG_{distal} signal in each figure was normalized such that 0 and 100% correspond respectively to the blood volume free of occlusion (no cuff attached) and that of complete arterial and venous occlusion with the band cuff at the proximal portion of the fingertip. The P_c and the PG_{distal} signals were continuously re-

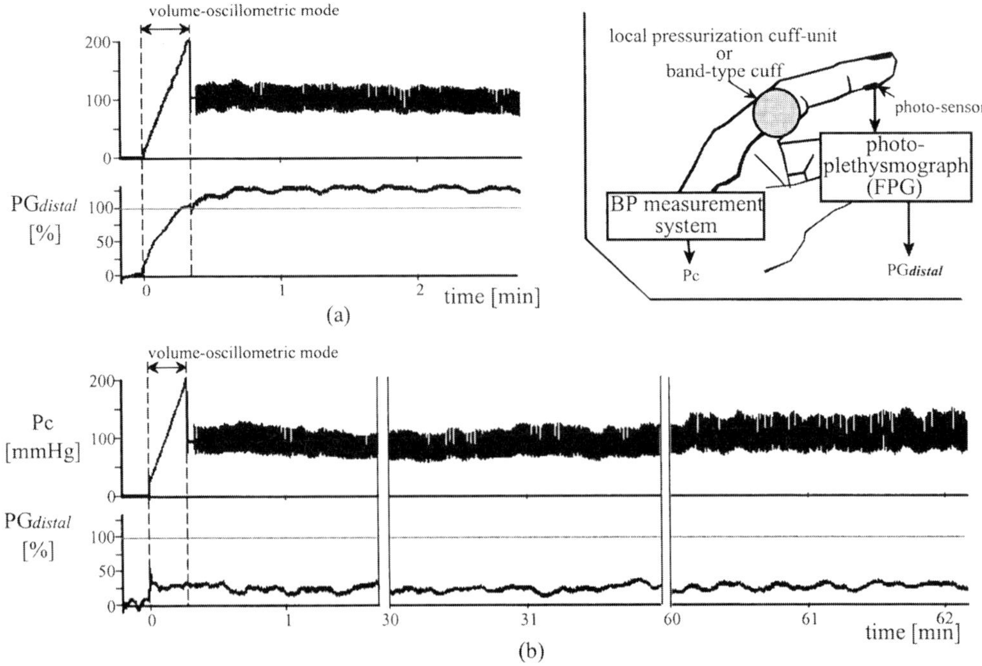

Fig. 5.9 Example of simultaneous recordings of the cuff pressure signal and a photo-plethysmographic signal by the FPG during indirect BP measurement with the band cuff and the present cuff-unit obtained in a healthy subject.

corded for about 3 min with the band cuff (Fig. 5.9a), and for more than 1 h with the present cuff-unit (Fig. 5.9b) observing the PG_{distal} level. In Fig. 5.9b, the recording is shown for three time intervals: 0–1.5, 30–32, and 60–62 min. Fig. 5.9 also indicates the changes of the P_c and the PG_{distal} during the volume-oscillometric mode at the beginning of the measurement.

It is clearly demonstrated that the PG_{distal} far exceeded 100% during the BP measurement with the band cuff. It reached almost 100% even under the volume-oscillometric mode. With the present cuff-unit, the PG_{distal} was less than about 35% throughout more than 1 h except transitionally at the beginning. Almost the same recordings were obtained in all the subjects. None of the subjects complained of discomfort during this prolonged study, nor did they experience the pain felt during the use of the band-cuff.

5.2.4.3 **Measurement Examples**

The accuracy of indirect BP measurement by local pressurization was evaluated by comparison with the reference measurement obtained with the conventional band cuff using 12 healthy subjects (1 woman and 11 men, 20–46 years old), and to the direct brachial arterial pressure measurement in one patient with dilated cardiomyopathy (66-year-old man) undergoing catheterization for medical treatment. They participated in this study with informed consent. Indirect BP measurement was made for more than 5 min in each of the 12 healthy subjects in a sitting position. The disk- and the band cuff were attached respectively on the root of the index and the middle finger of the left hand, which was held horizontally at the heart level. For the comparison between indirect and direct BP measurements, the patient lay in a supine position, and the disk cuff was attached on the root of the left index finger. An indwelling polyethylene catheter (20G-1-1/4 inch, Terumo, Japan) was inserted into the left brachial artery leading to a disposable pressure monitoring kit (UK 1000, Baxter International, USA).

To compare the indirect BP measurements by the use of the two types of cuffs, more than 3000 beats of SBP and DBP values were chosen in each subject to analyze the correlation coefficient, the mean and standard deviation of the difference between the BP data. Over the 12 subjects, the correlation coefficient ranged from 0.95 to 0.99 ($p<0.0001$) in both SBP and DBP values. The mean difference ranged from −1.45 to +1.84 mmHg with a standard deviation of 1.53–3.48 mmHg in both BP values over the 12 subjects. These results strongly indicate that the finger BP measurement by the local pressurization method is an appropriate substitute for that made by a conventional band cuff method.

During this comparison study, we tentatively let some subjects make flexion and extension movements of the finger concerned to confirm whether or not the pressure recording could be done by the band- and the disk cuff. Remarkable fluctuations and/or artifactual signals on the pressure record and often unmeasurable pressure data were obtained for the use of the band cuff. However, relatively stable pressure recording without any motion artifacts was made by the disk cuff, as shown in Fig. 5.10a, although the credibility of this recording may be unclear due to the fact that it is quite difficult to confirm experimentally the true intra-arterial pressure at the finger during such finger movements.

Fig. 5.10b shows an example of the simultaneous recordings of the direct brachial BP (upper record) and the indirect finger BP (lower record) with the present cuff-unit. The figure indicates a good agreement of the waveforms between the two measurements. The finger pressure tracks both direct brachial BP changes related to an arrhythmic cardiac cycle and normal cardiac cycles.

5.2.5 Blood-Pressure Measurement at the Posterior of the Thigh

5.2.5.1 Concept of the Development

Owing to the aging population, the requirement for home health-care or health monitoring during daily living is increasing. This health monitoring is considered as an effective approach to the early diagnosis and treatment of cardiovascular and other diseases, and for prevention and control of such diseases, and for the reduction of medical expenses.

It would be highly desirable if the monitoring could be done in a fully automated manner without the attachment of any biological sensors to the subject's

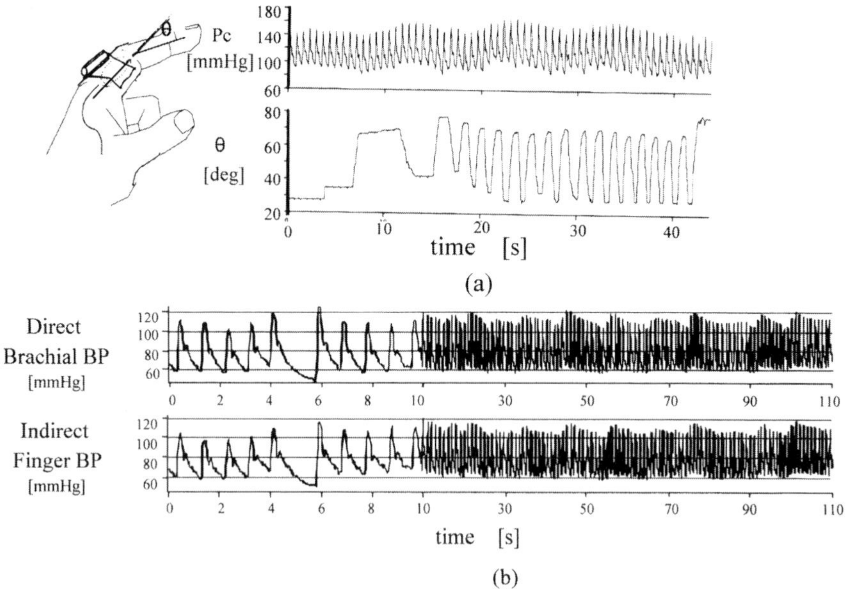

Fig. 5.10 a) Cuff pressure (Pc) recorded with angle of medial joint (θ). b) Simultaneous recordings of the direct brachial BP and the indirect finger BP obtained with the disk cuff.

body as well as any troublesome operations for the measurement. Thus, the subject does not need to be aware of the measurement and the physiological data can be unconsciously stored during ordinary daily living. With this in view, we note the studies by several investigators, who developed methods to record ECGs in the bathtub and on the toilet seat [34], and temperature distribution [35] and cardiopulmonary function in the bed [36]: all of these data are acquired in daily life.

In addition to these developments, we have also recently designed a monitoring system with a highly accurate weight measuring device installed in a lavatory floor around a toilet bowl, which allows the automatic measurement of body weight and weight of evacuation together with the ballistocardiogram as an index of cardiac ejecting function [37]. In recognition of the importance of measuring blood pressure (BP) during the use of the toilet, we have contemplated introducing BP measurement at the posterior of the thigh contacting with a toilet seat to improve this system. This section describes this development of BP measurement system from the toilet seat based on the volume-oscillometric method with local pressurization.

5.2.5.2 **Measurement System with Toilet Seat**

The measuring site is limited to the area where the skin contacts the toilet seat. Anatomically, there exists small arteries of perforation in the posterior of the thigh, which could be used as the site for BP measurement.

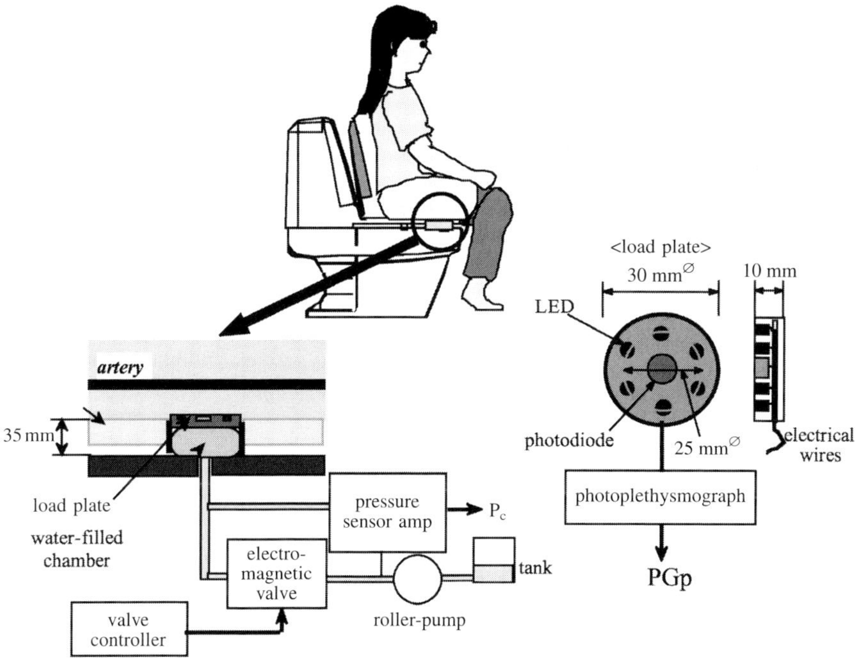

Fig. 5.11 Experimental set-up for BP measurement at the back of the thigh contacting the toilet seat.

A diagram of the experimental set-up for BP measurement is shown in Fig. 5.11. To detect the volume pulsation from the perforation arteries, a reflectance photoplethysmographic sensor is used, which consists of six high-luminance near-infrared LEDs (TLN-102, Toshiba, Japan) and a photodiode (SPD-102; photodetecting area, 50.24 mm^2, Sharp, Japan) embedded in an acrylic load plate of 30 mm diameter. The LEDs are arranged at equal separation to surround the photodiode placed on the center of the load plate, as shown in Fig. 5.11. The load plate can be pushed up and down by regulating the volume of distilled water in a water-filled chamber made of polyurethane membrane using a roller-pump, so that the local compression to the tissue is made by the load plate to transmit the pressure to the arteries concerned. An electromagnetic valve is set between the chamber and the roller-pump to maintain the chamber pressure (P_c), when necessary.

5.2.5.3 Measurement Examples

BP measurements with this experimental set-up have been successfully carried out in 10 healthy men (22–25 years old) while sitting on the toilet seat as usual. During the measurements, they were requested not to change their sitting position. Five to six repetitions of the measurements were made at a 2–3 min intervals in each subject. Altogether 54 measurement data were obtained for all of the subjects, to analyze the differences of BP values determined at the posterior of the thigh and at the wrist as reference.

Fig. 5.12 shows an example of the simultaneous recordings of the cuff or the chamber pressure (P_c) and the photoplethysmographic pulsation signal (PGp) obtained in a healthy subject. The upper record indicates the PGp signal (PGp_W) measured at the wrist by the device mentioned above, and the lower record is the PGp signal at the posterior of the thigh (PGp_T). The values of P_c for the wrist (P_{cW}) and for the thigh (P_{cT}) are compensated by the hydrostatic pressure difference between the measuring site and the heart to determine the BP value at the heart level. It is demonstrated that the pattern of the change in amplitude of the PGp_T signal following the gradual decrease in P_c is similar to that of the PGp_W signal, and the P_c values corresponding to the systolic end point (SEP; SEP_T for the former and SEP_W for the latter) and the maximum amplitude point (MAP; MAP_T and MAP_W) in each of the PGp signal are to be given respectively as the systolic (SBP; SBP_T and SBP_W) and the mean pressure (MBP; MBP_T and MBP_W).

The mean differences with standard deviations of these two BP values (SBP_W–SBP_T and MBP_W–MBP_T) in whole measurement data obtained from all of the subjects were -2.5 ± 6.4 mmHg and -3.0 ± 5.0 mmHg, respectively. Since the true BP values at the posterior of the thigh are unknown, the validity of this BP measurement from the toilet seat should be further investigated. However, this measurement technique appears to be feasible considering that the PGp_T signal shows the characteristic changes in amplitude according to the arterial tube law and the peculiar points (MAP and SEP) can be clearly identified from the PGp_T signal, as usually observed at other measuring sites such as the finger and the wrist by the volume-oscillometric method.

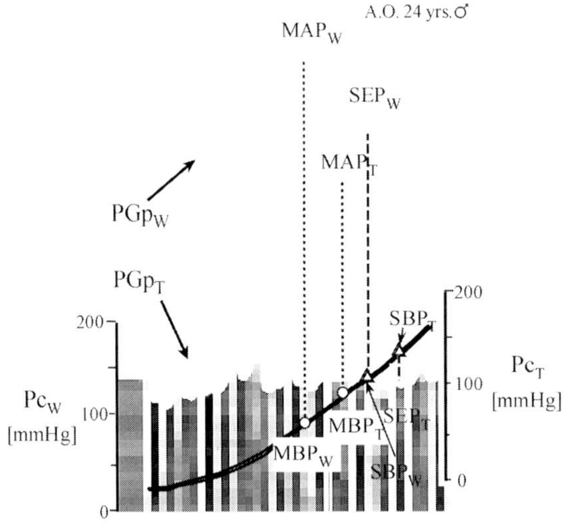

Fig. 5.12 Simultaneous recordings of the cuff or the chamber pressure and the photoplethysmographic pulsation signal obtained in a healthy subject.

5.3 Cardiac Output Measurement

5.3.1 Overview of Measurement Methods

Various methods for measuring cardiac output (CO) or stroke volume (SV) have been developed such as electromagnetic blood flowmetry, ultrasonic blood flowmetry, indicator dilution method (dye-dilution, thermal dilution, radioisotope dilution or nuclear stethoscope, and Fick method), echocardiography, nuclear cardiac imaging, electrical impedance cardiography, and so on [2, 3, 38]. Among all of these methods, there are at present only two methods capable of measuring CO or SV non-invasively simply without any traumatic procedures: echocardiography and electrical impedance cardiography.

Since its introduction into clinical cardiology about two decades ago, echocardiography has been widely used in clinics to evaluate both functional and structural features of the heart. This has made a major impact on clinical cardiac examinations. The advantages of echocardiography are that the technique is a safe way to acquire qualitative and quantitative data of ventricular size, wall motion and movements of cardiac valves, detection of the presence of pericardial effusion and intracardiac thrombi, as well as assessment of blood flow including forward (cardiac output) and backward flow (regurgitant flow) across cardiac valves. This method may therefore be superior to other non-invasive techniques for determining various cardiac functions, and is simply done by placing an echocardiographic transducer on the surface of the chest wall. However, the instrument is bulky, expensive, and inconvenient for portable use.

On the other hand, electrical impedance [39, 40] or admittance plethysmography [41, 42] provides a non-invasive method of estimating blood volume change and/or blood flow. Measurement of transthoracic electrical impedance (or admittance) gives the assessment of CO or SV and myocardial contractility. This technique has had a mixed history of acceptance, due probably to a stance that favors precise invasive techniques, the poor quality of early studies to validate the measurement of SV, and a lack of an appropriate and reliable electrical model of the thorax to calculate SV [3].

Nevertheless, impedance (or admittance) cardiography has many advantages. When it is properly applied, the technique provides a completely safe, relatively unobtrusive, inexpensive, convenient method for SV and contractility measurements on a beat-by-beat basis, requiring only the attachment of electrodes on the thorax. Also, the instrument is small enough to be portable. These characteristics make the technique ideally suited to behavioral or ambulatory studies [3]. No methods superior to this technique have yet been developed.

Although impedance cardiography has so many advantages, a disadvantage for practical application is the attachment of a tetra-polar band-electrode system used in standard practice. When the band-electrode system is used, two pairs of electrodes are placed around the subject's neck and abdomen to pass current and measure voltage. Attaching the band electrodes to a patient lying on a bed, for example, is not always easy, and its long-term as well as ambulatory use is often uncomfortable and sometimes causes dermatitis. These are serious problems for practical measurement.

Replacement of band electrodes with spot electrodes has therefore been studied by many investigators [43–50], and a variety of spot-electrode arrangements have been proposed. They have been based on comparisons of the SV or CO measurements with band-electrode arrays or other CO measurement methods. The current distribution on the thorax is an important factor that determines the utility of the spot-electrode array. However, this has seldom been measured for the proposed spot-electrode arrays, probably due to a lack of theoretical and analytical processes based on such measurement of current distribution in the thorax; a standard spot-electrode array has not yet been established.

We describe impedance or admittance cardiography and its measurement principle along with the evaluation of its accuracy. Then, the replacement of the band- by an appropriate spot-electrode array is discussed, based on the measurement of current distribution on the thorax.

5.3.2 Basic Principle and Evaluation of Accuracy

5.3.2.1 Basic Principle

The most commonly used technique of impedance cardiography is based on the method of Kubicek et al. [40]. As mentioned above, a tetra-polar band-electrode system is used to detect a biological impedance (or admittance) signal, in which the two outer electrodes placed on the neck and the abdomen impose sinusoidal current (I, 50–100 kHz, 2–4 mA_{rms}) and the two inner electrodes placed more than 3 cm apart from each current electrode detect the potential difference between a length L (V, Fig. 5.13 a). As the electrical impedance (Z, or the admittance, Y) is defined as $Z=V/I$ (or $Y=I/V$), the V signal corresponds to the Z signal when I is constant (or the I signal corresponds to the Y signal when V is constant). The pulsatile component superimposed on the Z signal (ΔZ) caused by each cardiac beat can be obtained by amplifying the Z signal (or also ΔY by amplifying the Y signal).

If the thorax is roughly assumed to be a two-compartment coaxial cylindrical model composed of the aorta and its surrounding thoracic tissues (Fig. 5.13 b), as modified by an original proposal given by Nyboer [51], the total thoracic impedance Z (or Y) after blood inflow into the aorta can be represented by an electrically parallel connection with the total impedance Z_0 (or Y_0) before blood inflow and the impedance component due to the increase in blood volume (ΔV) in the aorta (Z_b, or the admittance component Y_b) (Fig. 5.13 b). Therefore, Z can be expressed as $Z=Z_0Z_b/(Z_0+Z_b)$ (or $Y=Y_0+Y_b$). Accordingly, the impedance change before and after the blood inflow ($\Delta Z=Z_0-Z$; or the admittance change $\Delta Y=Y_0-Y$) is derived as

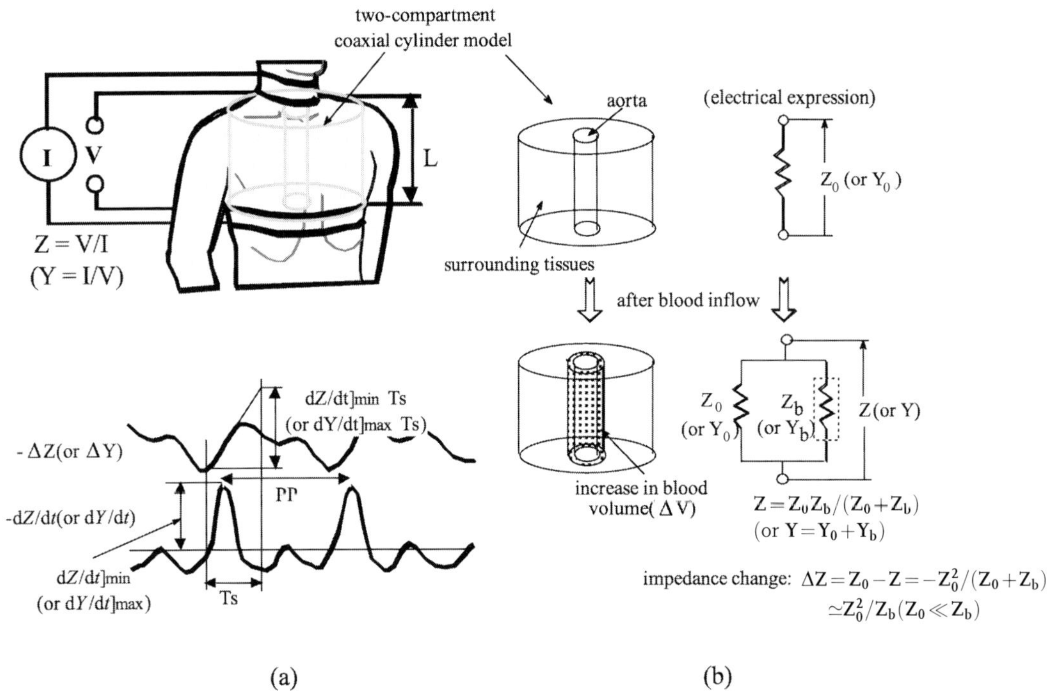

Fig. 5.13 Transthoracic impedance (or admittance) cardiography.

$$\Delta Z = Z_0 - Z = -Z_0^2/(Z_0 + Z_b) \simeq -Z_0^2/Z_b \quad (\text{if } Z_0 \gg Z_b) \tag{5.2}$$

or

$$\Delta Y = Y_0 - Y = Y_b \tag{5.2'}$$

A minus sign indicates a decrease in ΔZ. Since Z_b is given as $Z_b = \rho_b L^2/\Delta V$ (or $Y_b = \Delta V/\rho_b L^2$), ΔV can be obtained as the Nyboer's equation

$$\Delta V = \rho_b (L/Z_0)^2 \Delta Z \tag{5.3}$$

or

$$\Delta V = \rho_b L^2 \Delta Y \tag{5.3'}$$

where ρ_b is the blood resistivity and is a function of hematocrit [52–54], and ΔZ indicates the absolute value.

The ΔZ (or ΔY) signal is thought to reflect two sources of blood accumulation: inflow due primarily to ejection from the ventricle, and outflow due to runoff via the aorta. In order to obtain an estimate of forward flow due to left ventricular ejection, it is necessary to correct the ΔZ (or ΔY) signal for estimation of SV. Kubicek et al. [40, 53] proposed that the slope from the onset of the ascending limb of the ΔZ (or ΔY) curve continues during the ventricular ejection time (T_s) assuming that the volume of blood inflows into the aorta without outflow. The slope can be obtained from differentiating the ΔZ (or ΔY) signal (dZ/dt or dY/dt). Therefore, the corrected ΔZ (or ΔY) value (ΔZ^* or ΔY^*) is given as the minimum peak of dZ/dt ($dZ/dt]_{min}$, or the maximum peak of dY/dt ($dY/dt]_{max}$) times T_s ($dZ/dt]_{min} \cdot T_s$ or $dY/dt]_{max} \cdot T_s$), as shown in the ΔZ (or ΔY) record in Fig. 5.13 a. Replacing this corrected value into ΔZ (or ΔY) in the equations yields the Kubicek et al. [40, 53] formula for estimating SV

$$SV = \rho_b (L/Z_0)^2 dZ/dt]_{min} T_s \tag{5.4}$$

or

$$SV = \rho_b L^2 dY/dt]_{max} T_s \tag{5.4'}$$

Usually, ρ_b is in Ωcm, L in cm, Z_0 in Ω, $dZ/dt]_{min}$ in Ω/s, $dY/dt]_{max}$ in S/s, and T_s in s, and thus SV is calculated in mL. ρ_b should be corrected using the subject's hematocrit, but is given as 150 Ωcm in normal healthy subjects [40]. Also, the ventricular ejection time (T_s) is readily determined from the dZ/dt (or dY/dt) signal according to the method given by Lababidi et al. [55].

By noting the heart rate (HR in beats/min) over the interval in which SV is being determined, the cardiac output (CO in L/min) may be calculated from

$$CO = (SV/1000) HR . \tag{5.5}$$

Beat-by-beat or instantaneous CO may also be obtained by using the instantaneous HR ($HR_{instant}$) calculated from the interbeat pulse interval (PP in s, Fig. 5.13 a) as $HR_{instant} = 60/PP$.

Since the electrical impedance and admittance is a reciprocal relation, both methods provide similar information. However, it is obvious from Equation 5.4 that both the basal thoracic impedance (Z_0) and the dZ/dt signal are needed to estimate the SV value in the impedance method, while only the dY/dt signal is needed in the admittance method. From the viewpoint of measurement technique, the latter method is simpler and more practical than the former one [41, 42, 56].

5.3.2.2 **Evaluation of Accuracy**

Many studies evaluating CO measurement by the impedance method have been carried out by comparing this method with, for example, dye-dilution, radioisotope dilution, the Fick method [38, 57]. Although the results of comparison studies correlated well, demonstrating correlation coefficients of more than 0.9, there are also reports of poor correlation coefficients of less than 0.6 [38, 57]. Even among the studies which reported higher correlation coefficients, the regression equations are different. This implies a difficulty in the quantitative estimation of SV or CO by applying a simple model to the thorax.

We have also compared the admittance method and a thermal dilution method, focusing on tracking the changes in CO by the admittance method for longer periods of time using a within-subjects design [58]. A recently developed, commercially available continuous CO measurement device based on the thermal dilution method (Vigilance, Edwards Lifesciences, USA) was used as a reference measurement in this study. Nine men and two women patients (32–65 years old) without cardiac failures participated in this study with informed consent. They lay in a supine position on a bed in an ICU. The variables monitored were ECG, blood pressure (arterial catheterization), arterial oxygen saturation (pulse oxymetry), and CO using the Vigilance apparatus. The measurements were concomitant with the medical treatment that the subjects underwent. The monitoring time used for the comparison was 2.53–35.83 h (15.15 ± 10.61 h). The CO values calculated by the admittance method (CO_{admit}) were provided on a beat-by-beat basis, while those by the Vigilance (CO_{therm}) were from its output terminal (processing time for CO determination by the device was not specified in the technical specifications of the Vigilance).

Fig. 5.14 is an example of the trend chart during the first 10 h of 17.75 h monitoring, showing the simultaneous recordings of CO_{admit} and CO_{therm} values obtained in a patient (61-year-old man, 53 kg body weight). The changes in the values of CO_{admit} agree well with those of CO_{therm}. However, there were different delay times (3–8 min) between these two changes when abrupt changes in CO occurred as denoted by the arrows. Also, the changes in the values of CO_{therm} were smaller than those of CO_{admit} as indicated by the circles in this figure.

The delay as well as the smaller CO_{therm} values are probably caused by the device itself, because it takes some processing time for the determination of CO

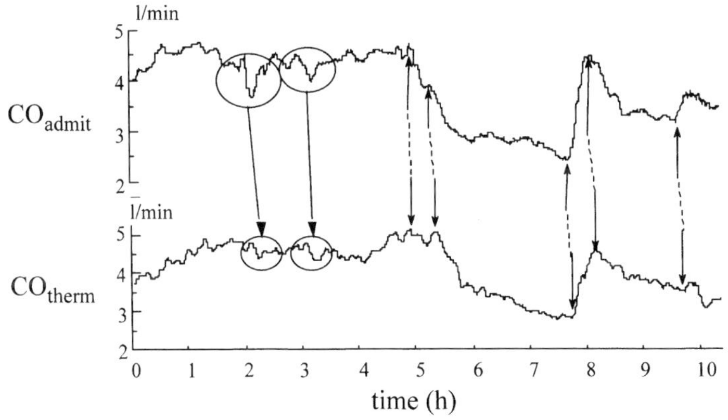

Fig. 5.14 Comparison of the admittance method and the thermal dilution method.

based on the thermal dilution method used in the device. In the admittance method, beat-by-beat recording of the CO_{admit} values can be done in real-time. From this viewpoint, tracking by the admittance method (and the impedance method) may be superior to the indicator dilution-based methods used.

Almost the same results were obtained in all the patients tested with respect to tracking by the admittance method. However, the difference between the values of CO_{therm} and CO_{admit} ($\Delta CO = CO_{therm} - CO_{admit}$) varied from subject to subject: ΔCO and the error ratio ($\Delta CO/CO_{therm}$ in %) ranged from +0.13 to +1.20 L/min (0.54 ± 0.30 L/min (mean ± SD)) and from +3.9 to +15.3% (10.1 ± 4.08%), respectively. As mentioned above, these results indicate that this method is difficult to use to estimate quantitatively the CO value as compared with that determined by the Vigilance as a reference measurement.

5.3.3 Replacement of Band Electrode by Spot-Electrode Array

As mentioned, we have tried to design an optimal spot-electrode array instead of the conventional band-electrode array based on the measurements of basal (static) impedance contour maps (Z maps) and pulsatile change of impedance contour maps (ΔZ maps) on the thorax. This approach may also provide useful information on the contribution of various organs in the thorax to the thoracic impedance as well as on the effect of each organ on the total impedance change.

5.3.3.1 Current Distribution on the Thorax

Description of Experimental System and Construction of Impedance Mapping

With these objectives, we have recently designed a 64-channel impedance mapping system, the details of which have been reported in the literature [59]. Briefly, this system consists of a current source (50 kHz sinusoidal signal of 2 mA_{rms}), voltage sensing units, an RS-232C interface, and a personal computer (Fig. 5.15).

The current passes through the subject's thorax between a pair of band (10 mm width; SDM121, NEC Medical Systems, Japan) or spot electrodes (disposable ECG electrode, 25 mm in diameter; SEC145, NEC Medical Systems, Japan). The resultant voltage drop between a reference spot-electrode and each of 64 selected detecting points (32 points on the front of the thorax and 32 points on the back) are measured with the voltage sensing units (F1–F32 for the front and B1–B32 for the back) (Fig. 5.15).

Each voltage sensing unit is composed of an electrode plate (10 mm in diameter) coated with gold, a preamplifier (amp), a rectifier, and an AC amplifier (AC amp) with a 1.6 s time constant. The output of the rectifier is proportional to the impedance (Z) signal without the 50 kHz component. Also, the output of the AC amp is the voltage signal as the pulsatile component of the impedance (ΔZ signal). Eight voltage sensing units with center-to-center separation of 50 mm, along with a 16-channel A/D converter (12-bit, 10 ms of sampling interval per channel),

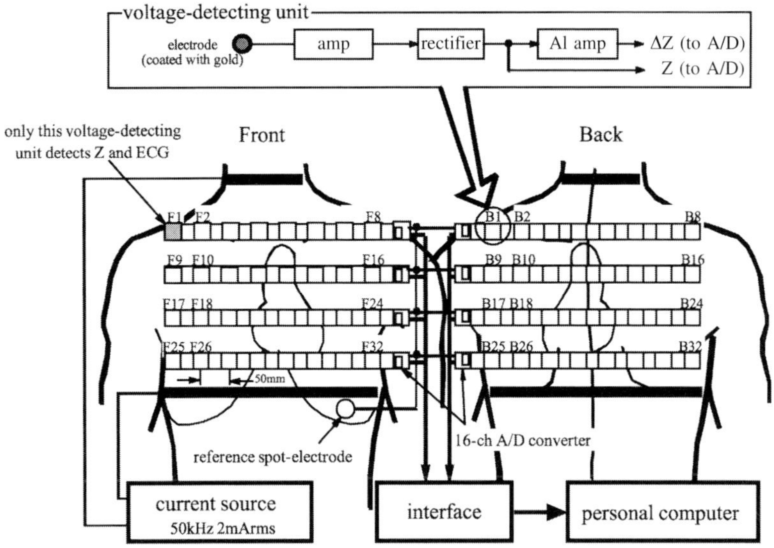

Fig. 5.15 Block diagram of 64-channel impedance mapping system.

are mounted on each of eight flexible circuit boards (25 mm×425 mm). Each unit has a measurement range of 0–144 Ω for the Z value and ±0.36 Ω for the ΔZ value, both within ±3% of error. Eight pairs of Z and ΔZ signals are led to the A/D converter on the same circuit board, and eight circuit boards are connected to the personal computer via the interface. One of the voltage sensing units (F1 in Fig. 5.15) is used to measure an ECG signal instead of ΔZ signal, in order to observe a synchronization of cardiac cycle time.

Four circuit boards were firmly attached at equal distances (53–77 mm for all the subjects used in the study) on the surface of the front of the thorax using a conductive adhesive sheet (25 mm×375 mm; NEC Medical Systems, Japan). Four more circuit boards were placed similarly on the back of the thorax. Care was taken to position the circuit boards such that the electrode plates of the top and the bottom boards were located at the level of clavicle and xiphisternum, respectively.

The basal impedance contour maps (Z maps) and its superimposed pulsatile component contour maps (ΔZ maps) were drawn from the 64 Z signals and 63 ΔZ signals, respectively. The contours were determined through linear interpolation as

$$Z(x, y) = (1 - y)\{(1 - x)Z(0, 0) + xZ(1, 0)\} + y\{(1 - x)Z(0, 1) + xZ(1, 1)\} \quad (5.6)$$

where $Z(x, y)$ is the impedance value at a location (x, y), and $Z(0, 0)$, $Z(1, 0)$, $Z(0, 1)$, and $Z(1, 1)$ are the impedance values measured at the points forming a

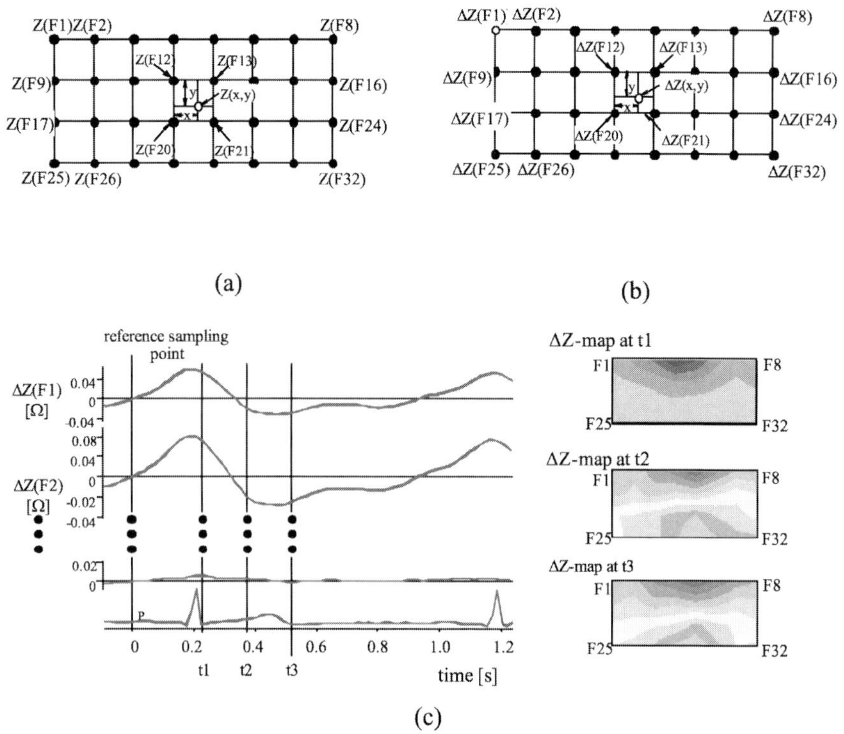

Fig. 5.16 a) Matrices of measurement points of basal impedance. b) Pulsatile component. c) Construction of Z maps and ΔZ maps.

rectangle surrounding the (x, y) point (e.g., F12, F13, F20 and F21 in Fig. 5.16). Fig. 5.16 shows the matrices of Z and ΔZ measurement points on the front side of the thorax. The missing ΔZ value at the point F1 is determined through linear extrapolation as

$$\Delta Z(\mathrm{F1}) = \Delta Z(\mathrm{F2}) - \Delta Z(\mathrm{F3})/2 + \Delta Z(\mathrm{F9}) - \Delta Z(\mathrm{F17})/2\,. \tag{5.7}$$

To construct the ΔZ maps, the ΔZ value at the time of the ECG-P wave is given as the reference value for each measurement point, and the increments from the reference value were used to draw the ΔZ maps at various time points (Fig. 5.16c).

Experimental Procedures and Results

Experimental procedures: Eleven normal men (20–25 years old, 47–68 kg body weight) participated in the study under their informed consent. They were asked to take a supine position on a bed. The data were acquired during breath holding at the resting expiratory level.

The Z maps were obtained for different positioning of spot electrodes for current injection, including the arrangements previously proposed by Penney et al. [43], Qu et al. [45], Gotshall et al. [48], and Woltjer et al. [50]. The Z maps with Kubicek's band-electrode array were obtained for reference. The Z maps thus obtained were statistically compared with those calculated for a cylindrical model of the thorax having uniform current distribution (ideal Z map). Then, an optimal spot-electrode array for current injection was developed through the obtained Z maps that approximated the cylindrical model of the thorax. Meanwhile, an appropriate spot-electrode array for voltage pick-up was investigated from the data that were based on the measurement of the ΔZ maps.

Results of Z maps. Fig. 5.17 shows several examples of the Z maps obtained with the Kubicek's band-electrode and the spot-electrode arrays for current injection. In this figure are shown the schematic representations of the previously proposed spot-electrode arrays for current injection, together with the present array that was finally determined through trial and error.

The band electrodes produce nearly uniform current distribution. Their Z maps are similar to those estimated with the cylinder of the thorax, which has a uniform current distribution. The Z maps obtained show, however, a slightly larger current density than those estimated near the transition level from the neck to the thorax. On the other hand, the five sets of Z maps shown in Fig. 5.17 clearly demonstrate little uniformity of current distribution in the thorax and undesirable current concentrations near the current injection electrodes.

While observing the Z maps, the pair of spot electrodes for current injection was relocated to various points on the subjects' bodies and optimum locations for the electrodes were sought through trial and error. It was found that the electrodes should be located as far from the thorax as possible in order to minimize the current concentration in the thorax. Indeed, one of the best locations is to place the source electrodes on the forehead and on the left medial knee. The Z map obtained with this arrangement as shown in an example of Fig. 5.17g is relatively similar to that obtained with the band-electrode array (Fig. 5.17a).

Each of these Z maps was correlated with the Z map calculated for the cylindrical model assuming uniform current distribution (ideal Z map). In all the subjects studied, the correlation coefficients ranged from 0.916 to 0.967 for the band-electrode array and from 0.920 to 0.962 for the present spot-electrode array with the 99.9% confidence level for both. The correlation coefficients for the other spot-electrode arrays ranged between 0.701 and 0.941. The Z maps with the band- as well as the present spot-electrode array correlate well with the ideal Z map. In addition to this, the correlation coefficients obtained for all the subjects with one of the current injection electrodes attached on the right medial knee ranged from 0.919 to 0.960, indicating slightly lower correlation to the ideal Z map than those with one electrode located on the left medial knee. Therefore, the left medial knee is a slightly superior electrode location for current injection.

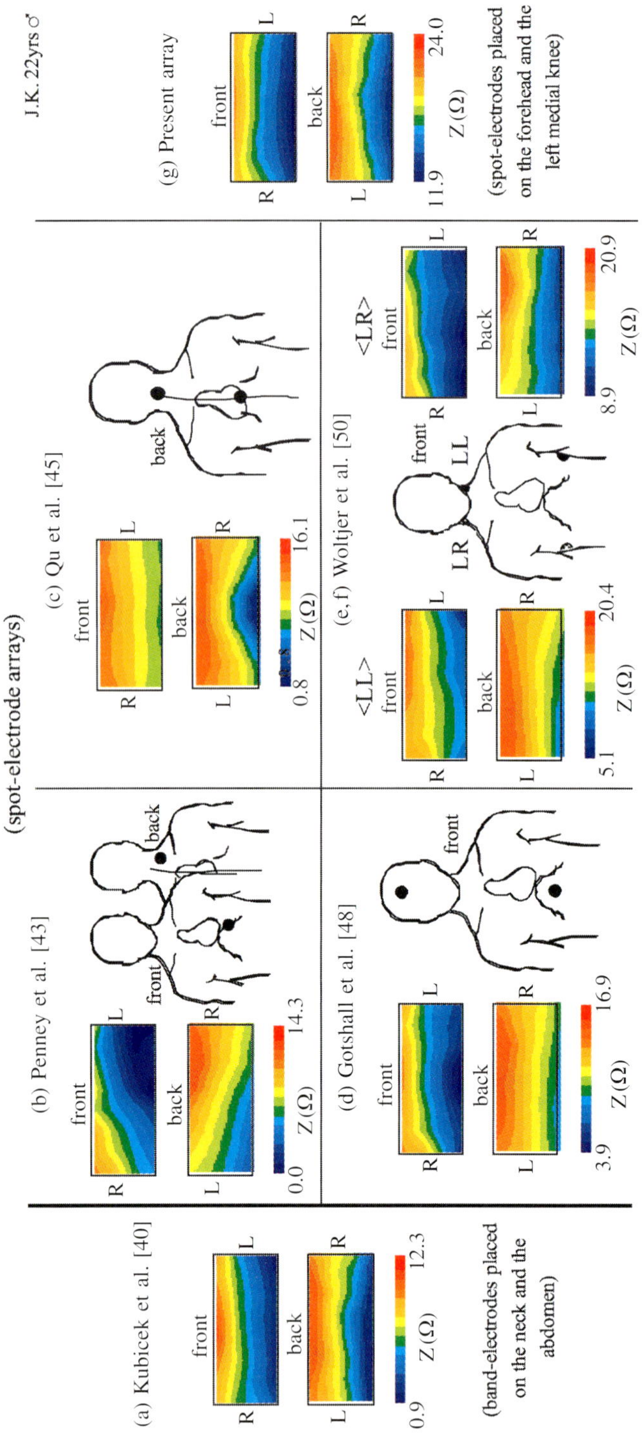

Fig. 5.17 Examples of the Z maps obtained with the Kubicek's band-electrode and the spot-electrode arrays for current injection.

Results of ΔZ maps: Fig. 5.18 shows the time sequence of ΔZ maps together with the ECG signal obtained in the same subject as Fig. 5.17. The presently proposed spot-electrode array was used for current injection. The time increment is 20 ms ($\Delta t = 20$ ms). The figure clearly represents that the ΔZ maps change in accordance with cardiac blood ejection cycle. It also shows little uniformity of the ΔZ values over the thorax. This lack of uniformity was probably dominated by the pulmonary blood circulation. The electrodes for voltage measurement, therefore, should be located so that the influence of the pulmonary blood circulation is avoided or minimized. From an anatomical viewpoint, appropriate locations for voltage measurement are the medial portion at the level of clavicle and the portion on the xiphisternum. Clear changes in the ΔZ values are observed along the medial line on the front in Fig. 5.18. The same was true for all the subjects. The same was also true when the band-electrode array was used for current injection.

As mentioned earlier, Fig. 5.18 supports the fact that the thoracic impedance waveform detected by the band electrodes can be affected by blood volume changes in various organs inside the thorax, and the thoracic impedance may be detected as the sum of the impedances contributed by each organ. The effect of each organ on the total impedance change has been quantitatively estimated

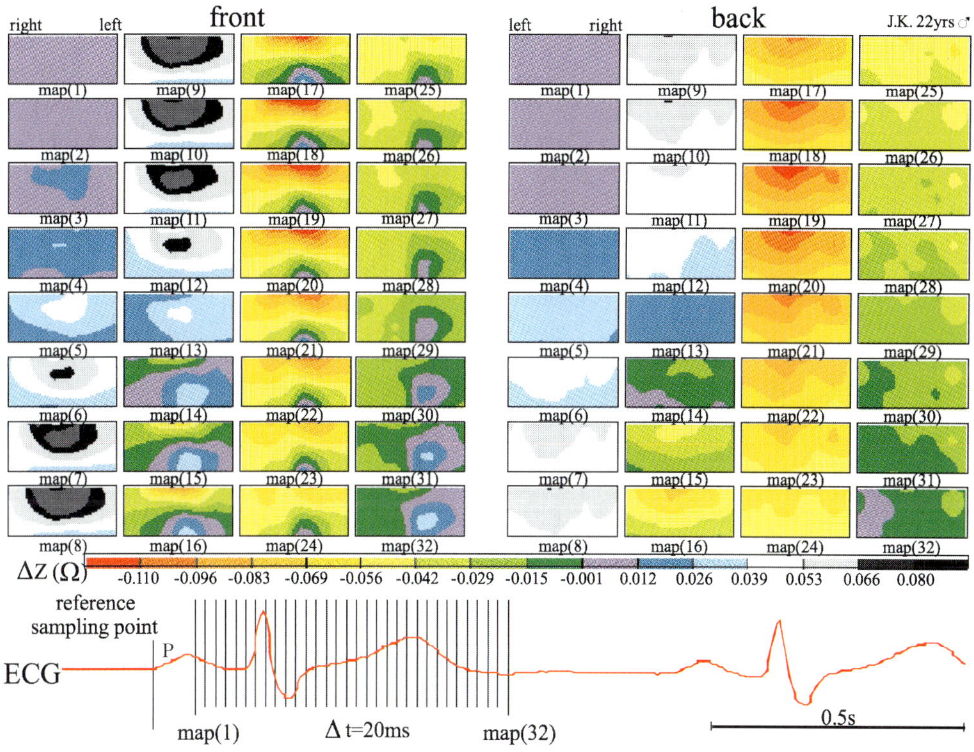

Fig. 5.18 Time sequence of ΔZ maps together with the ECG signal.

through animal experiments [60] and models [61, 62]. According to these studies, the effect of the volume change in the aorta and large arteries is less than 30%, while that in the lung is about 60%.

Although such quantitative estimation of the effect of each organ on the total impedance change has not yet been analyzed using the present mapping system, it appears from Fig. 5.18 that the impedance changes caused by the pulmonary blood circulation would produce a considerable part of the thoracic impedance changes. From this aspect, it could be acceptable to determine the electrode locations for voltage measurement so as to avoid the contribution of the pulmonary blood volume changes. However, it is not so easy to assess quantitatively a percentage contribution of the blood volume change in the aorta to the impedance change detected by the presently proposed electrode locations for voltage measurement.

5.3.3.2 New Spot-Electrode Array and Comparison with Band-Electrode Array

If the two-compartment cylindrical model of the thorax with the assumption of uniform current distribution is valid, the ΔZ maps would be also uniform especially with the blood accumulation into the aorta during systole. However, the ΔZ maps obtained showed little uniformity over the thorax (Fig. 5.18) due mainly to the contribution of the pulmonary blood circulation. This implies that the simple model of the thorax is not appropriate. Determination of an optimal spot-electrode arrangement for voltage pickup therefore becomes more difficult without the cylindrical model of the thorax. However, one can assume a partially cylindrical model for the frontal part of the thorax excluding the lungs, taking the thoracic anatomy into account. This assumption is based on the findings of nearly uniform cyclic changes of the ΔZ maps in association with the cardiac blood ejection cycle observed along the medial line on the front of the thorax. Considering this assumption, the locations of a new spot-electrode array can be proposed as the forehead and the left medial knee for current injection and the medial portion at the level of clavicle and the portion on the xiphisternum for voltage pickup in this study (Fig. 5.19).

Then, the new spot-electrode array thus determined and the Kubicek's band-electrode array were compared through the SV measurements. The two electrode array systems were alternated approximately every 10 s during SV measurements with an impedance cardiograph (current source, 2 mA_{rms} and 50 kHz sinusoidal wave; input impedance, 10 MΩ; measurement range of impedance (Z_c) and its first derivative (dZ_c/dt), 0–50 Ω and –5 to +5 Ω/s; frequency range of Z_c, 0–50 Hz (–3 dB); absolute error of Z_c, ±1.5%; time constant of dZ_c/dt, 0.01 s). The SV value was calculated according to the Equation 5.4 and the ρ_b value was tentatively given as 150 Ω cm in this study.

Fig. 5.19 shows sample recordings of dZ_c/dt signals, yielding the SV value of 60.6 and 52.4 mL for the new spot- and the band-electrode array, respectively. From such recordings, the mean values and standard deviations of Z_c, $dZ_c/dt]_{min}$, T_s, and SV over 10 cardiac cycles were analyzed in each of all the subjects. Particu-

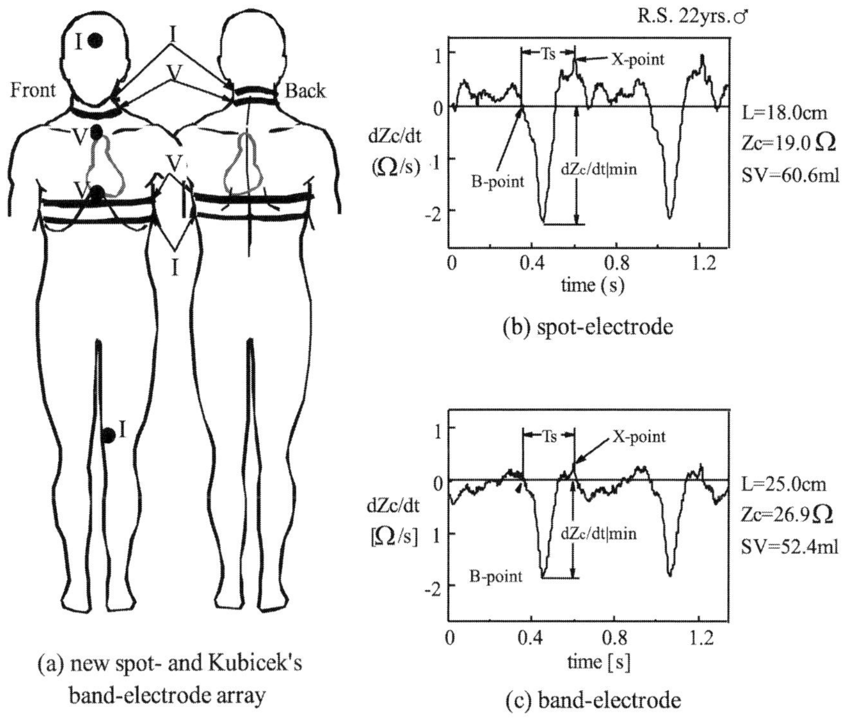

Fig. 5.19 Locations of a new spot-electrode array can be proposed as the forehead and the left medial knee for current injection and the medial portion at the level of clavicle and the portion on the xiphisternum for voltage pickup.

larly, the ratios of the SV values obtained with the spot- (SV_{spot}) and the band-electrode array (SV_{band}), SV_{spot}/SV_{band}, were given as the range 1.05–1.19 (1.129 ± 0.045) over 11 subjects. This indicates that the SV_{spot} values are 5–19% larger than the SV_{band} values.

This overestimation as compared with the SV_{band} values is consistent with the results reported by Raaijmarkers et al. [63], who analyzed the thoracic impedance (Z_c) as a nonlinear function of the distance (L) between the two voltage pickup electrodes using a 3D finite element model of the thorax. According to their results, the calculated SV value increases considerably as the position of the voltage pickup electrode on the neck is close to the level of the clavicle. One of the voltage pickup electrodes is placed at the level of the clavicle for the present electrode array, and accordingly, the estimated SV values are larger as is supported by the results of Raaijmarkers et al.

Another causative factor of the overestimation would be the contribution of blood volume changes of the lung to the total impedance changes measured with the band electrodes. The voltage potential at the xiphisternum level, where one of the voltage pick-up band electrodes is located, would be affected by the pulmonary blood volume changes, and thus would decrease following the cardiac blood ejection. However, the ΔZ maps in Fig. 5.18 show little changes in the ΔZ value near the xiphisternum, indicating less influence of the pulmonary blood circulation. This may imply that the band electrodes cause smaller voltage drops than the spot electrodes do. This means that the value of ΔZ_c (or $dZ_c/dt]_{min}$) measured with the band electrodes should be smaller than that with the spot electrodes. Due to an uncertain sensitivity of the impedance measurement to the distance L between the two band electrodes, such effect of the pulmonary blood circulation on this value cannot be exactly confirmed from the impedance data in the present experiments. Considering Equation 5.3 or Equation 5.4, however, the values of $\Delta Z_c/Z_c^2$ (or $dZ_c/dt]_{min}/Z_c^2$) measured with the band- and with the spot electrodes can be roughly compared with each other, provided that the same distance L is given in the two electrode arrays. The mean ratio of the values of $\Delta Z_c/Z_c^2$ (or $dZ_c/dt]_{min}/Z_c^2$) measured with the spot-electrode ($\Delta Z_c/Z_c^2]_{spot}$ (or $dZ_c/dt]_{min}/Z_c^2]_{spot}$) to those with the band-electrode array ($\Delta Z_c/Z_c^2]_{band}$ (or $dZ_c/dt]_{min}/Z_c^2]_{band}$) was 4.20 ± 1.48 (SD) (or 3.96 ± 1.18). This result is suggestive of less contribution of blood volume changes of the lung to the impedance change with the present spot-electrode array as compared with that with the band-electrode array.

In summary, the overestimation of the SV values determined by the new spot-electrode array is mainly due to less influence of the current concentration near the neck-thorax transition and to less contribution of the pulmonary blood circulation on the impedance change (ΔZ_c) as opposed to the data obtained with the band-electrode array. Accordingly, ΔZ_c (and thus dZ_c/dt) with the spot electrodes would be larger than that with the band electrodes. It is still uncertain whether or not the partial two-compartment cylindrical model for the frontal part of the thorax will hold, because the true SV value is unknown in the present comparison study. However, the present spot-electrode array may be a feasible means of providing a quantitative estimation of the SV value, considering the underestimation of the SV values measured with the band-electrode array as compared with those by the thermal dilution method, as mentioned above.

Besides this, the new spot-electrode array proposed here is a superior replacement with respect to the conventional band-electrode array. It is useful for clinical practice and for long-term monitoring of SV and CO. Further investigations to address the problems still remaining are to produce a more certain electrical model of the thorax, possibly through more detailed impedance mapping, and to carry out a comparison study with an accurate and reliable measurement method to provide a quantitative estimation of the SV or CO value.

5.4 Development of a Prototype Ambulatory Cardiovascular Monitoring System and its Applications

The availability of new technologies that enable cardiovascular hemodynamic variables including BP and CO to be monitored in a non-invasive, automated, and ambulatory fashion during prolonged observation periods should create exciting opportunities for studying new approaches to the developments of early diagnosis and treatment as well as prevention and control of cardiovascular and other diseases in modern clinics. This is also beneficial for normal subjects as a means of health care during daily life. If such monitoring could be made on a beat-by-beat basis, it would be much more desirable for detailed diagnosis and therapy of inpatients as well as outpatient cardiovascular function and circulatory autonomic regulation.

For ambulatory BP monitoring (ABPM), commercially available portable devices, based on the auscultation and/or cuff-oscillometric method, have been used often clinically. Due to the measurement principle, these devices do not allow continuous BP measurement on a beat-by-beat basis, but can only monitor BP every 10–15 min or, more practically, every 30 min because frequent measurements make the subject uncomfortable and disturb his/her sleep due to the arm-cuff occlusion. They are considered useful for understanding the temporal characteristics of the BP change, particularly for the evaluation and treatment of hypertensive patients [64]. Still, the process involved is not straightforward.

We have previously designed such ABPM systems. One allows the ambulatory monitoring of intermittent BP data with shorter time intervals by measuring BP in the finger with the volume-oscillometric method [24, 65]. Another assesses beat-by-beat BP data in the finger or a portion of superficial temporal artery with the volume-compensation method [31, 66]. A similar type of the latter system, based on the Peñáz's method, has been commercialized as Portapres [30]. These systems have been evaluated as useful and helpful techniques to better measure BP fluctuations in various laboratory and field tests.

On the other hand, Gonneke et al. [67] and Sherwood et al. [68] independently designed a system for ambulatory CO monitoring (ACOM) based on conventional Kubicek's impedance cardiography. As mentioned above, electrical admittance cardiography is considered practical to use than are other methods. Also, we have designed a voltage clamp method to obtain the bio-admittance signal [69], and pointed out that this method is suitable for a compact design of such ACOM device.

As described above, methods have been developed for monitoring BP and CO separately. Thus the simultaneous monitoring of BP and CO is needed for the determination of detailed cardiovascular function. With this in view, we have designed a prototype portable instrument capable of monitoring simultaneously cardiovascular hemodynamic variables including BP and CO on an intermittent basis [70] and on a beat-by-beat basis [32, 70]. The former instrument is based on a combination of the volume-oscillometric method and admittance cardiography, while in the latter, the volume-compensation method is employed.

This section therefore introduces these two systems with slight improvements based on the new techniques of both the local pressurization method for the BP measurement and the tetra-polar spot-electrode method for the CO measurement.

5.4.1 Intermittent Cardiovascular Hemodynamic Monitoring System

5.4.1.1 Description of the System

Fig. 5.20 shows a schematic block diagram of the ambulatory system for intermittent monitoring of cardiovascular hemodynamic variables by combining the volume-oscillometric method and admittance cardiography. The system consists of three units. The first is a BP sensing unit of a finger-cuff unit or a head-cuff unit for the BP measurement and tetra-polar spot electrodes for the detection of the admittance signal to determine CO, the second is a portable unit to control measurements, process signals, and store data with a microprocessor, and the third is a conventional personal computer system to reproduce and analyze data. Disposable ECG electrodes (SEC145, NEC Medical Systems, Japan) were used as the spot electrodes, which were firmly covered with an elastic surgical adhesive tape so as not to slip from the skin in the face of body movements during ambulatory monitoring. The size of the portable unit is 104 mm×64 mm×32 mm and 160 g weight including a 1.5 V alkaline battery (LR03, Panasonic, Japan) and a 3.6 V lithium thionyl chloride battery (TL-2100, Tadiran, Israel).

To minimize discomfort and to allow movement of the subject's hand during the BP measurement, the local pressurization technique was adopted on the finger or on the head on a segment of the superficial temporal artery. The finger-cuff unit is the same as described above, and the head-cuff unit is formed like a disk cap with a 25 mm diameter and 6 mm height made of black-colored plastic. A polyurethane film 0.05 mm thick is fixed on one side of this cuff unit contacting

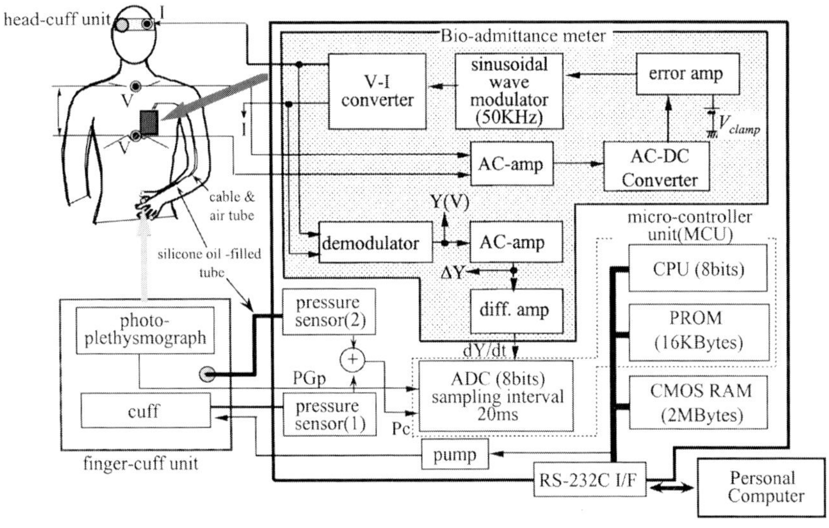

Fig. 5.20 Diagram of the ambulatory system for intermittent monitoring of cardiovascular hemodynamic variables.

with the skin, and a reflectance photoplethysmographic sensor is also attached to the inside of the film at its central part. The sensor's light source and photodetector are those used in the finger-cuff. The head-cuff unit is properly positioned just above the superficial temporal artery and is held in position by a headband with Velcro.

Both the air tube for the cuff pressure measurement and the silicone-oil-filled tube from the cuff for hydrostatic pressure measurement between the heart and the BP measuring site are led to the portable unit. The portable unit contains two pressure sensors (FPM-05PG, Fujikura, Japan and 1210B-005, IC Sensors, USA, respectively) and is carried by the subject in his/her breast pocket at the heart level. This unit can therefore measure the pressure at the heart level, to compensate for the hydrostatic pressure change due to the difference between the height of the measuring site and the heart.

As shown in Fig. 5.20, the portable unit is divided into three parts. The first is a signal-detecting circuit for the photoplethysmographic pulsation (PGp) and cuff pressure (P_c) controlled by a micro-air pump (P36B01, Oken Seiko, Japan); the second is a bio-admittance meter for the detection of the first derivative of the admittance signal (dY/dt); the third is the signal processing and recording system. The third unit performs A/D conversion of PGp, P_c, and dY/dt signals, all the necessary measurement procedures and data storage on a digital memory IC (CMOS RAM, 2 Mb) using a single-chip microcontroller unit (MCU; PC68HC711EQZ, Motorola, USA: ADC, 8 bits and 20 ms of sampling interval; CPU, 8 bits; PROM, 16 kb) with interactive software.

A voltage clamp method was successfully designed for the bio-admittance meter to measure the admittance signal (Y). This method can detect the admittance signal from the supplied current, which is controlled so as to maintain the voltage drop across the inner electrodes to equal a constant reference voltage, V_{clamp} [69].

Fig. 5.21 shows a timing chart of the sequential procedure and signal processing. Following a start signal, P_c is gradually increased by the micropump, and the mean (MBP) and systolic pressure (SBP) are determined according to the volume-oscillometric method. The maximum dY/dt ($dY/dt]_{max_i}$), ejection time (Ts_i), and inter-beat interval (PP_i) are determined from the detection of the positive and negative peaks of dY/dt signal on a beat-by-beat basis during the BP measurement, and their mean values of $dY/dt]^*_{max}$, T^*_s, and PP^*, with standard deviations, are processed in the CPU for each measurement. At this time, zero level of each dY/dt signal ($\bar{Y}_i$) is obtained from the mean electric level calculated from the integration of dY/dt during each cardiac cycle, as shown in the lowest part of this figure. If one of these three values, $dY/dt]_{max}$, T_s and PP, is out of range compared with the preceding processed value, this signal is deleted as an artifactual signal. All of these processed values for each measurement are stored in the CMOS RAM. Approximately 20 s are required for each measurement. The measurement interval can be preset at 2, 5, or 10 min.

After the monitoring, these stored data are reproduced by the computer to calculate heart rate HR ($=1/PP^*$), stroke volume SV ($=\rho_b L^2 dY/dt]^*_{max}\ T^*_s$; ρ_b is corrected by the subject's hematocrit according to our previous study [54], cardiac out-

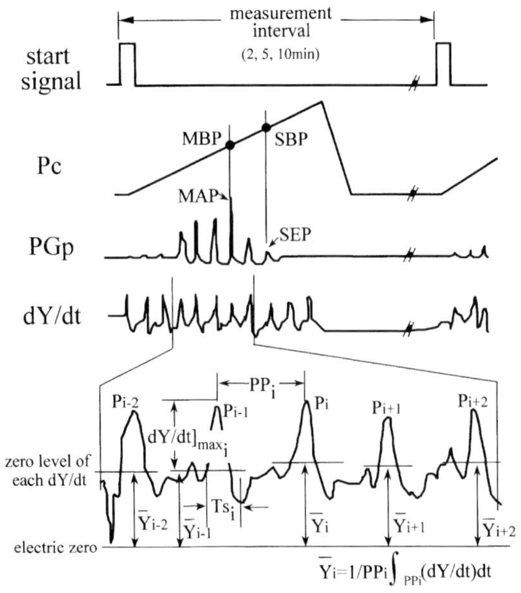

Fig. 5.21 Timing chart of the sequential procedure and signal processing.

put CO (=SV×HR), peripheral resistance R_p (=MBP/CO), rate pressure product RPP (=SBP×HR) as an index of cardiac oxygen consumption, and diastolic pressure DBP by Equation 5.1, if necessary. All of these time series data, including SBP and MBP, are displayed on a CRT monitor.

5.4.1.2 Monitoring Examples and Operational Evaluation of the System

In order to evaluate the operational function of the system, 24 h monitoring with a 5 min measurement interval was tried on 13 normal subjects (10 men and 3 women, 22–26 years old) during their daily life. An additional 15 patients (10 men and 5 women, 19–62 years old) undergoing drug treatment for cardiovascular diseases such as hypertension or pheochromocytoma were observed during their stay in hospital. This experimental study was done with all of the subjects' informed consent.

Fig. 5.22 shows 24 h trends of cardiovascular variables recorded with the system from two different normal subjects. The dynamic changes and short-term fluctuations of cardiovascular variables were well monitored in each subject, as were the diurnal changes in those variables. In the left side of this figure the decrease in oxygen consumption of the heart, expressed by the rate pressure product (RPP) during sleep (indicated by shaded region), was caused mainly by the decrease in only BP (SBP), while that in the right case it was caused by a decrease in both BP and HR. Furthermore, the decrease in the average BP during sleep would be due

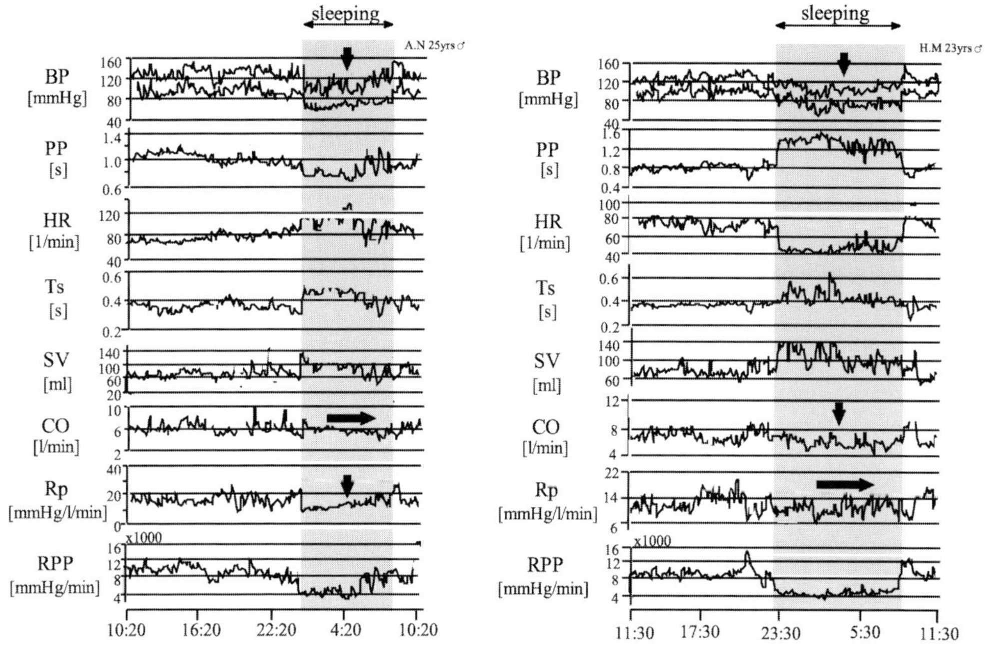

	SBP/MBP[mmHg]	CO[l/min]	Rp[mmHg/l/min]		SBP/MBP[mmHg]	CO[l/min]	Rp[mmHg/l/min]
awakening	133± 1.0/94± 0.8	5.9± 0.1	17.0± 0.3	* P < 0.00001	122± 0.8/93± 0.7	6.8± 0.1	14.1± 0.2
	*	NS	*	NS: not significant	*	*	NS
sleeping	109± 1.5/67± 1.1	5.2 ± 0.2	13.6 ± 0.4		101± 1.1/69± 1.1	5.5± 0.1	12.8± 0.3

Fig. 5.22 24 h trends of cardiovascular variables recorded with the system from two different normal subjects.

mainly to the decrease in the average R_p rather than to the reduction in the average CO in the left case, while it would be due particularly to the reduction in the average CO rather than to the decrease in the average R_p in the right case. The supporting statistics are shown in the table in the lowest part of this figure.

Fig. 5.23 also shows two examples of 24 h trends obtained in a patient with hypertension (left side) and a patient with pheochromocytoma (right side) during hospitalization, who had received a drug treatment. In both cases, no significant decrease in the average BP was observed during sleeping. There was no significant reduction in either the average CO or R_p during sleep in the left case, but there was a decrease in the average CO and an increase in the average R_p in the right. This observation of no significant decrease in BP could therefore be interpreted, if BP is simply determined by an Ohm's law in the cardiovascular system in these cases ($BP = CO \times R_p$).

These findings cannot be demonstrated with a conventional ABPM system. The present system may therefore be useful and available for better understanding of cardiovascular functional tendencies as well as of interactive relations among the cardiovascular hemodynamic variables.

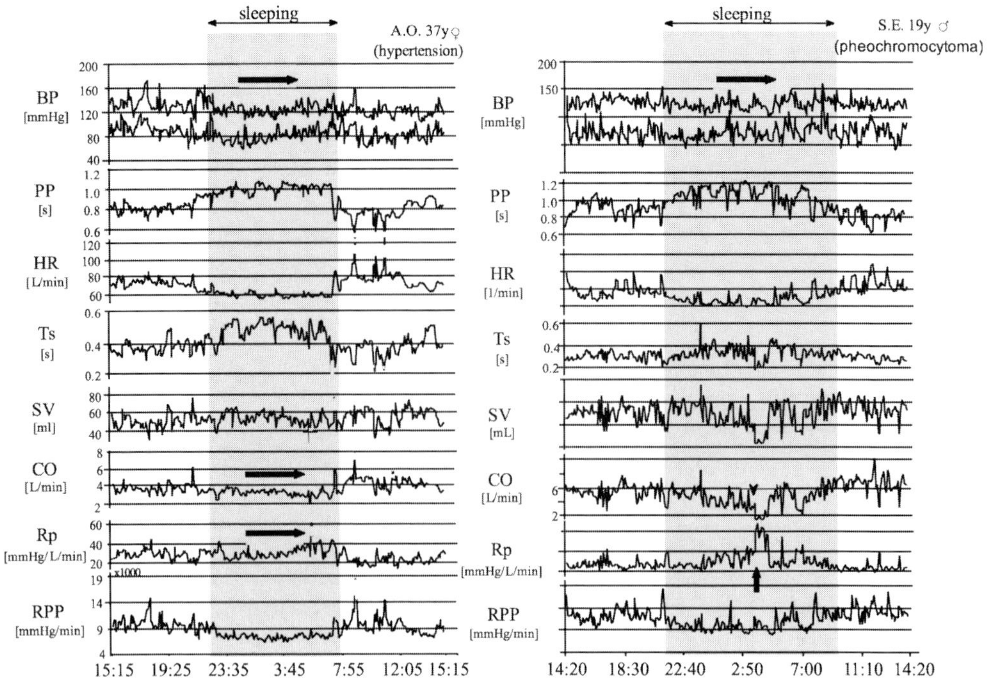

Fig. 5.23 Two examples of 24 h trends obtained in a patient with hypertension (left) and a hospital in-patient with pheochromocytoma (right), who had received a drug treatment.

Throughout the whole 24 h monitoring in all of the normal subjects and the patients tested in this study, satisfactory results were obtained. The subjects complained little about discomfort during monitoring. They insisted on the necessity for electrical wireless and tubeless connections between the portable unit and the sensing unit attached to the body. They were concerned about going without bathing during the study.

Sometimes data were unmeasurable. That was probably due to strong movements of the body that led to motion artifacts. However, such data were detected easily and rejected as artifact by observing trends. A percentage of the numbers of such unmeasured data points to those of the total data points during the 24 h monitoring (244 data points when the measurement interval is 5 min) ranged from about 2 to 7% ($3.2 \pm 1.5\%$) in this study.

5.4.1.3 Blood-Pressure Monitoring at Head Level

It is well known that BP at the heart level normally shows a higher level during wakefulness in the daytime and a lower level during sleep at night (Fig. 5.22). BP is regulated mainly by the baroreceptors of the carotid sinus. During a standing or sitting position, the BP level detected by the baroreceptors is less than that of the heart by about 15 mmHg. When the subject is in a supine position, the two pressures are nearly equal. Therefore, it may be that the diurnal change in BP at a position rostral to the heart would be smaller rather than that in BP at the heart level.

To investigate the validity of this hypothesis, we have carried out preliminary experiments with the monitoring system in 7 healthy subjects (6 men and 1 woman, 22–39 years old) with their informed consent. In this study, since the system enables BP to be measured at a level of superficial temporal artery, the BP at the head level was adopted as that at the upper position from the heart. Thus, a hydrostatic pressure difference (HSP) between the portable unit placed on the heart level and the BP measuring site (the portion of superficial temporal artery) was measured by the output of the pressure sensor (2) shown in Fig. 5.20. The measurement interval was 5 min and the subjects were requested to behave normally.

Fig. 5.24 is an example of 24 h trends found in the HSP and the cardiovascular variables obtained in a healthy subject. Peripheral resistance, R_p, and rate pres-

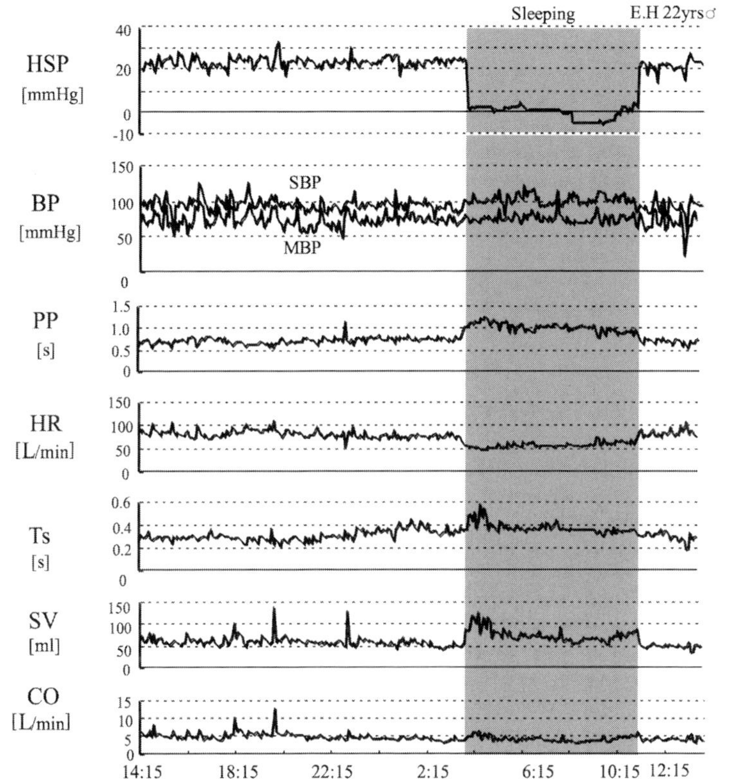

Fig. 5.24 Example of 24 h trends found in the HSP between the heart and the BP measuring site (head) and the cardiovascular variables obtained in a healthy subject.

sure product, RPP, are not shown. The HSP record shows almost 0 mmHg between 4:00 am and 11:00 am, indicating the subject lying down to sleep. Except for this time period, the HSP is 20–30 mmHg, indicating that the subject's upper body remained in an upright position. It is noted that a clear diurnal change of the BP at the head level is not observed, but there are diurnal changes of the other cardiovascular variables. The 24 h trends were similar to this in all the subjects tested.

The BP values at the heart level compensated for the measured HSP values were calculated in each subject to display the 24 h trends as illustrated in Fig. 5.25. This figure shows two examples of the BP trends at the head (BP_{head}, upper trace) and at the heart level (BP_{heart}, lower trace). A conventional moving average method applied for every five data points is made for better understanding of the pattern of the diurnal changes of both BP_{head} and BP_{heart}. It is demonstrated that there were no significant decreases in BP_{head} during sleeping, but the level of BP_{heart} is clearly lower during sleeping. Similar trends were obtained in all the subjects.

In order to estimate quantitatively the differences between the BP_{head} and BP_{heart} values, the MBP data during sleeping and the other period (awake period, or the period when the subjects were at least in an upright position) are sampled from the trends in each subject to give their means and standard deviations. Fig. 5.26 shows results obtained in two normal subjects. In this figure, open and closed circles indicate the MBP values at the heart (MBP_{heart}) and at the head lev-

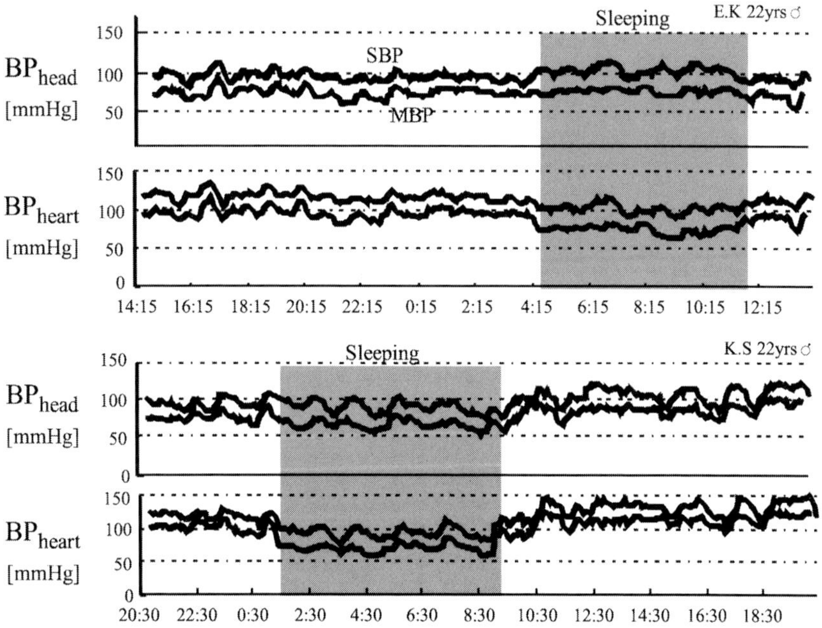

Fig. 5.25 Two examples of the 24 h BP trends at the head and at the heart level.

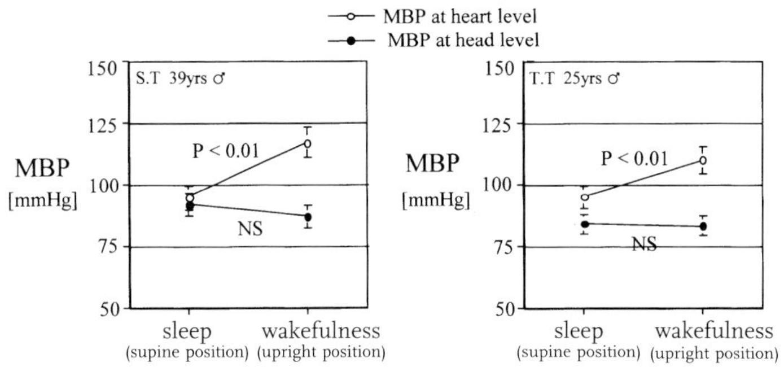

Fig. 5.26 Results obtained in two normal subjects comparing MBP during sleep and when awake.

el (MBP_{head}), respectively. It is clearly shown that the mean MBP_{heart} value during wakefulness is significantly higher than that during sleep ($p<0.01$), while there is no significant difference between the mean MBP_{head} values during wakefulness and sleep. The same results were obtained in the other five subjects.

From these results, the difference in BP_{head} between during wakefulness and sleep is small, and accordingly, the diurnal change of BP at the head level is less than that of BP at the heart. If this is true, BP at the head level might possibly be raised during sleep in case of no clear diurnal change of BP at the heart level, as shown, for example in Fig. 5.23. Also, if this is true, BP at the head level would be regulated to remain almost constant, so that the blood circulation in the brain perfused with constant pressure. These ideas are, of course, only speculative, but it is the fact through the present experiments that the diurnal change of BP at the head level is small compared with that at the heart level, though a mechanism of such BP regulation is unknown. Further experimental examinations are needed to verify this phenomenon during daily living using normal subjects as well as patients with cardiovascular disorders.

5.4.2
Beat-By-Beat Cardiovascular Hemodynamic Monitoring System

5.4.2.1 Brief Description of the System

The system designed in this study is based on the simultaneous measurements of BP and CO on a beat-by-beat basis using a combination of the volume-compensation method and admittance cardiography with a microprocessor-based digital technique. The present system is essentially the same as the previously developed system reported in the literature [32] except for a difference in the sensing unit for BP and CO measurement. A brief description of the system is given below.

As shown in Fig. 5.27, the system is composed of: the local pressurizing finger-cuff unit described above and the tetra-polar spot electrodes for the detection of admittance signal described above; a main portable unit; and a conventional personal computer for data retrieval and analysis. The portable unit has seven func-

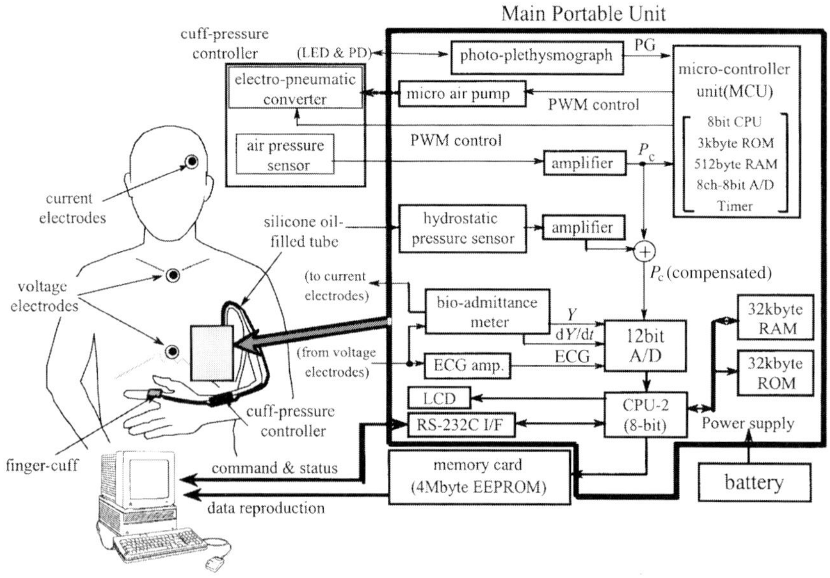

Fig. 5.27 Diagram of the system for monitoring ambulatory beat-by-beat cardiovascular hemodynamic variables.

tional roles: BP measurement, CO measurement, signal processing and control of each measurement using CPUs, data storage using a memory device, data display using an LCD, interactive communication between the unit and the computer using an RS-232 serial interface, and power supply using a lithium-ion rechargeable battery (7.2 V, 1250 mAh; NP-510, Sony, Japan) which allows more than 200 min continuous use. During monitoring, the subject carries the portable unit (72 mm×36 mm×126 mm and 480 g weight including the battery) in his/her breast pocket to determine the BP value at the heart level.

After the monitoring, the computer reads out the stored data in the memory device, analysis is done to display the cardiovascular variables. The following 13 variables are processed on a beat-by-beat basis: systolic (SBP), mean (MBP) and diastolic BP (DBP), ECG R-R interval (RR), instantaneous heart rate (HR), stroke volume (SV), cardiac output (CO), pre-ejection period (PEP), ventricular ejection time (T_S), pulse transit time (PTT), peripheral vascular resistance (R_P), rate pressure product (RPP), and respiration rate (Resp). The BP and the CO measurements, as well as the main functional roles of this portable unit, are briefly described as follows.

BP Measurement

The microcontroller unit (MCU) for automatic BP measurement based on the volume servocontrol is similar to that already described. The servocontrolled cuff pressure (P_c) processed by the MCU and given as the indirect BP waveform is led to a 12-bit A/D converter (sampling frequency of 100 Hz), and then processed in another 8-bit signal processor (CPU-2 shown in Fig. 5.27; TMPZ84C015BF, Toshiba Corp., Japan) to determine the values of SBP, MBP, and DBP on a beat-by-beat basis.

CO Measurement

Beat-by-beat SV or CO measurement using admittance cardiography is done by processing the signals from the bio-admittance meter in the portable unit (Fig. 5.20). Signal processing and the determination of SV and CO on a beat-by-beat basis are basically the same as before. In this meter, an ECG signal can be detected from the voltage pickup electrodes to provide as a timing signal for the signal processing. Also, a basal (total) admittance signal (Y) is taken from the meter to obtain respiration rate (Resp) by detecting a peak-to-peak time interval (T_{resp}) of the slow change in the Y signal due to inspiration and expiration (Fig. 5.28). The outputs from the meter, Y, dY/dt, and ECG signals, are sent to the CPU-2 via the A/D converter to perform the necessary processing. In order to eliminate the baseline fluctuation of the dY/dt signal, caused by normal respiration and body movements, a digital filtering function is provided in CPU-2 (high-pass filter, 0.8 Hz of cutoff frequency).

Signal Processing, Data Storage and Display

Fig. 5.28 shows a flow chart of the signal processing from the measured signals, contents of the stored data in the memory device, and calculation and data display in the personal computer system. The left side of this figure shows an example of simultaneous recordings of ECG, dY/dt, indirect BP, and Y, representing the variables to be determined and stored on a beat-by-beat basis.

Following an ECG-R wave, detections of a bottom peak followed by a maximal first peak value from the BP curve are made to give a DBP and an SBP value, respectively. An MBP value is determined from the integration of the BP curve during the same cardiac cycle. At this cycle, ECG-R to -R interval (RR), pulse transit time (PTT), pre-ejection period (PEP), and ventricular ejection time (T_s) are processed from a time interval between two peculiar points of ECG, BP and dY/dt waves. A maximum dY/dt ($dY/dt]_{max}$) is measured by the height from the baseline to a peak point of the dY/dt wave. The digital filter in the CPU-2 facilitates the determination of these variables from the dY/dt wave, whose filtered waveform is sufficiently stable for this signal processing except during sudden changes in the subject's posture and strong movements such as jumping and jogging.

Respiration interval is also determined from the Y signal which varies as a function of inspiration and expiration. Signal processing is executed by CPU-2, and 10 beat-by-beat datasets are stored in the memory card inserted into the portable unit (Fig. 5.28). At this time, the values of Y, $dY/dt]_{max}$, RR, T_s, SBP, and DBP are displayed on a beat-by-beat basis using the LCD for making the real-time observation.

A smartMedia (TC5832DC, 2 Mb, Toshiba Corp., Japan) is used as the memory device. Altogether 31680 beats of 10 datasets-with text data file format can be

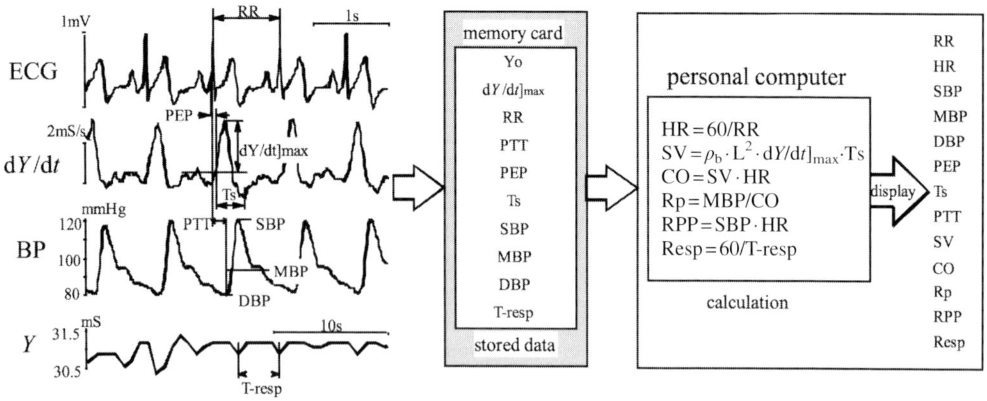

Fig. 5.28 Signal processing of measured signals, contents of stored data in memory card, and calculation and data display in the PC system.

stored. After the monitoring, this device is removed from the portable unit and set into a 3.5″ floppy disk adapter. This adapter is then inserted into the computer's floppy disk drive to reproduce the stored data. As shown in the right part of Fig. 5.28, the necessary calculation is executed by the computer using the reproduced data to display 13 processed variables on the CRT monitor. If required, some of necessary processed variables among them are properly selected to display. The present system enables us to observe these time-series during monitoring by connecting the portable unit to the computer with a communication cable.

5.4.2.2 **Monitoring Examples and Operational Evaluation of the System**

In order to evaluate the operational performance of the system as well as willingness to be subjected to the monitoring, ten men and two women (20–45 years old) participated in the laboratory and field testing, after giving their informed consent. They were requested to move freely and perform various normal activities, such as walking, going up/down stairs, desk work, driving a car, posture changes from sitting to standing or *vice versa*. They were asked not to make strong movements of the hand in the BP measuring site and to avoid severe body movement as much as possible. These events were excluded as either unreliable or artifactual: unusually abrupt or abnormal changes in BP and/or SV (CO); as well as sudden changes in Y and dY/dt, probably due to motion artifacts; slippage of the cuff unit from its original place or from a significant change in the vasomotor tone of the finger artery.

In the left part of Fig. 5.29 is shown an example of about 3 h trends for the ambulatory monitoring of beat-by-beat cardiovascular variables (16742 datasets) obtained in a normal male subject behaving normally, where his activities are given in the lowest part. About 130 numbers of beat-by-beat datasets (about 0.8% of the total) were confirmed as unreliable and/or artifactual and were deleted by the

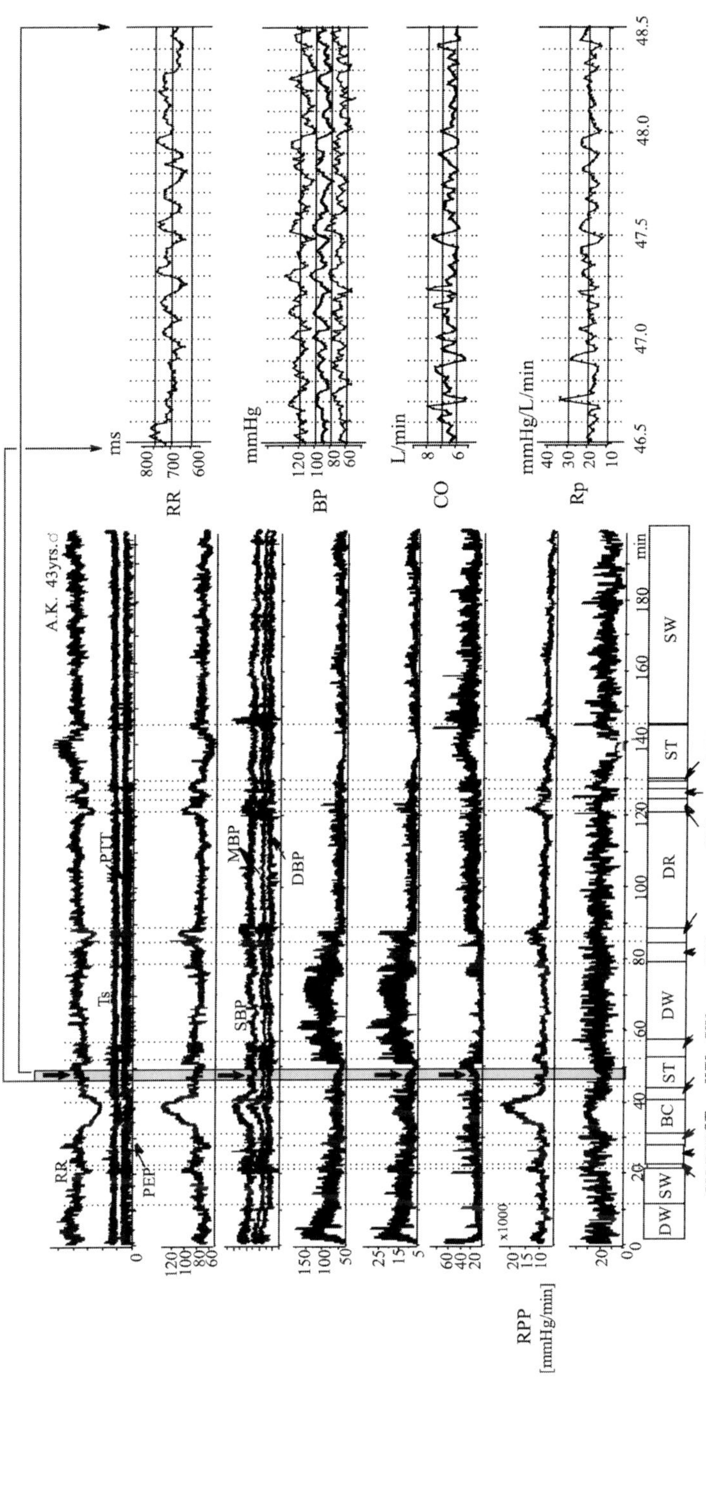

Fig. 5.29 Example of 3.3 h ambulatory monitoring in normal subject doing daily activities; right hand side shows expanded 2 min section for four variables. Various activities are indicated in the lowest part of the left figure: *ST*, sitting; *DW*, desk work; *SW*, standing work; *DR*, driving; *WK*, walking; *GU*, going up the stairs; *BC*, riding a bicycle. Other symbols see text.

manual editing, so that there were no such datasets in this chart. In spite of these data lacks, transient changes and interactive relations of the variables following various activities are clear from the trends.

The right side of Fig. 5.29 is a part of the trends expanded for the 2 min from 46.5 to 48.5 min during the sitting period, indicated by shading in the left trends, and showing only four variables of RR, BP, CO, and R_p, denoted by arrows in the shaded area. The beat-by-beat changes of these variables are clearly recorded, and about 10 s cyclic fluctuations in BP, so called Mayer waves, are clearly observed.

A total of 10 trials of monitoring lasting for 60–210 min were made on each of all the subjects with this system. Each trial was successful, there were no major functional or operational difficulties during laboratory and field tests. The results were satisfactory except for a small amount of bad data due mainly to motion artifacts. However, these bad data could be readily discovered with careful observation in a few time-series data points in which they occur. Overall, about 3% of the beat-by-beat datasets appeared as unreliable and/or artifactual. An additional software function for automatic editing of such bad data would make the system much easier to use.

The system appears ready to provide the essential parameters of RR (or HR), BP, CO (or SV), R_p, and Resp on a beat-by-beat basis for detailed assessment of the cardiovascular hemodynamic functions. The system involving monitoring PEP, PTT, T_s, and RPP may also be useful for further assessment of impaired circulatory functions. It creates an exciting opportunity to investigate the behaviors of these additional variables in clinical applications. In particular, the variables of PEP and T_s may be simple indicators of cardiac sympathetic activity [72–74].

However, several minor but practical problems were experienced during this study, which were concerned with the subjects' general notions about their continuous monitoring. Although the subjects complained little about discomfort due to the cuff occlusion and to the attachment of the spot electrodes, they insisted on the necessity of wireless data transfer between the portable unit and the sensing unit. In addition, the acoustic noise produced from the air-pump and the electro-pneumatic converter used for the portable unit at around 60 dB-SPL bothers the subjects during monitoring. These are unsolved technical problems at this time. With recent and future advances in microelectronics and micromechanics technologies, redesign of the system should be focused on improvements for more practical use. Although such technical problems to solve are still remain, the present prototype system offers an excellent means of creating many exciting opportunities for pursuing new approaches to cardiovascular hemodynamic studies.

5.4.2.3 Analysis of Autonomic Regulation During Various Physical Activities

Monitoring of beat-by-beat cardiovascular variables enables us to make an analytical evaluation of autonomic regulation of the cardiovascular system. We have tentatively analyzed the circulatory autonomic regulation during various physical activities by using the collected data from the field tests.

Analysis of Cardiac Baroreflex Function

In order to estimate a vagal activity to BP regulation, a baroreceptor–cardiac reflex sensitivity (BRS) was analyzed according to Bertinieri's method [75]. This baroreflex function is assessed to identify spontaneous sequences of three or more consecutive beats, in which the values of SBP progressively increase or decrease and the corresponding values of RR progressively lengthen or shorten in a linear fashion ($\gamma^2 \geq 0.8$; γ is regression coefficient). A slope between these consecutive beats of SBP and RR with $\gamma^2 \geq 0.8$ is a measure of BRS in this study.

Spectral Analysis

Spectral analysis was performed to assess the RR and R_p data with a fast Fourier transform (FFT). An FFT was applied to the datasets of 128 heartbeats, which were updated at every heartbeat (moving average FFT). A Hanning window was applied prior to each FFT calculation to prevent spectral leakage. Spectral powers of RR and R_p in the low-frequency band (0.04–0.12 Hz; PLF (RR) and PLF (R_p)) and power of RR in the high-frequency band (0.15–0.5 Hz; PHF (RR)) were calculated together with the ratio of RR powers in the LF and HF (LF/HF (RR)). It has been reported that LF/HF (RR) would be a better marker of sympathetic activity rather than PLF (RR) which includes both vagal and sympathetic activities, and PHF (RR) a good measure of vagal activity [76, 77]. PLF (R_p) is considered to be an estimate of the regulation of vasomotor tone via sympathetic activity, since the R_p value could possibly be influenced by changes in vasomotor tone.

The mean values of the above variables were obtained during a period in which the subject performed the same physical activity for more than 5–6 min. Then, the mean values of each quantity were compared among classified load levels of physical activities.

Transfer function analysis was also made between SBP as input and RR as output signal and between SBP as input and R_p as output signal using a computer algorithm. An FFT was applied for each signal in a data segment of 128 heartbeats overlapping by one-half during the selected period, yielding auto power and cross-power spectra to obtain the transfer functions. Gain and phase from SBP to RR or from SBP to R_p were calculated as a function of frequency. These gain and phase data were adopted only when the coherence was over 0.5. Then, a peak gain in the HF band between SBP and RR ($H_{SBP\text{-}RR}$ (HF)) and that in the LF band between SBP and R_p ($H_{SBP\text{-}R_p}$, (LF)) were assessed respectively as an index for vagal and sympathetic activity.

Results

Fig. 5.30 summarizes the mean values of various markers for the assessment of vagal and sympathetic activities against various kinds of physical activities except severe ones such as 'riding a bicycle' during normal daily living. On the left part of this figure is shown the markers of BRS, PHF (RR), PLF (RR), LF/HF (RR) and PLF (R_p), and on the right those of $H_{SBP\text{-}RR}$ (HF) and $H_{SBP\text{-}R_p}$ (LF), together with the ratio of numbers of acceptable data to total ones in the LF and HF band

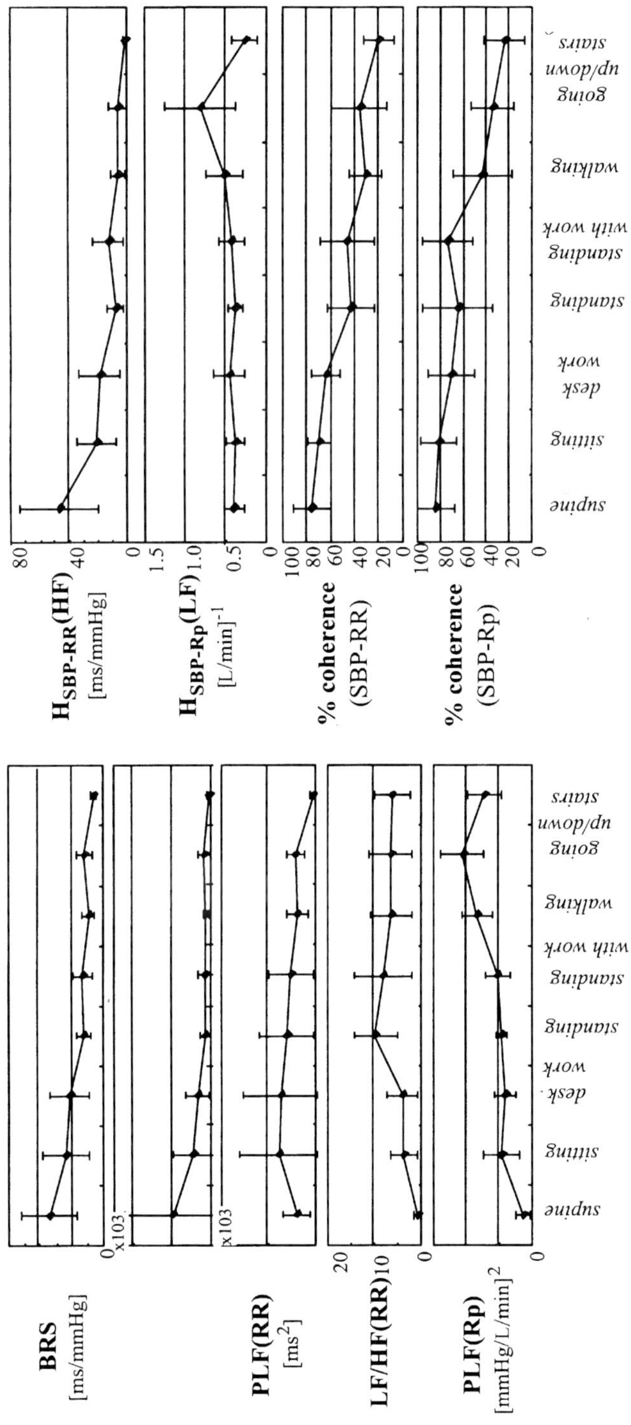

Fig. 5.30 Mean values of various markers for the assessment of vagal and sympathetic activities against various kinds of physical activities, with increasing demand from left to right.

(% coherence), which means that the interdependence between the input signal (SBP) and the output signal (RR or R_p) was acceptable under the 0.5 coherence criterion. Physical activities presented in the lowest part of this figure are classified by an load level order from easier (left) to harder (right). As expected from the physiological standpoint, the markers of BRS, PHF (RR) and $H_{SBP\text{-}RR}$ (HF) decrease, while those of PLF (R_p) and $H_{SBP\text{-}R_p}$ (LF) tend to increase, in accordance with a harder load level. However, there appears no clear tendency in the markers of PLF (RR) and LF/HF (RR).

Although the present study is a preliminarily examination with the prototype system using normal subjects, the results strongly indicate a feasible applicability to the assessment of circulatory autonomic regulation from the beat-by-beat cardiovascular variables monitored under daily activities in a non-invasive and ambulatory manner. A prospective research program should be made for clinical applications including such kinds of tests using patients with cardiovascular disorders.

5.5 Summary and Conclusions

This chapter describes particularly our current research on non-invasive BP and CO measurement techniques based on the volume-oscillometric and the volume-compensation method for the former and on the electrical admittance cardiography for the latter. A local pressurization technique for the BP measurement at the wrist, the finger, and the posterior of the thigh using a local compression cuff is proposed with successful and satisfactory results. This technique may provide a more useful and convenient substitute for a conventional cuff-sphygmomanometry using a band cuff wrapped around a biological segment. The instrument is compact and comfortable for the subject to use. It measures BP for long durations and can acquire frequent samples. In particular, a BP monitoring technique from a toilet seat could realize a fully automatic acquisition of a huge number of BP data without attachment of any sensors to the body and necessary measurement during daily life. This may be considerably available for health care monitoring in daily living at home.

Electrical impedance or admittance cardiography, which is considered at present as one of the best means of estimating SV and CO non-invasively, simply and continuously, is briefly reviewed. Replacement of a conventional tetra-polar band- to spot-electrode arrangement is attempted to address practical problems such as discomfort for the subject and frequent incidence of contact dermatitis caused by the band electrodes' attachment to the skin. An optimal spot-electrode array is newly proposed based on the measurement of current distribution on the thorax with an impedance mapping system. This electrode array is promising as a superior replacement to the band-electrode array, and prospective research is discussed to compare this technique with an accurate and reliable invasive method, as well as to improve a more precise electrical model of the thorax.

Two types of recent developments of non-invasive and ambulatory cardiovascular monitoring systems have been introduced. One design combines the volume-oscillometric method for BP measurement and the admittance cardiography for CO measurement on an intermittent basis. The other uses the volume-compensation method and admittance cardiography respectively to measure BP and CO on a beat-by-beat basis, taking advantage of both the local pressurization technique and the tetra-polar spot-electrode array. Successful and satisfactory monitoring results were obtained using the two systems.

The former system is found to be a promising means for better understanding of cardiovascular functional tendencies as well as interactive relations among the cardiovascular hemodynamic variables. Especially, it is useful and available for the assessment of causality of the diurnal change in BP, taking the fluctuations in both CO and peripheral vascular resistance (R_p) into account. Those fluctuations cannot be assessed with conventional ambulatory BP monitoring. The importance of BP measurement at head level is discussed after the introduction of preliminary examinations using normal subjects. These results suggest that the blood circulation in the brain might be a perfusion system with constant pressure, though further experimental examinations should be necessary to verify the physiological phenomenon under daily living using many normal subjects as well as patients with cardiovascular disorders.

The latter prototype system appears to be a useful and helpful means for more detailed data acquisition of transient changes as well as the beat-by-beat interrelations among the cardiovascular hemodynamic variables. Several technical problems to be solved are discussed, stressing the practical situation. Nevertheless, the present method may offer an excellent means of creating many exciting opportunities for pursuing new approaches to cardiovascular studies. In addition, the preliminarily experimental results obtained in normal subjects strongly suggest that the assessment of circulatory autonomic regulation could be made from beat-by-beat cardiovascular variables monitored non-invasively while subjects perform normal and ambulatory physical activities. These results suggest that prospective research program should be also designed for clinical applications using patients with cardiovascular and other diseases.

With recent advancements in microelectronics and micromechanics technologies, additional studies will be needed to further improve on these two systems for more practical use.

5.6 Acknowledgments

The author wishes to thank Associate Professor Shinobu Tanaka, Dr. Minoru Nakagawara, and Mr. Masamichi Nogawa for their invaluable assistance during the progress of the research programs on instrumentation, data analysis, and experimentation.

5.7 References

1 L.A. Geddes: *The Direct and Indirect Measurement of Blood Pressure*, Year Book Medical Publishers, Chicago, **1970**.

2 R.S.C. Cobbold: *Transducers for Biomedical Measurements: Principles and Applications*, John Wiley & Sons, New York, **1974**.

3 N. Scheiderman, S.M. Weiss and P.G. Kaufmann (eds.) *Handbook of Research Methods in Cardiovascular Behavioral Medicine*, Plenum Press, New York, **1989**.

4 T.G. Pickering and S.G. Blank: The measurement of blood pressure. In: N. Scheiderman, S.M. Weiss and P.G. Kaufmann (eds.) *Handbook of Research Methods in Cardiovascular Behavioral Medicine*, Plenum Press, New York, **1989**, 69–79.

5 E. O'Brien and K. O'Malley (eds.) *Blood Pressure Measurement*, Vol. 14: *Handbook of Hypertension* (Series Eds. W.H. Birkenhager and J.L. Reid), Elsevier Science Publishers, Amsterdam, **1991**.

6 K. Yamakoshi, H. Shimazu, M. Shibata and A. Kamiya: New oscillometric method for indirect measurement of systolic and mean arterial pressure in the human finger. Part 1 and Part 2, *Med. & Biol. Eng. & Comput.* **1982**, *20*, 307–313 and 314–318.

7 K. Yamakoshi, H. Shimazu, M. and T. Togawa: Indirect measurement of instantaneous arterial blood pressure in the rat, *Am. J. Physiol.* **1979**, *237*, H632–H637.

8 K. Yamakoshi, H. Shimazu, M. and T. Togawa: Indirect measurement of instantaneous arterial blood pressure in the human finger by the vascular unloading technique, *IEEE Trans. Biomed. Eng.* **1980**, *BME-27*, 150–155.

9 H. Ito, K. Yamakoshi and T. Togawa: Transthoracic admittance plethysmography for measuring cardiac output, *J. Appl. Physiol.* **1976**, *40*, 451–454.

10 R.S. Mackay, E. Marg and R. Oechsli: A quartz crystal tonometer, *IRE Trans. Bio-Med. Electron.* **1962**, *9*, 174.

11 G.L. Pressman and P.M. Newgard: A transducer for the continuous external measurement of arterial blood pressure, *IEEE Trans. Bio-Med. Electron.* **1963**, *10*, 73–81.

12 R.S. Mackay: The application of physical transducer to intracavity measurement with special application to tonometry, *Med. Electron. & Biol. Eng.* **1964**, *2*, 3–17.

13 J. Peñáz: Photoelectric measurement of blood pressure, volume and flow in the finger, *Digest 10th Int. Conf. Med. Biol. Eng.* **1973**, Dresden, Germany, 104.

14 K.H. Wesseling: Non-invasive continuous blood pressure wave form; measurement by the method of Penaz, *Scripta Medica* **1984**, *57* (Suppl. 6), 321–334.

15 K.H. Wesseling and J. Peñáz: Indirect blood pressure measurement in the finger; physiological background of a recent development, *Scripta Medica* **1986**, *59*, 203–206.

16 R.D. Boehmer: Continuous, real-time, noninvasive monitor of blood pressure; Penaz methodology applied to the finger, *J. Clin. Monit.* **1987**, *3*, 282–287.

17 H.W. Shirer: Blood pressure measuring methods, *IRE Trans. Bio-Med. Electron.* **1962**, *9*, 116–125.

18 P.D. Stein and E.F. Blick: Arterial tonometry for the atraumatic measurement of arterial blood pressure, *J. Appl. Physiol.* **1971**, *30*, 593–596.

19 S. Terry, J.S. Eckerle, D. Kornbluh, T. Low and C.M. Ablow: Silicon pressure transducer arrays for blood-pressure measurement, *Sens. Actuators* **1990**, *A21/A23*, 1070–1079.

20 K. Yamakoshi: Volume-compensation method for non-invasive measurement of instantaneous arterial blood pressure – Principle, methodology, and some applications, *Homeostasis* **1995**, *36* (2/3), 90–119.

21 H. Shimazu, H. Ito, H. Kobayashi and K. Yamakoshi: New idea to measure diastolic arterial pressure by the volume-oscillometric method in human fingers, *Med. Biol. Eng. & Comput.* **1986**, *24*, 549–554.

22 H. Shimazu, H. Kobayashi, H. Ito and K. Yamakoshi: Indirect measurement of the systolc, mean and diastolic arterial

pressure in the forearms and ankles of babies and children by the volume-oscillometric method, *J. Clin. Eng.* **1987**, *12*, 297–303.

23 K. Yamakoshi, P. Rolfe and C. Murphy: Current developments in non-invasive measurement of arterial blood pressure, *J. Biomed. Eng.* **1988**, *10*, 1–8.

24 Y. Imai, K. Abe, S. Sasaki, H. Sekino, K. Yamakoshi and K. Yoshinaga: A new automated volume-oscillometric device for monitoring ambulatory blood pressure in the human finger in man, *J. Hypertension* **1986**, *4*, S662–S664.

25 Y. Imai, M. Nihei, K. Abe, S. Sasaki, N. Minami, M. Minakata, S. Yumita, Y. Onoda, H. Sekino, K. Yamakoshi and K. Yoshinaga: A finger volume-oscillometric device for monitoring ambulatory blood pressure; Laboratory and clinical evaluations, *Clin. Exp. Hypertens. Theory Pract.* **1987**, *A9*, 2001–2025.

26 Y. Imai, K. Abe, Y. Miura, M. Nihei, S. Sasaki, N. Minami, M. Munakata, N. Taira, H. Sekino, K. Yamakoshi and K. Yoshinaga: Hypertensive episodes and circadian fluctuations of blood pressure in patients with phaeochromocytoma; Studies by long-term blood pressure monitorings based on a volume-oscillometric method, *J. Hypertension* **1988**, *6*, 9–15.

27 G.W. Mauck, C.R. Smith, L.A. Geddes and J.D. Bourland: The meaning of the point of maximum oscillations in cuff pressure in the indirect measurement of blood pressure. Part II, *J. Biomech. Eng.* **1980**, *102*, 28–33.

28 K. Yamakoshi: Non-invasive techniques for ambulatory blood pressure monitoring and simultaneous cardiovascular measurements, *J. Ambul. Monit.* **1991**, *4*, 123–143.

29 K. Yamakoshi, H. Shimazu, A. Kamiya, H. Ito and T. Togawa: Noninvasive automatic monitoring of instantaneous arterial blood pressure using the vascular unloading technique, *Med. & Biol. Eng. & Comput.* **1983**, *21*, 557–565.

30 B.P. Imholz, G.J. Langewouters, G.A. Van Montfrans, G. Parati, J. Van Goudoever, K.H. Wesseling, W. Wieling and G. Macia: Feasibility of ambulatory, continuous 24-hour finger arterial pressure recording, *Hypertens.* **1993**, *21*, 65–73.

31 S. Tanaka and K. Yamakoshi: Ambulatory instrument for monitoring indirect beat-to-beat blood pressure in superficial temporal artery using volume-compensation method, *Med. & Biol. Eng. & Comput.* **1996**, *34*, 441–447.

32 M. Nakagawara and K. Yamakoshi: A portable instrument for non-invasive monitoring of beat-by-beat cardiovascular hemodynamic parameters based on the volume-compensation and electrical admittance method, *Med. & Biol. Eng. & Comput.* **2000**, *38*, 17–25.

33 M. Nakagawara and K. Yamakoshi: Development of a finger cuff unit for a non-invasive blood-pressure monitoring system based on the volume-compensation method, *Jap. J. Med. Electr. & Biol. Eng.* **2000**, *38*, 283–290.

34 T. Togawa, H. Mizutani and T. Tamura: Physiological monitoring techniques for home health care, *Biomed. Sci. Instrum.* **1992**, *28*, 105–110.

35 T. Tamura, J. Zhou, H. Mizukami and T. Togawa: A system for monitoring temperature distribution in bed and its application to assess body movement, *Physiol. Meas.* **1993**, *14*, 33–34.

36 M. Ishijima: Cardiopulmonary monitoring by textile electrodes without subject-awareness of being monitored, *Med. & Biol. Eng. & Comput.* **1997**, *35*, 685–690.

37 K. Yamakoshi: Unconstrained physiological monitoring in daily living for health care, *Frontiers Med. Biol. Engng.* **2000**, *10*, 239–259.

38 T. Togawa, T. Tamura and P. Åke Öberg: *Biomedical Transducers and Instruments*, CRC Press, New York, **1997**.

39 J. Nyboer: *Electrical Impedance Plethysmography*, 2nd edn. C.C. Thomas, Springfield, Illinois, **1970**.

40 W.G. Kubicek, R.P. Paterson and D.A. Witsoe: Impedance cardiography as a noninvasive method of monitoring cardiac function and other parameters of the cardiovascular system, *Ann. N.Y. Acad. Sci.* **1970**, *170*, 724–732.

41 H. Ito, K. Yamakoshi and T. Togawa: Transthoracic admittance plethysmogra-

phy for measuring cardiac output, *J. Appl. Physiol.* **1976**, *40*, 451–454.

42 K. Yamakoshi, H. Shimazu, T. Togawa and H. Ito: Admittance plethysmography for accurate measurement of human limb blood flow, *Am. J. Physiol.* **1978**, *235*, H821–H829.

43 B. C. Penney, N. A. Patwardhan and H. B. Wheeler: Simplified electrode array for impedance cardiography, *Med. & Biol. Eng. & Comput.* **1985**, *23*, 1–7.

44 D. P. Bernstein: Continuous noninvasive real-time monitoring of stroke volume and cardiac output by thoracic electrical bioimpedance, *Crit. Care Med.* **1986**, *14*, 898–901.

45 M. Qu, Y. Zhang, J. G. Webster and W. J. Tompkins: Motion artifact from spot and band electrodes during impedance cardiography, *IEEE Trans. Biomed. Eng.* **1986**, *BME-33*, 1029–1035.

46 D. I. Boomsma, J. D. Vries and J. F. Orlebeke: Comparison of spot and band impedance cardiogram electrodes across different tasks, *Psychophysiology* **1989**, *26*, 695–699.

47 A. Sherwood, S. A. Royal, J. S. Hutcheson and J. R. Turner: Comparison of impedance cardiographic measurements using band and spot electrodes, *Psychophysiology* **1992**, *29*, 734–741.

48 R. W. Gotshall and W. R. Sexson: Comparison of band and spot electrodes for the measurement of stroke volume by the bioelectric impedance technique, *Crit. Care Med.* **1994**, *22*, 420–425.

49 B. J. M. Meer, H. H. Woltjer, A. M. Sousman, W. O. Schreuder, E. R. Bulder, M. A. J. M. Huybregts and M. J. M. Vries: Impedance cardiography – importance of the equation and the electrode configuration, *Intern. Care Med.* **1996**, *22*, 1120–1124.

50 H. H. Woltjer, B. W. G. Arntzen, H. J. Bogaard and P. M. J. M. Vries: Optimalisation of the spot electrode array in impedance cardiography, *Med. & Biol. Eng. & Comput.* **1996**, *34*, 84–87.

51 J. Nyboer: Electrical impedance plethysmography; A physical and physiological approach to peripheral vascular study, *Circulation* **1950**, *2*, 811–821.

52 L. A. Geddes and C. Sadler: The specific resistance of blood at body temperature, *Med. & Biol. Eng.* **1973**, *11*, 336–339.

53 W. G. Kubicek, F. J. Kottke, M. V. Ramos, R. P. Patterson, D. A. Witsoe, J. W. Labree, W. Remole, T. E. Layman, H. Schwening and J. T. Garamella, The Minesota impedance cardiograph – theory and applications, *Biomed. Eng.* **1974**, *9*, 410–416.

54 K. Yamakoshi, S. Tanaka and H. Shimazu: Electrical admittance cuff for non-invasive and simultaneous measurement of hematocrit, arterial pressure and elasticity using volume-oscillometric method, *Med. & Biol. Eng. & Comput.* **1994**, *32*, S99–S107.

55 Z. Lababidi, D. A. Ehmke, R. E. Durnin, P. E. Leaverton and R. M. Lauer: The first derivative thoracic impedance cardiogram, *Circulation* **1970**, *41*, 651–658.

56 K. Yamakoshi, T. Togawa and H. Ito: Evaluation of the theory of cardiac output computation from transthoracic impedance plethysmogram, *Med. & Biol. Eng.* **1977**, *15*, 479–488.

57 L. A. Geddes and L. E. Baker: *Principles of Applied Biomedical Instrumentation*, 3rd edn. John Wiley & Sons, New York, **1989**.

58 H. Maki, M. Nakagawara, K. Yamakoshi, K. Nakajima and T. Maekawa: Non-invasive and ambulatory monitoring system for cardiovascular hemodynamic variables; A special interest in comparison test for long-term continuous monitoring of cardiac output determined by invasive and present non-invasive method, *Jap. J. Med. Electr. & Biol. Eng.* **1995**, *33* (Suppl.), 123–124.

59 M. Nakagawara and K. Yamakoshi: Spot-electrode array for impedance cardiography determined from the measurement of current distribution on the thorax, *Jap. J. Med. Electr. & Biol. Eng.* **2000**, *38*, 238–245.

60 R. P. Patterson, W. G. Kubicek, D. A. Witsoe and A. H. L. From: Studies on the effect of controlled volume change on the thoracic electrical impedance, *Med. & Biol. Eng. & Comput.* **1978**, *16*, 531–536.

61 K. Sakamoto, K. Muto, H. Kanai and M. Iizuka: Problems of impedance cardiography, *Med. & Biol. Eng. & Comput.* **1979**, *17*, 697–709.

62 R.P. Patterson: Sources of the thoracic cardiogenic electrical impedance signal as determined by a model, *Med. & Biol. Eng. & Comput.* **1985**, *23*, 411–417.

63 E. Raaijmarkers, T. J.M. Faes, H.G. Goovaerts, J.H. Meijer, P.M. J.M. Vries and R.M. Heethaar: Thoracic geometry and its relation to electrical current distribution consequences for electrode placement in electrical impedance cardiography, *Med. & Biol. Eng. & Comput.* **1998**, *36*, 592–597.

64 M.A. Weber and J.I.M. Drayer (eds.) *Ambulatory Blood Pressure Monitoring*, Springer, New York, **1984**.

65 K. Yamakoshi, A. Kawarada, A. Kamiya, H. Shimazu and H. Ito: Long-term ambulatory monitoring of indirect arterial blood pressure using a volume-oscillometric method, *Med. & Biol. Eng. & Comput.* **1985**, *23*, 459–465.

66 A. Kawarada, H. Shimazu, H. Ito and K. Yamakoshi: Ambulatory monitoring of indirect beat-to-beat arterial pressure in human fingers by a volume-compensation method, *Med. & Biol. Eng. & Comput.* **1991**, *29*, 56–62.

67 H.M.W. Gonneke, J.C.D. Eco, H.A.M.K. Coert, J.P.V. Lorenz and C. Douglas: Ambulatory monitoring of the impedance cardiogram, *Psychophysiology* **1996**, *33*, 184–193.

68 A. Sherwood, J. McFetridge and S. Hutcheson: Ambulatory impedance cardiography; a feasibility study, *J. Appl. Physiol.* **1998**, *85*, 2365–2369.

69 K. Yamakoshi and M. Nakagawara: Voltage clamp method for the use of electrical admittance plethysmography in human body segments, *Med. & Biol. Eng. & Comput.* **1995**, *33*, 740–743.

70 K. Yamakoshi, M. Nakagawara and S. Tanaka: Recent advances in non-invasive and ambulatory monitoring of cardiovascular hemodynamic parameters, *Biocybern. & Biomed. Eng.* **1997**, *17*, 181–202.

71 S. Tanaka, M. Nakagawara and K. Yamakoshi: Assessment of diurnal change of blood pressure measured at superficial temporal artery using an ambulatory instrument for monitoring cardiovascular variables, *Proc. Biol. & Physiol. Eng.* **1999**, *14*, 447–450.

72 S.S. Ahmed, G.E. Levinson, C. J. Schwartz and P.O. Ettinger: Systolic time intervals as measures of the contractile state of the left ventricular myocardium in man, *Circulation* **1972**, *46*, 559–571.

73 J.T. Cacioppo, G.G. Berntson, P.F. Binkley, K.S. Quigley, B.N. Uchino and A. Fieldstone: Autonomic cardiac control. II. Noninvasive indices and basal response as revealed by autonomic blockades, *Psychophysiology* **1994**, *31*, 586–598.

74 R.M. Kelsey and W. Guethlein: An evaluation of the ensemble averaged impedance cardiogram, *Psychophysiology* **1990**, *27*, 24–33.

75 G. Bertinieri, M. Di Rienzo, A. Vavallazzi, A.U. Ferrari, A. Pedotti and G. Mancia: Evaluation of baroreceptor reflex by blood pressure monitoring in unanaesthetised cats, *Am. J. Physiol.* **1988**, *254*, H377–H383.

76 M. Pagani, G. Lombardi, S. Guzzetti, I. Rimoldi, R. Furtan, P. Pizzinelli, G. Sandrone, S. Dell Orto, E. Piccaluga, M. Turiel, G. Baselli, S. Cerutti and A. Malliani: Power spectral analysis of heart rate and arterial pressure variabilities as a marker of sympatho-vagal interaction in man and conscious dog, *Circ. Res.* **1986**, *59*, 178–193.

77 B. Pomeranz, R.J.B. Macaulay, M.A. Caudilli, I. Kutz, D. Adam, D. Gordon, K.M. Kilbom, A.C. Barger, D.C. Shannon, R.J. Cohen and H. Benson: Assessment of autonomic functions in human by heart rate spectral analysis, *Am. J. Physiol.* **1985**, *248*, H151–H153.

6
Sensors for Respiratory Monitoring

A. Johansson and B. Hök

6.1
Physiological and Clinical Relevance

In humans the diaphragm is responsible for most of the respiratory activity. During normal inspiration, both the abdomen and the rib cage move outward. The diaphragm also causes a rib cage expansion by elevating the rib margin. These actions cause an enlargement of the thoracic cage, and thus, a decrease in the alveolar pressure. The arisen pressure gradient causes gas to be transported into the lungs via the respiratory passageways. Expiration is mainly passive, but abdominal muscles might contract to facilitate expiration. In the alveoli, carbon dioxide and oxygen are exchanged between alveolar air and circulating blood, mainly by diffusion. The two gases are then transported to and from the various cells of the body via the circulatory system. Disturbances in this rather complex system often occur and supervision of respiration is of great importance in fields such as intensive care, postoperative care, anesthesia, and neonatal care. The monitoring of respiration is performed for two major reasons, namely to detect

- pathologic minute ventilation caused by injury or illness affecting respiratory rate and/or tidal volume, or
- hypoxemia caused by apnea or inadequate minute ventilation.

In clinical decision-making, respiratory rate and tidal volume are two of the most useful parameters to monitor. Respiratory rate is one of the vital signs, but is unfortunately often disregarded by the physician. Abnormalities are detected in two ways, either as an increased respiratory rate (tachypnea) or as a decreased respiratory rate (bradypnea). These states are seen in situations such as pneumonia, lower respiratory tract infection, pulmonary embolism, fever, anemia, congestive heart failure, metabolic abnormalities, shock, and head trauma. The physiologic response to injury or illness is often more accurately reflected in the minute ventilation than in the respiratory rate alone, and respiration monitoring systems should ideally follow both respiratory rate and tidal volume. Tidal volume is defined as the volume of air that passes into and out of the lungs with each respiration, normally about 500 mL. Furthermore, ventilation is defined as the quantity of air passing into the respiratory airways during a specific time interval, normally

Sensors in Medicine and Health Care
Edited by P. Å. Öberg, T. Togawa, and F. A. Spelman

ISBN: 3-527-29556-9

around 6 L per minute. Other severe respiratory disturbances are central or obstructive apneic events, for instance caused by opioid analgesics or neuromuscular blocking agents. Another high-risk patient group is premature infants where apneic events are sometimes seen as a consequence of the immaturely developed respiratory control system. In summary, the respiratory conditions causing hypoxemia can be divided into:

- central apnea, where all respiratory efforts cease,
- obstructive apnea, with respiratory efforts at a negligible tidal volume,
- low respiratory rate, and
- small tidal volume.

The monitoring of respiration is still often based on nursing observation. This will always be the base of respiratory monitoring, but as the human senses often are inaccurate, assistance from technical devices is needed. It is generally agreed that nursing observation is unreliable for respiratory rate measurement [1] and not sensitive enough for tidal volume estimation [2]. Of course, cost-effectiveness is another important factor to consider [3]. Throughout the years, several respiration sensors and systems have been developed for the purpose, but no satisfactory solution is yet seen solving all the issues of sensitivity, specificity, patient safety, and user friendliness. We have chosen to divide the presentation of the respiration sensors/systems into respiratory airflow detectors, measuring parameters directly correlated to the respiratory gas flow, indirect techniques, and blood gas monitors.

6.2 Sensors Based on Respiratory Airflow Detection

Any sensor intended for monitoring respiration must fulfil a number of general and specific criteria. Important general criteria are sensitivity, specificity (selectivity), reproducibility, and patient safety. More specific criteria are small size, minimal interference and obstruction, rapid response, and low cost. The demand on low cost could eventually be replaced by reusability, if cleaning/disinfection of the device could be performed without risk of transmitting infections.

Airflow due to respiration may be detected in a few locations, the preferred one being the nose/mouth. As shown in Fig. 6.1, the flow patterns of nose and mouth breathing differ significantly at inspiration and expiration. The flow distribution of inspired air has a multidirectional funnel shape, whereas expiration, especially the nasal one, has the form of a relatively well-defined directional jet, with a typical flow velocity of 2 m/s. Oral expiratory airflow velocities are typically an order of magnitude smaller. A deep breath corresponds to a volume of several liters, whereas a shallow one may be ten times smaller. Devices for detecting respiratory airflow velocity must therefore have an operating dynamic range (ratio between highest flow velocity and the smallest resolvable one) of three or more decades.

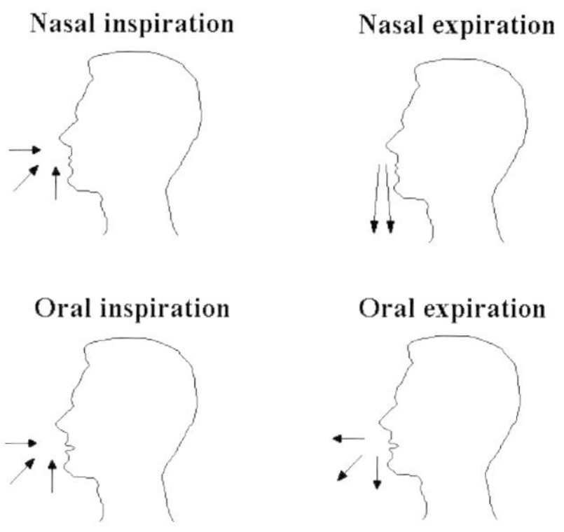

Fig. 6.1 Typical respiratory flow patterns of nose and mouth breathing during inspiration and expiration. Inspired air has a funnel shaped flow velocity distribution, whereas expiration, especially for the nasal case, has the form of a directional jet. Nasal expired flow velocities are typically an order of magnitude higher than oral flow velocities.

For local detection of respiratory airflow, there are a number of possible detection modalities.

- Gas flow velocity is normally measured by indirect techniques, for instance by the associated pressure gradients using *differential pressure measurements*, or *pressure fluctuations* due to the mixing effects, which may be detected by various means [4].
- A heated obstacle will be cooled by the flowing gas, and thus the temperature of the obstacle will be flow dependent *(hot wire anemometry)* [5, 6].
- Expired air is usually warmer than ambient air, thus it may be detected by *passive temperature sensing* [5, 6].
- Expired air is usually more humid than ambient air, therefore it may be detected by a *humidity sensor* [6].
- Expired air contains 'waste products', such as *carbon dioxide* (CO_2), which may be detected.

In Tab. 6.1, some basic characteristics of the methods described above are summarized. Approximate signal and background ranges are given in typical clinical conditions, and their ratio is calculated, with the maximum background range taken as normal. The ideal measurand should exhibit both high sensitivity and specificity. The specificity may in practice be the most critical parameter, since disturbance from various sources is a common problem in clinical use.

Tab. 6.1 shows that the signal/background ratios may vary considerably from one measurand to another. Devices based on flow velocity measurements suffer from the fact that other flow sources than respiration may occur in clinical situations. Temperature variations due to other origins are also likely, and likewise are humidity variations, although to a lesser degree. On the other hand, the high concentration of carbon dioxide in expiratory air compared with normal room air provides the basis for extraordinary high sensitivity and specificity. It should be noted that the numbers quoted in Tab. 6.1 are highly approximate, and should only be considered as indicative of the actual magnitude.

Tab. 6.1 Some basic properties of measurands for respiratory flow sensing

Basic measurand	*Signal range*	*Background range*	*Signal/ background*	*Sensitivity × specificity*
Airflow velocity	2 m/s	1 m/s	2	Low
Differential pressure	3 Pa	1 Pa	3	Low
Pressure fluctuations	3 Pa	1 Pa	3	Low
Hot wire anemometry	5 °C	2 °C	2.5	Low
Temperature	5 °C	2 °C	2.5	Low
Humidity	30%	5%	6	Moderate
Carbon dioxide	4%	0.05%	80	High

A number of actual devices for respiratory monitoring have been reported in the literature, and some of them have also reached a commercial phase.

6.2.1
Pressure and Acoustic Sensing Devices

The use of differential pressure sensing for flow measurements is widespread for gas flow measurements in general. It makes use of the simple relation based on Bernoulli's classical equation

$$\Delta P = \frac{\rho v^2}{2} \tag{6.1}$$

where ΔP represents the differential pressure between a stagnant and a neutral point with respect to the flow, ρ is the density, and v the flow velocity. The square law dependence of Equation (6.1) has the consequence that pressure-sensing devices will require much higher dynamic range than devices having linear response. On the other hand, small signal disturbances are suppressed, resulting in a stable and noise-free baseline.

Instead of true differential pressure measurements, it is also possible to measure pressure fluctuations due to mixing of expiratory and ambient air. For this purpose, true d.c. measurements are not necessary, and it is possible to use standard electret or silicon microphones. The frequency spectrum of these fluctuations is mostly contained within a bandwidth of a few hundred Hz, and the signal magnitude is well within reach for standard microphones. The square law dependence of Equation (6.1) will be valid also for this type of sensor.

Pressure and acoustic sensing devices have a number of attractive characteristics, which may be put to use in respiratory monitoring devices.

- Suppression of sub-threshold disturbances due to the square law response.
- Small size: silicon pressure-sensing devices of submillimeter size are commercially available, either in chip form, or as packaged units.
- Fast response.

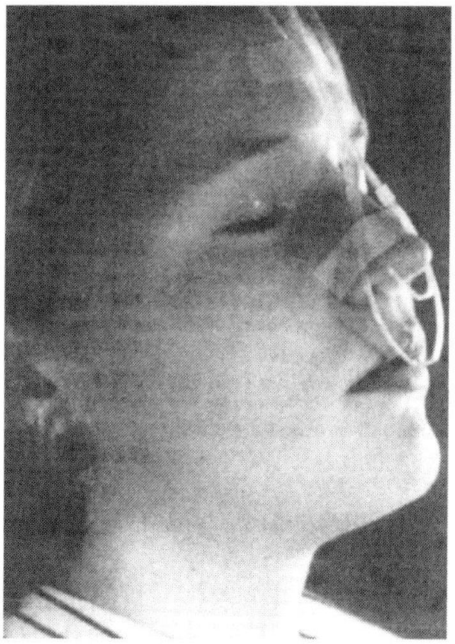

Fig. 6.2 Acoustic flow sensing device, using thin polymer tubing to pick up the acoustic signal at the subject's nostrils, and transmit it to remotely located microphone elements.

Hök et al. [7] designed an acoustic respiratory monitor with a signal pickup device (Fig. 6.2) using thin polymer tubing to transmit the acoustic signal to remotely located microphone elements. The device was subjected to several clinical testing procedures [8–10]. Using observation by experienced nurses as the 'gold standard', it was found that the majority of apnea alarms were true, and that false non-alarms were extremely rare. Involuntary mis-positioning of the pickup tubing caused the majority of false alarms. It should also be noted that these studies pointed at some fundamental difficulties in the clinical assessment of various techniques for respiratory monitoring. The use of expert observation as the 'gold standard' is controversial due to its subjectivity and the problem of properly defining what constitutes a true 'breath'.

6.2.2
Thermal Flow Sensors

The hot wire anemometry principle is widely used in measuring devices for airflow velocity, due to its simplicity, ruggedness, and wide dynamic range. For respiratory airflow detection, however, passive temperature-sensing devices are more common. A number of temperature-sensing devices may be used, including inert-metal resistive devices, like Pt-100 elements, thermistors with negative or positive coefficient, thermocouples, pyroelectric material, and so on [1, 5, 6]. Commercial products are available from several manufacturers, and include disposable sensor units for oral and nasal application. Other possible positions are the posterior pharynx, where both nasal and oral airflow pass [11], or integration in a facemask. Pyroelectric sensors with such design, using polarized polyvinylidene fluoride as sensor material [12, 13] have been reported.

Many thermal respiratory flow sensors suffer from slow response, and from the fact that the response time depends on the flow velocity. At appreciable flow velocities, the response time may be completely adequate, whereas it may be insufficient at low velocities. The solution to this problem is obviously to miniaturize the sensor, thereby reducing its thermal mass, and very thin pyroelectric foils represent one interesting approach. Another is, of course, to use silicon technology. Free-standing thin films may be fabricated by modified standard silicon processing, using bulk or surface micromachining.

6.2.3 Humidity Sensors

Expired air is almost completely saturated with water vapor, due to the humid mucus membranes of the airways. A classical and simple method for detecting expired air, known for centuries, is to place a glass mirror in front of the subject in order to cause condensation of the water vapor on the glass surface. Vegfors et al. [14] ingeniously designed a fiberoptic probe for the same purpose. Successive inspirations/expirations invariably cause droplets to condense and re-evaporate. The signal is derived from reflected light at the end surface of an optic fiber that will vary with the formation and evaporation of water droplets. A probe for nasal insertion was developed (Fig. 6.3) and adequate performance in laboratory conditions with voluntary test subjects has been reported [15]. An instrument based on this design is commercially available under the trade name of Optovent.

A capacitive humidity sensor with fast response has also been demonstrated [16]. The sensor was attached to the cheek and positioned a few millimeters in front of one nasal orifice. Using a digital filter, the baseline variation caused by the humidity in the environmental air could be minimized.

6.2.4 Carbon Dioxide Sensing, Capnometry

The alveolar volume concentration of carbon dioxide value is about 4%, which may be measured as the end tidal concentration of each breath. The early phase of expiration will involve mixing of air from the alveolar and dead space compartments. The large difference in CO_2 concentration between expired and ambient air results in superior sensitivity and specificity as earlier mentioned. In fact, due to its close relation to the basic respiration process itself, measurement of CO_2 concentration represents the ideal variable in respiratory monitoring.

Methods for measuring CO_2 concentration include mass and infrared spectroscopy [17, 18]. Most capnographs are based on infrared absorption at the 4.3 μm band. Usually black-body radiators are used as emitters, and pyroelectric devices, often in a differential arrangement as shown in Fig. 6.4, as detectors. Interference filters or diffractive gratings are used for wavelength dispersion. The optical path of the measuring cell is commonly several centimeters in order to obtain adequate resolution.

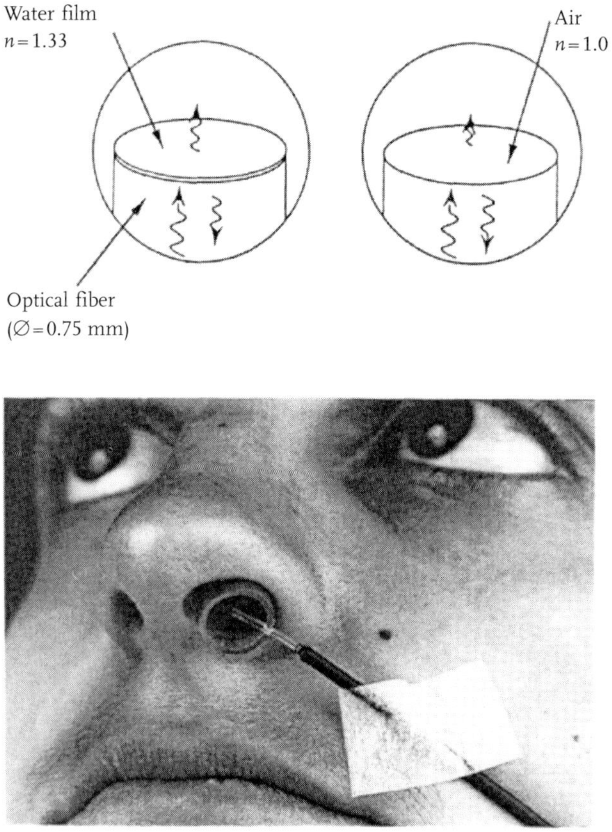

Fig. 6.3 A fiberoptic respiratory sensor based on the detection of droplet condensation and evaporation (courtesy of Optovent AB, Sweden).

The relatively large size of the IR-based CO_2 sensors makes it necessary to use a sampling line and a pump arrangement to transport the gas sample to the measuring cell. This is obviously a complication, both in terms of performance and from a practical point of view. The sampling line will introduce signal delay, and

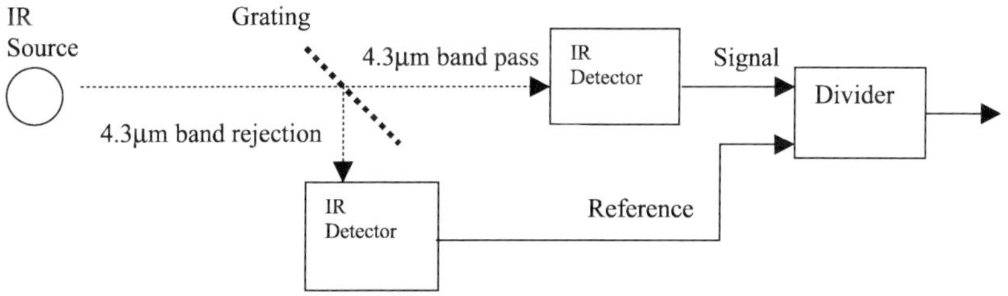

Fig. 6.4 Principle of CO_2 sensor using infrared absorption.

will be detrimental for the response time of the system. Nevertheless, acceptable values of a few hundred milliseconds have been achieved. The practical aspect is more serious. Continuous suction of air from the nasal area will inevitably cause mucus and liquid to be collected, blocking the sampling line. Nasal cannulae are commonly used in the patient end of the sampling line [17–20]. These are fully adequate in laboratory settings with careful control of subject behavior. In many clinical situations, however, the results are less convincing [19, 20]. For example, changes between mouth and nose breathing will cause disturbances.

Recently, Hök et al. [21–23] have suggested that CO_2 in air may be measured by its effect on the mean molecular weight, and on the velocity of sound. According to elementary thermodynamics, the sound velocity c is given by the equation

$$c = \sqrt{\frac{RT\gamma}{M}} \tag{6.2}$$

where R is the general gas constant=8.3 J/mol, T the absolute temperature, γ the ratio of specific heats at constant pressure and volume, respectively, and M the mean molecular weight.

Increasing concentration of CO_2 in air thus results in a higher value of M, and hence in a decreasing sound velocity. The basic principle being applicable to any molecular species differing from the mean value of air needs to be combined with adequate separation techniques [23] in order to be useful in medical applications.

Measurement of sound velocity may be performed in extremely small volumes of air. Recently, it has been shown that high resolution and fast response may be favorably combined [24]. The suggested sensor principle may be found useful in designing miniaturized CO_2 sensors with possible applications in respiratory monitoring.

6.3
Indirect Sensors of Respiration

As equipment around the airways is impractical and sometimes disturbing to the patient, it is advantageous to leave the face and the airways free. Devices requiring a facemask or a mouthpiece have also been shown to affect the breathing pattern by lowering the respiratory rate and increasing the tidal volume [25]. Furthermore, the flow-based sensors need more patient co-operation and may limit mobility. Numerous approaches have therefore been tried for monitoring respiration indirectly. In many instances, the clinical acceptance of the indirect sensors is a result of the practical advantages and not primarily the accuracy of the respiration record. The devices reviewed in this section all share the property of not measuring actual respiratory flow, but parameters involved in the respiratory drive, that is respiratory muscle activity, chest/abdominal motion, and intrathoracic pressure variations and effects thereof (Tab. 6.2).

Tab. 6.2 Devices for indirect sensing of respiration and underlying physiological mechanisms

Device type	*Respiratory muscle activity*	*Chest/ abdominal motion*	*Intrathoracic pressure variations*
Torso devices			
Strain gauge		×	
Respiratory inductance plethysmograph (RIP)		×	
Magnetometer		×	
Transthoracic impedance plethysmograph (TTI)		×	×
Photoplethysmographic (PPG) sensor			×
Mattress system		×	
Non-contact device		×	
Invasive sensor			×
Electrocardiographic (ECG) sensor			×
Electromyographic (EMG) sensor	×		
Pressure sensor		×	×

6.3.1 Torso Devices

Many systems based on the measurement of thoracic and/or abdominal motion during respiration have been developed. Of these, the transthoracic impedance plethysmograph (TTI), the respiratory inductance plethysmograph (RIP) [26], the magnetometer [27], and to some extent, the strain gauge are the most widely used. Respiration is caused by a rather complex interaction between the diaphragm and the respiratory muscles of the thorax and abdomen. However, Konno and Mead have demonstrated that the behavior of the respiratory system can be approximated to two degrees of freedom, with the volume change of the rib cage accounting for one and that of the abdomen for the other [28]. Thoracic and abdominal motion are often measured simultaneously.

- According to the Konno-Mead relation, tidal volumes can be monitored qualitatively or quantitatively after appropriate calibration.
- The co-ordination between the two compartments can be followed. The motion of the two compartments is usually phasic in non-pathological situations.
- The paradoxic motion of the rib cage and abdomen associated with obstructive apnea can be detected. This is more difficult for other indirect approaches.

6.3.1.1 Strain Gauges

A basic approach is to measure chest/abdominal circumference. Assuming a static abdomen, the chest circumference is nonlinearly related to the lung volume,

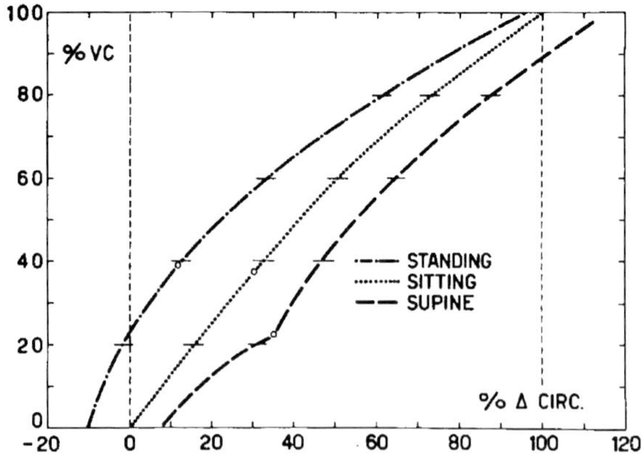

Fig. 6.5 Relation between lung volume (%VC) and chest circumference (%ΔCIRC) in three body positions. Lung volumes are expressed as percent of the vital capacity. From Agostoni et al. [31] with permission.

as the lung volume is proportional to the chest cross-sectional area. This relation further depends on the body position (Fig. 6.5). One of the earliest circumference monitors is the mercury-in-silastic strain gauge. The technique is based on measuring the resistance of a mercury column embedded in a stretchable tube placed around the thorax or the abdomen. When the gauge is stretched, the resistance of the mercury column increases as it becomes thinner and longer. A similar device uses stretch-sensitive (capacitive) coils [29]. Gauges are generally used in a qualitative fashion or for monitoring respiratory rate, but for the latter system, tidal volumes can be monitored within +15.2% to –16.9% of spirometrically derived values [30].

For use in electromagnetically noisy situations, a system based on measuring the loss of laser light intensity transmitted through a fiber loop has been presented [32]. When the chest/abdominal belt is stretched, the fiber loop becomes more bent and the transmitted intensity decreases (Fig. 6.6). If assuming the fiber to be bent in a circular loop of radius d, the intensity loss per unit curved length, α_c, is given by

$$\alpha_c = \frac{K_1}{\sqrt{d}} e^{-K_2 d} \qquad (6.3)$$

where K_1 and K_2 are optical fiber characteristics.

The main drawback of the strain gauges is that they measure circumference, and thus are nonlinear with respect to tidal volume. This fact and the somewhat difficult procedure of gauge positioning have limited their clinical acceptance.

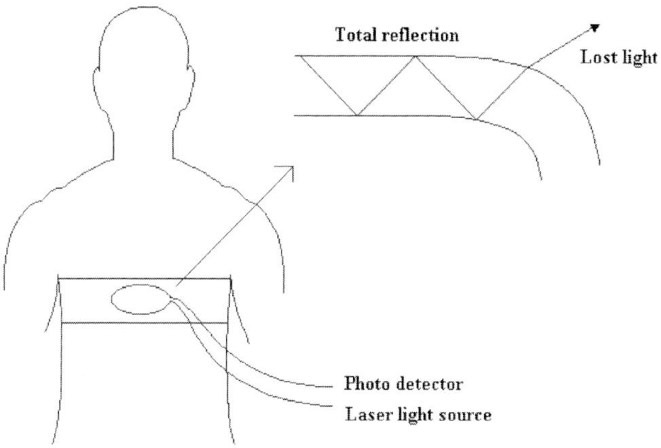

Fig. 6.6 Circumference monitor based on the measurement of laser light intensity transmitted through a fiber loop. Adapted from Davis et al. [32]. The detected intensity decreases during inspiration as the fiber loop is stretched, increased the amount of light lost.

6.3.1.2 Respiratory Inductance Plethysmography

In RIP [26], two elastic bands are positioned around the thorax and the abdomen, respectively. Each band includes an insulated coil positioned in a zigzag fashion (Fig. 6.7). The encircled cross-sectional areas vary with respiration, affecting the self-inductances of the coils and therefore the frequencies of two connected oscillators. The system output ΔV is signed according to the Konno-Mead relation

$$\Delta V = K_1 \Delta_{\mathrm{RC}} + K_2 \Delta_{\mathrm{ABD}} \tag{6.4}$$

where Δ_{RC} and Δ_{ABD} are the measured inductances of the rib cage and abdominal coils respectively, and K_1 and K_2 are coefficients relating respective compartment contribution to the tidal volume.

The constants in the relation are often determined by calibration to a spirometer or a pneumotachograph or by having the subject perform isovolumetric maneuvers, that is shifting air between the thoracic and abdominal compartment with a closed glottis. The main advantage of RIP is the characteristic of measuring cross-sectional area and not distance. This makes the output more linearly related to lung volume and less sensitive to distortion of the two compartments during motion or posture change. RIP has been found superior when comparing to other methods [33–35], and in normal subjects, the measured value remains within $\pm 10\%$ after calibration for 93% of the breaths [36]. RIP therefore is a reliable method for monitoring tidal volume in the spontaneously breathing patient. The technique is becoming more and more widely used in research and in clinics, but unfortunately, RIP is still relatively expensive, difficult to handle, and covers a large body area.

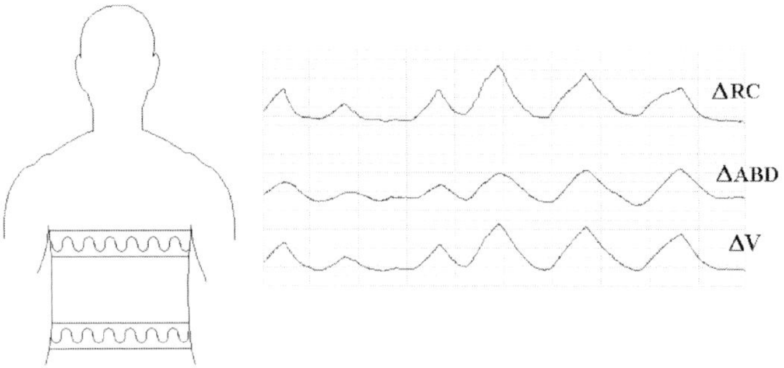

Fig. 6.7 In respiratory inductance plethysmography, variations in thoracic and abdominal cross-sectional areas induce inductance changes in the coils encircling the two compartments. Δ_{RC} and Δ_{ABD} are the measured inductances of the RIP rib cage and abdominal coils, and ΔV is formed as a calibrated linear combination of the two.

6.3.1.3 **Magnetometry**

Magnetometry was first presented by Mead et al. [27]. An alternating current is sent through a coil on one side of the chest. The generated magnetic field induces a current in another coil on the opposite side of the chest (Fig. 6.8). The magnetic field strength B of the exciter coil decreases by the cube of the distance d according to

$$B = \frac{K_1 I}{d^3} \tag{6.5}$$

where K_1 is the characteristics and I the current of the exciter coil.

In this way, respiratory movements can be sensed accurately by measuring the current induced in the receiver coil. As the distance variation is small in relation to the separation between the coils, a linear relation is often assumed. Under this assumption, tidal volumes can be sensed according to the Konno-Mead relation by

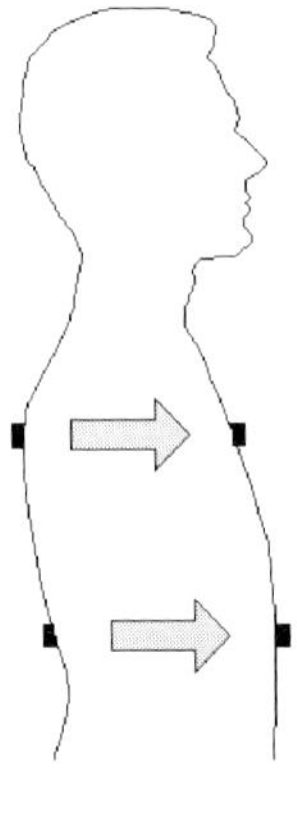

Fig. 6.8 In magnetometry, thoracic and abdominal diameters are measured by the induced current of two receiver coils sensing the magnetic fields induced by two coils on the back.

using another pair of coils positioned across the abdomen [37]. The magnetometer measures distance between the coils and not cross-sectional area. This makes it sensitive upon distortions of the body with respect to motion, posture variation and breathing pattern [37, 38]. For these reasons, magnetometers are rarely used today in clinical situations. Improvements can be made if adding coils sensing lateral movements.

6.3.1.4 Transthoracic Impedance Plethysmography

Atzler first suggested transthoracic impedance plethysmography/pneumography (TTI) [39]. TTI is the most widespread methodology in clinics for monitoring respiration. The technique employs standard skin surface potential electrodes. Part of its success is the simplicity of the transducer and the possibility to combine the measurement with that of cardiac rhythm by ECG, which most patients is monitored for. TTI is based on measuring the impedance variation (mainly resistive) across the thorax accompanying respiration. The impedance variation (ΔZ) per liter of inspired air (ΔV) is highly individual but normally in the area of 2–5 Ω, with maximum impedance at end-inspiration. An experimental relationship between body weight W and the impedance variation per liter of inspired air has also been shown [40]

$$\frac{\Delta Z}{\Delta V} = 453.23\ W^{-1.084}\ . \tag{6.6}$$

The origin of the impedance variation is still not fully understood but is believed to be due to the relation between air and blood in the thorax. Variations in the thoracic geometry and motion at the electrode–tissue interface are other explanations [41, 42, 80]. During measurement, a low-intensity alternating current (usually 20–100 kHz, 25–500 μA) is passed between two electrodes attached to the chest wall (Fig. 6.9). The potential between the two electrodes or, in some cases, between two separate electrodes is then measured. Both respiratory and cardiac synchronous components are seen in the detected potential. The relation between the two depends on electrode positioning. For maximum respiratory impedance changes, the electrodes should be positioned on a mid-axillary line at the height of the 6th rib [41].

Impedance systems are frequently used in respiratory rate and apnea monitors, often in combination with ECG. In practice, tidal volumes are not monitored by TTI and the possibility is debated. Some authors find it possible [35, 37, 40, 43, 44] and others regard it as more difficult [45, 46]. One conclusion is that TTI provides acceptable estimates of tidal volume during quiet breathing when averaging over a large number of breaths, but is not precise enough for individual breaths [37]. Unfortunately, TTI is sensitive to artifacts originating in the skin–electrode interface, cardiac pulsation (cardiogenic artifact), patient motion, and changes in posture. Improvement of the electrodes and optimization of the electrode configuration/position may reduce the susceptibility of error [47]. TTI is the only torso

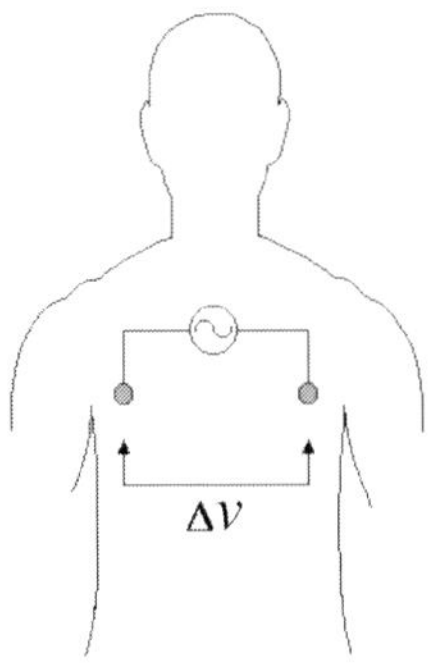

Fig. 6.9 In transthoracic impedance plethysmography, the potential ΔV between two skin surface potential electrodes is measured when passing a low-intensity alternating current between the two electrodes attached to the chest wall. In this manner, the impedance variations of the thorax caused by respiration can be followed.

device that cannot utilize the Konno-Mead principle. It therefore has difficulties in identifying paradoxic breathing.

6.3.1.5 **Photoplethysmographic Sensors**

The photoplethysmographic (PPG) technique is used in pulse oximetry (section on blood gas monitors) but has also been used for monitoring respiratory rate [48, 49]. These systems utilize a reflectance mode PPG sensor, emitting infrared light into the skin of the patient (Fig. 6.10). Infrared light is used as the absorption in blood and tissue for this wavelength region is low, causing a large penetration depth of the light. The amount of light reaching the photodetector depends on the skin blood volume and flow, as these strongly influence the absorption (μ_a) and scattering (μ_s) of the skin dermis. In addition to the cardiac synchronous pulsation, the detector signal includes respiratory synchronous variations and other components of lower frequencies. The respiratory variations are caused by the central interaction of the intrathoracic pressure and the circulation, and are transmitted to the skin, mainly via the venous system [50]. The PPG respiratory rate monitors extract the respiratory variations by digital filtering. The reflection mode is used as this makes it possible to attach the sensor anywhere on the skin. It has also been found that the PPG respiratory variations are best developed at locations demanding reflection measurements, such as the forehead or the forearm.

The obvious advantages of the PPG technique are the sensor simplicity and the possibility of simultaneous measurement of respiratory rate, heart rate, and arterial oxygen saturation. This combination is probably more useful in anesthesia/postoperative care than the usual combination of ECG and TTI. Negative drawbacks of the technique are the loss of signal quality during motion, a well-known problem in pulse oximetry, and disturbances in the complex respiratory signal transfer process. Obstructive apneic events appears somewhat different in the PPG than central events [51], but still, PPG shares the difficulty in classifying apnea type with most other indirect respiration monitors.

In healthy subjects, it has been shown that tidal volume information can be extracted using the PPG technique [52]. However, there is a risk that low-frequency

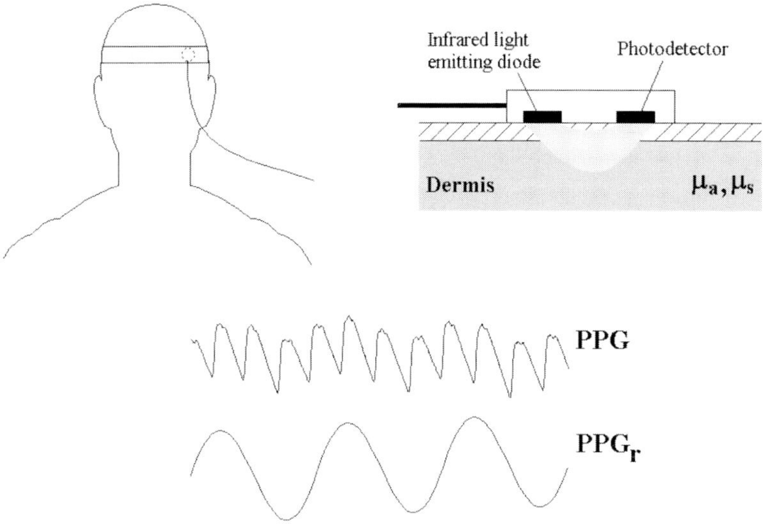

Fig. 6.10 Photoplethysmographic monitoring of respiratory rate. The PPG signal is the amount of light, back-scattered from the skin, as measured by a photodetector. The respiratory intrathoracic pressure variations are transmitted via the circulation to the skin where they effect the absorption (μ_a) and scattering (μ_s) of the dermis, and thus, the amount of incident light reaching the photodetector. Filtering techniques are used to extract the respiratory information (PPG_r) from the detector signal (PPG).

blood pressure waves may interfere during long-term registrations. Early clinical try-outs of the technique have given good results [53, 54], but monitoring respiratory rate/tidal volume using PPG is still at an early stage. It may become an important future complement to pulse oximetry.

6.3.2 Mattress Systems and Non-contact Devices

In some situations, such as measurements in the frail elderly, impaired, infants, or in the home, sensor positioning on the body is inappropriate as this limits mobility and/or may be difficult for the patient to handle. Several mattress systems have been developed for these situations. One approach is to detect respiratory motion by measuring the changes in air pressure inside an inflatable mattress [55–57]. The system in Fig. 6.11 includes a pressure transducer to measure the air pressure inside the mattress and a mechanical filter to protect the transducer from larger pressure changes caused by gross motion. The mattress system has been evaluated in healthy subjects with good result [58]. The mattress concept has also been tried out using modern pressure-sensing materials, converting weak mechanical signals to electrical [59]. One such material is the polymer polyvinylidene fluoride (PVDF).

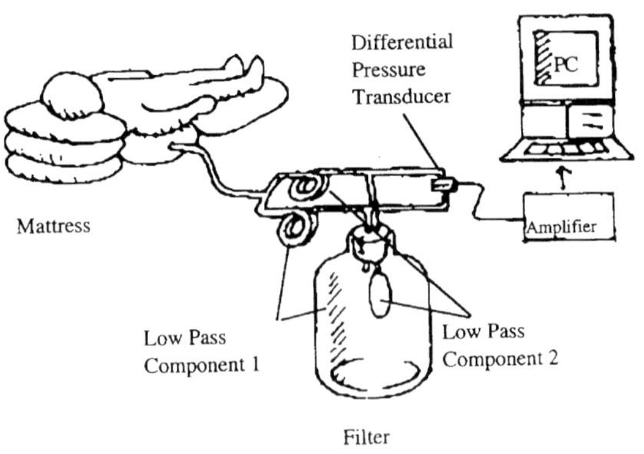

Fig. 6.11 Detection of respiratory motion by measuring the changes in air pressure inside an inflatable mattress (from Hernandez et al. [57] with permission from IOP Publishing). A mechanical filter is included to protect the transducer from large pressure changes caused by gross motion.

Other systems where patient mobility is maintained are optoelectric plethysmography, in which reflective optical markers positioned on the thorax and abdomen model respiratory motion [60], and capacitance respirometry, in which surface area changes of the abdomen and rib cage are detected [61]. The latter method utilizes the subject as one electrode of a capacitor and a second electrode at a fixed distance from the subject.

One step further is to measure respiration without any contact with the patient. This is advantageous for instance in the treatment of burn injuries. A non-contact system can be achieved by detecting the changes in microwave intensity and/or phase after scattering by the moving respiratory compartments [62, 63]. Characteristic of the non-contact devices is their sensitivity to motion other than respiration.

6.3.3
Invasive Sensors

Pleural pressure can be measured directly by using a small catheter placed into the pleural space, or estimated indirectly from the esophageal pressure as measured by an esophageal balloon. These techniques are highly accurate for monitoring respiration. Unfortunately, they are cumbersome for the patient and can only be used in laboratory settings. However, in clinical situations where a central venous line is demanded, there is a possibility of accessing the pleural pressure indirectly via the central venous pressure [64, 65]. These monitors include a filter for removing the cardiac pulsation of the central venous pressure. One drawback is the somewhat uncertain relationship between the central venous pressure and the pleural pressure [66].

6.3.4
Electrocardiographic Sensors

Besides TTI, skin surface potential electrodes can be used to extract respiratory parameters by deriving the respiratory sinus arrhythmia (RSA) of the ECG [67]. For normal subjects, the heart rate increases during inspiration and decreases dur-

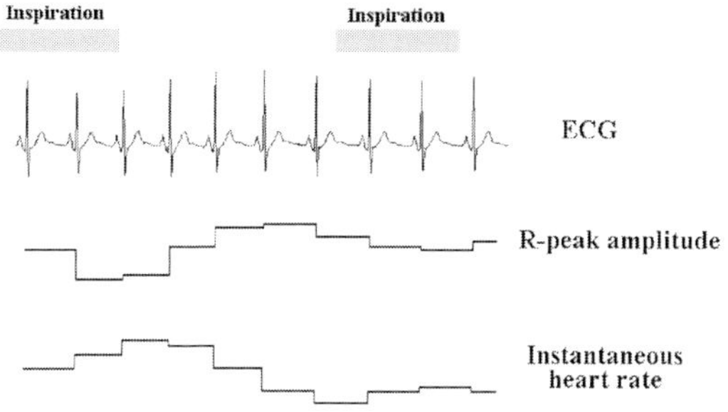

Fig. 6.12 Electrocardiographic (ECG) signal from a healthy subject. Both the variation in instantaneous heart rate (respiratory sinus arrhythmia, RSA) and the R-peak amplitude modulation can be extracted for monitoring respiration rate.

ing expiration. The underlying mechanism of the RSA is still not fully known but is primarily a vagal effect upon the sinus node of the heart. RSA is unfortunately not present in all subjects, and problems arise during cardiac arrhythmia. Another difficulty is when the respiratory rate is higher than half the heart rate, causing the 'sampling rate' to be inadequate (the Nyquist theorem). In this situation, both the R-peak and the T-wave of the ECG can be used to increase the frequency range [68]. In addition to the RSA, there exists a respiratory beat-to-beat variation of the ECG R-peak amplitude. This is to some part contributed to the thoracic impedance variation, but mainly to a variation in the heart's electrical axis during respiration (typically 3–6°) [68, 69]. Both the RSA and the amplitude modulation of the R-peak can be detected by appropriate placement of ECG electrodes. An advantage of these systems is the possibility of simultaneous respiration and cardiac monitoring. Whether tidal volume can be derived remains unclear, but for both the RSA and the R-peak amplitude modulation, a (nonlinear) tidal volume relation has been found [70]. Typical signals are shown in Fig. 6.12.

6.3.5 Electromyographic Sensors

Another attractive approach is to measure intercostal electromyography (EMG), in which the electrical activity of the diaphragm is sensed. The best signal quality is achieved when measuring the EMG in the esophagus, but for monitoring purposes standard skin potential electrodes are used. The EMG readings unfortunately have a low signal level and are easily disturbed by electrode motion or other electrical activity of the human body, primarily the ECG. Mathematical algorithms to subtract the unwanted components have been tried [71]. An advantage of the EMG method is the possibility of early detection of respiratory muscle fatigue [72], but the clinical application of the technique is still limited.

6.3.6 Pressure Sensors

Pressure sensors applied to the skin are used in several ways to extract respiratory information. Similar to the strain gauges, piezo-electric elements can be included

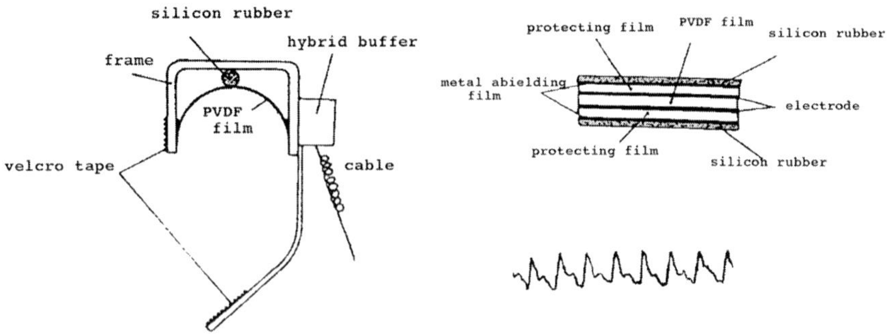

Fig. 6.13 A finger polyvinylidene fluoride (PVDF) sensor for picking up arterial pulsation, synchronous with the cardiac pulse and respiration (reprinted from Chen et al. [73], *Sensors and Actuators* 1990, *A21–A23*, 879–882, with permission).

in chest belts, sensing thoracic expansion and contraction. In other applications, one or more pressure capsules are attached to the chest for the purpose of detecting respiratory motion. A somewhat different approach is to pick up the peripheral pulse wave by using a pressure-sensitive material, such as polyvinylidene fluoride (PVDF) [73]. When picked up from a finger, the pressure signal includes variations synchronous with the cardiac pulse, but also, the respiratory rhythm. The latter is an effect of the interaction between the heart and the respiratory system, causing respiratory fluctuations in the arterial blood pressure. A finger PVDF sensor is presented in Fig. 6.13.

6.4 Blood Gas Monitors

After diffusion into the blood in the lungs, oxygen travels in blood bound to hemoglobin, or to a few percent, dissolved in the plasma. To assure adequate gas exchange in the lungs, blood gas transport, and tissue oxygenation, continuous arterial blood gas determination is a valuable guide. The partial pressure of oxygen in blood (pO_2) and the percentage of hemoglobin molecules saturated with oxygen (SO_2) are related by the curvilinear dissociation curve (Fig. 6.14). The normal levels of the two parameters are marked in the figure. The shape of the curve depends both upon temperature and pH. Traditionally, intermittent blood sampling served the purpose of assuring satisfactory blood gas levels. A breakthrough in the field was the development of transcutaneous measurement techniques, associated with less risk for the patient and continuous readings. In recent years, pulse oximetry has become an indisputable technique for monitoring hypoxemia and has to some extent replaced the transcutaneous electrodes. A common problem in pulse oximetry is the high rate of false desaturation alarms caused by patient motion. Recently, motion-resistant pulse oximeters have been presented, to some degree solving the problem. Since the introduction of pulse oximetry, the occurrence of hypoxemia has been significantly reduced.

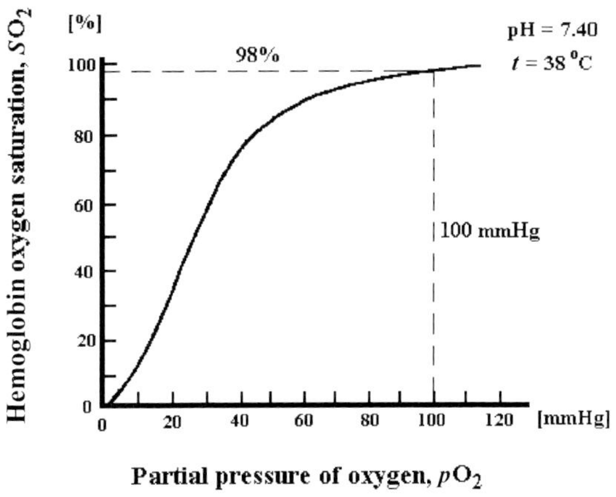

Fig. 6.14 The dissociation curve of hemoglobin. The normal levels of partial pressure of oxygen and hemoglobin oxygen saturation are marked.

6.4.1 Transcutaneous pO_2/pCO_2 Electrodes

Transcutaneous oxygen and/or carbon dioxide electrodes often accompany the monitors of respiratory rate/tidal volume. These devices are based on detection of the gases dissolved in plasma that cross the stratum corneum. The skin in adults is relatively thick and the tissue oxygenation of the outer skin layers is very low. These facts are overcome by heating of the skin to 43–44 °C. This gives rise to a better gas transport through the skin and a vasodilatation, increasing local blood flow. The oxygen electrodes are often of the classic Clark type with a platinum cathode, an Ag: AgCl anode and potassium chloride with a buffer solution as electrolyte. The transcutaneous pCO_2 electrode was developed soon after the pO_2 electrode. This electrode consists of a pH-sensitive glass electrode with an Ag: AgCl reference. The success of the pCO_2 electrode has been much less, primarily due to a slower response time. The transcutaneous electrodes have found their most widespread use in neonatal care. This is a high-risk patient group and transcutaneous measurement is facilitated by the infant's thin skin. In adults, the technique is less used and associated with a slight underestimation. An obvious risk with the technique is that of skin burns. The sensor must therefore be moved between skin sites every 4–6 h. The calibration of the device is relatively cumbersome and after repositioning, 10–20 min are required before the readings once again are stable. These impracticalities to some extent explain the rapidly increasing use of pulse oximeters instead of transcutaneous systems.

6.4.2 Pulse Oximeters

The absorption of light in blood and tissue varies strongly with the wavelength used. For the region 600–1300 nm, the absorption is low and light penetrates to significant depths in the skin. This wavelength region is often referred to as the optical window. The Beer-Lambert law describes the intensity of light passing through a non-scattering medium

$$I_x = I_0 e^{-ACx} \tag{6.7}$$

where I_x is the intensity at position x, I_0 the incident light intensity, A the absorption, and C the concentration of the absorbing material.

For human skin, the Beer-Lambert law is not valid due to the high scattering, primarily by blood and the dermis collagen fibers. When modeling light propagation in human skin, the radiative transport equation is used instead, often in the diffusion approximation. Other approaches are numerical such as the Monte-Carlo simulation.

The basic fact of oximetry is the different absorption characteristics of hemoglobin and oxyhemoglobin (Fig. 6.15). The absorption of deoxygenated hemoglobin is higher in the red region of the spectra, giving it its bluish appearance. Several early attempts were made to measure blood oxygenation non-invasively before the invention of pulse oximetry. The major breakthrough was when Aoyagi et al. [74] made the assumption that the change in light absorbed by the skin during systole is caused mainly by arterial blood. In this way, the arterial oxygen saturation (SaO_2) becomes related to the peak-to-peak ratio of the cardiac pulsation, detected by two different wavelengths. To overcome the dependency on incident light intensity, a normalization of the pulsatile components by their respective D.C. level is performed, resulting in the ratio

$$K = \frac{AC_R/DC_R}{AC_{IR}/DC_{IR}} \tag{6.8}$$

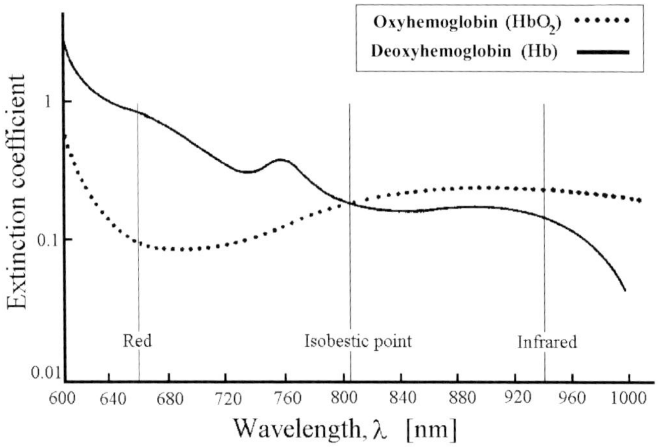

Fig. 6.15 Absorption of oxygenated and deoxygenated hemoglobin in the wavelength region 600–1000 nm.

where *AC* is the peak-to-peak value of the pulsatile component and *DC* is the baseline level of the detected intensity. The two wavelengths are often chosen in the red (R = 660 nm) and infrared regions (IR = 940 nm).

When the relative amount of oxygenated and deoxygenated hemoglobin in the blood changes, the normalized *K*-ratio in Equation (6.8) is affected as shown in Fig. 6.16. This slightly parabolic curve has been empirically determined and is relatively constant among individuals.

In practice, an infrared and a red light source (often LED components) are mounted together with a photodetector in a clip and positioned so that light is transmitted through a fingertip, a toe or an ear. There are also pulse oximeter sensors of reflection type, which can be positioned at any skin site of the body (Fig. 6.17). Pulse oximetry is regarded as one of the most significant technological advancements in the monitoring of patients in anesthesia, critical care, and postoperative care. The major advantage of the method is that blood oxygenation is provided continuously and non-invasively. This makes it possible to detect hypoxemia at an early stage and to notice undesired trends in the ventilatory gas exchange. The simplicity of the method is an important explanation of its rapid acceptance.

6.4.2.1 Limitations and Artifact Rejection

The pulse oximeter manufacturers often report an accuracy of ±2% (SD) above 70% saturation, and ±3% below 70% saturation [75]. Several factors influence the accuracy, one of the major being the sensitivity to motion. Others are difficulties in low perfusion conditions, ambient light disturbance, changes in hemoglobin concentration, presence of dyshemoglobins (or fetal hemoglobins), and venous perfusion variations. Ideas have been presented in order to reduce the sensitivity of pulse oximeters to motion. These assume that arteriovenous shunting causes the disturbance by inducing fluctuations in venous absorption during motion. The venous component is therefore removed by adaptive correlation canceling [76, 77]. Another suggestion is to use non-linear detection components and a separate normalization wavelength [78]. A third suggestion is to compare the distances be-

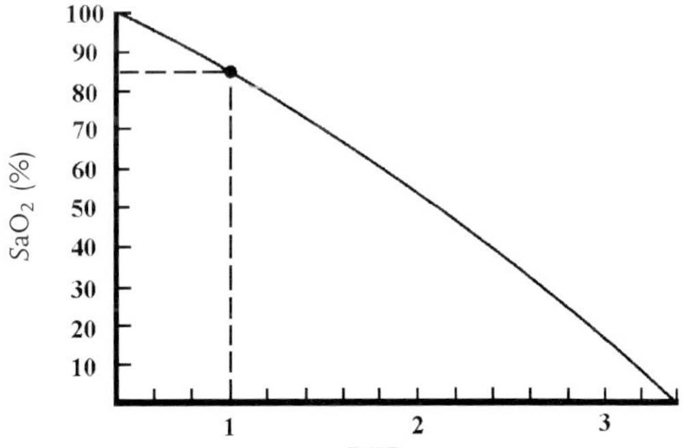

Fig. 6.16 The empirical relationship between the arterial oxygen saturation (SaO_2) and the normalized ratio of red to infrared (R/IR) signal.

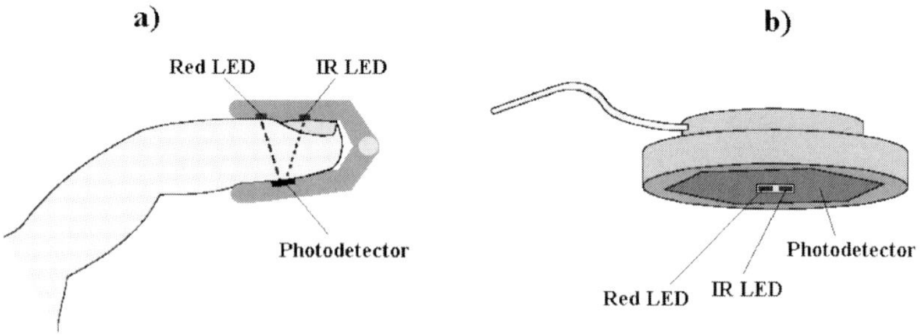

Fig. 6.17 Example of pulse oximeter sensors for the measurement of arterial oxygen saturation: a) finger-clip transmission mode sensor; b) reflectance mode sensor with a large detector area. Both sensors include a red and an infrared light emitting diode and a photodetector.

tween the cardiac pulses to the RR-intervals of the ECG. In this way, an annotation can be presented on the pulse oximeter display during motion [79].

6.5 Final Remarks

Respiration is highly non-deterministic, and a considerable variation in mean respiratory rate, tidal volume, and thoraco-abdominal partitioning is present between subjects. Other affecting variables are speech, sighs, and irregular respiration. These facts result in a fundamental difficulty in interpreting the response of a respiratory monitoring device, as the breath definition by itself is problematic. To some extent, this explains why there still is no completely satisfactory solution for monitoring respiration today. Another major issue is the lack of a 'gold standard' in the field. Direct observation may seem well suited for the task but turns out to be difficult as the human senses often are inaccurate and due to the complex interaction between the patient and the observer. This makes prototype evaluation somewhat arbitrary, and leads to more inappropriate devices introduced in the field. A more controlled prototype evaluation using a common reference is sought for. Some attempts to standardize the field of respiratory monitoring have been made, but no general standard is yet applied. Demands and recommendations are often stated as follows. A respiratory monitor *must*

- give information about tidal volume and/or respiratory rate,
- detect apnea,
- produce a low number of false alarms,
- cope with the clinical conditions, and
- be non-invasive.

Furthermore, the monitor *should*
- identify apnea type and
- be inexpensive.

Finally, it is *preferable* if
- sensor positioning around the patient's airways is avoided and
- arterial oxygen saturation is provided.

Today, no respiration monitor fulfils all of these demands, and a trend toward custom-made respiratory monitors is seen, for instance for monitoring in the neonatal period, during positive pressure ventilation, or in postoperative care. In this way, monitoring devices are chosen as to cope with the risks and patient needs in the specific situation.

6.6 References

1 D. HABORNE, Measuring respiratory rate, *Arch. Emerg. Med.* **1992**, *9*, 377–378.
2 B.J. SEMMES, M.J. TOBIN, J.V. SNYDER, A. GRENVIK, Subjective and objective measurement of tidal volume in critically ill patients, *Chest* **1985**, *87*, 577–579.
3 M.A. SACKNER, Monitoring of ventilation without a physical connection to the airway. In: Lung Biology in Health and Disease, Diagnostic Techniques in Pulmonary Disease, Marcel Dekker, NY **1980**, *16 (Part 1)*, 503–536.
4 E.O. DOEBELIN, Measurement Systems: Application and Design, McGraw-Hill, NY **1966**.
5 R.S.C. COBBOLD, Transducers for Biomedical Measurements: Principles and Applications, Wiley, NY **1974**, 57–111.
6 J.W. GARDNER, Microsensors: Principles and Applications, Wiley, NY **1994**.
7 B. HÖK, L. WIKLUND, S. HENNEBERG, A new respiratory rate monitor: development and initial clinical experience, *Int. J. Clin. Mon. Comput.* **1993**, *10*, 101–107.
8 T. GORDH, et al., Respiratory monitoring during postoperative analgesia, *J. Clin. Monit.* **1995**, *11*, 365–372.
9 L. WIKLUND, B. HÖK, K. STÅHL, A. JORDEBY-JÖNSSON, Postanesthesia monitoring revisited: frequency of true and false alarms from different monitoring devices, *J. Clin. Anesth.* **1994**, *6*, 182–188.
10 C. LARSSON, et al., Respiratory monitoring during MR imaging, *Acta Radiologica* **1999**, *40*, 33–36.
11 D.H. GORDON, W.L. THOMPSON, A new technique for monitoring spontaneous respiration, *Med. Instrum.* **1975**, *9*, 21–22.
12 A.M. CYNA, et al., Aura: a new respiratory monitor and apnoea alarm for spontaneously breathing patients, *Br. J. Anaesth.* **1991**, *67*, 341–345.
13 D. DODDS, J. PURDY, D. MOULTON, The PEP transducer: a new way of measuring respiratory rate in the nonintubated patient, *J. Acc. Emerg. Med.* **1999**, *16*, 26–28.
14 M. VEGFORS, et al., Presentation and evaluation of a new optical sensor for respiratory rate monitoring, *Int. J. Clin. Monit. Comput.* **1994**, *11*, 151–156.
15 C. LARSSON, P. STAUN, Evaluation of a new fibre-optic monitor for respiratory rate monitoring, *J. Clin. Monit. Comput.* **1999**, *15*, 295–298.
16 T. TATARA, K. TSUZAKI, An apnea monitor using a rapid-response hygrometer, *J. Clin. Monit.* **1997**, *13*, 5–9.
17 G. LENZ, W. HEIPERTZ, E. EPPLE, Capnometry for continuous postoperative monitoring of nonintubated, spontaneously breathing patients, *J. Clin. Monit.* **1991**, *7*, 245–248.
18 B.P. KAVANAGH, A.S. SANDLER, K.E. TURNER, Use of end-tidal PCO_2 as noninvasive measurement of arterial PCO_2 in extubated patients recovering from general anesthesia, *J. Clin. Monit.* **1992**, *8*, 226–230.
19 R.H. FRIESEN, M. ALSWANG, End-tidal PCO_2 monitoring via nasal cannulae in pediatric patients: accuracy and sources

of error, *J. Clin. Monit.* **1996**, *12*, 155–159.

20 M.H. Sanders, et al., Accuracy of end-tidal and transcutaneous PCO_2 monitoring during sleep, *Chest* **1994**, *106*, 472–483.

21 B. Hök, M. Tallfors, G. Sandberg, A. Blückert, Acoustic gas sensor with ppm resolution. Presented at Eurosensors XIII, The Hague, The Netherlands, **1999**.

22 B. Hök, M. Tallfors, G. Sandberg, A. Blückert, A new sensor for indoor air quality control. Presented at Eurosensors XII, Southampton UK, **1998**.

23 F. Granstedt, M. Folke, Y. Bäcklund, B. Hök, Gas sensor with electroacoustically coupled resonator, *Sensors Actuators* **2001**, *B78*, 161–168.

24 F. Granstedt, et al., New CO_2 sensor with high resolution and fast response. Presented at IEEE-EMBC, Istanbul, Turkey, **2001**.

25 R. Gilbert, J.H. Auchincloss, J. Brodsky, W. Boden, Changes in tidal volume, frequency, and ventilation induced by their measurement, *J. Appl. Physiol.* **1972**, *33*, 252–254.

26 M.A. Cohn, A.S.V. Rao, M. Broudy, et al., The respiratory inductive plethysmograph: a new non-invasive monitor of respiration, *Bull. Europ. Physiopath. Resp.* **1982**, *18*, 643–658.

27 J. Mead, N. Peterson, G. Grimby, J. Mead, Pulmonary ventilation measured from body surface movements, *Science* **1967**, *156*, 1383–1384.

28 K. Konno, J. Mead, Measurement of the separate volume changes of rib cage and abdomen during breathing, *J. Appl. Physiol.* **1967**, *22*, 407–422.

29 J.R. Brimacombe, A.G. Macfie, A. McCrirrick, The Extensometer – Potential applications in anaesthesia and intensive care, *Anaesthesia* **1991**, *46*, 756–761.

30 J. Brimacombe, Noninvasive monitoring of tidal volume with an extensometer: Laboratory and clinical studies, *Anaesth. Intens. Care* **1993**, *21*, 62–66.

31 E. Agostoni, P. Mognoni, G. Torri, F. Saracino, Relation between changes of rib cage circumference and lung volume, *J. Appl. Physiol.* **1965**, *20*, 1179–1186.

32 C. Davis, A. Mazzolini, D. Murphy, A new fibre optic sensor for respiratory monitoring, *Australas. Phys. Eng. Sci. Med.* **1997**, *20*, 214–219.

33 J.A. Adams, I.A. Zabaleta, D. Stroh, M.A. Sackner, Measurement of breath amplitudes: Comparison of three noninvasive respiratory monitors to integrated pneumotachograph, *Ped. Pulmonol.* **1993**, *16*, 254–258.

34 B.J. Semmes, M.J. Tobin, J.V. Snyder, A. Grenvik, Subjective and objective measurement of tidal volume in critically ill patients, *Chest* **1985**, *87*, 577–579.

35 K.P. Cohen, W.M. Ladd, D.M. Beams, et al., Comparison of impedance and inductance ventilation sensors on adults during breathing, motion, and simulated airway obstruction, *IEEE Trans. Biomed. Eng.* **1997**, *44*, 555–566.

36 T.S. Chadha, H. Watson, S. Birch, et al., Validation of respiratory inductance plethysmography using different calibration procedures, *Am. Rev. Respir. Dis.* **1982**, *125*, 644–649.

37 K. Ashutosh, R. Gilbert, J.H. Auchincloss, et al., Impedance pneumograph and magnetometer methods for monitoring tidal volume, *J. Appl. Physiol.* **1974**, *37*, 964–966.

38 D. Stagg, M. Goldman, J.N. Davis, Computer-aided measurement of breath volume and time components using magnetometers, *J. Appl. Physiol.* **1978**, *44*, 623–633.

39 E. Atzler, Dielektrographie, Handbuch der Biologischen Arbeitsmethoden, Urban und Schwarzenberg, Berlin Wien, **1935**, *vol. 5 (Part 8)*.

40 M.E. Valentinuzzi, L.A. Geddes, L.E. Baker, The law of impedance pneumography, *Med. Biol. Engng.* **1971**, *9*, 157–163.

41 L.A. Geddes, H.E. Hoff, D.M. Hickman, A.G. Moore, The impedance pneumograph, *Aerospace Med.* **1962**, *33*, 28–33.

42 R.V. Hill, J.C. Jansen, J.L. Fling, Electrical impedance pneumography: a critical analysis, *J. Appl. Physiol.* **1967**, *22*, 161–168.

43 R.D. Allison, E.L. Holmes, J. Nyboer, Volumetric dynamics of respiration as measured by electrical impedance

plethysmography, *J. Appl. Physiol.* **1964**, *19*, 166–173.

44 L. H. Hamilton, J. D. Beard, R. E. Carmean, R. C. Kory, An electrical impedance ventilometer to quantitate tidal volume and ventilation, *Med. Res. Eng.* **1967**, *6*, 11–16.

45 A. Grenvik, S. Ballou, E. McGinley, et al., Impedance pneumography – Comparison between chest impedance changes and respiratory volume in 11 healthy volunteers, *Chest* **1972**, *62*, 439–443.

46 W. G. Kubicek, E. Kinnen, A. Edin, Calibration of an impedance pneumograph, *J. Appl. Physiol.* **1964**, *19*, 557–560.

47 V. Sahakian, W. J. Tompkins, J. G. Webster, Electrode motion artifacts in electrical impedance pneumography, *IEEE Trans. Biomed. Eng.* **1985**, *BME-32*, 448–451.

48 L.-G. Lindberg, H. Ugnell, P. Å. Öberg, Monitoring of respiratory and heart rates using a fibre-optic sensor, *Med. Biol. Eng. Comput.* **1992**, *30*, 533–537.

49 K. Nakajima, T. Tamura, H. Miike, Monitoring of heart and respiratory rates by photoplethysmography using a digital filtering technique, *Med. Eng. Phys.* **1996**, *18*, 365–372.

50 A. Johansson, P. Å. Öberg, Estimation of respiratory volumes from the photoplethysmographic signal. Part II: a model study, *Med. Biol. Eng. Comput.* **1999**, *37*, 48–53.

51 T. Kimura, M. Kambe, T. Utumi, M. Yamakido, Volume pulse wave and respiratory diseases, *Jpn. J. Clin. Pathol.* **1994**, *42*, 413–417.

52 A. Johansson, P. Å. Öberg, Estimation of respiratory volumes from the photoplethysmographic signal. Part I: experimental results, *Med. Biol. Eng. Comput.* **1999**, *37*, 42–47.

53 A. Johansson, P. Å. Öberg, G. Sedin, Monitoring of heart and respiratory rates in newborn infants using a new photoplethysmographic technique, *J. Clin. Monit. Comput.* **1999**, *15*, 461–467.

54 L. Nilsson, A. Johansson, S. Kalman, Monitoring of respiratory rates in postoperative care using a new photoplethysmographic technique, *J. Clin. Monit. Comput.* **2000**, *16*, 309–315.

55 J. E. Lewin, An apnoea-alarm mattress, *The Lancet* **1969**, *September 27*, 667–668.

56 P. Chow, G. Nagendra, J. Abisheganaden, Y. T. Wang, Respiratory monitoring using an air-mattress system, *Physiol. Meas.* **2000**, *21*, 345–354.

57 L. Hernandez, B. Waag, H. Hsiao, V. Neelon, A new non-invasive approach for monitoring respiratory movements of sleeping subjects, *Physiol. Meas.* **1995**, *16*, 161–167.

58 B. Waag Carlson, V. J. Neelon, H. Hsiao, Evaluation of a non-invasive respiratory monitoring system for sleeping subjects, *Physiol. Meas.* **1999**, *20*, 53–63.

59 J. Siivola, New noninvasive piezoelectric transducer for recording of respiration, heart rate and body movements, *Med. Biol. Eng. Comput.* **1989**, *27*, 423–424.

60 A. Aliverti, R. Dellaca, P. Pelosi, et al., Optoelectronic plethysmography in intensive care patients, *Am. J. Respir. Crit. Care Med.* **2000**, *161*, 1546–1552.

61 R. E. Barrow, A. J. Vorwald, E. Domeier, Capacitance respirometry, *Arch. Environ. Health* **1969**, *19*, 579–585.

62 J. C. Lin, Noninvasive microwave measurement of respiration, *Proc. IEEE* **1975**, *October*, 1530.

63 G. Matthews, B. Sudduth, M. Burrow, A non-contact vital signs monitor, *Crit. Rev. Biomed. Eng.* **2000**, *28*, 173–178.

64 P. F. Meagher, R. E. Jensen, M. H. Weil, H. Shubin, Measurement of respiration rate from central venous pressure in the critically ill patient, *IEEE Trans. Biomed. Eng.* **1966**, *BME-13*, 54–57.

65 P. T. Walling, T. M. Savege, A comparison of oesophageal and central venous pressures in the management of transpulmonary pressure change, *Br. J. Anaesth.* **1976**, *48*, 475–479.

66 U. Hedstrand, S. Jakobson, U. Nylund, H. Sterner, The concordance of respiratory fluctuations in oesophageal and central venous pressures, *Uppsala J. Med. Sci.* **1976**, *81*, 49–53.

67 H. Bettermann, P. Engelke, P. van Leeuwen, C. Heckmann, Die Bestimmung der Atemfrequenz aus der Respir-

atorischen Sinusarrhythmie (RSA), *Biomed. Technik* **1996**, *41*, 319–323.
68 F. Pinciroli, R. Rossi, L. Vergani, P. Carnevali, et al., Remarks and experiments on the construction of respiratory waveforms from electrocardiographic tracings, *Comput. Biomed. Res.* **1986**, *19*, 391–409.
69 G. B. Moody, R. G. Mark, A. Zoccola, S. Mantero, Derivation of respiratory signals from multi-lead ECGs, *Comput. Cardiol.* **1985**, *12*, 113–116.
70 R. M. L. Whitlock, Respiratory rate measured from the electrocardiograph, *Dig. 9th Int. Conf. Med. Biol. Eng.* **1971**, Melbourne, Australia, 103.
71 J. Weinberg, C. Sinderby, L. Sullivan, et al., Evaluation of diaphragm electromyogram contamination during progressive inspiratory maneuvers in humans, *Electromyogr. Clin. Neurophysiol.* **1997**, *37*, 143–153.
72 A. Cohen, G. Zagelbaum, D. Gross, et al., Clinical manifestations of inspiratory muscle fatigue, *Am. J. Med.* **1982**, *73*, 308–316.
73 Y. Chen, L. Wang, W. H. Ko, A piezopolymer finger pulse and breathing wave sensor, *Sensors Actuators* **1990**, *A21–A23*, 879–882.
74 T. Aoyagi, M. Kishi, K. Yamaguchi, S. Watanabe, Improvement of the earpiece oximeter, *Proc. Abstr. 13th Ann. Meet. Japanese Soc. Med. Electron. Biol. Eng.* **1974**, Japan, 90–91.
75 K. K. Tremper, S. J. Barker, Pulse oximetry, *Anesthesiology* **1989**, *70*, 98–108.
76 S. J. Barker, N. K. Shah, The effects of motion on the performance of pulse oximeters in volunteers, *Anesthesiology* **1997**, *86*, 101–108.
77 C. Dumas, J. A. Wahr, K. K. Tremper, Clinical evaluation of a prototype motion artifact resistant pulse oximeter in the recovery room, *Anesth. Analg.* **1996**, *83*, 269–272.
78 J. H. Hayes, P. R. Smith, Artifact reduction in photoplethysmography, *Appl. Opt.* **1998**, *37*, 7437–7446.
79 M. F. Lewer, P. D. Larsen, J. M. Torrance, D. C. Galletly, Artefactual episodic hypoxemia during postoperative respiratory monitoring, *Can. J. Anaesth.* **1998**, *45*, 182–185.
80 A. Adler, R. Guardo, Impedance imaging of lung ventilation: Do we need to account for chest expansion?, *IEEE Trans. Biomed. Eng.* **1966**, *43*, 414–420

7
Sensors for Fetal and Neonatal Monitoring

P. Rolfe

7.1
Introduction

The clinical care of the fetus and the newborn baby has advanced significantly over the last two decades and sensor science and technology have played their parts in this. In fact, there has been a concentration of effort from multidisciplinary groups to develop new technologies suited to the task of improving the care of the fetus and the newborn baby that can be traced back to the mid 1960s [1–6]. As a result of this effort sensors have emerged for both diagnostic and therapeutic monitoring purposes and great ingenuity has been used to solve difficult problems at the interface between sensor and signal. Nevertheless, despite the successes, many problems still remain to be solved.

7.1.1
The Clinical Demands

There are certain similar general requirements for sensors to be used for the fetus and the newborn baby but the specific sensor designs and construction are distinctly different due to very different practical demands. These demands become clear when we consider the temporal progression from pregnancy, through labor and delivery, then beyond into the care of the ill newborn baby.

The fetus during pregnancy is completely concealed from direct vision, encased within the maternal abdomen and inside the uterine cavity. Obstetricians need technology with which to assess the health of the fetus throughout pregnancy, for example for detection of multiple fetuses, for the assessment of the growth of the fetus in order to detect intrauterine growth retardation (IUGR), and for detection of malformations. In order for this basic anatomical information about the fetus to be obtained ante-partum, remote sensing is needed. As the pregnancy progresses into the third trimester (6th to 9th month) there is an increasing need to collect evidence that the fetus is developing from the physiological systems viewpoint and in particular is able to respond appropriately to various stimuli. In the event that abnormal responses exist, then a clinical option is to initiate delivery of

Sensors in Medicine and Health Care
Edited by P. Å. Öberg, T. Togawa, and F. A. Spelman

ISBN: 3-527-29556-9

the baby, perhaps by performing a Cesarean section. This is not to be undertaken lightly and therefore the methods used for providing the data on which such a decision is taken need to be reliable.

Commencement of labor may signal the need for an acute phase of surveillance. There is a risk that the fetus may become distressed by the rigors of labor. As uterine contractions get underway the fetus remains physically inaccessible initially and therefore there remains the need for indirect methods of surveillance. Then as labor progresses the fetal head becomes available through the dilating cervix. However, even when the fetus becomes partly physically accessible, there are still significant problems in establishing reliable access and connection for measurement. At this time the midwife and obstetrician must look for fetal distress, again so that rapid intervention can be initiated. Thus, measures of acute change are needed, particularly to assess blood oxygenation and safeguard the brain. A proven indication of brain oxygenation to be used in a clinical setting is not yet available. Indirect measures are used, based on fetal heart rate, the electrocardiogram, or pH of scalp blood samples.

Monitoring the baby at the moment of birth can also present very different challenges. Babies distressed at birth may require immediate emergency care, perhaps involving cardiopulmonary resuscitation entailing artificial ventilation and cardiac massage. These babies are wet, small, and fragile. Any sensors that may be used to assist in rapid assessment of the clinical condition must be usable under these extreme conditions. As a minimum requirement heart rate and breathing efforts need to be monitored and there will usually be a need to determine deep body temperature to see if the baby is becoming hypothermic.

Any small preterm baby is likely to require at least some degree of surveillance in the first few days of life. In many cases this will only entail a day or two of observation to ensure that there are no breathing difficulties and that the baby can maintain a normal body temperature. In the minority of cases, however, more extensive and longer-term surveillance will be needed. These babies are those born perhaps 10 or more weeks early, weighing less than 1 kg, with poorly developed lungs and immature cardiorespiratory and thermoregulatory control mechanisms. Such babies may already have had difficulty during the critical transition at birth and may have required resuscitation. The assessment of cardiorespiratory status carried out during that initial resuscitation is an important part of this emergency procedure and monitoring of breathing efforts, blood oxygenation, and the physiological variables defining cardiac performance is essential. There will then be an ongoing need for the monitoring of these variables during the days or weeks of neonatal intensive or special care that follow. There will be concerns about possible brain hemorrhage – with bleeding into the CSF-filled lateral ventricles of the brain – and associated hypoxic-ischemic brain injury in the first day or two of life. Thus, methods for scanning the brain are needed. The precise causes for such hemorrhage remain unknown but the careful control of blood gas and pH levels, arterial blood pressure, and the use of minimally traumatic means for applying artificial ventilation are thought to be important in minimizing the likelihood of brain hemorrhage. Temperature control is of fundamental importance in the small preterm baby, as symbolized by the neonatal incubator and radiant warmer, and reliable monitoring is a key part of that.

7.1.2
General Sensor Requirements

Having considered the clinical demands and constraints it becomes clearer that there are at least some general practical and performance requirements for the sensors and the associated instruments that are needed for fetal and neonatal monitoring.

When carrying out any form of physiological measurement there is always the basic question concerning just how the variable of interest may be sensed conveniently, safely, reliably, and effectively. In a clinical situation the requirements for convenience and safety are paramount whilst of course not negating the effectiveness of the measurement. When considering very small newborn babies or the fetus the demands for safety and freedom from any encumbrance to the subject are very great. The attachment of inappropriately designed sensors to a very small newborn baby could constitute a significant disturbance to the baby. In the case of the fetus it is as much the issue of encumbrance to the mother as to the fetus that is important.

These factors suggest devices that are small and lightweight. Experience has also shown that it is not merely the size of sensor itself but also any connecting cables that will influence the applicability of the measurement. Thus small, lightweight, and very flexible cables or optical fibers connected directly to the sensor are needed.

It is obvious with invasive sensors that the device must be sterile and therefore the materials used for its construction must withstand appropriate sterilization. This might be one of a variety of methods for cold chemical sterilization, or high temperature autoclave or ionizing radiation. At the same time, sensors that are effectively non-invasive must still be clean and sterile. This means that they must withstand surface wipe-down with suitable chemical agents. With both invasive and non-invasive sensors there is also the need for any connecting cable or fiber to withstand an appropriate method of cleaning and sterilization.

7.2
Considerations of Safety and Convenience

Within the field of medical sensor science and technology there has been much debate about direct and indirect hazards associated with the clinical use of such devices. There are obvious issues in relation to the potential for electrical hazards with sensors directly connected to subjects. Thus, we see that when a particular sensor requires Ohmic contact with the patient then the use of floating front-end amplifiers, based on optically coupled or RF transformer-coupled electronics, is essential. In fact, these now tend to be the norm within all modern perinatal monitors regardless of whether or not the sensors need Ohmic contact. Furthermore, the use of sensors using fiber optic connection is acknowledged to be useful in providing some level of intrinsic safety advantage as compared with electrically connected sensors.

Also quite well-established are views concerning potential hazards of techniques in which some form of external energy is applied to the subject. Reasoning here has been based on experience with X-ray-based methods, including X-ray-computed tomography. The introduction of ultrasound techniques, initially for fetal heart rate measurement, fetal growth assessment, and neonatal brain scanning, then led to extensive debate about possible hazards to the developing fetus [7]. However, the clinical application of these methods was introduced prior to the conduct of detailed examination of possible hazards, largely because ultrasound was referred to as a non-invasive method. Experts in the field were already aware of the potential of ultrasound to produce undesirable biological effects, including cavitation, the possible establishment of standing waves in blood vessels, and thermal effects, but pressure from consumers was an important factor in ensuring that efforts were made to investigate these and other phenomena. The aim was to gain agreement on safety levels for ultrasound transducer power output. However, the matter of a risk/benefit ratio has also been debated. The National Council on Radiation Protection and Measurement recommended that users: 'should strive to obtain the most medically significant information possible while producing the least ultrasonic exposure to the patient' [8]. At the levels of power employed in currently available clinical instruments laboratory studies have failed to show hazards through these mechanical or thermal mechanisms. However, efforts continue to investigate the possibility of subtle damage at the gene level.

The consideration of safety issues surrounding the use of ultrasound led to a reappraisal of the implications of the use of the invasive/non-invasive terminology. It became accepted that the distinction between invasive and non-invasive devices or methods should not automatically lead to the assumption that the latter were less hazardous than the former. For example, optical methods using near infrared radiation were described as non-invasive and safe and were assumed to be fundamentally less hazardous than X-ray or ultrasound methods. Yet it is also known that at appropriate wavelengths clear interactions with cellular and sub-cellular components and phenomena do take place. An example is the selective absorption of near infrared energy, at around 830 nm, of the highly important intramitochondrial enzyme cytochrome aa_3. At the low power levels used for diagnostic purposes it is likely, although not yet proven, that such interactions do not lead to any specific hazard.

There is clearly an on-going need for not only new technologies to be evaluated for possible damaging biological effects but also for methods now in routine clinical use to be assessed as part of their on-going use. Despite the need for caution in interpreting the safety aspects of non-invasive methods it is certainly the case that they are generally less intrusive and more convenient clinically than their invasive counterparts. On the other hand, it is also the case that invasive devices will be positioned closer to the variable being measured and should, in principle, yield more accurate results. By their very nature non-invasive methods are less direct and may be subject to more confounding factors.

The matter of the intrusive or encumbering nature of a particular method is especially important with the fetus during birth and with very small, fragile, newborn babies. Making direct contact with the fetus during delivery as the cervix dilates is an important opportunity as far as monitoring is concerned. However, the procedure of examining the fetus in order to attach a sensor or withdraw a fetal scalp blood sample is intrusive both to the mother and her baby. Similarly, the very small, sick newborn baby is usually very fragile and in a precarious physiological state. The mere handling of such a baby for nursing and clinical procedures can have adverse physiological effects. The attachment of even non-invasive sensors can represent too much disturbance or encumbrance to the baby. Sensors may seem comparatively small and convenient at first sight but compared with the size and weight of a 500 g baby they actually may be too big and heavy. Perhaps the long-term aim should be to develop completely non-contact sensing methods. Indeed, some are already available and some groups are attempting to develop more.

7.3 Antepartum Fetal Assessment

During pregnancy the most obvious questions that must be addressed as part of fetal assessment are, firstly, is the fetus growing and developing normally and, secondly, is it responding appropriately to normal or any abnormal intrauterine stimuli to which it is subjected? Thus there are needs to assess the growth and development of the fetus in order that potentially serious conditions can be detected and possible clinical interventions carried out. Sensors and instruments have long played a role in this important phase of maternal and fetal care. Ultrasound methods for imaging and physiological measurement now represent a vital part of this phase of fetal surveillance. Other imaging techniques, including X-ray-computed tomography and magnetic resonance imaging (MRI), are also important but are not directly relevant to the topic of sensors covered by this book.

Assessing the gross clinical status of the fetus observing the fetal heart activity has been the most widely used historically and it is still central to modern technology for fetal surveillance. The monitoring and display of fetal heart rate alongside the temporal changes in intrauterine pressure associated with uterine contractions forms the basis of the commonly used method of cardiotocography.

7.3.1 Ultrasound Techniques

Although historically phonocardiography and electrocardiography had earlier associations with intrapartum fetal surveillance, especially for fetal heart rate monitoring, ultrasound methods subsequently found their place and are now of significant importance in ante-partum obstetric care [9, 10]. Of course, much of the sophistication of modern medical ultrasound apparatus lies in the instrumentation, especially in the software for signal processing, but the ultrasound sensors nevertheless represent a key part of the overall system. Such sensors evolved to meet the intended improvements in overall performance of the complete system. Particularly, this has been the case with regard to spatial resolution. Resolution is important in imaging systems as well as in those used for blood flow measurement or for anthropomorphic measurement.

7.3.1.1 Transducers

Depending upon the precise geometrical form of the ultrasound probe it may contain one or many individual piezoelectric elements, in many cases each element being used to emit as well as receive ultrasonic waves (Fig. 7.1). Most medical applications of ultrasound will use frequencies in the range 1–15 MHz, corresponding to wavelengths of 1.5–0.1 mm in soft tissues. Materials such as quartz, lead niobate, barium titanate ceramic, and lead zirconate titanate have been the most commonly used as the piezoelectric element, in the form of thin slices covered on opposing faces by a conductive layer or film. Thin polymer films exhibiting the piezoelectric effect are also available, such as polyvinylidene difluoride (PVDF). Depending upon the orientation of the crystal slice or polymer film in relation to atomic structure the element may exhibit a variety of deformation modes, such as thickness-expansion mode, volume-expansion mode, or thickness-shear mode.

Most piezoelectric materials exhibit a linear relation between induced surface charge, δQ, and the change in applied force, δF

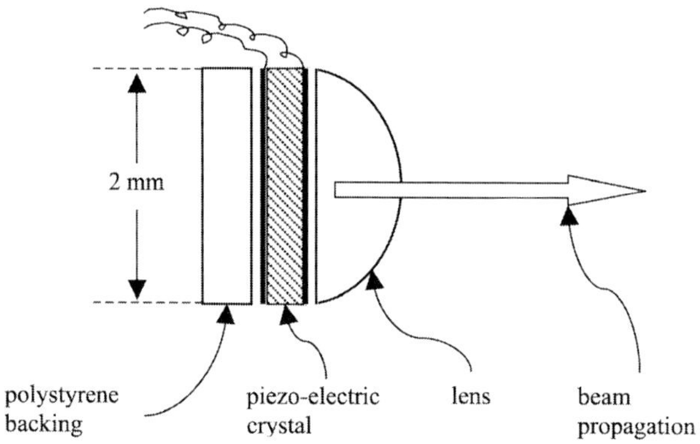

Fig. 7.1 Exploded view of a simple ultrasound transducer. The crystal is connected to drive or receive electronics via thin conductive films fixed to opposing faces. The polystyrene backing material ensures reflection of crystal energy to give forward propagation via the lens. Dimensions may be smaller or larger.

$$\delta Q = K_q \delta F \quad (7.1)$$

where K_q is the material charge sensitivity which varies, say, from 2.3×10^{-12} for quartz to 200×10^{-12} for lead niobate. Driving piezoelectric elements as transmitters or collecting the induced charge requires careful consideration of the element's capacitance and leakage resistance. Since the element has leakage resistance any induced charge will leak away, implying that the transducer can only be used for detecting time varying forces and cannot be used for steady forces. The capacitance will of course be proportional to element's area and inversely proportional to its thickness, with typical values being in the range 500 pF to 1000 pF for a crystal 1 mm thick and either 0.5 cm^2 or 1 cm^2 respectively. To produce drive circuits and charge amplifiers that can, for example, operate an element at 5 MHz with a repetition rate of 2 kHz in drive mode and, interleaved with bursts of transmission, collect the very small induced surface charge from resulting echoes, requires very careful electronic design [11]. Considerable research has been carried out to understand factors that influence ultrasound transducer design. Transducer size, shape and material, together with operating frequency, desired power and operational depth must all be considered when designing an assembly that is needed to fulfil a particular function. Focussing of the emitted beam, or the detection volume, is a critical design requirement and matching of crystal, lens and tissue properties. The characteristic impedance of each material, Z_m, needs to be considered. This is given by

$$Z_m = \rho c \quad (7.2)$$

where ρ is the material density and c is the speed of sound in that material. At boundaries between different materials, 1 and 2, there can be a proportion of the energy reflected, R, given by

$$R = [(Z_2 - Z_1)/(Z_2 + Z_1)]^2 \, . \quad (7.3)$$

Considering values of characteristic impedance for typical biological and other materials allows values of R to calculate. Total reflection occurs when $R=1$, for example at soft tissue and bone interface. Values of Z for typical materials are ($kg\ m^{-2}\ s^{-1}$): air, 4×10^{-4}; bone, 3.7–7.4; lung, 0.26; lead zirconate titanate, 30.

At plane surfaces so-called specular reflection takes place. Where there are objects having dimensions in the order of the ultrasound wavelength or less, scattering takes place and the scattered energy is radiated uniformly in all directions as spherical waves. The shape and spatial orientation of the scatterers will determine the pattern of the scattered waves.

The incident and reflected waves are also attenuated, according to the attenuation coefficient of the material (dB cm^{-1}), a_m, and part of this attenuation is due to absorption of the energy which is converted to heat [12]. Considering a simple case of a circular disc transducer element then the following expression can provide an indication of the steady state field

$$I_x/I_0 = \sin^2\{(\pi/\lambda)[a^2 + z^2)^{1/2} - z]\} \tag{7.4}$$

where I_x is the intensity at distance x, I_0 is the intensity emitted from the transducer of radius a. The beam exhibits a near field or Fresnel zone demarcated by the point at which the intensity reaches a maximum, beyond which lies the far field or Fraunhofer zone.

Focusing of the ultrasound beam can be achieved with plastic lenses. A typical arrangement is to have the piezo element with a polystyrene backing on one face with a lens on the forward face.

Ultrasound transducers have been developed to meet a wide variety of applications, but today there are basically three classes that are of importance for ante-partum clinical use: basic imaging systems [13]; simple Doppler probes for FHR measurement [14]; Doppler probes for blood flow measurement [15].

7.3.1.2 **Clinical Uses**

For imaging applications in ante-partum surveillance the hand-operated 2D static B-scanner was the first to become used widely in clinical practice. With this the ultrasound probe is constrained by two articulated arms so that it can only move in a specific 2D plane. Three synchro-resolvers are used to control two time base generators for a CRT-based display that receives the intensity-modulated signal from the probe. Real-time scanners then rapidly overtook this type of static scanner. In the mechanical sector real-time scanner, typically three single or multiple-element transducers are mounted on a wheel that is rotated. A commutation arrangement then allows each element to scan an image sector in turn. It is often the case that a single scan-head will contain different piezoelectric elements to allow the user to switch between operating frequencies to achieve optimum performance for a given clinical situation.

There is some difficulty in controlling the focussing characteristics of mechanical sector transducers and so array transducers have become popular. With the linear array transducer the piezoelectric elements are electronically addressed sequentially to generate a rectangular format image. The technology for constructing piezoelectric arrays with well-defined geometrical characteristics has led to array transducers and associated instrumentation capable of producing ultrasound images with high spatial and axial resolution and high sensitivity. To achieve high gray-scale and spatial resolution array transducers should be comprised of a large number of piezoelectric elements (96 or 128) spaced $\lambda/2$ apart to minimize grating lobes of acoustic sensitivity.

The evolution of the array transducer led to phased-arrays in which an assembly of crystals is driven electronically such that each element is activated in turn according to a chosen time sequence [16]. The effect is to produce a shaped or focused ultrasound beam that can be steered in a chosen direction.

Of course it is very straightforward to use ultrasound scanners simply to confirm pregnancy and to identify multiple fetuses. However, of more significant value is the assessment of the growth of the fetus. Fetal growth during pregnancy is perhaps one of the most obvious processes to monitor in order to assess the general clinical status of the fetus. Measuring the weight gain of the mother or increase in maternal abdominal circumference during pregnancy has been one of the most straightforward ways of deriving an indirect indication of fetal growth. However, the measurement of fetal size and its change through pregnancy is clearly more direct. This may be done simply using a tape measure with agreed reference points on the fetus as palpated through the maternal abdomen; for example, the crown–rump length. However, the advent of ultrasound in obstetrics was of great importance since it opened up the possibility of making more accurate and repeatable measurements of fetal size and growth. This was initially achieved by using pulse-echo ultrasound to measure the diameter of the fetal head [17]. A 2D scan is used to identify the fetal head and the scan plane adjusted until the midline echoes are seen bisecting the long axis of what is an ovoid. The distance between the skull echoes from the opposite sides of the head in the short axis, called the bi-parietal diameter (BPD) is then calculated using knowledge of the speed of ultrasound (1500 m s^{-1}). The normal fetus has a BPD of 27 mm at 14 weeks gestation, 80 mm at 30 weeks, and 95 mm at 36 weeks. Systems using swept gain to enhance progressively the echoes from deeper structures together with considerations of the typical dynamic range of around 100 dB have led to trade-offs between penetration depth, attenuation and operating frequency. Typically, attenuation will be 1 dB cm^{-1}. Before 14 weeks measurement of BPD is not possible and therefore other measurements have been made ultrasonically, for example crown–rump length [18].

Doppler ultrasound techniques have become very important in intrapartum surveillance. Of course the Doppler approach allows movement to be detected and quantified. The earliest application of this method was to detect fetal heart movement [14], which can be reliably found after about the 12th week of pregnancy. In addition to methods based on fetal electrocardiography or fetal phonocardiography, this has allowed processing and analysis techniques to be developed for the evaluation of fetal heart rate patterns during the ante-partum period.

Traditionally a mother's perception of the activity of her baby has been a useful indicator of fetal well-being. Recording by the mother of the number of movements of the fetus in a timed period is often carried out. Using a de-focused, broad beam Doppler ultrasound system it was shown that this system was more sensitive and reliable than either maternal records of movement or palpation of the abdomen by a clinician [19]. The fetal activity was said to be easily distinguished from the fetal heart Doppler signal. In addition to fetal limb and gross body movements there has been interest in using ultrasound techniques to detect and quantify fetal breathing movements [20].

In addition to allowing body movements and cardiac movements to be detected Doppler methods are used to measure fetal blood flow. Initially continuous wave Doppler instruments were used for blood flow assessment in a semi-quantitative way. Examination of flow in large blood vessels was possible but the CW method does not allow any depth discrimination. The pulsed Doppler instruments were developed to allow depth discrimination, by gating the receiver path to allow reflections at a chosen time, and therefore depth, to be extracted. Color Doppler flow mapping was a further extension, allowing the range-gated flow information to be superimposed upon an ultrasound image. Color flow Doppler of the fetal heart allows a variety of important observations to be made. These include structural and functional cardiac abnormalities such as: insufficiencies of atrioventricular valves; detection of turbulent high velocity jet in stenosis of semilunar valves; reverse flow in ascending aorta in atresia of aortic valves; turbulent jets and reverse flows in tetralogy of Fallot; bi-directional shunting of blood in ventricular septal defects [21].

7.3.2 Antepartum Electronic Fetal Monitoring

There are often needs to assess fetal status during pregnancy and this arises from previous medical history or from routine ultrasound or biochemical considerations early in pregnancy. The most obvious physiological variable upon which to base such an assessment is fetal heart rate (FHR). As mentioned above, ultrasound transducers may be used to derive FHR data. In addition to FHR there are other possibilities that may be offered by sensors and associated instrumentation.

7.3.2.1 Sounds and Movement

The practice of using the FHR to assess fetal status was derived from the very simple sensor in the form of the fetal stethoscope, or fetoscope. This is essentially a conical tube held in contact with the maternal abdomen, which allows a doctor or nurse to attempt to listen to sounds. Among the mixture of sounds heard, the fetal heart sounds, produced by valve opening and closing and by blood flow, may be perceived. When these heart sounds are clear it is then possible to calculate manually the mean heart rate, say over 30–60 s. It is also possible, with experience, to perceive changes in FHR, such as bradycardia. Use of a microphone in place of the simple fetoscope allows electronic recording of sounds detected from the maternal abdominal wall. The sounds detected include fetal heart sounds, thereby enabling the fetal phonocardiogram to be recorded [22]. However, the fetal heart sound is heard among a collection of other sounds. These may be due to movement of the fetus or the mother, maternal heart and breathing sounds, and also noises from the surrounding environment. Thus processing of the phono signal is essential in order to extract useful information from the fetus.

Although simple microphones were used widely in most commercially available fetal monitoring instruments, there was some attempt to improve the performance of phonocardiography by matching a suitable transducer to the mechanical properties of the tissues to which it was attached [23]. Such an approach was then used to achieve a transducer design with a wide bandwidth for fetal phonography with compliance-matched to the maternal abdominal wall [24]. Drawbacks arising from the limited low frequency response of piezoelectric crystals were overcome in a compliance matched transducer design based on inductive principles [25].

This transducer, called the INPHO, has a polypropylene membrane that can be stretched to an appropriate tension in order to achieve the required compliance. It has a frequency response flat to within 3 dB from 0.1 Hz to 200 Hz. It was shown that appropriate adaptive filtering can separate fetal heart sounds (>10 Hz) from fetal breathing movements (0.5–2.0 Hz) as well as eliminate the influence of maternal breathing movements [26]. It was also reported that the attachment of this transducer was best achieved using a double-sided adhesive disc rather than by means of belts or straps, since the latter interfere with compliance matching.

Although fetal phonocardiography had lost popularity in recent years, new work describes a low cost monitor based on phonocardiography and advanced signal processing [27]. The two-channel phonocardiographic device is said to provide performance for FHR variability monitoring comparable to that of ultrasound cardiography. The system was developed for home care use, offering an optimal trade-off between complexity and performance in a low-cost, stand-alone, battery-powered instrument. The developed system provided 83% accuracy compared with the simultaneously recorded reference ultrasound records [27].

7.3.2.2 The Fetomaternal Electrocardiograph

In addition to ultrasound and phonocardiography, it is of course possible to attach electrodes to the maternal abdominal wall to collect electrophysiological signals. The latter can comprise cardiac related signals, from both the fetus and the mother, as well as maternal electromyographic signals. For this reason the collection of ECG signals is often referred to as the fetomaternal ECG to distinguish it from the fetal ECG obtained from electrodes attached directly to the fetus, as described below. The fetal electrocardiograph provided the earliest form of electronic fetal heart monitoring [28]. When recorded from the maternal abdomen the extracted fetal ECG complex has an amplitude of 5–50 μV, with the maximum R-wave amplitude appearing between 21 and 24 weeks gestation. From 26 to 34 weeks the ECG magnitude is known to decrease, probably due to the presence of vernix, which has a high electrical resistivity [29]. Given the low signal level of the fetal ECG it is important to use low-noise Ag:AgCl electrodes. Skin preparation to de-grease and remove, at least partly, stratum corneum is essential, as is the use of contact gel. Positioning of the electrodes is usually a matter of trial and error although it is found that midline placement tends to minimize the magnitude of electromyograph signals. A complex approach developed to help eliminate the maternal ECG was developed by Schuler et al. [30]. With this an array of eight electrodes is placed circumferentially around the maternal abdomen. The signals from these are buffered and fed to a resistance mat analogue of the maternal abdomen. An optimum fetal signal is then found by exploring the mat with a probing electrode. The method is too cumbersome to use for routine clinical purposes but has some research value.

Apart from these general electrode requirements there is little more concerning the sensor part of the system. The most important remaining aspects are concerned with signal processing rather than sensor science and technology. Processing the collected signals from the maternal abdomen usually starts with filtering. The fetal ECG has maximum power at 40 Hz as compared with the maternal ECG at 25 Hz. Filtering cannot, however, completely eliminate the maternal ECG. In a subtraction method [31] a maternal QRS complex matching the maternal component within the combined fetal–maternal signal is discovered by switching around a number of maternal abdominal sites using a multi-electrode array [32].

7.3.2.3 **Uterine Activity**

Although uterine contractions do not become fully established until labor commences there are nevertheless sporadic contractions at various times during pregnancy and these act as a physiological stimulus to the fetus, in particular by producing changes in FHR. Manual detection of uterine activity is carried out by palpation or simply on the basis of maternal perception. Contraction of uterine muscles is felt as a hardening of the normally soft maternal abdomen. Thus sensors suitable for electronic recording of contractions were initially based on the 'probing' principle of fingers pressed into the maternal abdominal wall. Such a sensor developed by Reynolds et al. [33] was called a tocodynamometer. It consists of a spring-loaded plunger protruding from the flat face of a mounting case. The device is fixed to the maternal abdomen such that the plunger indents the abdominal wall. As the contraction stiffens the wall the plunger is pushed back into the case and the deflection is measured with a strain gauge.

An alternative type of sensor for uterine activity monitoring is the applanation transducer first described by Smyth [34]. This transducer presents a flat surface to be held in contact with the bulging maternal abdominal wall. With flattening, the intraabdominal pressure is balanced by the transducer and a force-sensitive spring-loaded central section responds to this pressure. A surrounding inactive guard-ring dissipates any bending forces. It is important for the central force-sensitive section to have very low compliance, that is, very small displacement for the range of pressures being measured. This kind of sensor is more comfortable for the mother than the plunger of the tocodynamometer.

The possibility of using the uterine electromyograph to derive indications of uterine activity has been studied for a number of years [35]. This is referred to as electrohystography (EHG). The method uses electrodes attached to the maternal abdomen to collect electrophysiological signals which include the maternal and fetal ECG signals as well as mixed EMG signals. The challenge has been, firstly to extract the uterine EMG signal from the composite signals and, secondly to establish a relationship between the EHG and uterine contractions or even intrauterine pressure itself. Single electrical bursts were found to correlate well with mechanical activity [36]. The use of the redundant wavelet packet transform has been found to allow very effective removal of the maternal ECG [37].

7.3.2.4 **Cardiotocography**

The recording of fetal heart rate (FHR) together with uterine activity, such as contractions, forms the basis of the widely used monitoring technique of cardiotocography. The interpretation of recorded patterns of FHR and uterine activity, made within the context of clinical signs and medical history, now represents the essence of so-called electronic fetal monitoring (EFM). As already indicated above, the FHR information can be derived from phonocardiograph signals representing heart valve action, or from the fetal ECG extracted from the mixed fetomaternal ECG signal picked up by electrodes on the maternal abdomen, or from ultrasonically-derived signals representing cardiac contractions or major vessel blood flow. The precise method of calculating the FHR is very important, however, if the full clinical potential of this variable is to be obtained. The clinical value of fetal heart rate actually arises from the variability of rate. This is because it provides a window into the physiological control mechanisms of the fetus. The presence of what are in effect fine adjustments to cardiac performance seen as changes in heart rate occurring from beat to beat is a useful indication that the fetus is in a good physiological state. Conversely, the absence of these fine adjustments, as evidenced by an essentially constant FHR – a 'flat' tracing – is suggestive of fetal distress. The FHR variability that may occur can be present between successive beats, as reflected in the so-called beat-to-beat heart rate, or between inter-beat periods separated in time. Whichever measure of variability is used it is important for the beat-to-beat time period to be measured accurately and this in turn means that a specific reference point in the cardiac cycle needs to be defined and detected precisely. The phonocardiograph signal, comprising as it does of heart valve action, does not offer sufficiently clearly defined and repeatable reference points. Ultrasonically-derived signals of cardiac movement or valve action or blood flow can, with appropriate signal processing, allow FHR variability indices to be derived with moderate repeatability. However, the high frequency components of the fetal ECG QRS complex offer the most useful reference points for variability analysis. The subject of FHR analysis is a large one and is outside the focus of this chapter on sensors. The reader is therefore referred to key texts on the subject, for example [38–40].

The essence of cardiotocography is to examine, firstly, baseline values of FHR and, secondly, to analyze patterns of FHR change in association with changes in uterine activity and with fetal limb and body movements. Thus, there are established values of baseline FHR to define, for example: marked bradycardia, FHR $<$ 100 beats min^{-1}; normal rate, FHR $=$ 120–160 beats min^{-1}; marked tachycardia, FHR $>$ 180 beats min^{-1}.

Patterns of change that are particularly useful are those termed acceleration and deceleration. Accelerations are transient increases in baseline FHR lasting less than 10 min. Important decelerations are those usually further sub-classified as: *early decelerations*, being those in which the onset, fall and recovery of FHR reflects the onset, peak, and fall of a uterine contraction; *late decelerations*, being those in which the deceleration has an onset, fall, and recovery, all of which are delayed in relation to the onset, peak, and fall of a contraction. In addition to these classifications there are so-called variable patterns and mixed patterns.

The patterns seen in the cardiotocograph are often judged by eye from the paper record of FHR and uterine activity. However, computer analysis has been the only possible means for detecting the subtle and complex changes likely to occur. A cardiotocogram-recorded ante-partum in the presence of spontaneously occurring contractions is termed the non-stress test and is commonly used. It is also the practice to apply external stresses, such as induced contractions or sounds or vibrations via the maternal abdomen, and to monitor the fetal response to these. This is the stress test. The benefit of the computer analysis of these tests has long been acknowledged, for example, in making non-stress test analysis more effective [41]. In this latter system an analysis is performed that minimizes monitoring time by alerting the operator when monitoring can be safely stopped because the fetal heart rate is normal. It also detects potentially sinister sinusoidal patterns of FHR and improves the quality of the test readings by quantifying signal loss. Sinusoidal FHR patterns are known to be associated with anemia, low umbilical cord pH, and large base excess [42] and this supports the idea that fetal heart rate can indeed be an early indicator of impending fetal compromise.

7.4 Intrapartum Monitoring

With the membranes ruptured and the cervix beginning to dilate there is the opportunity to gain direct access to the fetus. In most cases it is the fetal head that is the presenting part although in a small number of deliveries there is a breech presentation. This direct access has stimulated the development of a wide range of sensors that may be attached to the fetal scalp (or breech). The aim of monitoring the fetus during labor and delivery is to detect early signs that the fetus is not coping with the rigors of the birth process and in particular that it might be asphyxiated and as a result is at risk of serious brain damage. Even today, birth asphyxia, defined as a severe disturbance of oxygen supply to the fetus, has been said to be an unsolved problem of perinatal medicine [43].

7.4.1 Fetal Heart Rate and fECG Analysis

As with surveillance during pregnancy the response of the fetal cardiac system to stresses can provide an indication of the physiological and clinical status of the fetus. Thus, fetal heart rate (FHR) and heart rate variability are monitored during labor and delivery. In fact, in many cases the mother will already be connected to the cardiotocograph for monitoring and plotting the beat-to-beat heart rate and the uterine contractions. As described above, in many cases an ultrasound probe will be used for deriving the FHR information together with an external contraction sensor. However, the use of the fetal ECG is an important aspect of intrapartum monitoring.

7.4.1.1 **Direct Fetal ECG Electrodes**

With direct connection to the fetus now being possible it is clear that attachment of some kind of electrode directly to an appropriate fetal anatomical site is an obvious method for obtaining the fetal ECG signal with improved signal quality. A variety of direct fetal ECG electrodes has been evolved since the early 1960s. However, these mostly do not allow standard bipolar connections and so the ECG waveform must be interpreted accordingly.

One of the first electrodes for direct fetal ECG recording was an adaptation of a surgical clip, with two Ag:AgCl surfaced penetrating tines allowing contact with sub-epidermal tissue [44]. A further Ag:AgCl wire, which comes into contact with surrounding amniotic fluid, acts as the reference. A more popular electrode has been the 'spiral' design of Hon [45] in which a stainless steel helical needle, attached to a short cylindrical plastic body, is screwed into the fetal scalp. Sub-epidermal contact is made with this needle and a reference connection in the body makes contact with amniotic fluid. Versions with either single or double spirals have been used. Another popular design is the so-called 'Copeland' electrode. In this design a small plastic body holds a semi-circular needle which is attached a semi-flexible rotatable wire. The needle is initially concealed within the sensor body. After holding this assembly in contact with the fetal scalp, or presenting part, the needle is rotated, thereby piercing the skin. A reference wire is attached to the body for contact with amniotic fluid.

With these and other needle-like electrode designs the fetal tissue is penetrated and so there is the risk of infection or trauma. For this reason there have been attempts to design electrodes in which no skin penetration occurs. One such design comprises of a cup-like concave body, 15 mm in diameter, with a flat disc Ag:AgCl electrode at its center [46]. Suction is applied to the cup, thus holding the flat electrode in contact with the fetal tissue. A reference electrode is positioned on the rear of the suction cup. It has also been possible to use tissue adhesive in place of suction to allow secure attachment of the non-penetrating electrode [47].

The electrodes so far described have been assessed in various ways, including through the individual experiences of users in clinical practice. A randomized clinical trial was carried out in 780 patients to assess the performance of three types of fetal scalp electrode: Rocket-Rolon (double-helix spiral); Hewlett-Packard (double-helix spiral); and Surgicraft-Copeland (clip) [48]. This study found that the Copeland was least likely to become detached during monitoring, but that the Hewlett-Packard double-helix spiral was least likely to need replacement due to poor trace quality.

The extent to which FHR variability derived from a direct fetal ECG during labor may be analyzed is exemplified in work utilizing spectral analysis [49]. FHR variability was quantified by computing Fast-Fourier-transformed heart rate spectra at three frequency bands: low-frequency (LF) 0.03–0.07 Hz, mid-frequency (MF) 0.07–0.13 Hz and high-frequency (HF) 0.13–1.0 Hz. It was found that total FHR variability and MF FHR variability were lower in fetuses with cord arterial base deficit 8–12 mmol/L compared with the fetuses with normal cord arterial base deficit value.

7.4.1.2 Fetal ECG Waveform

For almost a century the ECG waveform in animals and in the human has been the subject of study, with the aim of identifying useful indicators of abnormality that may be used for diagnosis and monitoring. In fetal physiology and clinical care much has been achieved in terms of understanding the influences of hypoxia on the fetal ECG waveform. These influences on waveform are, however, quite subtle and have not been straightforward to apply in clinical monitoring. In indirect fetal electrocardiography one major obstacle has been the influence of the maternal ECG on the fetal ECG waveform. For example, Hon et al. noted apparent fetal ST segment depression that was in fact due to superimposition of the maternal T-wave [50]. The direct connection to the fetus improves the quality of the ECG and this has allowed some important progress. Much of the interest has focused on the ST-segment of the ECG, which represents the depolarization of the ventricles after ventricular contraction. In chronically instrumented fetal lamb studies it was shown that during 1 h of hypoxia there was an increase in the ST-segment with a rise in plasma lactate and a fall in pH [51]. As this line of research continued it became important that the direct fetal ECG electrodes were of optimum design to enable the subtle waveform changes to be recorded under clinical circumstances. In a prospective randomized study the technical quality of monitoring and the complications of use of spiral electrodes and Copeland electrodes were assessed [52]. The number of electrodes employed and the frequency of re-applications were significantly lower employing Copeland electrodes. The electrode signal was significantly better as assessed by the percentage of the duration of monitoring in which recording was not possible during the second stage of labor due to poor electrode signals. No differences were found in the frequencies of complications or subjective discomfort in the mother for the two types of electrodes. In another study an assessment was made of five commonly available scalp electrodes by examining their physical and electrical characteristics, together with a randomized clinical trial in which the ECG trace quality was assessed in 50 patients [53]. It was reported that the frequency response of Copeland electrodes was such that they attenuate the ECG signal more than the baseline noise and difficulties were experienced in obtaining optimum attachment and the long, semi-rigid design increased movement artifact resulting in significantly poorer quality ECG signals. The Hewlett-Packard double spiral electrode had a near ideal frequency response, but certain design features made it difficult to apply and remain secure so the clinical signals were of intermediate quality. The Corometrics and Cetro single spirals had the most stable attachment to the scalp and a near ideal frequency response, so produced significantly better signal quality in the clinical trial. In a further randomized trial with study a group of 106 patients comparison was made of the Copeland fetal scalp electrode (52 patients) and the Meditrace spiral single helix scalp electrode (54 patients) [54]. This study reported that the Meditrace spiral fetal scalp electrode was significantly easier to apply (unpaired t-test $p<0.02$) and it also obtained higher ratings for trace quality (unpaired t-test $p<0.02$). There were no serious neonatal injuries and no difference was found between the two electrodes in this regard.

The clinical value of ST-segment analysis has been assessed in a randomized trial of 4966 full-term deliveries [55]. Monitoring with cardiotocography combined with automatic ST-waveform analysis was found to increase the ability of obstetricians to identify fetal hypoxia and to intervene more appropriately, resulting in an improved perinatal outcome.

7.4.2 Intrauterine Pressure (IUP)

The sensors and methods described above for ante-partum monitoring of uterine activity are often used for intrapartum monitoring. However, direct measurement of intrauterine pressure (IUP) becomes feasible during labor and this can be performed by means of a fluid-filled catheter connected to a remote pressure transducer or by means of a catheter-tip transducer. There is little difficulty with either method in achieving the relatively low bandwidth need for recording the slow pressure changes that occur during a contraction. However, the catheter-tip transducer is less cumbersome than the liquid-filled manometer tubing and remote transducer arrangement which requires leveling to eliminate hydrostatic pressure errors.

The positioning of intrauterine catheters or transducers can present problems. The insertion of either device is usually only carried out after the membranes have ruptured. In this situation there are collections of amniotic fluid within the uterine cavity and pressure differences may well exist between these. It is possible to record different pressures if two separate transducers or fluid-filled catheters are inserted simultaneously. Blockage of the fluid-filled catheter can also occur although this can be overcome by attaching a latex balloon to the catheter-tip.

7.4.3 Fetal pH

7.4.3.1 Intermittent Sampling

When there is lack of oxygen the rates of metabolic process in which H^+ ions are normally consumed are reduced with the result that H^+ ions accumulate in the form of lactate or other organic acids. In 1961 Saling introduced the technique of sampling blood from the fetal scalp for remote laboratory biochemical analysis, especially to measure pH [56]. Micro samples are collected and therefore the laboratory analyzer must operate with such small volumes. This method became the standard technique for assessing fetal status during labor in high-risk cases and where suspicious fetal heart rate patterns were present. Blood gas analyzers using glass pH electrodes have been steadily refined over the decades since those early days.

The Saling method is clearly invasive and intrusive, and it must also be repeated for each measurement point. Following sampling of the blood, there are other options for the in vitro analysis in addition to use of a laboratory blood gas analyzer. Very rapid analysis may be possible with a pH sensor built into the blood sampling tube. The feasibility of combining a pH electrode with the fetal scalp blood sampling facility has been investigated [57]. The pH sensor was constructed using an ionophore contained within a polymer substrate. This was then mounted in a cylindrical body, 2 cm long and 8 mm in diameter, containing an electrical heater to hold the sensor at 37 °C. Calibration was carried out in sterile buffer before use. This concept was found to work well in practice, giving a very rapid result for pH. It was concluded that the electrode would need to be manufactured at sufficiently low cost to be a single use disposable in order to avoid problems of infection, including HIV.

The blood may be decanted into a cartridge containing a pH sensor. The cartridge pH analyzer from Corometrics was really an extension of the philosophy introduced by Professors Albert and Renate Huch with their so-called 'Trough' pH electrode. With both of these technologies, the sampled blood is analyzed in a small dedicated analyzer, which can be located close to the delivery room. With the cartridge analyzer, the blood sample is dispensed from the capillary into a disposable cartridge that contains a metal/metal oxide pH electrode. The analysis is quick, and costs about £8 for each disposable cartridge.

The third option is of course the conventional blood gas analyzer, which offers automation, and the measurement of blood gases and several derived parameters. There is a cost consideration, not only for capital purchase and running expenses, but for technical support to provide quality assurance and maintenance and repair. The precision and accuracy of pH measurement offered by laboratory blood gas analyzers is perfectly adequate for serial assessment of fetal blood samples, provided always that the sample is well taken and analyzed, without too much delay, using a calibrated machine. The performance of this type of machine, coupled with the associated procedures, sets the standard by which continuous fetal scalp pH measurement is judged. The precision and accuracy must be such that reliable decisions can be made, with indication for delivery at pH 7.15, normal being >7.25, and repeat measurement after 30 min when pH is between 7.20 and 7.25.

7.4.3.2 **Continuous Fetal pH Monitoring**

The Glass Electrode

Most experience with continuous fetal pH measurement has been gained with the glass pH electrode introduced in 1974 by Stamm and colleagues [58] and subsequently manufactured by Roche. The electrode is based upon the conventional pH-sensitive glass, with an internal reference electrode and a liquid junction between a further reference and the sample fluid. A Li-Ba-Si glass having low electrical resistance was used. The glass electrode tip, which is 1.3 mm in diameter and 1.0 mm long, is positioned within a 2 mm incision in the fetal scalp. This is made using the standard blade and holder for fetal scalp sampling. A conventional screw-in needle ECG button is used to hold the electrode in place. Continuous recordings are possible and the data usually combined with FHR and contractions on the cardiotocograph.

In order to overcome initial application problems, various techniques were developed for the application of the electrode [59]. Also, a modified button electrode was produced to ensure more secure attachment [60] and success rates improved from some 50% in earlier studies to around 80%. Electrode tip design remained a problem, in part related to the mechanical limitations of the glass itself and also to the electrode shape [61]. Nevertheless, this author reported good correlation between the continuous pH values and intermittent capillary micro blood samples ($r=0.83$, $n=237$, $y=1.04\times-0.30$).

The drift of the electrode, as judged by in vitro calibration before and after monitoring, may range from zero to 0.12 pH units. Drift is partly related to duration of monitoring. During long applications removal and re-calibration may be necessary, and this is not convenient. Drift is usually caused by the adsorption of proteins to the electrode surface, which may block partly the H^+ ion receptor sites in the glass membrane and the reference electrode. In the earlier electrode design abrupt changes in pH were sometimes seen, thought to be due to amniotic fluid entering into the incision. An enlarged shoulder shape around the reference was added which overcame the problem.

Taking out failed applications and cases of excessive drift, reported correlations with scalp capillary and arterial blood pH have been moderately good, with differences ranging from 0.04 to 0.17 pH units.

The simultaneous use of the fetal scalp glass pH electrode with a heated transcutaneous pCO_2 electrode was also reported [62], and it was thought to be useful in distinguishing between metabolic and respiratory acidosis. Of course this combination of the two techniques is more complex and encumbering.

Despite the value which some workers have reported for continuous pH monitoring with the glass pH electrode it was not used widely for routine clinical situations. This was probably due to a combination of less than perfect performance with the undoubted practical difficulties in clinical use [63].

Optical Sensors

Optical methods for biochemical and physiological measurement have been popular for some time. Of direct relevance here is an optical absorbance pH sensor developed by Peterson [64]. A pH-sensitive dye, in this case phenol red, is entrapped within a hydrogen ion-permeable membrane, and the optical absorbance change of the dye following pH change is measured via optical fibers. This approach can be used to construct sensors in various mechanical forms and sizes, including a needle form. Several groups developed fetal scalp pH sensors with optical fibers, and, as with the glass pH electrode, it has been possible to incorporate the sensor into the screw-type needle fetal scalp ECG electrode [65]. So far in vivo results have been reasonably encouraging, with present drift figures of approximately 0.10 pH during the course of monitoring. The drift is, perhaps, still too high, but there are good prospects for improvement. The response time of these devices is also considerably longer than that of glass electrodes; 3–4 min for the optical device compared with about 25 s for glass. The device also contains a needle with an incorporated glass bulb pH electrode, and this performs very well, with better in vivo characteristics than all of the other types, but it is extremely difficult to construct and would be a costly single use disposable.

The area of fetal scalp presenting during labor is not an ideal site at which tissue measurements might best correlate with arterial measurements. There are known to be circulatory changes in the fetal scalp during labor; the so-called 'tonsure effect' first described by O'Connor [66] and, secondly the inevitable disruptive effects which the invasive sensor must have on the cutaneous vasculature and micro-environment.

7.4.4
Transcutaneous Gases

Work in the 1960s on the measurement of oxygen at the skin surface [67, 68] led to the development of heated oxygen electrodes to be attached to the skin surface for the estimation of arterial pO_2 [69–71]. The method became known as transcutaneous gas monitoring (Fig. 7.2). It was found that when the skin surface is heated to approximately 42 °C, the cutaneous circulation is maximally vasodilated. Under these conditions the skin surface gas partial pressure correlates closely with arterial levels. Electrochemical oxygen sensors were first developed for this purpose, and, subsequently it was shown that the basic method employing thermally induced vasodilatation can also be used for measurement of other gases, particularly CO_2, but also N_2 and N_2O. Transcutaneous gas monitoring is used widely in neonatal intensive care (see below), the relatively thin neonatal skin allowing good correlation. The method was also refined for intrapartum fetal monitoring, with both electrochemical sensors [72] and also with mass spectrometry [73]. The attachment of the fetal scalp transducer was achieved both with tissue adhesive placed in a disposable attachment ring, but also with light suction applied to an annular groove around the front surface of the device.

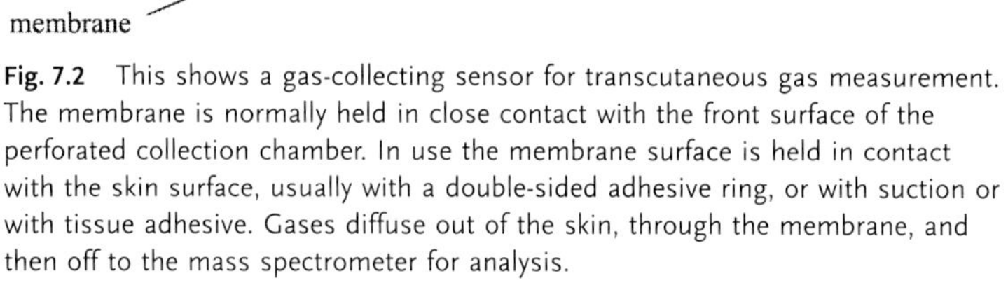

Fig. 7.2 This shows a gas-collecting sensor for transcutaneous gas measurement. The membrane is normally held in close contact with the front surface of the perforated collection chamber. In use the membrane surface is held in contact with the skin surface, usually with a double-sided adhesive ring, or with suction or with tissue adhesive. Gases diffuse out of the skin, through the membrane, and then off to the mass spectrometer for analysis.

Cutaneous blood flow is an important factor in determining the relationship between the skin surface and arterial gas levels. The method must therefore be used carefully with patients undergoing intensive care who may be in shock, because the skin surface gas levels will bear no sensible relationship with arterial values. The method cannot provide reliable information in subjects undergoing cardiac surgery with hypothermia.

7.4.5
Pulse Oximetry

The non-invasive determination of oxygen saturation with optical methods has been possible, with varying degrees of success, since the early part of the 20th century. The characteristic optical absorption spectra of oxy- and de-oxyhemoglobin are the basis of such optical oximetry. Practical realization of this method has required the use of small light sources and detectors together with appropriate processing. The pulse oximetry sensor therefore is very simple, often being just two LEDs and a photodetector attached to a piece of adhesive tape. Simplistically, a tissue segment is analyzed using the so-called cuvette model [74] in which a number of discrete wavelengths of interrogating light pass through the tissue segment before being received by the detector. The measured attenuation at each wavelength, together with knowledge of the extinction coefficients of HHb and O_2Hb at each of the wavelengths, then allows sO_2 to be determined. The length of the path taken by the light also affects the degree of absorption but if measurements are made at two wavelengths and a ratio taken then path length can be eliminated.

This basic method of oximetry suffered the drawback that it required a calibration at zero sO_2. This was overcome with the invention of pulse oximetry [75]. With this method use is made of the photoplethysmogram (PPG) that represents the variations in optical absorption seen throughout the cardiac cycle. The PPG is essentially a blood volume pulse, produced as arterial blood fills the tissue segment. It is assumed that the volume pulse is due only to arterial blood and this enables a simplified calculation of sO_2. The saturation is calculated from the ratio of optical absorbance at two different wavelengths. Examination of the absorption spectra of oxy- and de-oxyhemoglobin shows that if one wavelength is in the red at around 660 nm and the other in the infrared at around 890 nm then the maximum difference in absorption between HHb and O_2Hb can be utilized.

Pulse oximetry was first used in the newborn baby (see below) but eventually, in the late 1980s, attempts were made to apply it in the fetus during labor and delivery [76–78]. The design of suitable sensors has been a major undertaking due to the limitations of access to an appropriate anatomical site. In the newborn or adult it is relatively straightforward to use the transmission mode with, for example, the light emitters on one side of a section of tissue, such as a finger or hand, and the detector placed on the opposite side. Achieving this arrangement in the fetus *intrapartum* is clearly not possible and the reflection mode, with emitters and detector placed adjacent to each other, has been used to date.

Attachment of pulse oximetry probes to the fetal presenting part requires some dexterity and an inevitable degree of intrusion. Typically the sensor is placed on the fetal cheek or forehead. Methods of attachment were initially based on earlier experience of fetal scalp blood sampling and attachment of electrocardiograph electrodes, pH electrodes, and transcutaneous gas sensors. In this earlier work the potential problems presented by vernix, hair or blood had been confronted and largely analyzed. However, the propagation of light through fetal tissues and the possible interfering materials presents additional problems. These are very similar to those facing fetal near infrared spectroscopy and some had been addressed prior to the growth of fetal pulse oximetry.

The fetal head may become edematous with so-called caput formation and the scalp tissue is often swollen and 'boggy', and likely to accumulate de-oxygenated venous blood. A pulsatile photoplethysmograph signal is sometimes difficult to recover from this environment and even when it is, there may be substantial errors in the SpO_2 values obtained derived.

Models for photon propagation through the fetal and neonatal head has been examined using Monte Carlo and finite element methods. These and other studies have established some understanding of the relationships between emitter-detector spacing and photon paths through a variety of structures comprising of layers or shells with regions of absorbing and scattering material. Not surprisingly, increasing emitter–detector spacing increases the depth of penetration. Furthermore, the wavelength of the emitted light also influences penetration, infrared having greater penetration than red. This is relevant to the design of fetal pulse oximetry sensors.

Clinical studies of fetal pulse oximetry have been carried out since the late 1980s. These initially helped to establish the practical procedures for sensor attachment and, as these began to be implemented, there was a slow but continued increase in the acceptance of the method despite the recognition of limitations. The first question to be answered was whether or not the oxygen saturation values derived from pulse oximeters accurately reflect fetal blood values. This is not a straightforward matter. Standards for comparison are fetal scalp blood samples measured with a laboratory CO-oximeter. If such samples are taken from scalp tissue adjacent to the pulse oximeter sensor then the comparison is at least based on determinations made within the same tissue. However, this does not necessarily guarantee reliable comparison with central blood oxygen saturation. This difficulty has long been debated.

It has been shown, both by modeling of photon pathways and by means of animal studies, that accumulation of venous blood in the scalp beneath the pulse oximeter sensor can lead to a 30% underestimate of SpO_2. The analysis showed that this was due to increased relative penetration of underlying tissues by the infrared light. The effect was reduced by using emitters at 735 nm and 890 nm instead of the more usual 660 nm and 870 nm [79].

The second and equally important question is whether or not fetal SpO_2 monitoring can improve fetal surveillance by enabling timely detection of fetal hypoxic episodes. The establishment of a threshold value of 30% saturation appears to have led to a positive view on the usefulness of intrapartum pulse oximetry in this regard [80]. A multi-center study was carried out and the conclusion from this was that low fetal arterial oxygen saturation of $<30\%$ for 10 min or longer correlate significantly with low scalp pH values and have a predictive value concerning fetal outcome.

There are likely to be on-going efforts to improve the performance of reflection-mode sensors, perhaps to establish optimum emitter–detector spacing and choice of wavelengths. At the same time, transmission-mode sensors using helical screw-in needles containing emitters with a detector placed on the skin surface above the needle, may find a place if they can be optimized and be shown to be superior to the reflection mode sensors under real clinical conditions. Naturally, such transmission sensors have the disadvantage of being invasive.

7.4.6 Fetal Near Infrared Spectroscopy (NIRS)

A development of oximetry that emerged in the late 1970s was near infrared spectroscopy, NIRS. It was observed that penetration of tissues was significantly greater by light at wavelengths of about 900–1000 nm than it was by light in the visible or UV parts of the em spectrum. This opened up the possibility of measuring aspects of oxygenation in deeper tissues including the brain. In addition to the measurement of hemoglobin oxygenation there was the prospect of measuring the redox state of the respiratory enzyme cytochrome oxidase, or cyt aa3. Possible application of this method in the newborn was examined eagerly and then eventually intrapartum fetal surveillance with NIRS was also proposed [81] and examined in laboratory studies [82], and in clinical studies [83].

As is described below the method of NIRS requires that appropriate wavelengths of light be transmitted into tissue and emerging light be collected and analyzed to yield certain measures related to hemoglobin oxygenation. The practical challenge in applying the method to intrapartum monitoring is how best to interrogate the fetal tissues in a reliable meaningful way, given the restricted access. There have been progressive developments of sensors, or probes, together with methods of analysis that have allowed intrapartum NIRS measurement to be made with steady improvements. The first approach was to use two separate probes developed for transmission measurements in the neonatal head. These initially contained a 45° mirror to turn the light from the incoming or to the outgoing glass fiber bundle. Later probes replaced the mirror with a glass rod with a right angle bend (Fig. 7.3). This has lower insertion loss than the mirror.

The modified neonatal probes were then further modified to make them of lower and smoother profile for easier insertion and placement on the fetal head. This was achieved using clear acrylic molded into a flat elongated disc, tapered around its perimeter. However, clinical studies showed that these probes were subject to movement artifact, especially during uterine contractions [84]. It was also found to be difficult to achieve placement on opposite sides of the fetal head. The probes tended to be placed almost side-by-side. This raised the question about the pene-

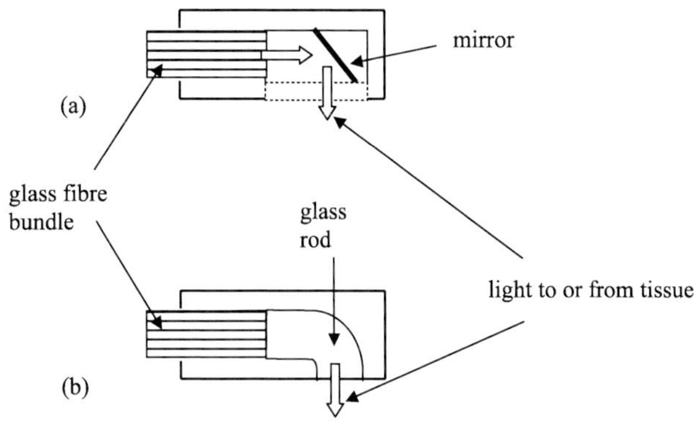

Fig. 7.3 Probes used for NIRS monitoring. Light is conveyed to or from the tissue using: a) a mirror; b) formed glass rod.

tration pattern of light that is achieved with different spatial arrangements. With probes placed effectively side-by-side the penetration is only superficial. It was found in modeling studies that for application in the fetal and neonatal head the probes must be at least 2.5 cm apart to interrogate cerebral tissue [74]. In this configuration propagation through the tissue from transmitter to receiver may be considered to be multiple scattering that is the 'backscatter' or 'reflection' mode. True 'transmission' mode is difficult to achieve in the human fetus.

Even though the 'backscatter' mode was found to give interpretable physiological data from human studies, the use of two separate probes was found still to be sensitive to movement, as the two probes moved relative to each other. For this reason several arrangements were investigated to fix the inter-probe distance mechanically. In one arrangement the two probes were fixed to either side of a hinged body. This effectively constrained the probes, thereby achieving semi-fixed inter-probe spacing whilst allowing some flexibility and thus minimizing the possibility of mechanical trauma to the fetus or to the mother. This arrangement was tested clinically but was found to be too rigid. This was largely due to the glass fiber bundles with their external spiral wire protective sheath, but also due to the hard acrylic material. These problems were overcome by using smaller diameter glass fiber bundles fixed at the patient end by incorporation within a silicone rubber molding. A further feature was added whereby low-level negative pressure was applied to the probe body in order to fix it to the fetal head. This probe design has allowed successful clinical studies to be performed [85, 86].

Clinical experience with NIRS has shown that changes in fetal brain oxygenation occur during labor. Several patterns of change have been found. A common pattern seen in early studies was that of a simultaneous decrease in HHb, HbO_2, and Hb_{tot} in synchronism with uterine contractions. This cannot be explained by physiological considerations and was concluded to be artifactual, caused by inter-probe spacing being modulated by uterine pressure changes. The second pattern, commonly seen in most reported studies, is termed the divergent pattern [87].

With this there is an increase in HHb and a decrease in HbO_2 in synchronism with uterine contractions, while Hb_{tot} changes are variable. This pattern suggests a slight induced hypoxic change during each uterine contraction. Other general observations have been made including: progressive hypoxia with on-going labor: reversal of postural hypoxia following maternal oxygen administration [86].

7.5 Neonatal Monitoring

7.5.1 Temperature Monitoring

Maintaining safe body temperature is a significant challenge for the small, preterm, immature baby. Incidence of hypothermia (a core temperature of $<36\,^{\circ}C$) is associated with increased mortality and morbidity. The high body surface area to volume ratio disadvantage in the small preterm baby, increases proportionately the heat losses (and therefore energy demands) by evaporative, radiative, conductive, and convective mechanisms as compared with older children and adults. Typically these babies will be cared for within a carefully controlled thermal environment, achieved with either warm air incubators or radiant warmers. It is therefore important to have a complete picture of thermal status of the baby from the moment of birth. Temperature measurement and monitoring can contribute to the creation of this.

At the moment of birth, measurements of the baby's core temperature (rectal or esophageal temperature) are useful to detect infection and to assess the likelihood of subsequent hypothermia [88]. An indication of core temperature from body surface-mounted probes can be achieved but this method has a long response time and is not suitable for rapid post-partum assessment. Direct rectal measurements, either with a glass thermometer or with an electronic sensor (thermistor or thermo-couple) are used. Under controlled environmental conditions the fall in rectal temperature over the first 15 min post-partum can indicate the baby's ability to respond to thermal stress, and a continuous recording over this period is therefore useful. The interpretation of this period of rectal temperature recording is simplified if there is also a simultaneous recording of the environmental temperature immediately around the baby, since this may not be stable. Continuous or semi-continuous monitoring of rectal and environmental temperature at delivery can then be extended into the following hours or days of intensive care.

The peripheral temperature is best measured with a loop thermistor sensor taped around the great toe. Minimal tension should be applied. Otherwise cutaneous blood flow will be reduced and falsely low readings will result. All surface-mounted temperature measurement sensors must be protected from radiant energy, for example that emitted by overhead warmers (this can be done with silver foil) otherwise a falsely high reading will be obtained. Also, the sensor should not be covered with a thick layer of insulating tape or other material because the skin surface temperature being measured will rise towards core temperature. Use of a perforated double-sided adhesive disc over the sensor, with the connecting cable lightly taped down to the skin a few cm away, is a successful approach.

Esophageal, auditory canal, or axillary temperatures may all be used in place of rectal temperature for single or continuous measurements. Of these, axillary measurement is the easiest in small babies, and it lends itself well to continuous monitoring with thermistor or thermocouple sensors attached as described above.

A further method for obtaining core temperature measurements non-invasively is the so-called zero temperature gradient [89, 90]. With this method a skin-mounted temperature sensor is covered by a thermal insulator. This makes the skin temperature rise towards the core temperature. A second sensor is positioned on the surface of the insulator, and this, too, is further covered by a small electrical heating coil. The heater increases the temperature of the skin surface further, until there is no gradient between the two temperature sensors, which may take 15–20 min after the device has been first applied. At this time both temperature sensors will be recording the core temperature. This method has been evaluated in babies under special environmental conditions, and good correlation has been found between the zero temperature gradient method and rectal temperature. However, the basic method can only function if ambient temperature is lower than the baby's core temperature and this is often not the case in incubator-nursed babies. Modification to include an electronic cooling element in the sensor can overcome this problem.

7.5.2 Breathing

In the small preterm baby the immaturity of respiratory control mechanisms, together with inadequate supply of the vital surfactant material in alveoli, means that the mechanics and regulation of breathing and gas exchange in the lungs become a significant focus of attention within clinical care and in research. The continuous surveillance of the baby's breathing efforts has become almost a traditional symbol of neonatal care and sensor developments have been a key part of the history and the practice. Babies under care may stop breathing for short or long periods of time; these are called apneic attacks. The depth and frequency of breathing can also become periodic. The assessment of the impact of such features of breathing requires detection of apneic attacks and monitoring of breathing patterns. However, the reliable, and clinically useful, monitoring of breathing patterns and the detection of apnea are still difficult to achieve in all preterm babies and especially in the very small babies. Problems arise from the difficulties of designing appropriately small, sensitive, reliable, and unencumbering sensors and connecting cables and leads. Over the last three decades many sensing methods have been employed. These include those that based on measurement of: transthoracic electrical impedance; diaphragm electromyograph; thorax and abdomen dimensional changes; direct and indirect pressure changes; body movements; breath sounds; air flow [91, 92].

Monitoring of trans-thoracic electrical impedance to extract the so-called impedance pneumogram is conveniently combined with ECG monitoring using the same electrodes. Even with care in application, though, at best this method still only provides detection of apnea and calculation of respiratory rate. The amplitude of the impedance change may correlate with tidal volume but only under very carefully controlled conditions, including attention to electrode positioning. Further significant improvements are also still needed in the impedance monitors themselves to deal with movement artifacts.

Abdominal surface electrodes have also been used to pick up the diaphragm EMG, and processing of this was found to produce a signal that correlated with tidal volume [93].

Dimensional changes of the chest and abdomen associated with breathing can be recorded with strain gauges, inflatable jackets [94], and inductive straps [95]. None of these methods is ideally suited for the very smallest babies because of the encumbrance of the necessary straps or bands which must be placed around the chest or abdomen. Magnetometers capable of measuring the relative displacement between two small coils placed on the chest or abdomen, have also been used for both apnea alarms and to record depth and frequency of breathing [92, 96]. Changes in curvature of the chest or abdominal wall associated with breathing can be detected by a small plastic 'pressure capsule' [97]. This technique has been found to be acceptable as an apnea detector, and is improved in terms of reliability by using two capsules, one to detect thoracic cage movement and the other abdominal movements.

There is an interest in assessing lung mechanics in babies under intensive care. Measurement of airways resistance and pulmonary compliance has been carried out by measuring air flow with miniature pneumotachographs, and respired volume and trans-pulmonary pressure. Recording of intraesophageal pressure changes in these procedures [98] can also be useful for more general respiratory monitoring in some very small babies. Although a thin latex rubber balloon (wall 0.045–0.075 mm, length 3.5–5.0 cm, 0.76 cm diameter) attached to a flexible plastic catheter can be used to make accurate measurements of the pressures, catheter-tip pressure transducers have also been used satisfactorily. Furthermore, a liquid filled, open-ended catheter or feeding tube is almost as good for quantitative pressure monitoring. This latter arrangement has been found to be adequate for apnea monitoring in the very small baby. In fact the feeding tube can be modified to incorporate small wire contacts to enable either EMG detection or electrical impedance monitoring of breathing via the esophagus. This is more convenient and more reliable than using surface electrodes on the chest or abdomen.

Apnea detection without any direct attachment to the baby first became feasible with the introduction of Lewin's air-filled multi-compartment mattress [99]. In this device the baby's weight re-distribution during breathing shifts air between separate compartments in an inflated mattress. The air flow is detected by a heated thermistor bead. This monitor has been used widely for several decades, although it lacks the sensitivity needed for the very smallest babies. Mattress-type apnea monitors using the same approach of detecting the weight re-distribution produced by breathing were improved by using very sensitive, thin, pressure-sensitive pads. These monitors (e.g., Draeger Medical Ltd; Eastwood and Son Ltd) also perform well with the very small babies, although it is important to position the pad such that cardiac movement during apnea does not prevent the alarm being activated.

Chest movements can still continue in obstructive apnea and therefore will not be detected by the simple apnea monitors described above, since none of these detects respired air flow. Small thermistors or thermo-couples can be used to detect nasal gas flow, but this approach is not always entirely reliable and cannot be used easily as a clinical routine.

7.5.3
pH and Blood Gases

7.5.3.1 Invasive Sensors

Invasive measurement of the concentrations of key biochemicals in blood can often provide the most direct, and immediate, indication of clinical condition. Improvement or deterioration in a baby's cardiorespiratory status could be detected quickly by means of continuous arterial O_2, CO_2, and pH measurement if suitable sensors were available. Several attempts have been made over the years to develop such sensors for routine clinical use, with only limited success. To date, use of invasive blood and pH sensors is not widespread.

Intravascular sensors are most conveniently incorporated into flexible catheters or cannulae, with dimensions appropriate for insertion into either a peripheral or central vessel. In critically ill newborn babies it is common practice to insert a catheter into the umbilical artery, and sometimes the vein, to access the central vessels.

Arterial pO_2 may be most easily measured electrochemically and there is a substantial background history in the development and use of such sensors. The basic membrane-covered amperometric oxygen electrode needs to be attached to a suitable catheter and selection of materials and mechanical design is of utmost importance to ensure safe and reliable operation. Since Clark's invention of the membrane-covered oxygen electrode many groups have developed catheter-tip devices [100–108] (Fig. 7.4). Each of these designs employs a particular approach for solving the problem of attaching the sensor and the membrane safely and securely to the catheter tip, and these have been reviewed [108]. The methods include: 'O'-ring fixation; screw-cap; snap-cap; push-on collar; solvent casting or dip-coating. In order to achieve a smooth external surface, which is important in relation to hemocompatibility, membrane attachment by 'dip-coating' appears to have certain advantages. This method also allows membrane thickness to be varied, and thus alter sensor response time, simply by varying the number of coats applied and the viscosity of the polymer solution. This approach is possible with a number of relevant polymers, such as the polyurethanes, polystyrene, and the poly vinyl chlorides (Fig. 7.5).

The effort to produce intravascular sensors for specific use in the newborn baby began in the mid-1960s. The aim was to incorporate the sensor(s) into 5 FG (1.67 mm diameter) catheters being used routinely for neonatal clinical care. Most interest was with O_2 sensors and the devices produced largely used Ag cathodes and Ag:AgCl reference anodes. As oxygen molecules are electrochemically reduced a current proportional to O_2 concentration is produced according to the following relationship derived from Fick's law and Faraday's law

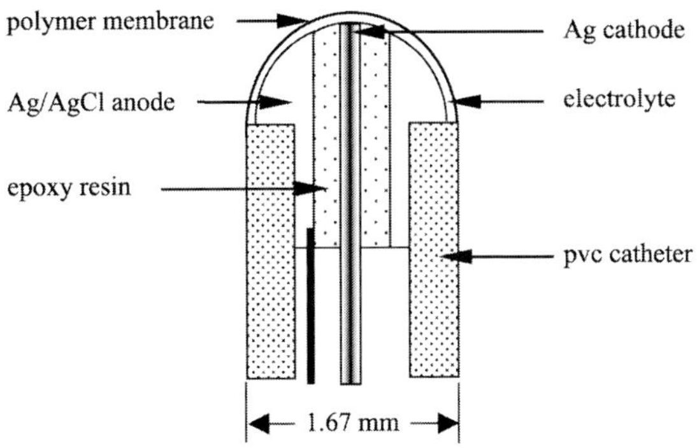

Fig. 7.4 Diagram of a catheter-tip amperometric oxygen sensor. The polymer membrane is dip-coated from a solution. Electrolyte may be gel or crystalline.

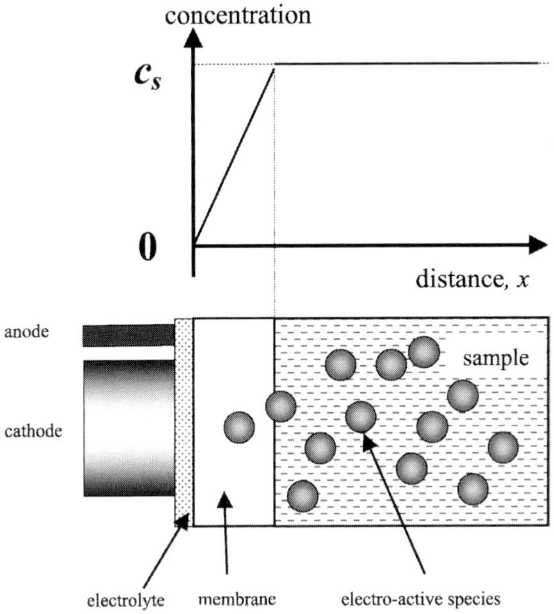

Fig. 7.5 Operation of membrane-covered amperometric sensor. The species concentration gradient from the sample bulk, through the membrane, to the cathode surface is also shown.

$$I_d = n(FAD_s/a)c_s \tag{7.5}$$

where a is the membrane thickness, A is the cathode area and D_s is the diffusion coefficient of the electro-active species, s, in this case O_2, in the membrane. The variables F, n, and c should be described as well for student readers.

The cathodes were large, approximately 150 μm diameter, by comparison with the earlier microelectrode designs which may only be 1–10 μm. Electrolyte was either dispersed over the anode–cathode surface as a powder or in gel form. This assembly then needed to be covered by a gas permeable membrane that could be attached securely to a supporting catheter. The choice of membrane material was restricted in all earlier devices due to the fact that plasticized poly vinyl chloride catheters were used almost exclusively. For this reason many pO_2 sensor designs also employed PVC (polyvinylchloride) membranes. The PVC could be dissolved in a suitable solvent (for example, cyclohexanone) and then the sensor assembly dip-coated to produce a membrane of a thickness determined by the viscosity of the solution and the number of successive coatings deposited. Typically membranes were 25 μm thick.

The limitation of plasticized PVC as a membrane material is that the material swells and thickens as it becomes hydrated. The membrane thickness therefore changes thus leading to a change in sensor sensitivity and response time. The plasticizers used for both the membrane and the supporting catheters are generally poorly bound to the base material and they leach out into the blood stream. This makes the catheter less flexible and is an additional cause of sensor instability since plasticizer content in the membrane influences membrane oxygen diffusion coefficient.

Alternative membrane materials have been used for intravascular pO_2 sensors including a polystyrene/butyl rubber co-polymer [106] low density polyethylene [108] and polyetherurethane [109]. With all materials there is the challenge of achieving secure attachment of the membrane to the sensor and catheter. Solvent casting of membranes to catheters of the same base material, such as PVC to PVC, produces the best attachment. In the case of the polystyrene/butyl rubber co-polymer the fixation to the PVC catheter is not a chemical bond but is probably merely a mechanical fit. The polyethylene membrane was heat welded to a polyethylene supporting tube and this achieved very secure attachment, as well as a rapid response time, around 3 s, and stable performance. Use of polyetherurethane as a membrane material was also combined with a move from the PVC catheter to use PU as the catheter material as well. This provided excellent attachment of the membrane, by solvent casting, as well as eliminating all of the problems associated with plasticizer leaching.

Invasive optical sensors have also been developed for blood oxygen monitoring, and the first approach towards this of relevance to neonatal application was with fiberoptic oximeter catheters originally developed for arterial, mixed venous or intracardiac oxygen saturation measurement in adults. As already described above, the absorption spectra of HbO_2 and HHb allow the relative proportions of these two components to be measured, thus giving oxygen saturation

$$sO_2 = HbO_2/(HHb + HbO_2) \times 100\% \,. \tag{7.6}$$

Enson et al. [110] described an invasive oximeter comprising of a cardiac catheter containing a glass fiber-optic bundle that could be inserted through a needle. This design was further developed by commercial organizations, such as Oximetrics, using polymer optical fibers contained within polyurethane catheters. Good performance was achieved with these devices in newborn babies [111]. A combined fiber-optic oximeter and electrochemical pO_2 sensor has also been reported [112].

Oxygen in blood may also be measured with a sensor based on fluorescence quenching [113]. A fluorescent dye, such as perylene dibutyrate, is adsorbed to organic beads contained within a hydrophobic gas permeable membrane. The dye is excited with blue light (468 nm) and it emits radiation at 514 nm (green) and pO_2 is calculated from the Stern-Volmer relationship

$$pO_2 = \text{const}\{(I_{\text{bl}}/I_{\text{gr}}) - 1\}^m \tag{7.7}$$

Monitoring other gases with invasive sensors has also been attempted. Most interest has been in the measurement of pCO_2, with an electrochemical potentiometric pH electrode surrounded by a CO_2 permeable membrane, the Stow-Severinghaus electrode [114] being the most popular approach. The use of a glass pH electrode can present fabrication problems, and metal/metal oxide pH sensors have been used, with antimony, iridium, or palladium [115]. It is also possible to measure CO_2 with an amperometric sensor in which the electrolyte is non-aqueous, for example dimethyl sulphoxide.

A fundamental problem with intraarterial sensors, and indeed with conventional catheters, is that of hemocompatibility. As is well known, a device placed within the vascular space will tend to have proteins adsorbed to its surface. Factors influencing this adsorption process particularly include the material surface chemical and physical properties, such as charge, topography, and hydrophobicity/hydrophilicity [116]. Examination by scanning electron microscope of surfaces of conventional catheters and sensors after several days of implantation generally shows substantial deposits of plasma proteins, often with fibrin networks and red cells adhered to these [117]. This is potentially hazardous, since protein and fibrin casts may break away and be swept into the vascular tree, possibly blocking small blood vessels. There is a further risk that thrombus may be formed.

The earliest attempts to improve the hemocompatibility of intravascular sensors consisted mainly of trying to produce 'smooth' surfaces. It was realized that the problems lay more with the larger surface area of the catheter than it did with the small area of sensor membrane. Also, studies showed that the area around the catheter blood sampling eye was most likely to be the initiation point for protein adsorption and subsequent events. This was thought to be due both to rough spots around the sampling eye as well as to stagnant blood within the catheter sampling lumen. This led to a so-called 'end hole' catheter and sensor configuration [118]. In this design the catheter blood sampling lumen terminates with an end-sampling orifice, facing into the on-coming blood flow. The anode and cathode of the pO_2 electrode form annular rings around the sampling hole. This arrangement has been found to give freedom from fibrin or clot formation at the catheter tip.

A further strategy to improve hemocompatibility has been to employ surface treatments for the catheter and the sensor. This approach has included the use of hydrogels to produce a hydrophilic surface, thought to resist protein adsorption, and also the attachment of heparin derivatives to prevent thrombus formation [119]. These two methods tackle different parts of the chain of events that may occur when blood comes into contact with the catheter and sensor surfaces.

A further strategy to improve hemocompatibility has been to use synthesized polymers having biochemical properties mimicking those of the red blood cell membrane [120, 121], since this membrane is compatible with plasma proteins, other blood cells and the vascular endothelium. A co-polymer has been developed in which the phospholipid properties have been created with a side chain and mechanically robust backbone added using *n*-butyl methacrylate [122]. This material was evaluated for use as a sensor membrane coating and it was found that protein adsorption was extremely low [123]. The performance of an intravascular pO_2 sensor constructed with such material was evaluated and found to be superior to sensors with heparin-treated surfaces [124, 125].

The experiences with electrochemical pO_2 sensors and optical sO_2 sensors for umbilical artery insertion soon led to the development of so-called multi-parameter devices. These have comprised of permutations of electrochemical and optical sensors for pO_2, sO_2, pCO_2, and pH. A dual sensor comprising of an amperometric pO_2 electrode and a Stowe-Severinghaus pCO_2 electrode was developed [126]. The pCO_2 electrode was composed of a glass pH electrode covered by a CO_2-permeable membrane and this was positioned at the tip of a 1.7 mm diameter supporting catheter. The cathode of the pO_2 electrode was arranged such that its active surface was flush with the lateral surface of the catheter wall, 2 mm back from the tip, and was covered by a gas and ion permeable membrane. This amperometric oxygen electrode was operated with a remote reference electrode. Although this device was shown to perform well in in vitro tests, it was not used to any significant extent clinically.

A triple sensor probe, developed from a similar device intended for use in adults, has three optical sensors for pH, pCO_2 and pO_2. This has followed a long period of research by several groups to perfect dyes-suitable stable oxygen quenching and pH-dependent fluorescence.

The use of oxygen saturation for the control of arterial oxygen levels in the small preterm baby is not entirely satisfactory from the theoretical viewpoint. The 'S' shape oxyhemoglobin binding curve means that at relatively high levels of oxygenation very small changes in saturation accompany very large changes in oxygen partial pressure. The prevention of hyperoxemia in terms of oxygen partial pressure is therefore very demanding of precision and accuracy in oxygen saturation measurement. This situation is worsened by the significant 'left-shifting' of the binding curve by fetal hemoglobin. Thus if one's clinical approach is to achieve control of arterial pO_2 by O_2 saturation monitoring it is essential, firstly, to determine the precise position of the oxyhemoglobin binding curve and, secondly, to ensure that the precision and accuracy of the oximeter is adequate.

7.5.3.2 Non-Invasive Techniques

The transcutaneous techniques described above have been used widely for continuous monitoring of pO_2 and pCO_2 in the ill newborn, and can thereby complement intermittent arterial blood sampling. The relationships between skin surface and arterial values of pO_2 and pCO_2 are influenced by epidermal thickness and gas permeability, as well as many other factors. The effect of the heating of the transducer has a large effect on cutaneous blood flow but also on the oxyhemoglobin dissociation curve. In the very small preterm baby this leads to skin surface pO_2 values which may be approximately 10% above arterial pO_2 in the range 8–12 kPa (60–90 mmHg). The theoretical work by Severinghaus [127] and Lübbers [128] together with more recent clinical results has emphasized the fact that the skin surface–arterial pO_2 relationship is nonlinear, the skin surface pO_2 being significantly lower than arterial pO_2 above approximately 13 kPa (100 mmHg). The electrical heating of skin surface gas sensors is critically important, and can be problematical. On the one hand adequate heating of the skin is necessary to induce the maximal vasodilatation which is essential for the pO_2 measurement to be essentially independent of moderate fluctuations in cutaneous perfusion. On the other hand the small preterm baby's skin can be very sensitive to the required heating and burning can occur. To reduce the risks of causing skin burns the sensor needs to be moved to a new site every 1–2 h, depending upon the size of the baby. A skin surface temperature of 42 °C is probably optimal in very small babies, but it is often not appreciated that the sensor temperature which is set on the associated monitor may be as much as 2 °C higher than the skin surface temperature achieved beneath the sensor. This temperature gradient between the sensor and the skin surface also varies according to the precise design of the sensor.

Skin surface pCO_2 monitoring is by no means as widely used as skin surface pO_2 monitoring. Generally the sensor temperature is less critical for pCO_2 monitoring, although maximal vasodilatation must still be achieved. Once again it is essential for individual centers to establish their own protocols for this method to produce useful information.

When both skin surface pO_2 and pCO_2 are to be monitored there is a certain practical advantage in employing a single sensor which combines means for both pO_2 and pCO_2 measurement. This reduces the number of devices which must be attached to the relatively small exposed surface area of the baby, and furthermore it is obviously much less demanding in terms of sensor calibration, membrane changing, and application site rotation.

There is now widespread use of pulse oximetry for non-invasive estimation of arterial oxygen saturation. This method, originally developed for use in adults [129, 130], utilizes the pulsatile blood volume change in a digit or palm to calculate oxygen saturation by employing spectrophotometric principles. The earliest pulse oximetry sensors were not ideal for use in the small newborn baby, being too large and cumbersome. However, current probes, employing LEDs and photodetectors, are much improved. The method has become popular because it is very much easier to use than conventional transcutaneous gas monitoring. The sensors do not need to be heated, and the optical sensors are very much more stable than the electrochemical sensors for pO_2 and pCO_2 measurement. However, pulse oximetry should be used with caution in small preterm babies due to the limitations imposed by the shape of the oxyhemoglobin binding curve at high levels of oxygenation.

The precision and accuracy of the various commercially available pulse oximeters has been examined by several groups. Systematic differences have been reported. For example, the Nellcor N-100 pulse oximeter was found to overestimate consistently the true S_aO_2 [131]. However, the Ohmeda 3700 Pulse oximeter has been reported to read consistently lower than the Nellcor N-100 [132].

7.5.4 Cardiac Monitoring

7.5.4.1 Blood Pressure

Monitoring of arterial blood pressure in the sick newborn baby is important, since it can provide valuable data with which cerebral perfusion may be evaluated. In some neonatal intensive care units it is still common to insert an umbilical artery catheter for blood gas and pH measurement, and arterial pressure measurement via the catheter is then useful in its own right, as well as providing an important guide to catheter patency. These catheters are usually 5 FG (1.67 mm diameter) but in very small babies it is necessary to use smaller catheters, such as 4 FG, and the problem of catheter lumen blockage due to thrombus formation is increased. The arterial pressure waveform (phasic) can be monitored reliably through small bore catheters by means of low volume-displacement pressure transducers, connected either directly to the Luer fitting of the catheter, or at 60 cm distance attached by saline-filled manometer tubing. Care must be taken to ensure air bubbles are not allowed to collect in the system otherwise the frequency response will be limited and the phasic signal attenuated. Ideally, the phasic pressure waveform is displayed on a monitor screen so that smoothing and damping of the waveform can be easily detected. The Luer fitting transducers have been commercially available for many years [91] and now single use disposable transducers have become widely used. The use of conventional blood pressure transducers was greatly simplified by the introduction of disposable pressure domes, obviating the need for the cumbersome and time-consuming sterilization of the unit. Catheter patency can be improved by continuous infusion through the catheter by means of 'intraflow' devices or similar. In spite of these precautionary measures, arterial catheters must still be regarded as potential hazards through thrombus formation, and improved catheter materials are needed. It is now well established that plasticized PVC catheters have far worse hemocompatibility characteristics than the more recently introduced material polyurethane. In spite of this PVC catheters continue to be used in many centers.

Non-invasive arterial pressure measurement is clearly desirable. Many systems are now available based on the use of limb encircling cuffs and the so-called oscillometric principle (Chapter 4). With this technique mean arterial pressure is first measured from a measurement of the cuff pressure at which cardiac-related pulsations in the cuff reach maximum amplitude. Subsequently systolic and diastolic pressures may be obtained. These arterial pressure monitors are now widely used for semi-continuous monitoring. However, the cuff needs to be matched to the size of the baby's limb, otherwise systematic errors will be introduced. Furthermore, repeated automatic inflation of the cuff is not to be recommended in very small fragile babies with poor peripheral perfusion. A non-invasive technique for continuous arterial pressure measurement has been described, but this is not yet available for routine use [133].

7.5.4.2 Neonatal ECG and Heart Rate

The initial resuscitation and stabilization of the preterm baby immediately after birth is aided by the use of quick and efficient means for heart rate monitoring. This is best achieved through the use of the ECG, which is the most likely signal to be able to give a reliable heart rate indication. Approaches used for the attachment of ECG electrodes in this situation include conventional adhesives, clips, clamps, and even needles. However, the baby is often wet immediately after birth. Electrodes can be attached by means of suction with a relatively low negative pressure (around 80 mmHg) applied in an annular groove around a soft plastic molding containing an Ag:AgCl electrode [1]. This level of suction is tolerated by the skin of the baby for the period of resuscitation. Three of these electrodes may be placed on the resuscitation table and the wet baby laid directly on them for almost immediate use.

In the late 1960s and early 1970s, as neonatal intensive care was developing rapidly, there was little appreciation of the need for ECG electrodes matched to the needs of the small, fragile preterm baby. As a result, the electrodes used were often adult electrodes, some 25 mm diameter, almost inflexible, with heavy connecting cables. This led to inefficient monitoring as electrodes became easily detached and often a baby's thin skin was damaged by the mechanical forces and inappropriate adhesives. Gradually manufacturers attempted to improve electrodes for the neonate, using smaller, more flexible electrodes and alternative adhesives.

At that time there was an interesting attempt to eliminate the electrode attachment and contact problem entirely. An electrode system was developed that was comprised of a thin metallized plastic plate on which to place the naked baby and thereby eliminate individual fixed electrode connections. This approach was unencumbering for the baby and convenient for nursing staff, but signal reliability was poor. Perhaps this approach could now be improved by using modern electronic and computing techniques.

The evolution of neonatal ECG electrodes was slow and chemical adhesive attachment was for a long times something of a compromise. Aggressive adhesives are more likely to keep the electrode in place, but they cause skin trauma. Milder adhesives lead to more frequent electrode detachment that produces false alarms. Electrodes constructed with the natural gum *Kuraya* were then found to be very successful [134]. The gum has very good properties and is skin compatible. The moistened gum adheres well to even the smallest babies, and the incorporated thin silver film electrode does not need the normal electrode conductive gel or cream. The gum must be moistened periodically, especially when the baby is nursed under a radiant warmer, but the incidence of skin trauma is very low.

The cables and leads for connecting electrodes on babies to monitors have also needed steady evolution. The most appropriate scheme is to have a relatively robust cable connected to the monitor, with a small connector block at its end for attachment of three or four lighter individual electrode leads. Very fine, light, and flexible electrode leads have some advantages for the smallest babies, but they can become tangled.

An alternative approach for the elimination of electrode contact problems is to take advantage of nasogastric tubes whenever they are in place. Polymer feeding tubes can be constructed easily to incorporate small conductive 'pick-up' regions, using wire or ring elements, and very reliable esophageal ECG monitoring can be achieved.

As described above in fetal monitoring the calculation of FHR variability can have some clinical value, and there has been considerable interest in studying neonatal HR variability. An example of this is in the study of the autonomic control of heart rate and respiration during the neonatal period in infants with prenatal exposure to cocaine [135]. Recordings of the neonatal ECG were obtained from 15 cocaine-exposed and 13 non-exposed full-term neonates at 2 weeks of age. The power spectrum of the R-R intervals was computed from the ECG to obtain the total power (0–2 Hz), and spectral power in the high-frequency (HFP, 0.3–2 Hz), mid-frequency (MFP, 0.1–0.2 Hz), and low-frequency (LFP, 0.03–0.1 Hz) bands. Cocaine-exposed neonates showed enhanced heart rate variability reflected by an increase in spectral power across all frequency bands. It was concluded that infants exposed to cocaine in utero show differences in the modulation of heart rate, reflecting an increase in both vagal and sympathetic influences. Current ECG monitors have the option for calculation and display of short-term and long-term heart rate variability but it is not obvious if this feature is being used routinely.

7.5.5 Cerebral Monitoring

In the late 1960s and early 1970s there was increasing interest in the study of neonatal cerebral hemodynamics due to the growing importance of hypoxic-ischemic brain injury in small preterm babies in terms of both mortality and morbidity [136]. Many methods were being developed for the purpose of elucidating the underlying pathophysiology. This included Doppler ultrasound, B-mode ultrasound imaging, X-ray CT, MRI, and MRS, positron emission tomography, electrical impedance plethysmography, electrical impedance tomography, and non-invasive intracranial pressure monitoring [137].

The clinical importance of hypoxic-ischemic brain injury in small preterm infants places great emphasis on the need to monitor different aspects of cerebral physiological and biochemical status.

The perfusion of cerebral tissues, and thereby the oxygenation of brain cells, is influenced by arterial pressure, cerebrovascular resistance and intracranial pressure (ICP). Raised ICP can lead to a reduction in cerebral blood flow and therefore knowledge of ICP together with measurement of arterial blood pressure can provide a guide to cerebral perfusion. In a mature infant there are cerebral autoregulatory mechanisms that can ensure the maintenance of near normal cerebral blood flow in the face of significant changes in arterial pressure and ICP.

7.5.5.1 Intracranial Pressure Monitoring

Concerns regarding the possibility of hypoxic-ischemic brain injury occurring in the small, preterm baby led to the evaluation of ICP monitoring as a potential clinical tool and as a research methodology. The purpose of ICP monitoring has been to gain information about the adequacy of cerebral perfusion. ICP together with arterial blood pressure can give a guide to total cerebral blood flow. Elevated ICP thus indicates possible reduced cerebral perfusion and oxygen delivery.

Both invasive and non-invasive methods have been used. Invasive ICP measurement in children and adults has been long established [138, 139]. It can be carried out with an indwelling cannula attached to an external pressure transducer, or with a catheter-tip pressure sensor, or with a screw pressure sensor. The principles for these invasive methods were established in children for the management of hydrocephalus. With a saline-filled catheter-manometer tube arrangement the leveling of the pressure transducer with the tip of the catheter is important to avoid errors due to hydrostatic pressure. Use of a transducer attached directly to the catheter eliminates most of this source of error. Catheter-tip transducers eliminate this problem entirely and one such sensor, originally designed for invasive use in adults, is based on a pneumatic control system that employs fiber optics. A flexible diaphragm carries a small mirror which reflects a beam of light from one fiber to a one of a further pair of fibers. An error signal from the reflection fibers feeds the pressure controller so that the diaphragm is held flat. Under these conditions internal pressure of the sensor is equal to the surrounding fluid pressure.

Clinical invasive ICP monitoring in newborn infants has been performed by several groups. The sites for measurement are the subarachnoid space, the lateral ventricles or the subdural space. Levene and Evans [140] used a cannula attached to an external pressure transducer by manometer tubing to measure subarachnoid pressure. Potential problems of invasive methods include infection and leakage of cerebrospinal fluid.

Non-invasive ICP monitoring in newborn babies has been performed with sensors attached to the anterior fontanelle, or 'soft spot'. The fontanelle is convex when the ICP is above atmospheric pressure and sensors based on the applanation principle may then be used to estimate ICP. The basis of applanation is that of flattening the convex fontanelle with a suitable sensor. Under ideal conditions of flattening the internal pressure inside the fontanelle is balanced by the application of the external pressure sensor. Additional bending forces are dissipated by an inactive guard-ring.

An applanation transducer originally designed for intrauterine pressure measurement, the APT-16 (Hewlett-Packard), was used for neonatal ICP monitoring [141–143]. In this device there is a spring-loaded plunger at the center of a circular guard ring. The plunger has very low displacement in response to the fontanelle force so that its front face and the guard ring remain flat under the range of pressures being measured. Although this device has been found to produce measurements of ICP that correlate well with invasive measurements it is not ideal, being large and quite cumbersome for use in very small babies. Also, although it has been found possible to hold the sensor in place using elastic-net bandage and thereby perform continuous measurements, in clinical practice it is not really practicable for long periods. Furthermore, the sensor zero is angle-dependent as the plunger assembly moves under gravity.

A much simpler, cheaper, smaller, and lightweight plastic sensor has been produced [144]. This is comprised of a 10 mm diameter circular, polyurethane disc, 4 mm thick, having a thin polyurethane membrane attached to the active front surface. There is a pressurized gas inlet that fills an annular groove in the front face of the disc, thereby pressurizing the membrane. There is a gas outlet from a central orifice. When the sensor is affixed to the fontanelle the membrane experiences the fontanelle pressure on one side and the pressure supply on the other. If the internal pressure exceeds the fontanelle pressure the membrane will be pushed away from the sensor surface, thereby uncovering the central gas escape port and reducing the sensor internal pressure. The system stabilizes when internal sensor pressure is equal to the fontanelle pressure. This sensor can be attached easily to small fontanelles using either collodian or double-sided adhesive rings.

The Ladd fiber-optic sensor has also been used for non-invasive measurement by attachment to the anterior fontanelle. It has been shown that it is important that attachment is carried out with minimal external pressure otherwise the reading of ICP will be falsely high [145]. The recognition of the importance of the application pressure with the Ladd sensor led to a proliferation of methods and subsidiary devices for standardizing fixation. These have included a spring-loaded device for intermittent measurements [146], adhesive-disc ECG electrodes for continuous monitoring [147], a plastic frame with a screw to adjust the force of a spring held over the sensor [145], and self-adhesive foam strip applied over the sensor [148].

7.5.5.2 **Cerebral Blood Flow Measurement**

ICP monitoring has generally been carried out to assess perfusion of the brain and, as already indicated, a more direct indication is given by measuring cerebral blood flow (CBF). When, in the early 1970s, intraventricular hemorrhage and hypoxic-ischemic brain injury emerged as the major clinical problem in neonatal medicine there were many efforts made to develop and use techniques that could safely monitor CBF in small sick babies. At that time the standard methods applicable in adults using analysis of the clearance of injected or inhaled radioactive gases such as ^{133}Xenon, ^{79}Krypton, based on the original N_2O clearance method of Kety and Schmidt [149], were not widely accepted for use in the ill newborn. Nevertheless, this method, using ^{133}Xenon did provide important data on normal values for CBF as well as values in babies having had brain hemorrhage. It requires the arterial injection of the tracer dissolved in saline. Around 0.5–1.0 mCi/kg body weight have been used, giving a radiation dose of 20 mRad. Then a radiation counter placed over the head picks up the activity-time curve. An estimate of the arterial input function is obtained by collecting an activity-time curve over the right side of the chest and a two-compartment model is used for the analysis [150].

Basic Doppler ultrasound instruments can be used to obtain a qualitative indication of short-term pulsatile changes of arterial or venous blood flow velocity. Small ultrasound sensors can be positioned, either hand-held or affixed to the skin, to direct a beam of ultrasound at vessels of interest including the carotid arteries, the jugular veins, the anterior cerebral arteries, the sagittal sinus, and the middle cerebral artery. To derive quantitative or semi-quantitative measures of volumetric blood flow (mL per 100 g per minute) one must know the angle between the ultrasound beam and the direction of blood flow, the internal diameter of the blood vessel and the distribution of blood flow across the vessel diameter: the velocity profile. Some of these problems can be overcome when the Doppler ultrasound system is combined with a real time ultrasound scanner, in the Duplex systems. In the very small preterm baby it is important to be sure that the resolution of the Doppler ultrasound system is adequate for the very small blood vessels of interest. The ultrasound scanner allows beam-vessel angle to be determined, as well as enabling individual vessels to be visualized and the depth selectivity of range-gated systems to be adjusted optimally. Nevertheless in the very small baby lateral and depth resolution of the ultrasound system may be inadequate to separate flow in adjacent vessels.

Improvements to resolution and to the methods of processing the Doppler signals continue to be made. In the absence of quantitative measures there have been attempts to derive indices from the blood velocity waveform. Much of the early work was based on the derivation of the pulsatility index, PI [151]. This is calculated as

$$\mathrm{PI} = (V_s - V_d)/V_s \qquad (7.8)$$

where V_s is the systolic velocity and V_d is the diastolic velocity. This index is independent of beam-vessel angle. More recently other methods of analysis have been developed and assessed. One technique referred to as compensatory fuzzy neural networks (CFNN) has been proposed as a pattern recognition technique for intelligent detection of Doppler ultrasound waveforms of abnormal neonatal cerebral hemodynamics [152].

There has been interest in measurements of the transcephalic electrical impedance, measured with four small, chlorided EEG-type electrodes, to derive indices relating to CBF or cerebrovascular resistance and to detect intracerebral bleeding. The impedance is determined by the relative proportions of the main components: scalp, skull, CSF, brain tissue, blood. Accumulation of blood in the lateral ventricles produces a change in the electrical impedance, and this may be detected by an impedance monitor, or may be displayed as an image using electrical impedance imaging techniques [153]. With each cardiac cycle there is a small reduction in the electrical impedance of the head due to the pulsatile blood volume increase. By means of computer processing techniques this signal can be monitored for several days continuously [154]. The pulsatile impedance signal correlated with other non-invasive estimates of cerebral blood flow [155].

7.5.5.3 Principles and Use of Near Infrared Spectroscopy (NIRS)

Near infrared spectrophotometry (NIRS) for in vivo use has been developed from principles first applied in instruments for in vitro laboratory analysis [156]. Uses in the newborn baby and in the fetus have, in fact, represented the major clinical focus of the technique. The design and operation of the probes used for brain monitoring in these subjects is not merely a question of providing convenient and safe optical interfaces with fetal or neonatal tissues. The design of the probes and the underlying theoretical operation of the complete instruments are intimately linked. The theory of in vivo NIRS has been described in chapter 2 of this volume and so only part of this is repeated here.

Consideration of a simple cuvette model leads to the use of the Lambert-Beer law linking the chemical concentration of an optically absorbing species, [C] μmol, incident electromagnetic energy intensity, I_0, emerging intensity, I, cuvette dimension, L cm, and molar extinction coefficient, ε, as follows

$$I = I_0 e^{-\varepsilon[\mathrm{C}]L} . \tag{7.9}$$

Then transmittance, T, is derived by taking logs

$$T = \log(I/I_0) = -\varepsilon[\mathrm{C}]L . \tag{7.10}$$

The attenuation, A, in Optical Density units (ODs), is given by

$$A = \log(I_0/I) = \varepsilon[\mathrm{C}]L . \tag{7.11}$$

It is then possible to determine in absolute terms the chemical concentration of the absorbing species, [C], from a measurement of the attenuation of the interrogating beam and a knowledge of ε and the cuvette internal dimension, L. The product ε[C] is often replaced by the quantity referred to as the absorption coefficient, μ_a. However, μ_a actually appears in an expression based on natural logarithms

$$\ln(I_0/I) = \mu_a L . \tag{7.12}$$

Use of a single wavelength, λ, is adequate for a single absorber, but with, say, three absorbers, HHb, O_2Hb, and oxidized cytochrome aa_3, then three wavelengths are required.

The extinction coefficients for all three absorbers must be known for each of the three wavelengths and these coefficients arise from the absorption spectra for the absorbing compounds.

In the case of the three absorbers O_2Hb, HHb, and oxidized cytochrome aa_3 with the use of three interrogating wavelengths, λ_1, λ_2, and λ_3, we assume that the total attenuation at each wavelength is equal to the sum of the attenuations of the individual absorbers. Thus

$$A_{\lambda_1} = (\varepsilon_{1,\lambda_1}[\mathrm{C}]_1 + \varepsilon_{2,\lambda_1}[\mathrm{C}]_2 + \varepsilon_{3,\lambda_1}[\mathrm{C}]_3)L$$
$$A_{\lambda_2} = (\varepsilon_{1,\lambda_2}[\mathrm{C}]_1 + \varepsilon_{2,\lambda_2}[\mathrm{C}]_2 + \varepsilon_{3,\lambda_2}[\mathrm{C}]_3)L$$
$$A_{\lambda_3} = (\varepsilon_{1,\lambda_3}[\mathrm{C}]_1 + \varepsilon_{2,\lambda_3}[\mathrm{C}]_2 + \varepsilon_{3,\lambda_3}[\mathrm{C}]_3)L$$

Extending this theory to application in tissue must bring in the matter of scatter. Multiple scattering in tissue leads to an extended pathlength by photons as they propagate through the tissue. This difference between physical (geometrical) pathlength, L, and optical pathlength, L_0, has a profound effect on calculations of chemical concentration made using the simple Lambert-Beer law above. This could be addressed if L_0 were known, or a correction for this effect could be made if a path length factor, ζ, could be applied to the physical pathlength measurement

$$L_0 = \zeta L \tag{7.14}$$

The propagating energy is attenuated due to the fact that some photons are scattered out of the transmit-receive path. Attenuation due to scattering is given by

$$\ln(I_0/I) = \mu_s L \tag{7.15}$$

μ_s being the scattering coefficient, which is the likelihood of a scattering event taking place per cm of travel. Different tissues give different patterns of angular distribution for the propagating photons. This is influenced by the dimensions of the scattering centers relative to the wavelength of the incident energy. Examples of such scatter patterns are, firstly, where the re-radiated energy has an isotropic pattern and, secondly, where the re-radiated energy continues predominantly in the same direction as the incident energy.

The degree of deflection of photons by scattering tissues is described by the term g, the mean cosine of the scatter angle, also termed the anisotropy parameter. The anisotropy coefficient, g, is given by

$$g = \int_{-1}^{1} \cos\theta f(\cos\theta)\mathrm{d}(\cos\theta) \tag{7.16}$$

for forward scatter, g=1, for isotropic scatter, g=0, and for backscatter, g=–1.

Patterns of propagation through tissues are influenced by the relative contributions to total attenuation of absorption and scatter. This is reflected in the term albedo, Λ, given by

$$\Lambda = \mu_s/(\mu_s + \mu_a) \ . \tag{7.17}$$

The mathematical description of propagation and scattering in tissue can be approached with either transport theory or analytical theory. The latter is the most rigorous approach, starting from Maxwell's equations, but its complexity has limited its application and transport theory has been more widely adopted.

Of direct relevance was Schuster's theoretical analysis of the interactions between radiation and water droplets in foggy atmosphere [157] based on intensity transfer in the medium. Henyey and Greenstein's research on diffuse radiation in the galaxy is also important [158], as is Ishimaru's work on the scattering of light by turbid media [159, 160]. Theoretical consideration of the propagation of light through tissue usually starts with the radiative transport equation [161]. This is equivalent to the Boltzmann equation, used in connection with the kinetic theory of gases and in neutron transport theory. Derivation of the transport equation is outside the scope of this chapter but has been reviewed [162, 163]. It is written as

$$\hat{s} * \nabla L(r, \hat{s}) = -\mu_t L(r, \hat{s}) + \mu_s \int_{4\pi} P(\hat{s}, \hat{s}') L(r, \hat{s}) d\varpi' + S(r, \hat{s}) \tag{7.18}$$

where $L(r, \hat{s})$ is the radiance ($Wm^{-2}sr^{-1}$) at position r, direction $\hat{s}$, μ_t is the attenuation coefficient ($\mu_t = \mu_a + \mu_s$), μ_a is the absorption coefficient (cm^{-1}), μ_s is the scattering coefficient (cm^{-1}), $P(\hat{s}, \hat{s}')$ is the scattering phase function which is the scattering contribution from $\hat{s}'$ into the $\hat{s}$ direction, and $d\varpi'$ is an elementary solid angle about $\hat{s}' \cdot \hat{s}$ and $\hat{s}'$ are unit vectors in the direction of observation and incidence respectively. $S(r, \hat{s})$ is a source term. The scattering phase function may be described by the Henyey-Greenstein phase function [158]

$$f(\theta) = \frac{1}{4\pi} \frac{1 - g^2}{(1 + g^2 - 2g \cos\theta)^{3/2}} \tag{7.19}$$

Methods whereby analytical solutions of the transport equation may be obtained have also been reviewed [164] but exact solutions have only been found for very restricted cases.

Useful qualitative concepts emerge by considering what happens to the pattern of optical propagation when the concentration of scattering particles is progressively increased from zero. Initially, there is no scattering and the incident beam is attenuated purely by absorptive events. Next, with low scatterer concentration, photons undergo a single scattering event before emerging with the same net direction as the incident beam, suffering attenuation largely by absorption. With further increases in scatterer concentration, photons may experience progressively more scattering events, the mean optical pathlength will steadily increase, and attenuation becomes progressively more influenced by the loss of photons scattered out of the view of a detector. Finally, with relatively high scatterer concentration, photons experience a very large number of scattering events, mean optical pathlength is significantly increased, attenuation by scattering becomes dominant and backscatter is very clear. Twersky partly developed the understanding of this process but assumed that the multiple scattering was forward-directed [165]. He included absorption and scattering terms in what is in effect a modification of the Lambert-Beer law:

$$\frac{I}{I_0} = e^{-\mu_a cd}[e^{-\mu_s d} + q(1 - e^{-\mu_s d})] \,. \tag{7.20}$$

In this the fraction of the EM wave scattered by one scatterer, which subsequently arrives at the detector, is given by q. However, as described above, beyond a certain level of increasing scatterer concentration the likelihood of backscatter increases significantly.

Owing to the difficulties of finding exact solutions to the transport equation approximations based on diffusion theory have been developed and used. This can be approximated by the diffusion equation

$$\frac{1}{c}\frac{\partial}{\partial t}\phi(r,t) - D\nabla^2\phi(r,t) + \mu_a\phi(r,t) = S(r,t) \,. \tag{7.21}$$

The diffusion coefficient D is given by

$$D = \frac{1}{3\{\mu_a + (1-g)\mu_{tr}\}} \,. \tag{7.22}$$

A further term, the reduced scattering coefficient (transport scattering coefficient), $\mu_s{}'$, is introduced and is defined as

$$\mu_s' = (1-g)\mu_s \,. \tag{7.23}$$

This is used in the diffusion approximation of photon transport in tissue because it appears in the diffusion equation as a single term, and g and μ_s cannot be separated.

In-Vivo Multivariate Analysis

In order to achieve the equivalent of in vivo multivariate analysis, using optical probes and their associated instrumentation, several lines of research were pursued. The Lambert-Beer law was modified by the addition of a scattering term, G

$$A = \log\frac{I}{I_0} = \varepsilon[\mathrm{C}]L\zeta + G \,. \tag{7.24}$$

Subsequently, the approach of calculating absolute values of changes in concentration, $[\Delta\mathrm{C}]$, from changes in absorption, ΔA, was used. This requires modification of the set of simultaneous equations as follows

$$\begin{aligned}
\Delta A_{775} &= (\varepsilon_{1,775}\Delta C_{O_2Hb} + \varepsilon_{2,775}\Delta C_{HHb} + \varepsilon_{3,775}\Delta C_{cyt})\zeta_{775}L \\
\Delta A_{845} &= (\varepsilon_{1,845}\Delta C_{O_2Hb} + \varepsilon_{2,845}\Delta C_{HHb} + \varepsilon_{3,845}\Delta C_{cyt})\zeta_{845}L \\
\Delta A_{904} &= (\varepsilon_{1,904}\Delta C_{O_2Hb} + \varepsilon_{2,904}\Delta C_{HHb} + \varepsilon_{3,904}\Delta C_{cyt})\zeta_{904}L
\end{aligned} \tag{7.25}$$

Using a matrix inversion approach gives: $\Delta[C] = E^{-1}A/(\zeta L) = \tilde{N}A/(\zeta L)$ where the matrix $\tilde{N}$ contains what are termed the NIR coefficients. It is of course essential to know the $\varepsilon_{i,j}$ values. These have been determined in vitro using laboratory spectrophotometers [166–168]. The subscripts in Equation 7.25 refer to the wavelengths of IR light used.

Time-Resolved NIRS

Approaches based on time-resolved spectroscopy (TRS) [169, 170] have perhaps been of greater interest. These comprise of time domain (TD) and frequency domain (FD) methods; see the review of the theoretical basis for their operation in the work by Arridge et al. [171]. In this latter work a so-called differential pathlength factor, or DPF, is used instead of the path length factor, ζ, in the present chapter.

With TD NIRS a short light pulse (about 2–5 ps) is delivered to the sample and is then typically detected with a streak camera [172]. The family of photon paths produced by scattering leads to a broadening of the pulse, as seen in the temporal point-spread function (TPSF). The time, $\bar{t}$ at which the maximum detected intensity occurs relative to the input pulse is the mean arrival time of photons and this is used, together with velocity of light in vacuo, c_v, and tissue refractive index, n_t, to calculate mean optical pathlength

$$\overline{L_0} = \frac{c_v}{n_t}\bar{t}\,. \tag{7.26}$$

Since the measurement of time is the basis of this approach it is also referred to as the 'time-of-flight' method. Thus attenuation within a scattering tissue is given by

$$A = (\mu_a + \mu_s)\frac{c_v}{n_t}\bar{t}\,. \tag{7.27}$$

The time domain methods have been used for basic research but cannot be used for clinical monitoring owing to the large size and cost of the apparatus. For this reason there has been much interest in frequency domain methods.

Frequency Domain NIRS

In frequency-domain spectroscopy the interrogating energy is intensity modulated (IM) and the detected energy exhibits a phase shift, Φ, as compared with the modulating signal, due to the propagation delay, as well as attenuation due to absorption and scattering. The detected intensity will be of the form

$$I = I_{dc} + I_{ac} \sin(2\pi \nu t - \Phi) \ . \tag{7.28}$$

Measurement of Φ can allow optical path length to be calculated since $L_0 = \Phi c_v / 2\pi \nu n_t$ where ν is the modulating frequency, n_t is the refractive index of the tissue, and c_v is the speed of light in vacuo. The approach is also referred to as phase modulation [173, 174].

Extension of the basic frequency domain method into what is termed spatially resolved NIRS then allows the absorption and scattering coefficients to be calculated and form these the chemical concentrations. They derived expressions for phase shift, Φ, the steady state photon density, U_{DC}, and the amplitude of the photon density oscillations, U_{AC}. These may then be used to calculate values for μ_a and μ_s'

$$\ln(rU_{DC}) = -r\left(\frac{\mu_a}{D}\right)^{1/2} + \ln\left(\frac{S}{4\pi \nu_t D}\right) \ . \tag{7.29}$$

$$\ln(rU_{AC}) = -r\left(\frac{\nu_t^2 \mu_a^2 + \omega^2}{\nu_t^2 D^2}\right)^{1/4} \cos\left[\frac{1}{2}\tan^{-1}\left(\frac{\omega}{\nu_t \mu_a}\right)\right] + \ln\left(\frac{SA}{4\pi \nu_t D}\right) \tag{7.30}$$

$$\Phi = r\left(\frac{\nu_t^2 \mu_a^2 + \omega^2}{\nu_t^2 D}\right)^{1/4} \sin\left[\frac{1}{2}\tan^{-1}\left(\frac{\omega}{\nu_t \mu_a}\right)\right] \tag{7.31}$$

In these expressions r is the distance between source and detector, S is the source term, and A is the modulation of the source (A=AC/DC). These expressions are linear functions of r and the slopes of these functions, Φ, exhibit dependence upon μ_a and μ_s'. The slopes derived from these equations may be termed M_{dc}, M_{ac}, and M_Φ. By making measurements at two or more values of source-detector spacing it is possible to determine values for M_{dc}, M_{ac}, and M_Φ. Expressions for μ_a and μ_s' can then be derived

$$\mu_a = -\frac{\varpi}{2\nu} \frac{M_{dc}}{M_\Phi} \left(\frac{M_\Phi^2}{M_{dc}^2}\right)^{-1/2} \tag{7.32}$$

$$\mu_s' = \frac{M_{dc}^2}{3\mu_a} - \mu_a \ . \tag{7.33}$$

Values for [HHb] and [O_2Hb] can then be calculated if μ_a and μ_s' are derived for at least two wavelengths. It is also possible to derive μ_a and μ_s' from data of relative phase shift and relative demodulation [175].

As will be apparent, this approach requires optical probes in which sources and detectors are combined in specific geometrical arrangements; often referred to as multi-site or multi-distance probes [176]. The probes can be constructed by using optical fibers with sources and detectors situated in the instrument. Alternatively, the probe can contain semiconductor sources and detectors. The spatially resolved method is based on the solution of the diffusion approximation for a highly scattering medium [170]. These authors provide a solution for the case of a semi-infinite half-space geometry for an input δ-function, giving the following expression for reflected intensity, $R(\rho, t)$

$$R(\rho, t) = (4\pi D c_t)^{-3/2} \frac{1}{\mu_s'} t^{-5/2} e^{-\mu_a c_t t} e^{-\left(\frac{\rho^2 + (\mu_s')^2}{4 D c_t t}\right)} . \tag{7.34}$$

Matcher et al. [177] showed that, by making the approximation $\mu_a + \mu_s{}' \approx \mu_s{}'$, the following relation can be derived

$$\mu_a \mu_s' \approx \frac{1}{3}\left(\frac{\partial A}{\partial \rho} - \frac{2}{\rho}\right) . \tag{7.35}$$

The wavelength dependence of $\mu_s{}'$ is then used and the gradient of attenuation with distance measured at several wavelengths to derive the spectral shape of μ_a. Hemoglobin saturation can then be calculated.

Clinical Use of NIRS

The potential application of NIRS to neonatal monitoring was presented in 1984 by Crowe et al. [178] and in 1985 by Rea et al. [81] and Wickramasinghe et al. [179]. Rea et al. emphasized then the need for safe, convenient, non-invasive methods with which cerebral oxygenation could be monitored in at risk neonates during the first few days of life. They speculated that NIRS might eventually emerge as a very useful method. They evaluated the relevance of the reported use of NIRS for monitoring cyt aa3 [180] to the needs in neonates. The possibility of being able to monitor brain intracellular oxygenation non-invasively was perhaps the most exciting attraction of NIRS when Jöbsis first published his work in 1977 [180]. Despite this excitement, the feasibility of in vivo cyt aa_3 monitoring remained controversial until relatively recently.

Soon after this the first preliminary observations were made in human neonates [181], using the instrumentation produced by Jöbsis and interest grew rapidly [182–185]. Wyatt et al. [184] described how indices of cerebral oxygenation and hemodynamics in sick newborn infants could be quantified. They reported changes of cyt aa_3 in seven babies in response to changes in S_aO_2 (0.002–0.014 µmol/100 g per 1%) and P_aCO_2 (0.02–0.33 µmol/100 g per kPa). They concluded that alterations in cerebral oxygen availability at the mitochondrial level could be observed and quantified. They used a hemoglobin oxygenation index (O_2Hb–HHb) that was found to change linearly with S_aO_2.

Other reports concluded that at that time with the available instrumentation changes in cyt aa_3 redox state in preterm infants following desaturation was inconsistent and insignificant [186]. This study examined the influences of falls in arterial O_2 saturation on cerebral cyt aa_3 redox state and found that oxidation of the enzyme was as likely to occur as was reduction.

The effect of intermittent positive pressure ventilation (IPPV) and continuous negative extra-thoracic pressure (CNEP) ventilation on CBV in preterm infants was studied with NIRS [187]. With CNEP, CBV was found to decrease, suggested to be due to increased venous drainage. CBV also fell with IPPV but HHb increased and O_2Hb fell, suggesting that in this case venous drainage had decreased.

NIRS has been used to derive estimates for CBF and CBV [188]. For CBF determination the arterial oxygen saturation must be changed swiftly. Then the Fick technique can be applied to the cerebral NIRS wash-out/wash-in data. There are practical difficulties in applying this method clinically. The estimate of CBV has been compared with measurements made using jugular venous occlusion plethysmography [189] and good correlation was found. Cyclical fluctuations in CBV have been reported in newborn babies [190]. This observation was studied retrospectively in 20 preterm infants [191]. In 19 of the 20, cyclical changes of CBV were found within a frequency range of 2–4.7 cycles/min which is comparable with the results of Doppler studies on fluctuations in CBF velocity.

NIRS used in neonatal intensive care has been reviewed [192]. The ability of the method to provide continuous long-term data on cerebral oxygenation and hemodynamic variables was emphasized. The research data demonstrated significant changes in the hemodynamic variables with patient activity and clinical interventions in both premature and term infants.

7.6 Conclusions

This review of sensors used for fetal and neonatal physiological measurement is not complete. Topics such as cerebral electrophysiological measurement, including EEG and evoked responses have not been dealt with. This is partly because the sensor aspect is really only a matter of simple electrode design. But it is also due to the wide breadth and extensive depth of the subject matter which cannot be dealt with fully in such a chapter. The field covers all possible means available for physiological measurement, each subtopic of which is generally only dealt with in isolation in an adult context. Although this field continues to evolve there are some problems in terms of sensor technology that have been difficult to solve over the past 30 years or so. An example of this is non-contact neonatal monitoring. Also, there is now more of an acceptance that comprehensive assessment of measurement systems is essential and this must include clinical trials. The fetus and the newborn will continue to demand special attention medically and from those engaged in research and development of sensors and instruments, including investigators in both academia and industry.

7.7
References

1 Rolfe, P. Monitoring in newborn intensive care. *Biomed. Eng.*, **1975**, *10*, 339–404, 413.

2 Rolfe, P. (ed), *Non-Invasive Physiological Measurements. Vol I.* Academic Press, London, **1979**.

3 Rolfe, P. (ed), *Fetal and Neonatal Physiological Measurements.* Pitman Medical, Tunbridge Wells, **1980**.

4 Rolfe, P. (ed), *Non Invasive Physiological Measurements, Vol II.* Academic Press, London, **1983**.

5 Rolfe, P. (ed), *Fetal Physiological Measurements.* Butterworths, Guildford, **1986**.

6 Rolfe, P. (ed), *Neonatal Physiological Measurements.* Butterworths, Guildford, **1986**.

7 Mannor, S. M., Serr, D. M., Tamari, I., Meshorer, A. and Frei, E. H. The safety of ultrasound in fetal monitoring. *J. Obstet. Gynecol. Brit. Commw.* **1972**, *113*, 653–661.

8 NRCP. *Biological Effects of Ultrasound*, National Council on Radiation Protection and Measurement, Report no. 74, Bethesda. NRCP Publications, **1983**.

9 Leopold, G. R. Ante-partum obstetrical ultrasound examination guidelines. *J. Ultrasound in Med.* **1986**, *5*, 241–242.

10 Leerentveld, R. A., Gilberts, E. C. A. M., Arnold, M. J. C. W. J. and Wladimiroff, J. W. Accuracy and safety of transvaginal sonographic placental localization. *Obstet. Gynecol.* **1990**, *76*, 759–762.

11 Brunner, E. How Ultrasound System Considerations Influence Front-End Component Choice. *Analogue Dialogue* **2002**, *36*(3), May–July.

12 Ginter, S. Numerical simulation of ultrasound thermotherapy combining nonlinear wave propagation with broadband soft-tissue absorption. *Ultrasonics* **2000**, *37*, 693–696.

13 Kremkau, F. W. *Diagnostic Ultrasound: Principles, Instruments and Exercises*, 3rd edn. WB Saunders, Philadelphia, **1989**.

14 Bishop, E. H. Obstetric uses of the ultrasonic motion sensor. *Am. J. Obstet. Gynecol.* **1966**, *96*, 863–867.

15 Veille, J. C. and Cohen, I. Middle cerebral artery blood flow in normal and growth-retarded fetuses. *Am. J. Obstet. Gynecol.* **1990**, *162*, 391–396.

16 Wells, P. N. T. Ultrasound Imaging. In: *Non-Invasive Physiological Measurements: 2.* P. Rolfe (ed.). Academic Press, London, **1983**, 313–352.

17 Campbell, S. The prediction of fetal maturity by ultrasonic measurement of biparietal diameter. *J. Obstet. Gynaecol. Br. Commonw.* **1969**, *76*, 603–609.

18 Robinson, H. P. Sonar measurements of fetal crown-rump length as a means of assessing maturity in the first trimester of pregnancy. *Br. Med. J.* **1973**, *4*, 28–31.

19 Lauersen, N. H., Hochberg, H. M. Automatic detection of fetal movement by Doppler ultrasound during non-stress testing. *Int. J. Gynaecol. Obstet.* **1982**, *20*(3), 219–222.

20 Dawes, G. S. Breathing before birth in animals and man. An essay in developmental medicine. *New Engl. J. Med.* **1974**, *290*, 557–559.

21 Gembruch, U., Chatterjee, M. S., Bald, R., Redel, D. A., Hansmann, M. Color Doppler flow mapping of fetal heart. *J. Perinat. Med.* **1991**, *19*(1/2), 27–32.

22 Hammacher, K. Neue Methode zur selektiven Registrierung der fetalen Herzschlagfrequenz. *Geburtsh. Frauenheilk.* **1962**, *22*, 1542.

23 Schwartz, R. S., Reeves, J. T., Sodal, I. E. and Barnes, F. S. Improved phonocardiogram system based on acoustic impedance matching. *Am. J. Physiol. (Heart Circ. Physiol.)* **1980**, *238*, H604–H609.

24 Talbert, D. G., Lyn Davies, W., Johnson, F., Abraham, N., Colley, N., and Southall, D. P. Wide bandwidth fetal phonography using a sensor matched to the compliance of the mother's abdominal wall. *IEEE Trans. Biomed. Eng.* **1986**, *BME-33, 175–181.*

25 Goovaerts, H. G., Rompelman, O., and van Geijn, H. P. A transducer for detection of fetal breathing movements. *IEEE Trans. Biomed. Eng.* **1989**, *BME-36*, 471–478.

26 Goovaerts, H.G., van Geijn, H.P., Rompelman, O., Mantel, R. and Swartjes, J.M. Recording fetal breathing movements with a passive transducer based on an inductive principle. *Med. Biol. Eng. Comput.* **1991**, *29*, 358–364.

27 Varady, P., Wildt, L. Fetal phonocardiography with a novel approach. *Orv. Hetil.* **2001**, *142*(36), 1971–1976.

28 Cremer, M. Über die direkte Ableitung der Aktionsströme des menschlichen Herzens vom Oesophagus und über das Elektrokardiogramm des Fötus. *Münch. Med. Wschr.* **1906**, *53*, 811.

29 Roche, J.B., Hon, E.H. The fetal electrocardiogram. V. Comparison of lead systems. *Am. J. Obstet. Gynecol.* **1965**, *92*, 1149.

30 Schuler, G., Puddicombe, P., Park, P. Fetal ECG improvement due to selective signal addition. *Am. J. Obstet. Gynecol.* **1968**, *101*, 1120.

31 Hon, E.H., Hess, O.W. Instrumentation of fetal electrocardiography. *Science* **1957**, *125*, 553.

32 Sureau, C. and Trocellier, R. Un probleme technique d'electrocardiographie foetale. Note sur l'elimination de l'electrocardiogramme maternel. *Gynec. Obstet.* **1961**, *60*, 43.

33 Reynolds, S.R.M., Heard, O.O., Bruns, P. and Hellman, L.M. A multi-channel strain-gauge tokodynamometer: an instrument for studying patterns of uterine contractions in pregnant women. *Bull. Hosp. Joint Dis.* **1948**, *82*, 446.

34 Smyth, C.N. The Guard-ring Tocodynamometer. Absolute measurement of intra-amniotic pressure by a New Instrument. *J. Obstet. Gynaec. Br. Emp.* **1957**, *64*, 59.

35 Marque, C., Duchene, J. Human abdominal EHG processing for uterine contraction monitoring. *Biotechnology* **1989**, *11*, 187–226.

36 Gondry, J., Marque, C., Duchene, J., Cabrol, D. Electrohysterography during pregnancy: preliminary report. *Biomed. Instrum. Technol.* **1993**, *27*(4), 318–324.

37 Leman, H., Marque, C. Rejection of the maternal electrocardiogram in the electrohysterogram signal. *IEEE Trans. Biomed. Eng.* **2000**, *47*(8), 1010–1017.

38 Hon, E.H. The classification of fetal heart rate. 1. A working classification. *Obstet. Gynecol.* **1963**, *22*, 137.

39 Caldeyro-Barcia, R., Medez-Bauer, C., Poseiro, J.J. Control of human fetal heart rate during labour. In *The Heart and Circulation in the Newborn and Infant*, (D.E. Cassels ed). Grune and Stratton, New York, London, **1966**, 7–36.

40 Dawes, G.S., Houghton, C.R.S., Redman, C.W.G. and Visser, G.H.A. Pattern of the normal human fetal heart rate. *Brit. J. Obstet. Gynaecol.***1982**, *89*, 276.

41 Pardey, J., Moulden, M., Redman, C.W. A computer system for the numerical analysis of nonstress tests. *Am. J. Obstet. Gynecol.* **2002**, *186*(5), 1095–1103.

42 Kang, A.H., Boehm, F.H. The clinical significance of intermittent sinusoidal fetal heart rate. *Am. J. Obstet. Gynecol.* **1999**, *180*(1), 151–152.

43 Schneider, H. Birth asphyxia – an unsolved problem of perinatal medicine. *Z. Geburtshilfe Neonatal.* **2001**, *205*(6), 205–212.

44 Hon, E.H. Instrumentation of fetal heart rate and fetal electrocardiography. II. A vaginal electrode. *Am. J. Obstet. Gynecol.* **1963**, *86*, 772.

45 Hon, E.H., Paul, R.H. and Hon, R.W. Electrode evaluation of the fetal heart rate. XI. Description of a spiral electrode. *Obstet. Gynecol.* **1972**, *40*, 362.

46 Goodlin, R.C. and Fabricant, S. A new fetal scalp electrode. *Obstet. Gynecol.* **1970**, *35*, 646.

47 Schmidt, S., Langner, K., Rothe, J., Saling, E. A new ECG electrode concept for the conduction of fetal heart action potentials without penetration of the skin. *Z. Geburtshilfe Perinatol.* **1982**, *186*(5), 242–243.

48 Needs, L., Grant, A., Sleep, J., Ayers, S., Henson, G. A randomized controlled trial to compare three types of fetal scalp electrode. *Br. J. Obstet. Gynaecol.* **1992**, 99(4), 302–306.

49 Rantonen, T., Ekholm, E., Siira, S., Metsala, T., Leino, R., Ekblad, U., Valimaki, I. Periodic spectral components of fetal heart rate variability reflect the changes in cord arterial base deficit val-

ues: a preliminary report. *Early Hum. Dev.* **2001**, *60*(3), 233–238.

50 Hon, E.H. and Lee. The fetal electrocardiogram. I. The electrocardiogram of the dying fetus. *Am. J. Obstet. Gynecol.* **1963**, *87*, 804.

51 Greene, K.R., Dawes, G.S., Lilja, H. and Rosén, K.G. Changes in the ST waveform of the fetal lamb electrocardiogram with hypoxemia. *Am. J. Obstet. Gynecol.* **1982**, *144*, 950.

52 Nickelsen, C., Weber, T., Parnell, C., Nim, J., Kemp, A.M., Junge, I. Cardiotocographic monitoring of deliveries. A prospective comparative study of 2 types of electrodes. *Ugeskr. Laeger* **1989**, *151*(7), 440–442.

53 Westgate, J., Keith, R.D., Curnow, J.S., Ifeachor, E.C., Greene, K.R. Suitability of fetal scalp electrodes for monitoring the fetal electrocardiogram during labour. *Clin. Phys. Physiol. Meas.* **1990**, *11*(4), 297–306.

54 Peres, M.F., Fisher, C.C., Leader, L.R. Continuous intrapartum fetal monitoring: a prospective randomized comparison of 2 types of fetal scalp electrode. *Aust. NZ J. Obstet. Gynaecol.* **1993**, *33*(4), 367–370.

55 Amer-Wahlin, I., Hellsten, C., Noren, H, Hagberg, H., Herbst, A., Kjellmer, I., Lilja, H., Lindoff, C., Mansson, M., Martensson, L., Olofsson, P., Sundstrom, A., Marsal, K. Cardiotocography only versus cardiotocography plus ST analysis of fetal electrocardiogram for intrapartum fetal monitoring: a Swedish randomised controlled trial. *Lancet* **2001**, *358*, 534–538.

56 Saling, E. *Geburtsh. Frauenheilk.* **1961**, *21*, 905.

57 O'Dowd, M., Martin, M.J., Wheble, A., Gillmer, M.D.G. and Rolfe, P. Ion-selective sensors for assessment of the fetus. *J. Biomed. Eng.* **1988**, *10*, 165–170.

58 Stamm, O., Latscha, U., Janecek, P. and Campana, A. Kontinuierliche pH-Messung am kindlichen Kopf post partum und sub partum. *Z. Geburtsh. Perinat.* **1974**, *178*, 36.

59 Flynn, A.M., Kelly, J. Continuous tissue pH (tpH) in the human fetus during labour. In Rolfe, P. (ed) *Fetal and Neonatal Physiological Measurements.* Pitman Medical, London, **1980**, 364–371.

60 Hochberg, H.M. New instrument developments for fetal pH monitoring. *Arch. Gynecol.* **1978**, *226*, 79–84.

61 Henner, H. Continuous tissue pH – measurement on the fetus during delivery. In Rolfe, P. (ed) *Fetal and Neonatal Physiological Measurements.* Pitman Medical, **1980**, 372–377.

62 Nickelsen, C., Thomsen, S.G. and Weber, T. Continuous acid-base assessment of the human fetus during labour by tissue pH and transcutaneous carbon dioxide monitoring. *Br. J. Obstet. Gynaecol.* **1985**, *92*, 220–225.

63 Nickelsen, C. and Weber, T. The current status of intrapartum continuous fetal tissue measurements. *J. Perinat. Med.* **1991**, *19*, 87–92.

64 Peterson, J.I., Goldstein, S.R., Fitzgerald, R.V., and Buckhold, D.K. Fibre-optic pH probe for physiological use. *Anal. Chem.* **1980**, *52*, 864–869.

65 Chatterjee, M.S., Hochberg, H.M. Continuous intrapartum measurement of tissue pH of the human fetus using newly developed techniques. *J. Perinat. Med.* **1991**, *19*(1/2), 93–96.

66 O'Connor, M.C., Hytten, F.E., Zanelli, G.D. Interpretation of intrapartum falls in fetal oxygenation as measured by transcutaneous pO_2 ($TcpO_2$) electrode (Draeger Transoxode). In Rolfe, P. (ed) *Fetal and Neonatal Physiological Measurements.* Pitman Medical, London, **1979**, 378–384.

67 Evans, N.T.S., Naylor, P.F.D. The systemic oxygen supply to the surface of the human skin. *Respir. Physiol.* **1967**, *3*, 21–37.

68 Evans, N.T.S., Naylor, P.F.D. The oxygen tension gradient across human epidermis. *Respir. Physiol.* **1967**, *3*, 38–42.

69 Huch, R., Huch, A., Rolfe, P. Transcutaneous measurement of pO2 using electrochemical analysis. In *Non-invasive physiological measurements,* Vol 1 (P. Rolfe ed). Academic Press, London, **1979**, 313–331.

70 Eberhard, P., Hammacher, K., Mindt, W. Methode zur Kutanen Messung des

Sauerstoffpartialdruckes. *Biomed. Technik.* **1973**, *6*, 216.

71 Eberhard, P., Mindt, W. Skin sensors for continuous oxygen monitoring of newborns. *Biotelemetry* **1977**, *4*, 48–76.

72 Huch, A., Huch, R., Schneider, H., Rooth, G. Continuous transcutaneous monitoring of fetal oxygen during labour. *Br. J. Obstet. Gynaecol.* **1977** *(Suppl 1)*, 1–37.

73 Rolfe, P., Burton, P. J., Crowe, J. A., et al. Foetal scalp mass spectrometer blood-gas transducer. *Med. Biol. Eng. Comput.* **1982**, *20*, 375–382.

74 Rolfe, P. In Vivo Near Infra-Red Spectrophotometry. *Ann. Rev. Biomed. Eng.* **2000**, *2*, 315–354.

75 Aoyagi, T., Kishi, T., Yamaguchi, K., Watanabe, S. Improvement of ear-piece oximeter. In *Proc. 13th Jpn. Medical Electronics and Biological Engineering Conf.* Osaka, **1974**, 90–91.

76 Peat, S., Booker, M., Lanigan, C., Ponte, J. Continuous intrapartum measurement of fetal oxygen saturation. *Lancet* **1988**, July 23, *ii* (8604), 213.

77 Johnson, N., Lilford, R. J. Continuous intrapartum measurement of fetal oxygen saturation. *Lancet* **1988**, Aug 27, *ii* (8609), 517.

78 Gardosi, J., Carter, M., Becket, T. Continuous intrapartum monitoring of fetal oxygen saturation. *Lancet* **1989**, Sep 16, *ii* (8664), 692–693.

79 Nijland, R., Jongsma, H. W., Nijhuis, J. G., Oeseburg, B. Accuracy of fetal pulse oximetry and pitfalls in measurements. *Eur. J. Obstet. Gynecol. Reprod. Biol.* **1997**, *72*(Suppl), S21–S27.

80 Kuhnert, M., Seelbach-Goebel, B., Butterwegge, M. Predictive agreement between the fetal arterial oxygen saturation and fetal scalp pH: results of the German multicenter study. *Am. J. Obstet. Gynecol.* **1998**, *178*, 330–335.

81 Rea, P. A., Crowe, J., Wickramasinghe, Y. A. B. D., Rolfe, P. Non-invasive optical methods for the study of cerebral metabolism in the human newborn: a technique for the future? *J. Med. Eng. Technol.* **1985**, *9*, 160–166.

82 Schmidt, S., Decleer, W., Gorissen-Bosselmann, S., Eilers, H., Pringle, K., Helledie, N., Rolfe, P., Krebs, D. Laser-spectroscopy of Induced Hyperoxia – An Experimental Study in the Lamb. *Biomedizinische Technik.* **1990**, *35*, 185–189.

83 Faris, F., Rolfe, P., Thorniley, M., Wickramasinghe, Y., Houston, R., O'Brien, S. Non-invasive optical monitoring of cerebral blood oxygenation in the fetus and newborn. *J. Biomed. Eng.* **1992**, *14*, 303–306.

84 Faris, F., Doyle, M., Wickramasinghe, Y. A. B. D., Houston, R., Rolfe, P., and O'Brien, S. A non-invasive optical method for intra-partum fetal monitoring: Preliminary clinical studies. *J. Biomed. Eng.* **1994**, *16*, 287–291.

85 Rolfe, P., Wickramasinghe, Y., Thorniley, M., Faris, F., Houston, R., Zhang, K., Yamakoshi, K., O'Brien, S., Doyle, M., Palmer, K., Spencer, S. Fetal and neonatal cerebral oxygen monitoring with NIRS: theory and practice. *Advances in Perinatal Medicine* (S. Sakamoto, Y. Takeda eds). Excepta Medica, Amsterdam, **1992**, 269–273.

86 Doyle, M., O'Brien, P. M. S., Wickramasinghe, Y. A. B. D., Houston, R. J. F., and Rolfe, P. Near infrared spectroscopy used to observe changes in fetal cerebral hemodynamics during labour. *J. Perinat. Med.* **1994**, *22*, 265–268.

87 Schmidt, S., Schwartzler, C. M., Sierra, F., Meyer-Wittkopf, M., Rolfe, P. Blood volume changes and oxygenation during labor – a laser spectroscopic analysis. *Z. Geburtshilfe Neonatol.* **2001**, *205*, 33–37.

88 Miller, D. L. and Oliver, T. K. Body temperature in the immediate neonatal period: the effect of reducing thermal losses. *Am. J. Obstet. Gynecol.* **1966**, *94*, 964–969.

89 Fox, R. H. and Solman, A. J. A new technique for monitoring the deep body temperature in man from the intact skin surface. *J. Physiol.* **1971**, *212*, 8P–10P.

90 Togawa, T. Medical thermometer making use of zero heat flow method. *Rep. Inst. Med. Dent. Eng.* **1973**, *7*, 75–83.

91 Rolfe, P. Monitoring equipment for the neonate. *Br. J. Hosp. Med.* **1976**, *1*, 189–205.
92 Rolfe, P. Neonatal critical care monitoring. *J. Med. Eng. Tech.* **1986**, *10*, 115–120.
93 O'Brien, M. J., van Eykern, L. A., and Prechtl, H. F. R. Monitoring respiratory activity in infants – a non-intrusive diaphragm EMG technique. In *Non-Invasive Physiological Measurements, Vol 2* (P. Rolfe ed), Academic Press, London, **1983**, 131–177.
94 Milner, A. D. The respiratory jacket. *Lancet* **1970**, *i*, 80.
95 Milledge, J. S. and Stott, M. Inductive plethysmography – a new respiratory transducer *J. Physiol.* **1977**, *267*, 4P–5P.
96 Rolfe, P. A magnetometer respiration monitor for use with premature babies. *Biol. Med. Eng.* **1971**, *6*, 402–404.
97 Wright, B. M. and Callan, K. A new respiratory recording and monitoring system. In *Proc. Int. Soc. Amb. Mon.* Academic Press, London, **1979**, 329–335.
98 Beardsmore, C. S., Helms, P., Socks, J. and Silverman, M. Some problems of oesophageal pressure measurement in the newborn. In *Fetal and Neonatal Physiological Measurements*. Ed. P. Rolfe, Pitman Medical Ltd., London, **1981**, 78–85.
99 Lewin, J. E. An apnoea alarm mattress. *Lancet* **1969**, *ii*, 667.
100 Kreuzer, F., Nessler, C. G. Method of polarographic *in vivo* continuous recording of blood oxygen tension. *Science* **1958**, *128*, 1005–1006.
101 Kreuzer, F., Nessler, C. G. Method for polarographic continuous in vivo recording of blood oxygen tension. *Physiologist* **1958**, *1*, 44.
102 Charlton, G., Read, D., Read, J. Continuous arterial pO_2 in normal man using a flexible micro-electrode. *J. Appl. Physiol.* **1963**, *18*, 1247.
103 Krog, J., Johansen, K. Construction and characteristics of Teflon covered polarographic electrodes for intravascular oxygen determination. *Rev. Sci. Instrum.* **1959**, *30*, 108.
104 Said, S. I., Davis, R. K., Crosier, J. L. Continuous recording *in vivo* of arterial blood pO_2 in dogs and in man. *J. Appl. Physiol.* **1961**, *16*, 1129–1132.
105 Koeff, S. T., Tsao, M. U., Vadrey, A., and Wilson, J. L. Continuous measurements of intravascular oxygen tension in normal adults. *J. Clin. Invest.* **1962**, *41*, 1125–1129.
106 Goddard, P. J., Keith, I., Marcovitch, H., Roberton, N. R. C., Rolfe, P., Scopes, J. W. The use of a continuously recording intravascular oxygen electrode in the newborn. *Arch. Dis. Childh.* **1974**, *49*, 853–860.
107 Parker, D., Key, A., Scopes, J. W., and Marcovitch, H. A disposable catheter tip transducer for continuous measurement of blood oxygen tension in vivo. *J. Biomed. Engng.* **1971**, *6*, 313–318.
108 Rolfe, P. Arterial oxygen measurement in the newborn with intra-vascular transducer. In *Medical Electronics Monographs No. 18–22* Hill, D. W., Watson, B. W. (eds). Peter Peregrinus, Stevenage, **1976**.
109 Zhang, S. PhD thesis: *Development of intravascular sensors with improved hemocompatibility.* Keele University, **1996**.
110 Enson, W., Briscoe, W. A., Polanyi, M. L., Courand, A. *In vivo* studies with an intravascular and intracardiac reflection oximeter. *J. Appl. Physiol.* **1962**, *17*, 552–558.
111 Wilkinson, A. R., Phibbs, R. H., and Gregory, G. A. Continuous measurement of oxygen saturation in sick newborn infants *J Pediatr.* **1978**, *93*, 1016–1019.
112 Parker, D. Continuous measurement of blood oxygen tension and saturation. In: *Monitoring of Vital Parameters during Extra-corporeal Circulation*, (H. P. Kimmich ed). Karger, Basel, **1981**, 23–28.
113 Peterson, J. I., Fitzgerald, R. V., Buckhold, D. K. Fibre-optic probe for *in vivo* measurement of oxygen partial pressure. *Anal. Chem.* **1984**, *56*, 62–67.
114 Stow, R. W., Randall, B. F. Rapid Measurement of the tension of carbon dioxide in blood. *Arch. Phys. Med. Rehabil.* **1957**, *38*, 646–650.
115 Coon, R. L., Lai, N. J. C., and Kampine, J. P. Evaluation of dual-function Pt and pCO_2 in-vivo sensor. *J. Appl. Physiol.* **1976**, *40*, 625–629.
116 Rolfe, P., Martin, M., William, D., Walters, R., Yamakoashi, K., Tanaka, S.

In vivo sensor-hemocompatibility. In: *Advances in In vivo Sensors.* TURNER, A. J. P. (ed). **1992** Cranfield Press, Cranfield, 167–178.

117 ROLFE, P. MONITORING. In *Practical Perinatal Care: The Baby under 1000 g.* G. LEVITT, D. HARVEY, R. COOKE eds). Butterworth-Heinemann, Oxford, **1999**, 83–100.

118 ROLFE, P. Catheters. UK Patent 1,527,801, **1976**.

119 KIM, S. W., FEJIEN, J. Surface modification of polymer for improved blood compatibility. *Crit. Rev. Biocompat.* **1985**, *1*, 229–260.

120 HAYWARD, J. A., CHAPMAN, D. Biomembrane surfaces as models for polymer design: the potential for hemocompatibility. *Biomaterials* **1984**, *5*, 135–142.

121 ISHIHARA, K., UEDA, T., NAKABAYASHI, N. Preparation of phospholipid polymers and their properties as polymer hydrogel membranes. *Polym. J.* **1990**, *22*, 355–360.

122 ISHIHARA, K., TSUJI, T., KUROSAKI, T., NAKABAYASHI, N. Haemocompatibility on graft copolymers composed of poly(2-methacryloyloxyethyl phosphorylcholine) side chain and poly(n-butyl methacrylate) backbone. *J. Biomed. Mater. Res.* **1994**, *28*, 225–232.

123 ZHANG, S., ROLFE, P. Blood-compatible material based on cell membrane mimicry for intravascular sensors. *Cell. Eng.* **1996**, *1*, 103–108.

124 ZHANG, S., ROLFE, P., TANAKA, S., ISHIHARA, K. Development of a hemocompatible pO_2 sensor with phospholipid based copolymer membrane. *Biosensors Bioelectr.* **1996**, *11*, 1019–1029.

125 ZHANG, S., WRIGHT, G., KINGSTON, M. A., ROLFE, P. Improved performance of an intravascular pO_2 sensor incorporating a poly(MPC-co-BMA) membrane. *Med. Biol. Eng. Comput.* **1996**, *34*, 313–315.

126 PARKER, D., DELPY, D., and LEWIS, M. Catheter-tip electrode for continuous measurement of pO_2 and pCO_2. *Med. Biol. Eng. Comput.* **1978**, *16*, 599–600.

127 THUNSTROM, A. M., STAFFORD, M. J. and SEVERINGHAUS, J. W. A two temperature, two pO_2 method of estimating the determinants of $tcpO_2$. In *Continuous Transcutaneous Blood Gas Monitoring* (A. HUCH, R. HUCH and J. F. LUCEY, eds). Alan R. Liss Inc, New York, **1979**, 167–182.

128 LUBBERS, D. W. Cutaneous and transcutaneous pO_2 and pCO_2 and their measuring conditions. In *Continuous Transcutaneous Blood Gas Monitoring*, (A. HUCH, R. HUCH and J. F. LUCEY eds). Alan, R. Liss Inc, New York, **1979**, 13–31.

129 YOSHIYA, I., SHIMADA, Y. and TANAKA, K. Spectrophotometric monitoring of arterial oxygen saturation in the fingertip. *Med. Biol. Eng. Comput.* **1980**, *18*, 27–32.

130 YOSHIYA, I., SHIMADA, Y. Non-invasive spectrophotometric estimates of arterial oxygen saturation. In *Non-invasive Physiological Measurements* P. ROLFE ed). Academic Press, London, **1983**, 252–286.

131 BUCHER, H. U., FANCONI, S., BAERCKERT, P. and DUC, G. Hyperoxemia in newborn infants: detection by pulse oximetry. *Pediatrics* **1989**, *84*, 226–230.

132 THILO, E. H., SCHMIDT, J., ANDERSON, D. and LUCKEY, D. Saturation by pulse oximetry: are different brands' results interchangeable? *Clin. Res.* **1992**, *40*, 133A.

133 ROLFE, P., KANJILAL, P. P., MURPHY, C. and BURTON, P. J. Continuous non-invasive beat-by-beat blood pressure (BP) measurement in the newborn. In *Proc. 3rd Int. Symp. Continuous Transcutaneous Monitoring*, **1987**. Eds. HUCH A., Huch, R., Rooth G., Plenum Press, New York.

134 CARTLIDGE, P. H. T. and RUTTER, N. Karaya gum ECG electrodes for the preterm infant. *Arch. Dis. Childh.* **1987**, *62*, 1281–1282.

135 REGALADO, M. G., SCHECHTMAN, V. L., KHOO, M. C., BEAN, X. D. Spectral analysis of heart rate variability and respiration during sleep in cocaine-exposed neonates. *Clin. Physiol.* **2001**, *21*(4), 428–436.

136 LOU, H. C., LASSEN, N. A. and FRIIS-HANSEN, B. Impaired autoregulation of cerebral blood flow in the distressed newborn infant. *J. Pediatr.* **1979**, *94*, 118–121.

137 ROLFE, P., PERSSON, B. and ZETTERSTROM, R. An appraisal of techniques for studying cerebral circulation in the newborn. *Acta Paed. Scand.* **1983**, *Suppl.* 5–13.

138 GUILLAUME, J. and JANNY, P. Manometric intracranienne continui: interet de la

methode et premiers resultats. *Rev. Neurol. Psychiatr.* **1951**, *84*, 131–142.

139 Lundberg, N. Continuous recording and control of ventricular fluid pressure in neurosurgical practice. *Acta Psychiatr. Neurol. Scand.* **1960**, *36*(Suppl 149), 1–193.

140 Levene, M. I. and Evans, D. H. Continuous measurement of subarachnoid pressure in neurosurgical practice. *Arch. Dis. Childh.* **1983**, *58*, 1013–1015.

141 Wealthall, S. R. and Smallwood, R. Methods of measureing intracranial pressure via the fontanel without puncture. *J. Neurol. Neurosurg. Psych.* **1974**, *37*, 88–96.

142 Robinson, R. O., Rolfe, P., and Sutton, P. Non invasive method for measuring intra-cranial pressure in normal newborn infants. *Dev. Med. Child. Neurol.* **1977**, *19*, 305–308.

143 Salmon, J. H., Hajjas, W., and Bada, H. S. The fontogram: a noninvasive pressure monitor. *Pediatrics* **1977**, *62*, 996–999.

144 Rochefort, M. J., Rolfe, P., and Wilkinson, A. R. New fontanometer for continuous estimation of intracranial pressure in the newborn. *Arch. Dis. Childh.* **1987**, *62*, 152–155.

145 Horbar, J. D., Yeager, S., Philip, A. G. S., and Lucey, J. F. Effect of application force on noninvasive measurements of intracranial pressure. *Pediatrics* **1980**, *66*, 455–457.

146 Hill, A. Intracranial pressure measurements in the newborn. *Clin. Perinatol.* **1985**, *12*, 161–177.

147 Philip, A. G. S. Noninvasive monitoring of intracranial pressure. A new approach for neonatal clinical pharmacology. *Clin. Perinatol.* **1979**, *6*, 123–137.

148 Hill, A. and Volpe, J. J. Normal pressure hydrocephalus in the newborn. *Pediatrics* **1981**, *68*, 623–629.

149 Kety, S. S. and Schmidt, C. F. The nitrous oxide method for the quantitative determination of cerebral blood flow in man: theory procedure and normal values. *J. Clin. Invest.* **1948**, *27*, 476.

150 Greisen, G., Frederiksen, P. S., Mali, J., Friis-Hansen, B. Analysis of cranial 133-Xenon clearance in the newborn infant by the two-compartment model. *Scand. J. Clin. Lab. Invest.* **1984**, *44*(3), 239–250.

151 Bada, H. S., Hajjar, W., Sumner, D. S., and Bashiru, M. Non-invasive monitoring of cerebrovascular changes in neonatal asphyxia and intra-ventricular hemorrhage. In *Fetal and Neonatal Physiological Measurements*, P. Rolfe ed). Pitman Medical, London, **1980**, 334–339.

152 Seker, H., Evans, D. H., Aydin, N., Yazgan, E. Compensatory fuzzy neural networks-based intelligent detection of abnormal neonatal cerebral Doppler ultrasound waveforms. *IEEE Trans. Inf. Technol. Biomed.* **2001**, *5*(3), 187–194.

153 Tarassenko, L. and Rolfe, P. Imaging distributions of electrical resistivity; an alternative approach. *Electron. Lett.* **1984**, *20*, 547–575.

154 Gronlund, J. U., Jalonen, J., Korhonen, I., Rolfe, P., Valimaki, I. A. Multivariate autoregressive modelling combined with transcephalic electrical impedance: method to relate neonatal systemic circulation and respiration to cerebral circulation. *Med. Biol. Eng. Comput.* **1995**, *33*(3 Special Issue), 458–463.

155 Gronlund, J. U., Kero, P., Korvenranta, H., Aarimaa, T., Jalonen, J., Tuominen, J., Valimaki, I. A. Cerebral circulation assessed by transcephalic electrical impedance during the first day of life – a potential predictor of outcome? *Early Hum. Dev.* **1995**, *41*(2), 129–145.

156 Rolfe, P. In Vivo Near Infra-Red Spectrophotometry. *Ann. Rev. Biomed. Eng.* **2000**, *2*, 315–354.

157 Schuster, A. Radiation through a foggy atmosphere. *Astrophys. J.* **1905**, *21*, 1–22.

158 Henyey, L. G. and Greenstein, J. L. Diffuse radiation in the galaxy. *J. Astro. Phys.* **1941**, *93*, 70–83.

159 Ishimaru, A. *Wave Propagation and Scattering in Random Media*. Academic Press, New York, **1978**.

160 Ishimaru, A. Wave Propagation and Scattering in Random Media. IEEE/OUP Series on Electromagnetic Wave Theory, IEEE Press, New York, **1997**.

161 Chandrasekar, R. Radiation Transfer. Clarendon Press, Oxford, **1950**.

162 Duderstadt, J. J. and Hamilton, L. J. Nuclear Reactor Analysis. Wiley, New York, **1976**.

163 Case, M. C. and Zweifel, P. F. Linear Transport Theory. Addison-Wesley, New York, **1967**.

164 Patterson, M. S., Wilson, B. C. and Wyman, D. R. The propagation of optical radiation in tissue I; Models of radiation transport and their application. *Lasers Med. Sci.* **1991**, *6*, 155–168.

165 Twersky, V. Absorption and multiple scattering by biological suspensions. *J. Opt. Soc. Am.* **1970**, *60*, 1084–1093.

166 Van Assendelft, O. W. Spectrophotometry of Hemoglobin Derivatives. Vangorcum Press, Assen **1970**.

167 Rea, P. A., Crowe, J., Wickramasinghe, Y. and Rolfe, P. Non-invasive optical methods for the study of cerebral metabolism in the human newborn: a technique for the future? *J. Med. Eng. Technol.* **1985**, *9*(4), 160–166.

168 Wray, S., Cope, M., Delpy, D. T., Wyatt, J. S., Reynolds, E. O. R. Characterisation of the near infra-red absorption spectra of cytochrome aa_3 and hemoglobin for the non invasive monitoring of cerebral oxygenation. *Biochim. Biophys. Acta* **1988**, *170*(933), 184–192.

169 Chance, B., Leigh, J. S., Miyake, M., Smith, D. S., Nioka, S., Greenfeld, R., Finander, M., Kaufmann, K., Levy, W., Young, M., Cohen, P., Yoshioka, H., Boretsky, R. Comparison of Time-resolved and -unresolved Measurements of Deoxyhemoglobin in Brain. *Proc. Natl. Acad. Sci. USA* **1988**, *85*, 4971–4975.

170 Patterson, M. S., Chance, B., and Wilson, B. C. Time resolved reflectance and transmittance for the noninvasive measurement of tissue optical properties. *Appl. Opt.* **1989**, *28*, 2331–2336.

171 Arridge, S. R., Cope, M. and Delpy D. T. The theoretical basis for the determination of optical pathlengths in tissue: temporal and frequency analysis. *Phys. Med. Biol.* **1992**, *37*(7), 1531–1560.

172 Delpy, D. T., Cope, M., van der Zee, P., Arridge, S. R., Wray, S., Wyatt, J. S. Estimation of optical pathlength through tissue from direct time of flight measurement. *Phys. Med. Biol.* **1988**, *33*(12), 1433–1442.

173 Chance, B., Maris, M., Sorge, J., Zhang, M. Z. A phase modulation system for dual wavelength difference spectroscopy of hemoglobin deoxygenation in tissues. *Proc. SPIE* **1990**, *1204*, 481–491.

174 Weng, J., Zhang, M. Z., Simons, K. and Chance, B. Measurement of Biological Tissue Metabolism Using Phase Modulation Spectroscopic Techniques. *Proc. SPIE* **1991**, *1431*, 161–170.

175 Fishkin, J. B., So, P. T. C., Cerussi, A. E., Fantini, S., Franceschini, M. A., and E. Gratton. Frequency-domain method for measuring spectral properties in multiple-scattering media: methemoglobin absorption spectrum in a tissuelike phantom. *Appl. Opt.* **1995**, *34*(7), 1143–1155.

176 Fantini, S., Franceschini, M. A., Maier, J. S., Walker, S. A., Barbieri, B. and Gratton E. Frequency-domain multichannel optical detector for non-invasive tissue spectroscopy and oximetry. *Opt. Eng.* **1995**, *28*, 32–42.

177 Matcher, S. J., Kirkpatrick, P., Nahid, K., Cope, M. and Delpy, D. T. Absolute quantification methods in tissue near infrared spectroscopy. *Proc. SPIE* **1995**, *2389*, 486–495.

178 Crowe, J., Rea, P. A., Wickramasinghe, Y. and Rolfe, P. Towards non-invasive optical monitoring of cerebral metabolism. *Proc. 2nd Int. Conf. Fet. Neonat. Physiol. Meas.* Oxford, UK, **1984**.

179 Wickramasinghe, Y., Crowe, J., Rolfe, P. Optical method adaptable for cerebral monitoring in the newborn. *Med. Biol. Eng. Comput.* **1985**, *23*, 468–469.

180 Jöbsis, F. F. Non-invasive, infra-red monitoring of cerebral and myocardial oxygen sufficiency and circulatory parameters. *Science* **1977**, *198*, 1264–1266.

181 Brazy, J. E., Lewis, D. V., Mitnick, M. H., and Jöbsis, F. F. Non-invasive monitoring of cerebral oxygenation in preterm infants: preliminary observations. *Pediatrics* **1985**, *75*, 217–225.

182 Crowe, J., Rea, P. A., Wickramasinghe, Y. and Rolfe, P. Towards non-invasive monitoring of cerebral metabolism. In *Neonatal Physiological Measurements*, (P. Rolfe ed). Butterworths. **1986**, 150–156.

183 Ferrari, M., De Marchis, C., Giannini, I., Nicola, A., Agostino, R., Nodari, S., Bucci, G. Cerebral blood volume and hemoglobin oxygen saturation monitoring in neonatal brain by near infrared spectroscopy. *Adv. Exp. Med. Biol.* **1986**, *200*, 203–212.

184 Wyatt, J. S., Cope, M., Delpy, D. T., Edwards, A. D., Wray, S. C., Reynolds, E. O. R. Quantification of cerebral oxygenation and hemodynamics in sick newborn infants by near infrared spectrophotometry. *Lancet* **1986**, *ii*, 1063–1066.

185 Brazy, J. E. Cerebral oxygen monitoring with near infrared spectroscopy. Clinical application to neonates. *J. Clin. Monit.* **1981**, *7*, 735–743.

186 Wickramasinghe, Y. A, Rolfe, P., Palmer, K., Spencer, S. A. Investigation of neonatal brain cytochrome redox state. *Dev. Brain Res.* **1995**, *89*, 307–308.

187 Palmer, K. S., Spencer, S. A., Wickramasinghe, Y. A. B. D., Wright, T., Southall, D. P., Rolfe, P. Effects of positive and negative pressure ventilation on cerebral blood volume of newborn infants. *Acta Paediatr.* **1995**, *84*, 132–139.

188 Edwards, A. D., Wyatt, J. S., Richardson, C. E., Delpy, D. T., Cope, M., Reynolds, E. O. R. Cotside measurement of cerebral blood flow in ill newborn infants by near infrared spectroscopy. *Lancet* **1988**, *11*, 770–771.

189 Wickramasinghe, Y. A. B. D., Livera, L. N., Spencer, S. A., Rolfe, P., Thorniley, M. S. Plethysmographic validation of near infra-red spectroscopic monitoring of cerebral blood volume. *Arch. Dis. Childh.* **1992**, *67*, 407–411.

190 Livera, L. N., Wickramasinghe, Y. A. B. D., Spencer, S. A., Rolfe, P., Thorniley, M. Cyclical fluctuations in cerebral blood volume. *Arch. Dis. Childh.* **1992**, *67*, *62–63*.

191 von Siebenthal, K., Beran, J., Wolf, M., Keel, M., Dietz, V., Kundu, S., Bucher, H. U. Cyclical fluctuations in blood pressure, heart rate and cerebral blood volume in preterm infants. *Brain Dev.* **1999**, *21*, 529–534.

192 Adcock, L. M., Wafelman, L. S., Hegemier, S., Moise, A. A., Speer, M. E., Contant, C. F., Goddard-Finegold, J. Neonatal intensive care applications of near-infrared spectroscopy. *Clin. Perinatol.* **1999**, *26*, 893–903.

8
Body Motion Analysis

T. Tamura

8.1
Introduction

Human motion and its measurement are of great interest. Motion analysis has been used to study gait, as well as shoulder, arm, trunk, and head movements. Gait analysis focuses on the lower limbs and quantifies the patterns of motion and forces during walking in the study of kinematics and kinetics, respectively. The practical measurement of gait kinematics was impossible until optical technology evolved to enable easy and repeatable recordings of motion. Kinematic and kinetic measurements allow the definition of normal patterns, and several key variables that are thought to characterize gait have been identified. In addition, the human activities of daily living involve motion. This chapter discusses the measurement of body motion and gait analysis.

Motion can be measured in two ways: directly, by sensors attached to the human body or indirectly, by optical methods. Both types of measurement have advantages and disadvantages. Direct measurement is simple and inexpensive, but less accurate, and it is difficult to attach the sensors, while indirect measurement is accurate but can only be used in limited areas and is expensive. In choosing a method, one must consider the aim of the study and select suitable sensors and transducers.

This chapter reviews recent developments in sensors, transducers, and measurement systems. It also discusses motion-related topics, such as force, EMG, and energy consumption.

8.2
Direct Measurement

The study of human movement can be divided into measuring motion and related parameters, such as force and electromyogram activity. Motion measurements include both linear motion and rotation. The three main types of contact transducers are goniometers, accelerometers, and magnetic tracking devices.

Sensors in Medicine and Health Care
Edited by P. Å. Öberg, T. Togawa, and F. A. Spelman

ISBN: 3-527-29556-9

8.2.1 Goniometry

A goniometer is a device that measures directly the angle between two limb segments using a simple manual protractor or a potentiometer. The goniometer is usually attached to the body and measures the angular displacement of a joint. A simple goniometer consists of a rotational potentiometer, as shown in Fig. 8.1. A single-axis lightweight rotational potentiometer with a specific resistance range is placed so that its axis is coaxial with that of the protractor. A resistance is attached to one arm of the protractor and a slide is attached to the other arm. Angular changes are recorded as voltage changes. The advantages of this device are its low cost and ease of use. However, actual joint motion does not consist of simple rotation around one fixed axis, but has more degree of freedom [1]. To measure precisely joint motion, a complex arrangement of goniometers is used to measure the true kinematic angles of the joint [2]. Such goniometers have been used in motion studies of the knee [3] and hip [4] joints.

The flexible goniometer shown in Fig. 8.2 consists of plastic endblocks that are connected at each end by a conductive piece of rubber [5–7]. The voltage output is proportional to the angle between the endblocks. The strain gauge housing inside the conductive rubber changes its electrical resistance in proportion to the change in the angle between the longitudinal axes of the endblocks. When the sensor is attached to the body, the motion angle can be obtained without disturbing movement. The axis of the goniometer does not need to be perfectly aligned with the axes of the joint. In addition, it is not easily damaged under large deformations. The biaxial version of this device allows one to simultaneously measure movement in up to two planes, such as the sagittal and frontal planes. During mea-

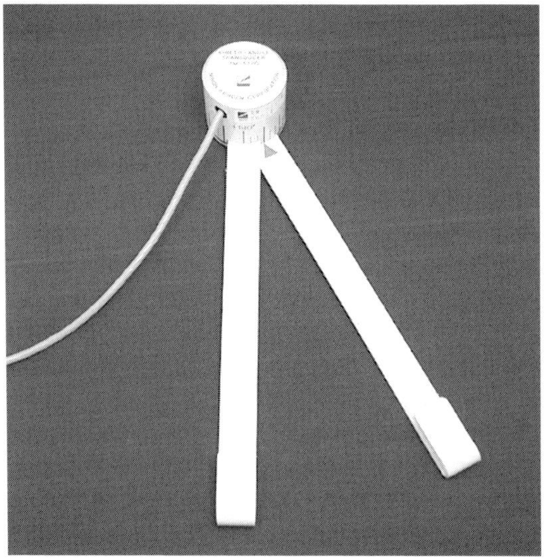

Fig. 8.1 An electrogoniometer.

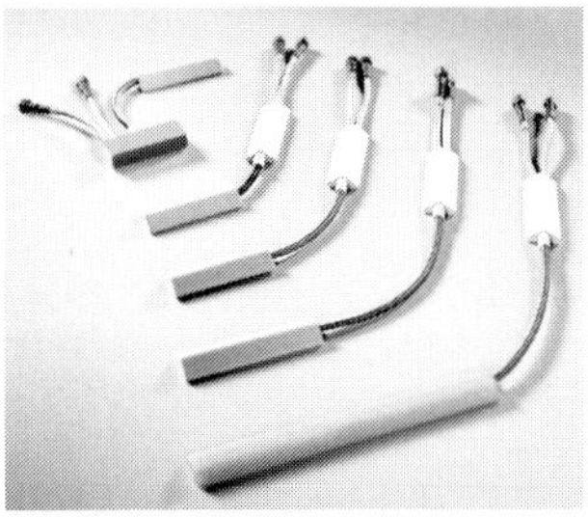

Fig. 8.2 Flexible electrogoniometer.

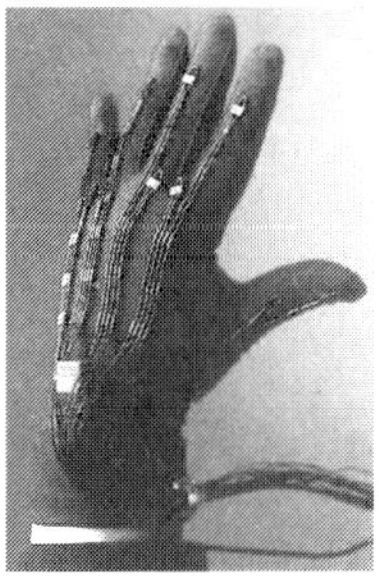

Fig. 8.3 Data-glove.

surements, a data logger stores the data for downloading to a PC using an amplifier and A/D converter.

An optical fiber goniometer can also be used to measure angles in fingers and hands (Fig. 8.3). This device is mainly used as a virtual-reality interface. The CyberGlove [8] is a fully instrumented glove that accurately measures up to 22 joint angles from the bending of the fibers. The ratio of light intensity between the ascending and descending fibers is proportional to the angle of the joint. The resolution, repeatability, and linearity are 0.5°, 1°, and 0.6% maximum nonlinearity over the full joint range, respectively.

8.2.2 Accelerometry

Accelerometers measure acceleration. While the acceleration of an object can be derived from the velocity (first derivative) or displacement (second derivative), differentiation of the signal usually causes noise. Direct measurement of the acceleration is often more convenient and easier. Translational acceleration can be measured simply as the force acting on a mass, while angular acceleration can be measured as the torque appearing in a body with a moment of inertia at the axis of rotation. However, when both translational and rotational motions exist, the situation is more complex, and apparent forces, such as centrifugal and Coriolis forces, may appear, as described in later.

The various types of accelerometers have different functional ranges, both in terms of frequency and magnitude. The maximum acceleration in voluntary human body motions does not exceed a few g, which is within the sensitivity of commercial accelerometers. Silicone micromachining technology has led to the development of very small and inexpensive, but sensitive, accelerometers. The correct type of accelerometer must be selected for the specific application.

Beam-type accelerometers are the most sensitive in the acceleration range of body motion. An elastic beam is fixed to a base at one end and a mass, called the seismic mass, is attached to the other end, as shown in Fig. 8.4a. When the seismic mass is accelerated, a force proportional to the mass times the acceleration occurs, and the beam bends elastically in proportion to the force. To avoid resonant oscillation of the seismic mass after a transient input, an adequate damping coefficient must be built into the mechanical system. Instead of a beam, a diaphragm, spring, or any other elastic material can be used in the accelerometer.

To determine the amplitude and direction of acceleration in 3D space requires either three translational accelerometers or a triaxial accelerometer. Fig. 8.4b shows a triaxial accelerometer in which three beams with seismic masses are assembled so that the beams are sensitive in perpendicular directions.

The displacement of the seismic mass can be detected by different methods, including those based on piezoresistance, piezoelectric effects, or capacitance. The semiconductor strain-gauge type of accelerometer is widely used to measure motion in humans and animals. Tab. 8.1 lists uniaxial and triaxial accelerometers that are small, inexpensive, sensitive, and available for human and animal studies.

A piezoresistive element is traditionally a strain gauge bonded to a mass-loaded cantilever. As the beam bends in response to acceleration, the material is stressed and its resistance changes. Four individual sensing elements are arranged in a Wheatstone bridge configuration, which provides output, while canceling cross-axis, temperature, and other spurious inputs. With micromachining technology, piezoresistors can be implanted into the support beams connecting the seismic mass to the frame or support structure.

The piezoelectric type of accelerometer is commonly used when only the a.c. component of acceleration needs to be measured. A polarization voltage appears in the piezoelectric material that is proportional to the deformation. The polarity

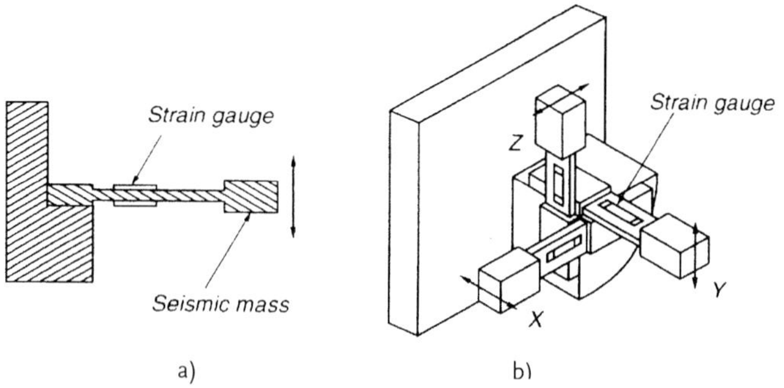

Fig. 8.4 Accelerometers: a) uniaxial; b) triaxial.

Tab. 8.1 Accelerometers for use in motion analysis

Manufacture	*Type*	*Component no.*	*Axes*	*Dimensions (mm)*	*Sensitivity (mV/G)*	*Frequency response (Hz)*	*Full scale range (G)*	*Cross axis sensitivity (%)*	*Non-linearity (%FSD)*	*Temp. drift (%FSD)*	*Supply voltage (V)*	*URL*
Akebono Brake	Piezoresistive	–	3	11.43×11.43×5.6	*X*, *Y*:5 *Z*:4	DC-200	±2	5	±0.5	±0.3	–	www.akebono-brake.co.jp/english/
ICSensors	Piezoresistive	3028	1	7.6×7.6×3.6	1.75	DC-350	±2, ±5, ±10	3	±1	±0.02	–	www.msiusa.com/icsensors/
ICSensors	Piezoresistive	3022	1	15.2×12.4×4	8	DC-250	±2, ±5, ±10	3	±1	±0.04	–	
ICSensors	Piezoresistive	3326	2	10×10×3.6	80	DC-2000	±25	5	±1	–	5	
Analog Devices	Capacitive	ADXL202E	2	5×5×2	312	DC-500	±2	2	±0.2	±0.1	3–5	www.analog.com/iMEMS/
Analog Devices	Capacitive	ADXL210	2	10×10×5.5	100	DC-500	±10	2	±0.1	±0.02	3–5	
Crossbow	Capacitive	CXL10LP3	3	33×28.7×11.5	500	*X*,*Z*:30 *Y*:7	±1.5	3	±1	±0.4	3.3	www.xbow.com/
Hokuriku	Piezoelastic	ACS010B	3	10.5×13.5×6.0	10	1-200	±10	20	±5	–	–	www.hdk.co.jp/top_e.htm
Hokuriku	Piezoelastic	ACM010B	3	11.0×13.9×10.0	200	1-200	±10	20	±2	–	3–7	
WACOH	Piezoelastic	PEA304	3	20.5×12.5×5.0	500	0.5-500	±4	5	±2	–	5	www.wacoh.co.jp/index-j.htm
WACOH	Piezoresistive	RPG	3	14×11.4×5.5	*X*, *Y*:4, *Z*:3	DC-200	±3	5	±1	*X*, *Y*:±4 *Z*:±3	–	
WACOH	Capacitive	MC300	3	29×16×13.5	30	DC-15	±1	10	±10	–	3–5	
Microstone	Piezoelastic	MA3-04Ac	3	20.5×12.5×5	500	0.8-1000	±4	10	±2	–	5	www.microstone.co.jp/index2.htm (Japanese)

Tab. 8.1 (cont.)

Manufacture	*Type*	*Component no.*	*Axes*	*Dimensions (mm)*	*Sensitivity (mV/G)*	*Frequency response (Hz)*	*Full scale range (G)*	*Cross axis sensitivity (%)*	*Non-linearity (%FSD)*	*Temp. drift (%FSD)*	*Supply voltage (V)*	*URL*
Japan Resistors	Piezoelastic	ASP-32N	3	16×16×7	7	–	±10	10	±2	–	–	www.jrm.co.jp/english/index.html
Silicon Designs	Capacitive	Model 1210	1	9×9×3	800	DC-400	±5	3	±1	±0.2	5	www.silicondesigns.com/Prod.html
Silicon Designs	Capacitive	Model 1010	1	9×9×3	–	DC-600	±5	3	±1	±0.3	5	
Endevco	Piezoelastic	Model 65	1	15.2×20.3×5.33	10	2–10000	150	5	±1	–	3–36	www.endevco.com/
VTI Hamlin	Capacitive	SCA600	1	10.5×11.5×5.2	1330	DC-400	±1.5	5	±30 mg	–	5	www.vti.fi/en/index.html
VTI Hamlin	Capacitive	SCA610	1	10.5×11.3×5.2	1330	DC-50±30	±1.5	5	±30 mg	–	5	www.vti.fi/en/index.html
VTI Hamlin	Capacitive	SCA320	2	12.7×12.7×6.4	750	DC-115	±1.5	5	±60 mg	–	5	www.vti.fi/en/index.html
Motorola	Piezoelastic	MMA1270D	1	10.5×10.7×3.6	250	DC-250	±8	5	3	–	5	e.www.motorola.com

Sensitivity is ratiometric to supply voltage (V_S). V_S is about 4.75–5.25 V

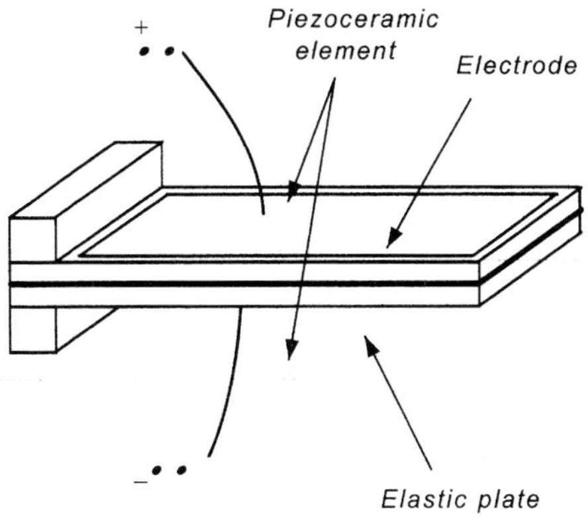

Fig. 8.5 Bimorph.

of the polarization voltage depends on the molecular structure of the material. Fig. 8.5 shows an example of a beam containing two piezoelectric elements of different polarities designed to produce a double or differential output. This configuration is called a bimorph.

The terminal voltage measures the charge that develops because the piezoelectric element has a capacitance, and hence the terminal voltage is proportional to the stored charge. More accurately, a charge amplifier measures the generated charge, as shown in Fig. 8.6.

When the input capacitance, which includes the capacitance of the piezoelectric element, and the stray capacity are represented by C_i, then the following relationships are valid

$$Q = C_i V_i + C(V_i - V) \tag{8.1}$$

$$V = -AV_i \tag{8.2}$$

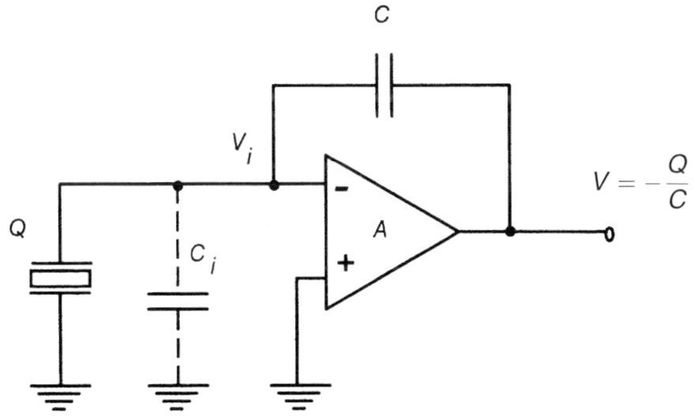

Fig. 8.6 A charge amplifier.

where Q is the generated charge, and V_i, V, and A are the input and output voltages and the gain of the operational amplifier, respectively. If $A \gg 1$, and $AC \gg C_i$, then

$$V = -Q/C \tag{8.3}$$

This means that the output voltage is proportional to the generated charge, regardless of the input capacitance. While the actual circuit has a finite time constant that is mainly determined by the current leaking from the feedback capacitor, very long time constants can be achieved by using a high quality capacitor. One commercial charge amplifier has a time constant to the order of 10^5 s (Kistler 5011, Winterthur, Switzerland). Another advantage of using a charge amplifier is that the sensitivity is not affected by the stray capacitance of the connecting cable or the capacitance of the piezoelectric element.

Very small, highly sensitive accelerometers have been developed using micromachining technology. Fig. 8.7 shows the typical configuration of an accelerometer with capacitive detection. A seismic mass supported by a beam is formed by silicone fabrication technology, and the displacement of the seismic mass is detected by the capacitance change between the electrode on the mass and the fixed electrode. Very high sensitivity, exemplified by a working range of ±0.1 g, can be attained in a device that has a seismic mass of 14.7 mg, an electrode gap of 7 μm, an 8.3×5.9×1.9 mm chip, and a resonance frequency of 126 Hz [9].

By introducing an electrostatic servocontrol operation, as shown schematically in Fig. 8.8, the force appearing in the seismic mass caused by acceleration can be balanced by an electrostatic force, so that the seismic mass stays at the equilib-

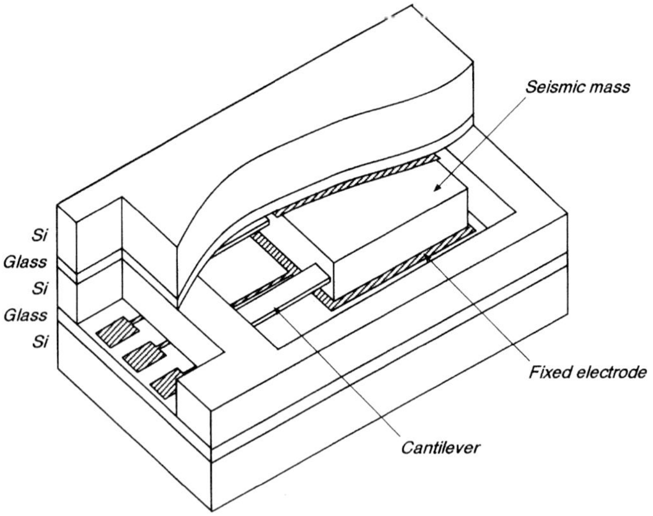

Fig. 8.7 A capacitance accelerometer.

rium point. This method is advantageous in that nonlinearity and fracture, due to a large displacement of the seismic mass, can be avoided and adequate damping can be applied by adjusting feedback circuit parameters [10].

In order to measure the acceleration of a body, the accelerometer must be rigidly attached to the body. To determine the 3D motion of a single body requires a minimum of six single-axis accelerometers. For accuracy, a minimum of nine single-axis devices per body segment is recommended [11].

From a biomechanical point of view, the attachment of the sensor warrants serious attention. There are significant differences between the outputs of skin-mounted and bone-mounted accelerometers [12]. However, accelerometry is simple and has been used to determine the spatial position of limb segments [13] and in gait analysis [14, 15]. The acceleration of the hips, shoulder, and upper-body joints can be calculated by combining the signals obtained from a triaxial accelerometer and a force plate [16]. For gait analysis, a triaxial accelerometer is rigidly attached to the back, close to the center of gravity, and the obtained acceleration signals are subjected to either frequency or time-frequency analysis [17, 18]. The difference between level walking and climbing stairs has been classified using wavelet transform [17]. A neural network has been used to classify gait according to inclination and speed of walking [18]. Gait-phase detection using accelerometers can be attempted with rule-induction algorithms [19]. The acceleration data is valid for measuring angular error and angular velocity [20]. When using an accelerometer in inclinometry, the angular error is small (1.3°), the reproducibility is high (0.2°), and the inherent angular noise is small (0.04°) and independent of device orientation.

A wearable accelerometric system has been developed [21]. The Wearable Accelerometric Motion Analysis System (WAMAS) (Fig. 8.9) enables real-time quantitative body motion analysis in non-laboratory settings. The WAMAS is a tool for diagnosing and treating movement disorders that is midway between a simple ob-

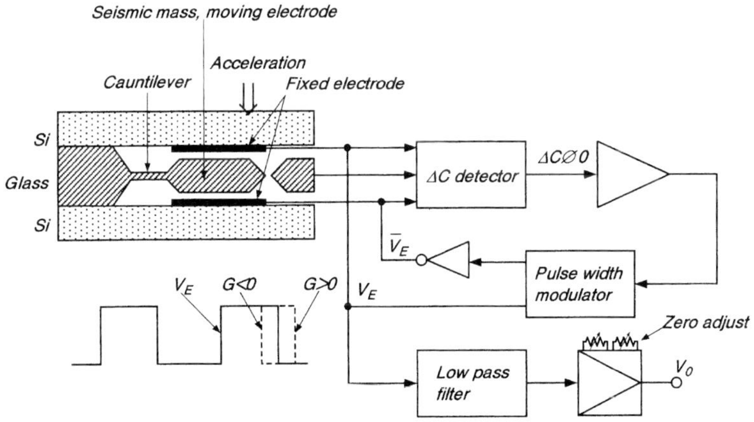

Fig. 8.8 An accelerometer with servocontrolled operation [10].

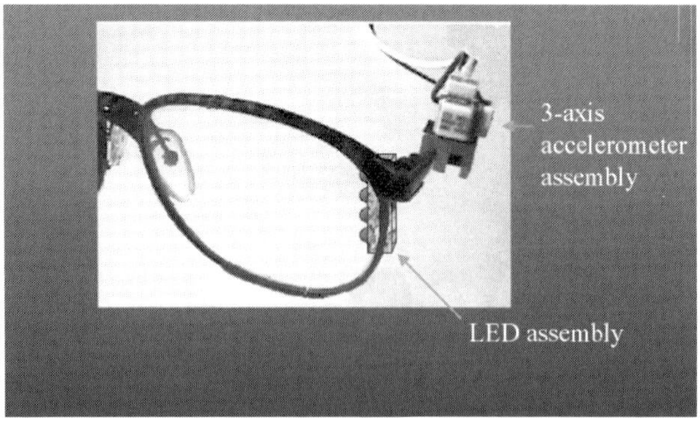

Fig. 8.9 Wearable accelerometer.

servational estimation of the risk of falling and consequent injury, and expensive laboratory-based gait analysis. It can provide unattended on-site quantitative records of balance status in the homes of outpatients; it is also suitable for use in outlying clinics that are remote from central laboratories. If physical- or occupational-therapist time with each patient is limited, the motion analysis system can act as a surrogate therapist, monitoring a patient's performance and compliance with a course of therapy. The wearable accelerometric instrument is useful: as a diagnostic tool to quantify hitherto qualitative measures of balance, as a biofeedback device during therapy, and as a fall-prevention aid, or what might be called a 'balance orthosis', for the institutionalized and community-living elderly who are prone to falling.

8.2.3
Gyroscope

Essentially, a gyroscope is a spinning wheel mounted on a movable frame. When the wheel is set spinning, it tends to retain its initial orientation in space, regardless of the external forces applied to it. When the axis is forced to change its direction, a torque proportional to the angular velocity of the axis of inclination arises, which can be used to detect angular velocity. An example of this type of transducer is the dynamically tuned gyroscope (DTG).

Typically, micromachined gyroscopes are specialized vibrating accelerometers that measure Coriolis forces. The basic operating principle of all vibratory gyroscopes relies on the generation and detection of a Coriolis force (Fig. 8.10). A very basic vibratory gyroscope consists of a proof mass mounted on a suspension that allows the proof mass to move in two orthogonal directions. To generate a Coriolis force, the proof mass must be in motion. To this end, the proof mass is electrostatically forced to oscillate in a direction that is parallel to the chip surface. If the gyroscope chip is rotated about the axis perpendicular to the chip surface, then a Coriolis force causes the proof mass to be deflected in the second direc-

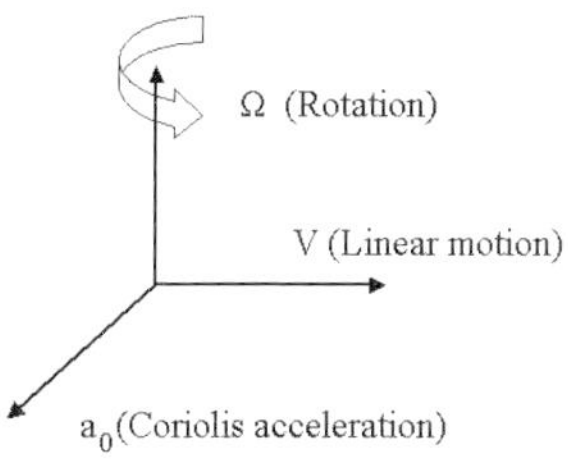

Fig. 8.10 The principle Coriolis-force detection.

tion. The amplitude of this oscillatory deflection is proportional to the rate of rotation, so that capacitive sensing, as in the case of the accelerometer discussed above, can be used to produce a voltage proportional to the angular rotation rate. Although this appears straightforward, building an effective micromachine gyroscope requires a great deal of finesse.

Usually, angular velocity is detected directly using a transducer called a tuning fork, which consists of four piezoelectric elements or bimorphs (Fig. 8.11). Two of the bimorphs are for driving and the other two are for sensing. The driving elements are attached to the base and connected to the sensing elements so that they are perpendicular to each other. The driving element bends resonantly through electric excitation, thereby swinging the sensing element. With zero angular velocity, the sensing element does not produce a signal. However, a bending motion is induced when the element rotates due to the Coriolis force and a signal that is proportional to the angular velocity is induced in the sensing element. While translational acceleration, with a component in the direction normal to the surface of the sensing element, may also induce a signal, this can be compensated by employing the differential operation of two pairs of elements. A commercial model (Angular Rate Sensor, Watson Industries, Eau Claire, MA, USA) has the following specifications: measurement range $\pm 30^\circ$/s or $\pm 10^\circ$/s, resolution 0.04°/s or 0.1% of the maximum angular velocity, precision 2%, nonlinearity less than 1% of the maximum angular velocity, frequency response 50 Hz, power consumption 150 mW, and weight 50 g [22]. A model with 2 or 3 axes is also available, as is a small, lightweight, quick-response device (Murata, Kyoto, Japan) [23]. Micromachined silicon rings have been designed to replace the conventional spinning wheel or bimorph. The deep-trench-etching process enables the production of tiny micromachined rings with maximum precision. A new ring-shaped micromachined resonator is resistant to external shocks

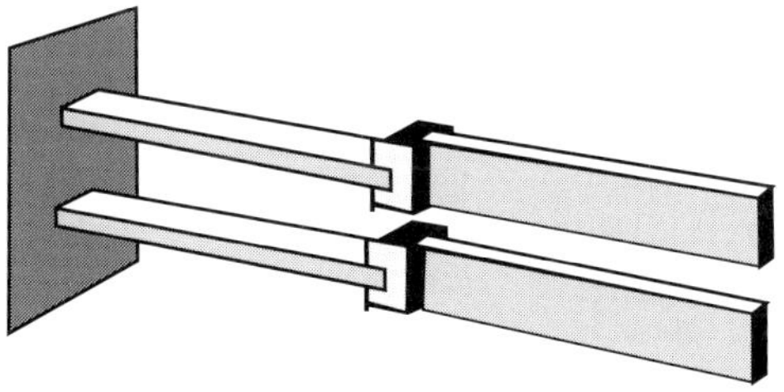

Fig. 8.11 Angular velocity transducer.

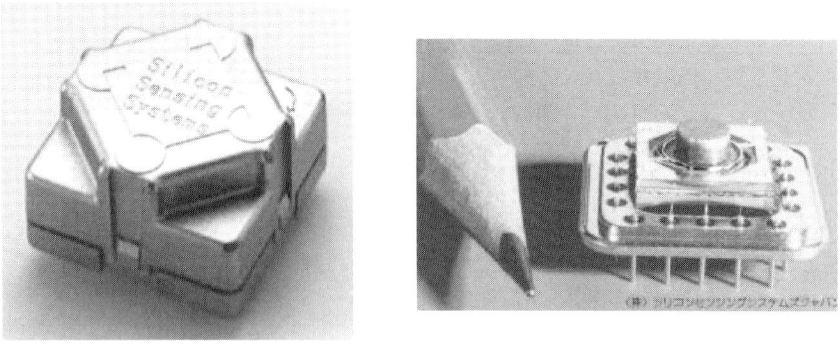

Fig. 8.12 Rate gyroscope.

and vibrations over a wide range of temperatures and a resolution of the order of 0.05 °/s over a 10 Hz bandwidth is attained in a model with a measurement range of ±100 °/s [24] (Fig. 8.12).

Used in conjunction with a three-axis accelerometer, a three-axis gyroscope results in a six-degrees-of-freedom inertial measurement unit. Silicon gyroscopes with sensitivity in the 0.1 °/s range over a 100 Hz bandwidth can be fabricated [25].

Real-time detection of angles based on the rate of gyroscope precession was limited due to the zero frequency offset when the device was stationary. A high-pass filtered piezoelectric gyro attached to the leg and hip was used to measure flexion extension angles and to determine walking speed [26]. To prevent offset, accelerometers are used to compensate for the error in the integral of the rate gyroscope. The mean differences between a rate gyroscope–accelerometer system and a reference goniometer in paraplegic standing trials for knee angle were 2.1 and 2.3 °/s, respectively, and for knee velocity were 2.3 and 11.8 °/s, respectively [27]. The rate gyroscope–accelerometer system is more accurate than using an accelerometer as an inclinometer. Kinematic data are also increasingly used to control neural prostheses and for functional electrical stimulation (FES) [28]. A combination of accelerometers, magnetic sensors, a rate gyroscope, and a strain gauge has been used in real-time control of FES [29].

8.2.4
Magnetic Tracking Methods

Magnetic tracking devices consist of reference and receiver parts that track the position of electromagnetic receivers in a magnetic field. The reference device is the transmitter that produces the reference magnetic field. The position of the receiver is measured in relation to an axis system established by the transmitter. The receiver can measure both rotation and displacement. If the transmitter is fixed with respect to a global axis system, absolute measurements can be made, but if the transmitter is located on one body segment, only relative values can be obtained. Fig. 8.13 shows simple examples. In Fig. 8.13 a, a magnetic field is generated by one coil and detected by another. In Fig. 8.13b, a permanent magnet and a Hall element are combined. Both methods use magnetometers. Since the magnetic flux density obeys the inverse square law for electromagnetic fields, the accuracy decreases as the distance between the transmitter and receiver increases.

Simple magnetometers are also used to monitor respiratory movements of the chest wall or abdomen [30–35].

In order to measure relative position and orientation in 3D space, a system consisting of both a triaxial magnetic dipole source and a triaxial magnetic sensor can be used. In a system developed by Raab et al. [36], both the source and the sensor consist of three orthogonally arranged coils wound around a ferrite core. A controlling computer excites the three source coils sequentially and signals from the three coils in the sensor are sent to the computer, as shown in Fig. 8.14. Then, the position and orientation of the sensor relative to the source, which has six degrees of freedom, can be computed. Such a system is commercially available (3 Space, Polhemus, Colchester, VT, USA) [37]. The sensor is small enough to attach to a subject, although the measurement is strongly affected by noise. The system performs as specified when the sensor is within 75 cm of the reference. Operation over a range of up to 1.5 m is possible with slightly reduced performance. The sampling rate is 120 Hz. Although environmental noise and drift affect the accuracy of measure-

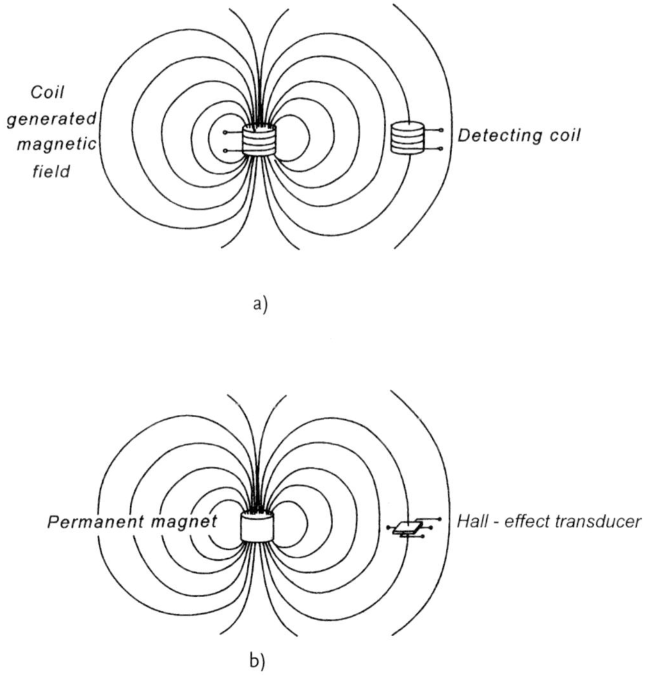

Fig. 8.13 Displacement measurement by magnetometers: a) magnetic-field generated; b) the Hall effect.

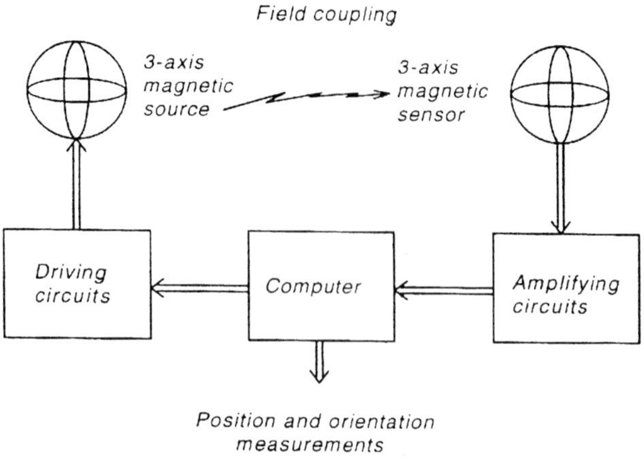

Fig. 8.14 A method for determining relative position and orientation.

ment, the simple inexpensive Polhemus sensor is commonly used to measure tremor. The calibration method has been described by Day et al. [38]. Positional and orientation data have been collected and calibrated using a locally linear model that is based on the position of the measurement. After calibration, the average position and angular errors were less than 1.8 cm and 1.2° up to 1.8 m from a long-range transmitter, respectively. For the standard transmitter, even after calibration, the errors increased sharply when the sensor was more than 1.2 m from the transmitter. Up closer, the post-calibration errors were less than 1.2 cm and 1.2°, respectively, while at distances of up to 1.8 m they were below 5 cm and 4°. Noise filtering could further reduce these errors. The authors do not recommend use of the standard transmitter at distances greater than 1.2 m due to orientation-based effects. A typical application of this system is to attach the sensor to a helmet and track the wearer's line-of-sight [39]. The system has also been used to measure various body movements, such as wrist motion [40] and lumbar spine mobility [41]. Arm position can be reconstructed from the angles of rotation of joints using the spatial tracking system [42]. In addition, tremors can be classified and the magnetic tracking system has been used as a screening tool [43].

8.2.5 Clinometer

A clinometer is an angle-sensing device that uses gravity as a reference. The common principle of measuring angle is based on either the potentiometric or capacitive differences of sensors. Each sensor is half filled with a dielectric liquid and an inert gas. Plates or a tilt tube, etched to form either a potentiometer or two variable capacitors, are used to provide an electrical reference for the horizontal position. The sensors are described as resistive and may use standard bridge circuitry to produce an angular readout. Rotation of the sensor produces a linear change in either resistance (single axis glass sensors) [44–46] or capacitance signal (AccuStar electronic clinometer, Shaevitz Sensor) [47], thereby producing the corresponding output (Fig. 8.15). The viscosity and conductive properties of the electrolyte can be suited to wide-ranging applications. Femoral anteversion (AV) has been determined by ultrasound (US). The angle of tilt of the US transducer, which represents the angle of AV, was measured with an attached clinometer [48]. An electromagnetic sensor has also been used as a clinometer and to monitor posture during daily life activities [49].

In addition, an earth-magnetic sensor has been used to evaluate movement disorders. A combination of EMG monitoring and earth-magnetic sensors improves the accuracy of body position monitoring. Monitoring the 3D orientation of body parts can be applied to long-term ambulatory recordings during daily-life activities. The 3D sensor combines miniature sensors of the earth's gravity and magnetism. It measures 60×50×10 mm and draws less than 1 mA of current from a 5 V battery. The non-horizontal direction of the earth's magnetism and body-part torque are corrected mathematically. This allows measurement of the 3D orientation of a body part in terms of vertical inclination and horizontal azimuth [50].

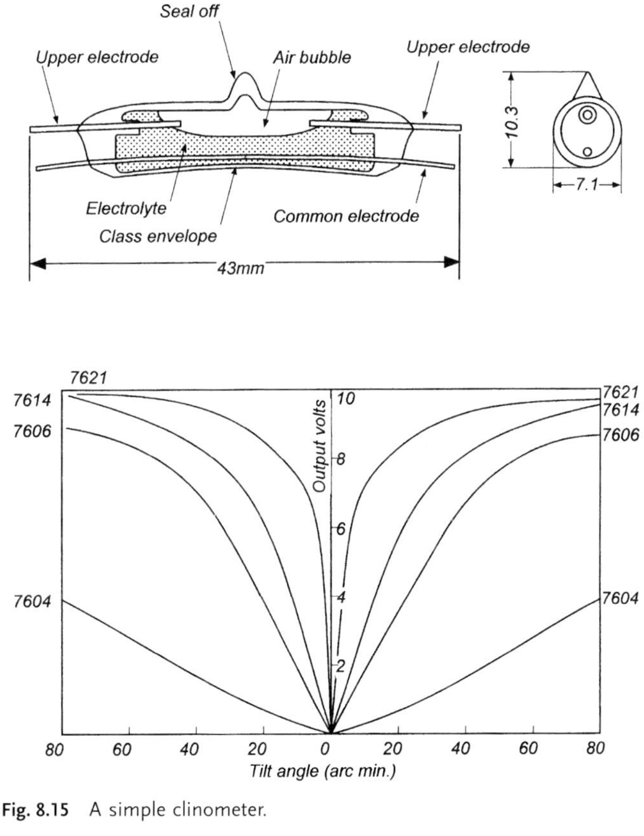

Fig. 8.15 A simple clinometer.

8.2.6
Velocity Measurement by Ultrasound

A portable, low cost, real-time, direct US ranging system (DURS) has been developed for the quantitative clinical evaluation of gait in real-time [51]. The DURS estimates the instantaneous forward velocity of the center of mass of a body, from which parameters such as gait speed, cadence, step time, step length, peak-to-peak variation, and time to achieve steady state walking may be calculated.

The DURS shown in Fig. 8.16 consists of three components: a transponder worn by the subject, a base-unit, IR emitter/US receiver, and a laptop computer.

The DURS operates by emitting high intensity infrared (IR) pulses from the base unit at a frequency of 22 Hz. Each IR pulse triggers the transponder to emit an US pulse back to the base unit. The base unit measures the time difference between the emission of an IR pulse and the arrival of a US pulse at the base unit. The US pulse travels at the speed of sound in air. The speed of sound in air is dependent on temperature and can be computed using the formula

$$\mathrm{speed(m/s)} = 331 + 0.6 \times \mathrm{temp}\,(^\circ\mathrm{C})\,. \tag{8.4}$$

The speed of sound at room temperature (22 °C) is approximately 344 m/s. By calibrating for the speed of sound in air, the time difference can then be converted into a measurement of the distance between the base and transponder units. Since distance measurements are taken at a rate of 22 Hz, a measure of how distance changes over time (i.e., velocity) is obtained. For static distance measurements, the principle has been exploited in inexpensive, commercial, distance-ranging devices.

The accuracy of the DURS was compared with that of the CODA 3 motion analysis system as described later. The velocity profiles obtained from the DURS and CODA 3 systems are very similar. Both devices accurately measure the periodic fluctuation in the forward velocity of the body trunk that results from the rising and falling of the center of mass during normal gait. The gait speed determined

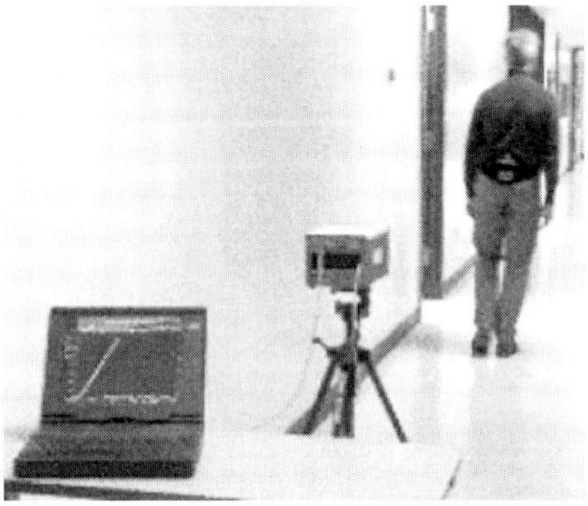

Fig. 8.16 An ultrasound velocity monitoring system [51].

with the DURS was consistently within 3% of that determined with the CODA 3 system. The current system can accurately measure at a distance of 15 m.

8.2.7
Footswitches

Footswitches are a convenient and inexpensive method of measuring temporal gait. Walking speed, cadence, stride length, and limb-support times can be obtained from the times of floor–floor contact and traversed distance. Two basic types of footswitches are available: compression-closing and force-sensitive resistor (FSR) switches. These are worn as thin insoles in the subject's shoes or taped to the bottom of bare feet. Compression-closing switches [52] consist of two thin pieces of brass shim separated by a compressible foam rubber insole. In the contact areas, conductive rubber cylinders are inserted into holes in the insole. When pressure is applied, the insole compresses and the conductive rubber cylinder contacts the pieces of brass on each side of the insole and closes an electrical circuit. The FSR (Fig. 8.17) consists of two thin layers of flexible plastic with printed circuits on the inner surfaces, separated by a thin layer of double-sided adhesive [53]. When pressure is applied, carbon on one surface contacts a metal pattern on the other surface, creating an electrical circuit.

Another FSR foot switch consists of a 25 mm diameter membrane switch with a 15 mm sensor area on a 100 mm flexible tail. The entire switch sensor is only 1 mm thick and responds reliably to 50 g pressure anywhere in the sensor area [54]. The small size of the sensor allows flexibility in placement on the subjects' foot, making it much easier to accurately define gait events in clinical subjects as the sensors can be placed on any part of the foot that comes in contact with the floor.

Contacts are usually placed on the heel, 5th metatarsal, 1st metatarsal, and big toe areas to indicate when these areas of the foot are bearing weight. The heel section is separated from the forefoot section so that one pair of switches can accommodate a range of shoe sizes. The advantage of discrete switches is that different sizes are not required to fit a large range of foot sizes. The disadvantage is in getting reliable data because of difficulties in consistently placing the switches at the proper locations under the feet. Typical footswitch activation delay times are about 1–2% of the gait cycle. Footswitch software can be used that compensates for this delay.

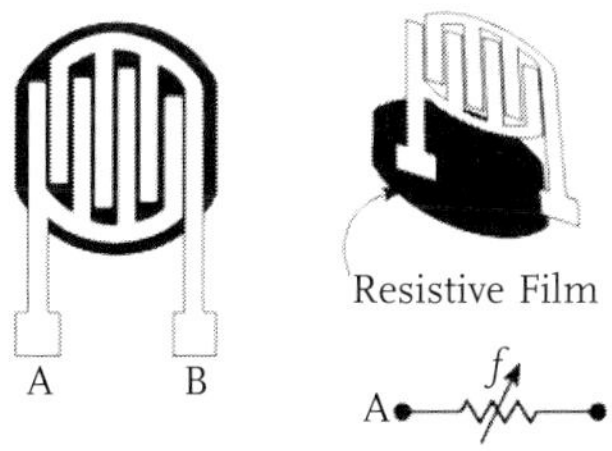

Fig. 8.17 An interlink force-sensing resistor.

The Footswitch Stride Analyser is a computer-based instrument that calculates temporal gait parameters based on footswitch data averaged over a measured distance [55]. The subject wears a small battery-powered microprocessor recording unit (data logger), which stores the footswitch data. In a footswitch system, a pair of light beams is placed along the line of a walkway. The light beams are placed apart at set distances. A light-sensitive switch worn on the subject's arm triggers the recorder when the subject passes the lights at the beginning and end of the measured walkway.

8.3 Non-contact (Optical) Measurements

The first optical method used was a photographic method that employs hand digitization of 16 mm cinefilm [56]. In this method, markers are placed on a subject to identify the skeleton. Three movie cameras are placed orthogonal to a walkway and the subject is photographed while walking. After the film is processed, it is analyzed frame-by-frame and the researcher digitizes the points by hand, by directly entering the x–y co-ordinates into a computer. Then joint-angle rotation and stride characteristics are plotted or shown graphically. The disadvantages of this method are that it is slow, the film is expensive, and it is labor intensive.

The optical methods currently used in biomechanical analysis involve real-time measurement of the position of a discrete light source or markers fixed to the subject using video. Three video methods are available: passive, active, and pattern-matching. The markers are either active or passive.

A passive marker is reflective, and is illuminated by a light source close to the measurement camera. The markers are usually made from lightweight plastic spheres, and with material such as silver Scotch-Lite tape. The light source is generally a ring of light emitting diodes (LEDs) around the lens, which is strobed to the field rate of the camera. In clinical applications, most measurements are made in an environment with relatively well-controlled ambient lighting. Monochromatic IR light and IR video cameras are commonly used to minimize disturbance to the subject. Owing to its passive nature, the trajectory of each marker must be identified with a marker label and then tracked throughout the examination. When makers are lost from view, the proper identification can be lost. The position of each marker can be determined by an intelligent tracking-algorithm.

LEDs are commonly used as active markers. An active marker can be identified automatically by fast, sequential illumination. If many LEDs are driven sequentially, the spatial co-ordinates of each LED can be determined more easily than in a situation in which many reflecting markers appear simultaneously, because, in the sequential operation, only one light spot is seen at any one time. A multiplexing frequency is chosen to scan for all markers. Synchronization results from the inherent time skew in multiplexing. The disadvantages of active markers are the need for an electric power supply and subject cabling. The cabling can interfere with the subject and narrow the viewing angle.

There are two different types of light sources/markers: one-dimensional, linear image sensors and 2D sensor arrays. Both types of sensor detect the intensity of focussed images of the markers and convert them into digital form. The detector directly converts the transition between light and darkness at the edge of the image into image position co-ordinates. The number of detectors across the image determines the inherent resolution of the sensor. Since the image of a marker normally covers several detectors, its center can be determined by averaging either the intensities or the outline of the image. This spatial averaging is allowed by the accuracy of the opto-electronic system.

Automated motion measurement systems are also available and are more convenient for data analysis. In these systems, the spatial co-ordinates of specific positions on the body are measured and stored in a computer. Automated systems use either passive or active markers. VICON (Oxford Metrics) is a video-based system in which reflective markers are placed on the subject and detected by the system via IR strobes that are placed around the camera lens [57]. The subject walks and at least two cameras can detect markers. To allow 3D calculations, VICON offers two types of camera: one with a sampling rate of 50 frames/s and one with a sampling rate of up to 200 frames/s that uses a rotating shatter to freeze movement. The markers range in diameter from 4 to >50 mm.

The Selspot system (Selcom Selective Electronics, Sweden) consists of IR LEDs that are placed on the subject as skeleton markers [58]. The diodes are pulsed at different rates by a control unit. Two cameras are placed along the walkway, so that both cameras can see each marker. Since each marker pulses at its own rate, the light detector in the camera samples each marker and can calculate its x–y co-ordinates. The data are then transmitted to a computer for 3D calculation.

A one-dimensional image sensor can be used to determine the position of a light spot. Two types of image sensors are used. As shown in Fig. 8.18a, the MOS-type image sensor has a photodiode array. Each photodiode is connected to the output port via a switching transistor driven sequentially by a scanner. A charge-coupled device (CCD) image sensor with a photodiode array is also used (Fig. 8.18b). The outputs of all the photodiodes are simultaneously transferred to the CCD, which then sequentially transfers the outputs from each photodiode to the common output port. A 2D position-sensitive detector can determine directly the position of a light spot in 2D space. Fig. 8.19 shows two different configurations of position-sensitive detectors [59]. Fig. 8.19a shows a tetralateral position-sensitive detector in which the photoconductive layers are on one side of the device, and Fig. 8.19b shows a duolateral position-sensitive detector, which has two photoconductive layers on each side of the device. While the distortion of the image is small in the duolateral type, the tetralateral type is advantageous because there is less dark current. Both types of position-sensitive detectors are available commercially (Hamamatsu Photonics, Hamamatsu, Japan).

The position of a light marker in 3D space can be determined by using three one-dimensional position sensors. As shown in Fig. 8.20, if a cylindrical lens projects a target light spot onto a one-dimensional position sensor placed in the focal plane of the lens, so that the sensor axis is orthogonal to the nodal axis of the

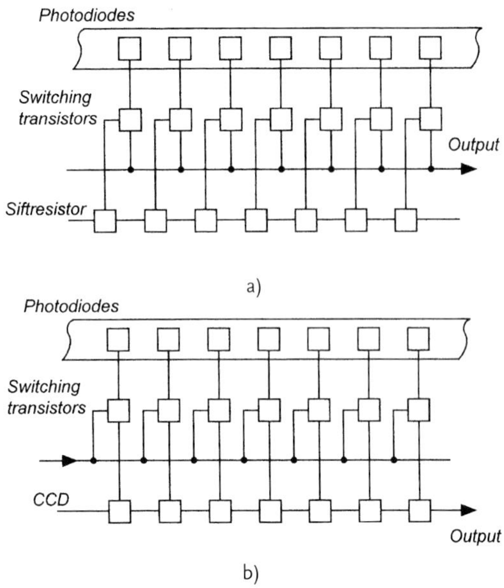

Fig. 8.18 Two types of image sensor: a) an image sensor in which the output of photodiodes is scanned by switching transistors; b) an image sensor in which a CCD transfers the output from the photodiodes sequentially to a common output port.

lens, this defines a plane that includes the target light spot and the nodal axis of the lens. By using three sensors with cylindrical lenses aimed in different directions, the position of the target can be computed as the intercept of these three planes [60]. Using CCD-type one-dimensional position sensors with 2048 elements, a resolution of 1/4000 was attained in a 2.2×0.6×2.2 m measurement field at a viewing distance of 5 m.

The CODA (Charnwood Dynamics, Leicestershire, UK) [61] scanner contains three position-sensitive sensors, which are mounted on a rigid frame within the unit. The sensors and associated DSP processors located in a host PC use advanced cross-correlation techniques to accurately locate the positions of reactive markers. The sensor outputs are used by the software to calculate the x, y, and z co-ordinates of the markers in real-time by triangulation. This means that one

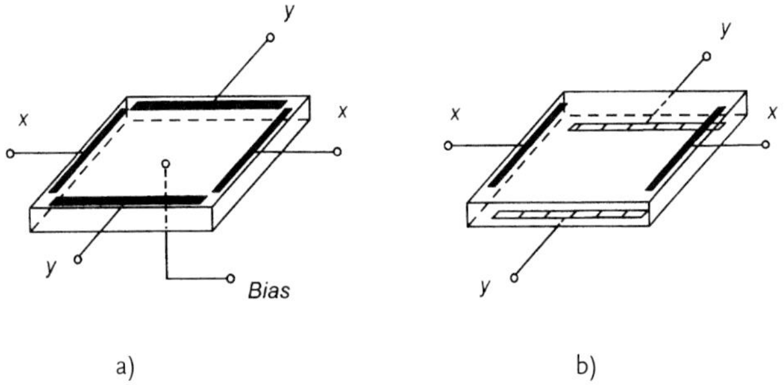

Fig. 8.19 Two configurations of a position-sensitive detector.

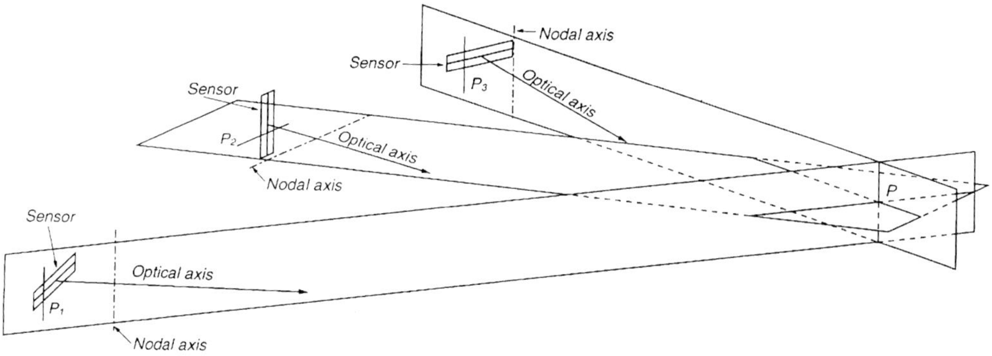

Fig. 8.20 Method for measuring the position of a light marker using three one-dimensional position sensors.

scanner can collect full 3D data. Up to 6 scanners can be used together for a larger field-of-view.

The OPTOTRAK (Northern Digital, Waterloo, ON, Canada) [62] is another non-contact motion measurement system that tracks small IR markers attached to a subject or object. Unlike video tracking systems that employ reflective markers, OPTOTRAK employs active marker technology. True real-time 3D/6D data rates and higher accuracy are two of the main advantages of this technology. The position sensor can identify the IR light emitted from markers, while ignoring interference from reflections and ambient lighting. The 3D positions of markers are accurately determined in real-time for viewing and processing. Simultaneously, data may be saved to file for *post hoc* analysis.

High-resolution digital cameras have revolutionized the motion-capture industry with extreme resolution, unprecedented high frame rate, upgradeable functionality, and ease of use. The motion analysis motion-capture system, The Eagle, has a resolution of 1.3 million pixels and a digital pixel rate of 600 million pixels per second [63].

The optical method commonly uses either one- or 2D image sensors. More precise analysis of human movement requires 3D calculation. Photogrammetric calculation transforms 2D co-ordinates taken from the image plane of one or several cameras into 3D co-ordinates. If two pictures of the same scene are captured by two cameras with different perspective centers, it is possible to derive 3D positions by knowing the orientation parameters of the individual cameras and by measuring the 2D co-ordinates of common image points in each plane. Thus, the position and orientation of the camera and the principal distance between two cameras are required for camera calibration. A control frame that creates a free, net, orthogonal, 3D co-ordinate system is commonly used. The control frame is captured by each camera and the position of camera is determined inversely. Direct linear transformation (DLT) is a popular algorithm for such calculations. The parameters in the equations are found by combining general mathematical form with measurements from a system of known 3D points on calibration objects. The following equations are used

$$x = \frac{b_{11}X + b_{12}Y + b_{13}z + b_{14}}{b_{31}X + b_{32} + b_{33}z + 1} \quad \text{and} \quad y = \frac{b_{21}X + b_{22}Y + b_{23}Z + b_{24}}{b_{31}X + b_{32}Y + b_{33}Z + 1} \tag{8.5}$$

where x and y are the 2D image/source co-ordinates, X, Y, and Z are the 3D square co-ordinates, and b_{11} to b_{33} are 11 parameters representing the internal and external positions, orientations, and distortions of each camera [64].

A least-squares method is usually used to optimize the redundant equations in these formulae. DLT uses linear corrections for the various lens-distortion parameters. For higher precision and resolution, a nonlinear transformation protocol, called Direct Non-Linear Transformation, has been proposed [65]. Moreover, polynomial equations can be used to compensate for optical distortions [66].

Improved system accuracy requires the use of markers and their accurate placement. All current systems attempt to estimate the motion of bone using measurements made with surface-mounted markers. In these systems, major inaccuracies arise from the motion of the soft tissues relative to the bones. These errors are called skin movement artifacts. To solve this problem, the relative movements of surface-mounted markers and bone were measured by X-ray and the error caused by skin movement was detected [67]. Furthermore, some groups attempted to decrease the error at the knee joint by using a simple knee band, which decreased the error [68]. Another method uses three markers on the truck instead of measuring joint position [69]. When more than three markers are attached to the trunk their relative position and orientation are determined from these markers and joint position. The position and orientation of joints can be extrapolated from the positions of the markers [70], as shown in Fig. 8.21.

Another problem arises when markers are hidden during movement. One way to avoid this is to use several cameras; another is to use more than three markers. The skin movement artifact can be diminished as well as the angle-measurement rate.

In template-detectors, intensity variations across the image are compared with those of a known template for pattern matching. A sharp contrast between the bright image of the marker and the background is essential.

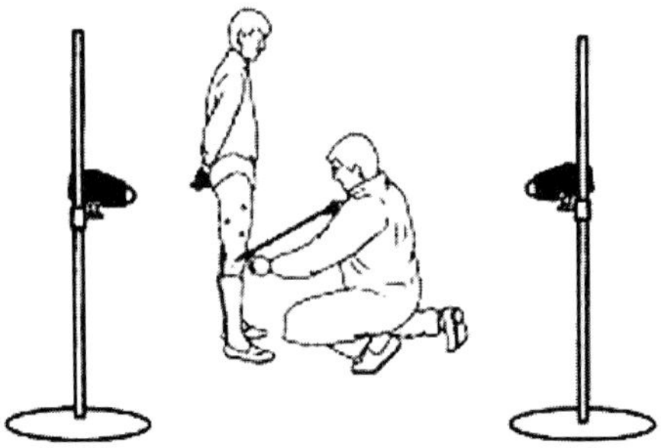

Fig. 8.21 Femur anatomical landmark calibration using a pointer.

8.4 Force Measurements

Muscles produce forces and may cause body movements that in turn cause a reactive force from the environment, such as the reaction from the floor when a subject walks on it. The environment passively exerts many forces on the body; these are due to natural phenomena or are artificial in origin. Force measurements are required when these forces are studied quantitatively.

A human body standing on the ground exerts a gravitational force on the ground. An equal and opposite reaction force is exerted on the body by the ground. When the body moves, the force exerted on the ground or body varies in strength and direction due to the acceleration and deceleration of the body. Thus, the force exerted on the ground, which is called the ground force, provides information about body motion. Ground-force measurements have been used in clinical gait analysis, in stabilometry, and to evaluate athletic capacity in sports medicine.

A transducer placed on the ground or attached to the foot is used to measure the ground force. While these two forces have the same magnitude, their directions are opposite, so that the data obtained appears different, because a transducer on the ground measures the force with reference to a fixed co-ordinate system, whereas a transducer attached to the foot measures the force with reference to a moving co-ordinate system.

8.4.1 Force Plate

A force plate is commonly used to record the ground-force reaction and provides the three orthogonal components of the ground-force reaction vector, the co-ordinates of the center of pressure in the plane of the force plate and the associated spin moment about a vertical axis.

A force plate or force platform, which consists of a plate supported by force transducers, is installed on the ground to measure ground force, as shown in Fig. 8.22. A simple force plate has transducers that only measure vertical forces, and outputs the vertical component and point of application of the ground force. The more commonly used force plates have transducers that measure one vertical and two horizontal (shear) components of the force, so that the strength, direction, point of application, and torque of the ground force can be obtained.

Each component of the ground force is determined from the transducer outputs in the following manner. If the positions of four transducers in the horizontal x–y plane are (x_1, y_1), (x_2, y_2), (x_3, y_3), and (x_4, y_4), and the transducer outputs, which include two horizontal components and one vertical component, are (F_{x1}, F_{y1}, F_{z1}), (F_{x2}, F_{y2}, F_{z2}), (F_{x3}, F_{y3}, F_{z3}), and (F_{x4}, F_{y4}, F_{z4}), then the x, y, and z components of the ground force are given by

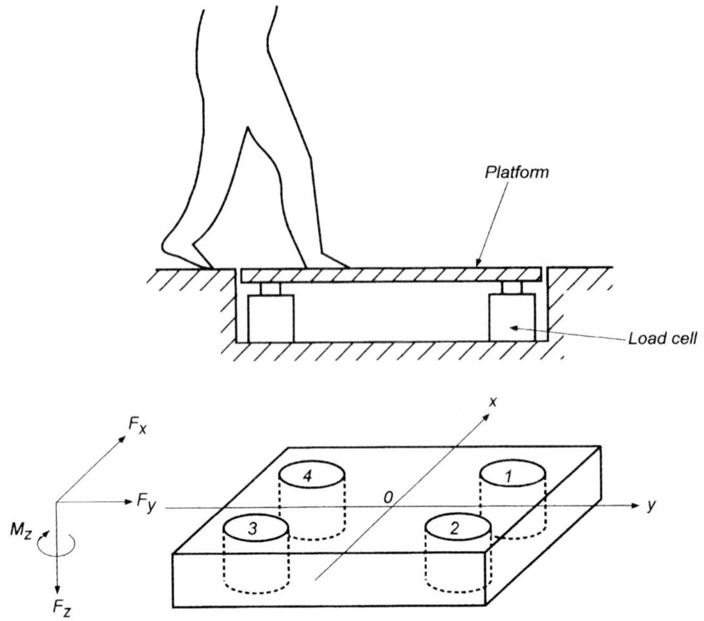

Fig. 8.22 The force plate.

$$F_x = \sum_{i=1}^{4} F_{xi},\ F_y = \sum_{i=1}^{4} F_{yi},\ F_z = \sum_{i=1}^{4} F_{zi} \tag{8.6}$$

The point of application of the ground force, $(\bar{x}, \bar{y})$, is given by

$$\bar{x} = \frac{\sum_{i=1}^{4} f_{zi} x_i}{F_z} + \frac{F_x}{F_z}\Delta z,\ \bar{y} = \frac{\sum_{i=1}^{4} F_{zi} y_i}{F_z} + \frac{F_y}{F_z}\Delta z \tag{8.7}$$

where Δz is the depth of the transducer elements measured from the ground surface level. The torque referring to the vertical axis at the point of application is given by

$$M_z = \sum_{i=1}^{4} F_{yi}(x_i - \bar{x}) - \sum_{i=1}^{4} F_{xi}(y_i - \bar{y})\,. \tag{8.8}$$

The force transducers in a force plate are either of the strain-gauge type or of the piezoelectric type. Fig. 8.23 shows an example of a strain-gauge type of force plate [71, 72]. The platform is supported by four metal pipe pylons to which strain gauges are attached in order to detect the compression, bend, and twist of the pylon.

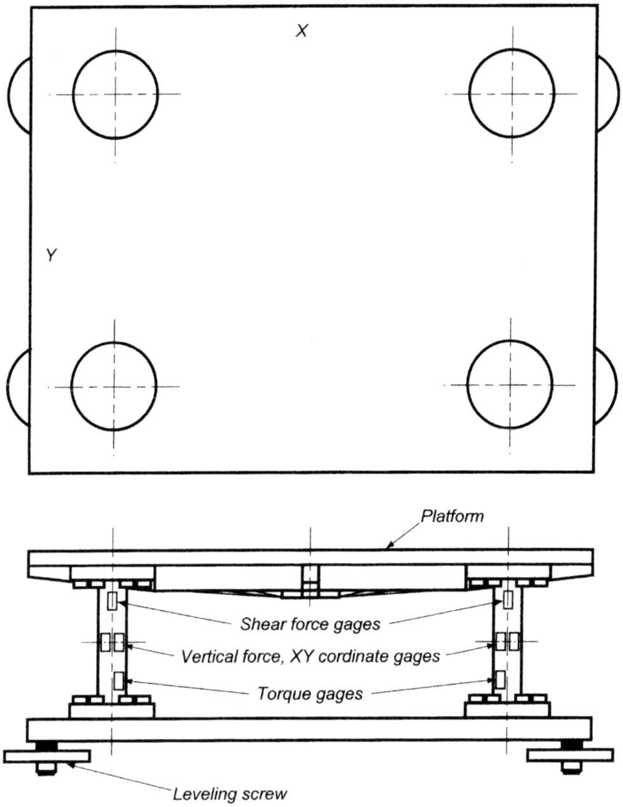

Fig. 8.23 A strain-gauge type force plate.

Fig. 8.24 shows a piezoelectric type force transducer developed by Kistler Instruments (Winterthur, Switzerland) [73]. It consists of three pairs of quartz disks sandwiched between the steel base and the top of the transducer, and provides output corresponding to the three orthogonal components of the applied force. While a piezoelectric transducer cannot measure static force, a fairly long time constant can be realized using charge amplifiers. In practice, a Kistler force plate attains a time constant of up to 1000 s. Typically, a Kistler force plate has sensitivities of 5 and 10 mN for the horizontal and vertical force components, respectively, a measurement range of –10 to 10 kN, and a resonance frequency of 800 Hz (Kistler, Type 9281 B).

The length and width of a force plate are typically 600 and 400 mm, respectively. A walkway, along which many force plates are installed, is required to measure the ground forces of many steps in gait analysis. For example, eight force plates arranged in line were used in a study that estimated mechanical work during walking and running [74].

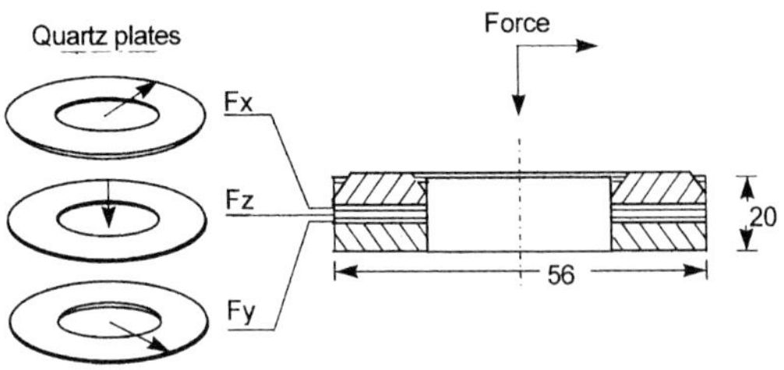

Fig. 8.24 A piezoelectric force transducer.

8.4.2
Stabilometers

Clinical examination of posture stability is called stabilometry, and the force plate designed for stabilometry is called a stabilometer. Simple stabilometers only measure the locus of the point of application of the ground force by using triangularly arranged vertical force transducers, as shown in Fig. 8.25.

If the co-ordinates of the transducers in the x–y plane are $(0, a)$, $(-b, -c)$, and $(b, -c)$, and their vertical force components are F_1, F_2, and F_3, then the x and y co-ordinates of the application point of the ground force, $(\overline{x}, \overline{y})$, are given as

$$\overline{X} = \frac{b(F_3 - F_2)}{F}, \quad \overline{Y} = \frac{aF_1 - c(F_2 + F_3)}{F} \tag{8.9}$$

where $F = F_1 + F_2 + F_3$.

When a subject is standing still, the point of application of the ground force stays directly below the center of gravity of the body. However, when the body is accelerated in a horizontal direction, the point of application deviates from the point directly below the center of gravity. Thus, the locus of the point of application does not simply reflect the excursion of the center of gravity. If the center of gravity has a hor-

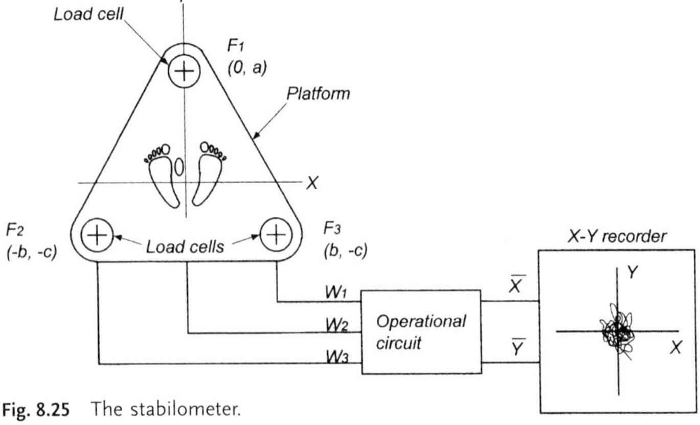

Fig. 8.25 The stabilometer.

izontal acceleration a, and the height of the center of gravity from the ground level is h, then the deviation of the application point of the ground force from the point directly below the center of gravity is ah/g. In an extreme situation, such as jumping forward, deviations of around 30 cm can occur [75].

In order to study the characteristics of the posture control system in humans, the response to externally applied disturbances has been measured using moving force plates. For example, a motor-driven force plate was used to apply pseudo-random acceleration to a standing human subject [76].

8.4.3 Instrumented Shoe

The ground force acting on the foot can be measured by attaching instruments to the foot. The shoe designed for such measurements is called an 'instrumented shoe'. The instrumented shoe is advantageous in foot-force measurement because measurements can be performed during many steps as in natural locomotion without restricting the subject to a walkway where force plates, or other instruments, such as a video camera, are installed.

Fig. 8.26 shows an instrumented shoe [77]. It has a load cell attached at the toe and another at the heel. Each load cell consists of an end-support spring element on which strain gauges are mounted, and provides output corresponding to the anterior-posterior and medial-lateral shear, axial compression, and torque. Kljajic and Krajnik used an instrumented shoe with eight or nine strain-gauge type force transducers [78]. Each transducer consists of a steel beam with a protuberance in the middle that contacts the ground. By employing many transducers, the shoe not only provides the vertical component of the ground force, but also the point of application of the vertical force along the shoe.

Miyazaki and Iwakura reported a simple instrumented shoe in which two vertical-load transducers were attached at the metatarsal part and the heel [79]. Each transducer consisted of a stainless steel beam with strain gauges. A capacitive transducer in which a foam-rubber sheet was sandwiched between two copper sheets so that the capacitance between the copper sheets varied when the rubber sheet was compressed by the applied load, as shown in Fig. 8.27, was also studied [80]. Although the sensitivity might change when the load was localized but not uniform, the observed error was limited to ±10%.

A pediatric insole system using a FSR is also available. The FSR is attached to a custom insole. Interlink circular FSRs (18 mm outer diameter, 15 mm active-sensing diameter, 0.4 mm overall thickness) are flexible, durable, reliable, inexpensive and have overload tolerance and electrical simplicity. The conductive polymer sensors provide a logarithmic resistance drop with increasing load [81].

Many of the earlier studies that were used to obtain in-shoe pressure measurements were classified as discrete devices. Nicol developed a flexible capacitance transducer array to monitor in-shoe pressures. This system refines the commercially available PEDAR insole (Novel Electronics) [82].

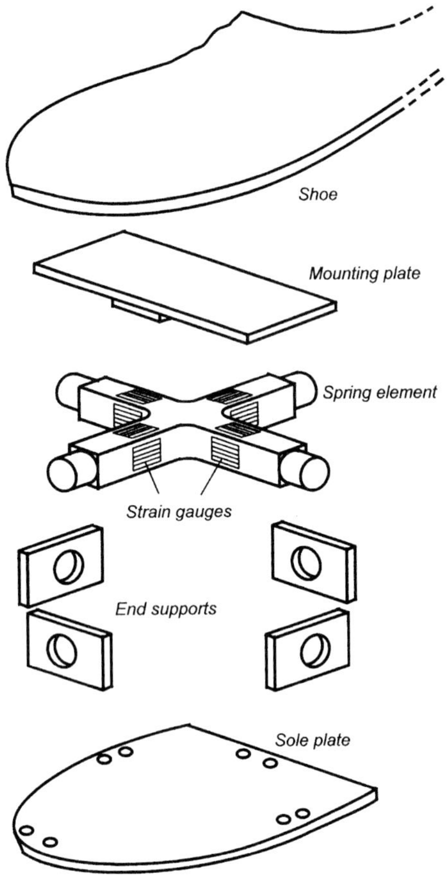

Fig. 8.26 An instrument shoe [77].

Currently available insoles have 99 sensors for adults with an average sensor area of 1.5 cm^2. Their day-to-day reliability appears to be excellent each element may be calibrated individually.

The F-scan insole (Tekscan, Boston, MA, USA and Nitta, Osaka, Japan) [83–85] is a very thin (0.15 mm), disposable FSR insole with a maximum of 1260 elements, which can be trimmed to fit various shoe sizes. It consists of two printed circuits separated by a conductive ink layer. The insole must be protected from possible damage and the clinician must ensure that the floor/insole interface does not create a slip hazard for the wearer.

The problem with discrete devices is that the individual sensors are positioned at pre-determined anatomical positions and can migrate secondary to shear stress at the foot/shoe interface [86, 87]. PEDAR and the F-scan have been compared in a series of bench tests and in dynamic testing with four human subjects walking on a treadmill (Fig. 8.28). The absolute error of the F-scan is much larger than that of PEDAR [88]. The validity and reliability of the F-scan insole were less satisfactory for certain measurements.

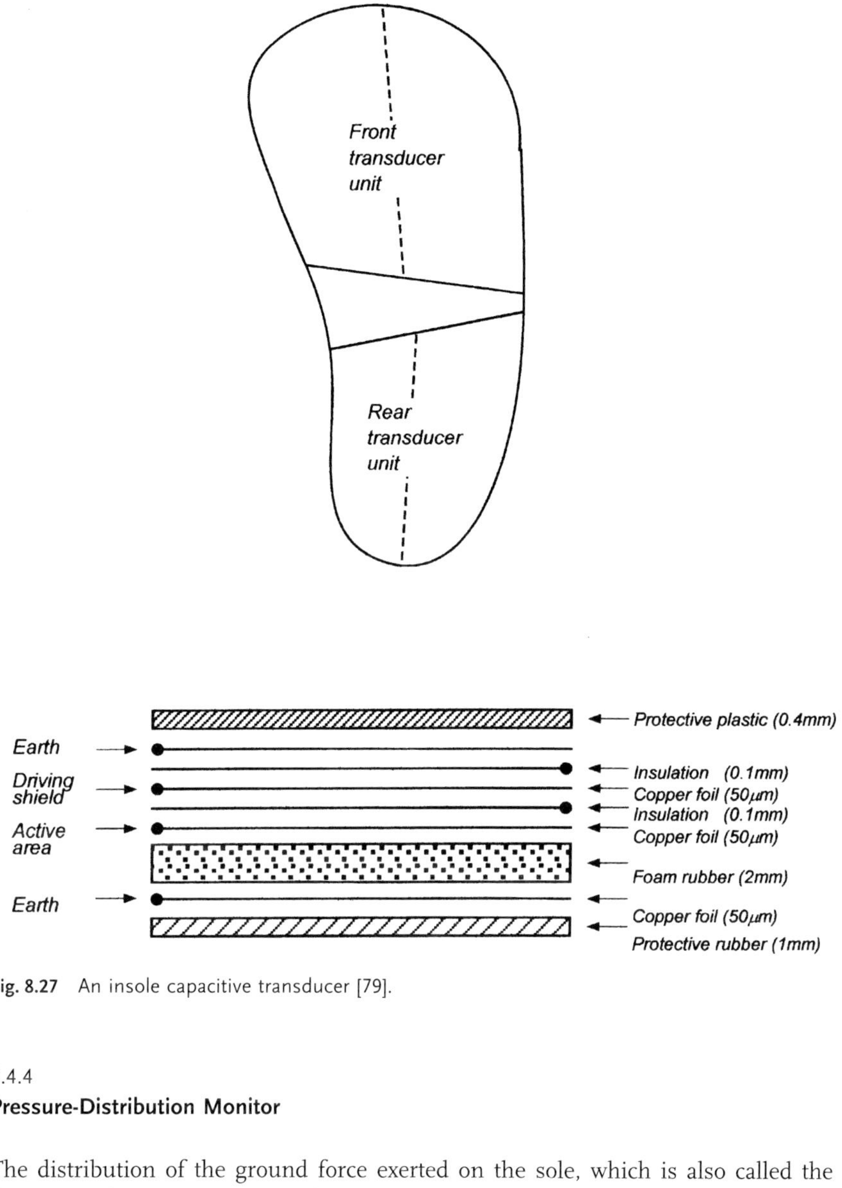

Fig. 8.27 An insole capacitive transducer [79].

8.4.4 Pressure-Distribution Monitor

The distribution of the ground force exerted on the sole, which is also called the foot–ground pressure pattern, has been measured to investigate the function of the foot in postural control of the body [89]. Early studies used a printing technique that provided a pressure-dependent footprint. One technique, called the kinetograph, consists of a corrugated rubber mat and inked fabric that recognizes pressure difference as density differences. Many studies have used similar printing techniques. Quantitative force distribution can be estimated from the obtained footprint by using a calibrated pressure-sensitive sheet [90]. Prescale [91] pressure-

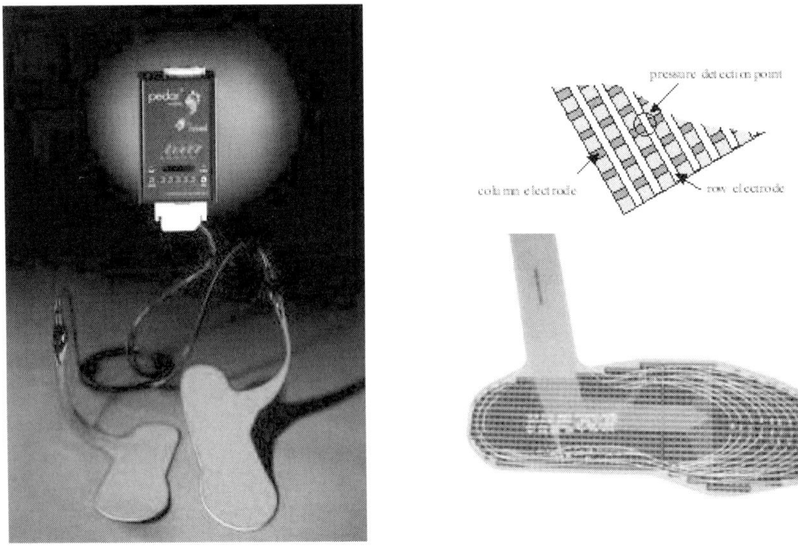

Fig. 8.28 Computerized insole sensors: a) PEDAR; b) F-scan.

detecting sheets consist of two sheets: one is coated with microcapsules containing a color-producing agent and the other is coated with a color-developing agent (Fig. 8.29). With pressure, the microcapsules rupture and the color of the sheet corresponds to the pressure applied.

A direct visualization technique called the barograph has been developed. It consists of a black rubber mat covered with many small pyramidal protrusions that is placed on a glass plate. When viewed from below, the footprint appears as a matrix of black dots. In order to enhance the contrast of the image, a white opaque fluid is introduced between the mat and the glass. Many modifications of the original barograph have been made using different materials and shapes for the deformable mat, as well as different optical-observation techniques. Video-to-computer interfacing was also introduced to facilitate data processing and presentation [92]. A photo-elastic sheet has been employed for optical detection of the force distribution [93]. As shown in Fig. 8.30, the discrete applied loads are measured as a result of the contact between many hemispherical solids and a layer consisting of the reflector, photo-elastic sheet, and polarizers, which are placed on a glass plate. When viewed from below, a circular interference pattern appears and the diameter of the circle is a function of the force exerted by the solid on the

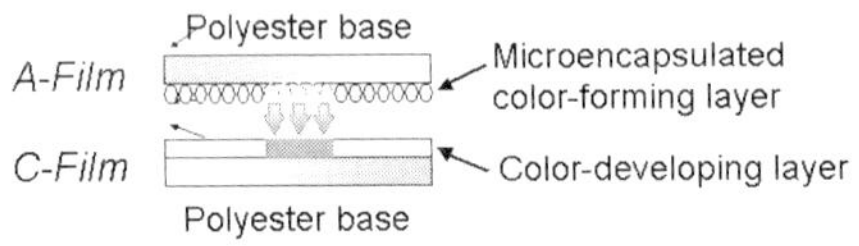

Fig. 8.29 The principle of Prescale.

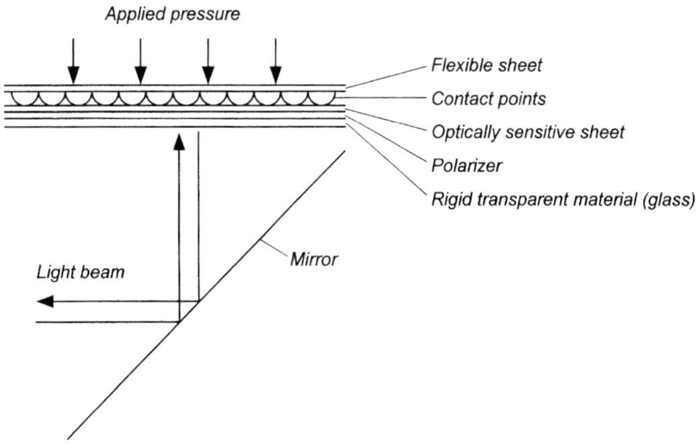

Fig. 8.30 A method of direct visualization of the foot/ground pressure pattern.

layer. Computer processing produces quasi 3D plots of the ground-force distribution [94].

The Harris mat [95] (Fig. 8.31) consists of a rubber mat with a repeated pattern of small projections. The mat is inked and progressively increasing forces cause a high concentration of ink to be deposited at the locations of the highest pressure, where all the layers of the mat are compressed. When an aluminum foil sheet is placed on top of an insole sheet with pyramidal projections, the applied pressure during gait deforms the sheet. An optical scanner measures the height of the deformations in the foil and scales the result in pressure units [96].

Ground-force distribution has also been measured using a matrix-force plate consisting of many cells. The force plate has 128 strain-gauge type transducers, each 14 mm square, arranged in a 16×8 matrix [97, 98]. Nicol reported the use of

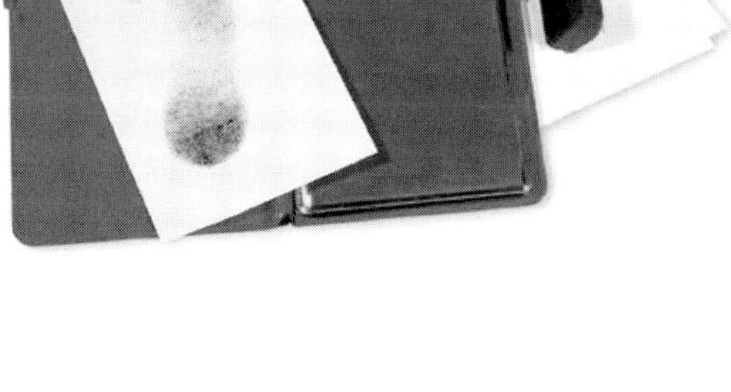

Fig. 8.31 The Harris mat.

a capacitive force transducer that was made from a compressible elastic dielectric material [99]. Square transducers measuring between 3×3 and 30×30 mm were used in a matrix arrangement and matrices of up to 256×128 (32768) transducers were evaluated. A commercially available force plate that contains 960 capacitance points and is scanned by a computer 30 times per second is available for clinical application [100]. Polyvinylidene fluoride (PVDF) [101] is the most popular piezo-electric polymer material for shock sensors. PVDF can be used as a stress gauge, and for pedobarography.

The instrumented walkway shown in Fig. 8.32 can be used to determine stride characteristics without attaching any device to the subjects. Sensors are placed on the floor, not on the subject. A gait mat automates the spatial and temporal measurements of gait. The gait mat consists of an array of switches along the walkway that are covered by a protective mat. As a subject walks along the mat, the switches close underfoot, enabling the computer to calculate the timing of each switch closure. Such walkways are portable, can be laid over any flat surface, require minimum set-up and test time, and do not require the placement of any devices on the patient.

The standard gait mat (GaitRite, CIR systems) [102] and electronic walkway contains six sensor pads in a roll-up carpet and produces an active area of 61 cm wide and 366 cm long. In this arrangement, the active area forms a grid of 48×288 switches placed on 1.27 cm centers. On the other hand, GaitMat (Gait Mat, EQ) [103] uses pressure-sensitive switches arranged in 38 rows of 256 switches and has a walking surface of 60×3.84 m. The switches are on 15 mm centers, giving a spatial resolution of 15 mm to the system.

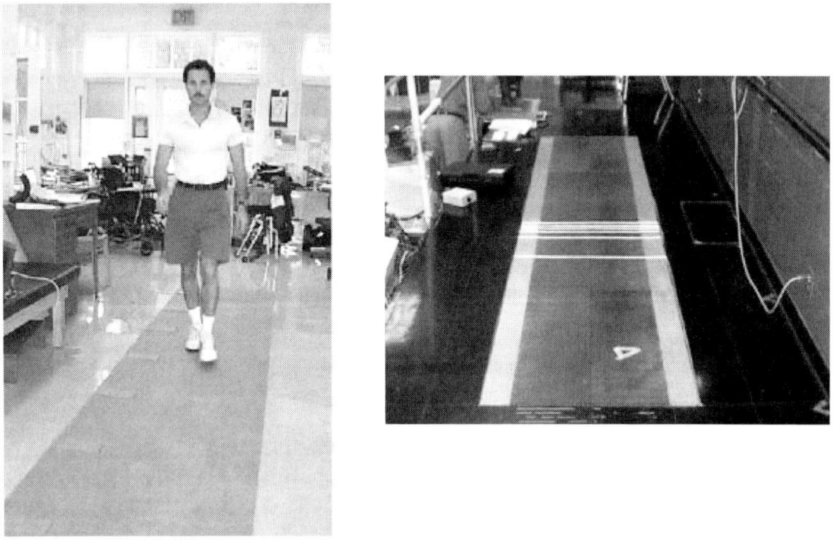

Fig. 8.32 The configuration of a gait mat [102].

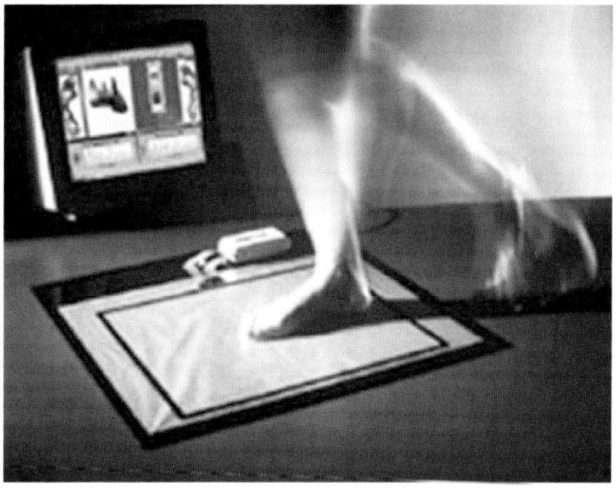

Fig. 8.33 A pressure mat for continuous gait analysis [104].

As the patient ambulates across the walkway, the system captures the relative arrangement, geometry, and applied pressure of each footfall as a function of time. The application software controls the functionality of the walkway, processes the raw data into footfall patterns, and computes temporal and spatial parameters. The software stores each test by patient and supports a variety of reports and analyses. These systems can be used to test patients with or without shoes, including those using assistance devices and ambulatory aids, such as crutches, canes, or walkers.

The advantages of such devices are the elimination of any gait-encumbering attachments, low cost, and portability. The disadvantages include limitations on spatial resolution due to the finite size of the switches and on temporal resolution due to restricted scan rates.

As shown in Fig. 8.33, a pressure mat is placed on the center of walkway and used as a force plate that the subject steps on. This provides a quick and easy way of obtaining a plantar-pressure distribution. The pressure mat is factory-calibrated and comes with software that produces color pressure pictures, force and pressure versus time plots, etc. The principle of the sensor is the same as that of the insole pressure distribution system. EMED [104] uses capacitive sensors and F-Mat and Big mat use FSR sensors. EMEDs are high-resolution sensor platforms for the fast diagnosis of dynamic foot function and foot deformities. The measurement method is based on calibrated capacitive sensors. The signals, produced from a maximum of 4000 pressure sensors, are displayed as conformational color pictures on a monitor or as printouts on a color printer. EMED system platforms are available in different sizes and with various local resolutions.

8.5 Related Measurements

8.5.1 Electromyogram

Electromyography is the recording of the electrical signals produced by muscles during activity. An electromyogram (EMG) is the recorded signal, which is the summation of the motor unit action potential in a localized area of muscle. These measurements can be either invasive or non-invasive. In the latter, surface electrodes are attached to the skin over the appropriate muscle, while in the former, needle- or fine wire-electrodes are inserted into the muscle. The EMG of deep muscles can only be obtained reliably with intramuscular wire electrodes to avoid cross-talk from more superficial muscles.

Surface electrodes are classified into two types: passive and active. A typical passive electrode is the Beckman Ag:AgCl type. A pair of electrodes can be placed over a muscle as desired. Sizes range from 7 to 20 mm in diameter. A conductive electrode gel is required, which is attached to the skin with double-sided adhesive tape.

Active electrodes amplify the signal directly. This reduces the electrical noise that is often picked up by passive electrode lead wires. A high-impedance differential amplifier with a high common mode rejection ratio is used. Delsys manufactures a soluble differential active surface electrode that is reported to reduce cross-talk.

A wireless electrode has been developed. The wireless electrode consists of an electrode terminal, an amplifier, a high-pass filter, a built-in stabilized reference circuit, a transmitter, and a power supply. It is 22.5×49×14.5 mm and weighs 25 g with a button battery [105].

EMG acquisition systems are also classified into two types: cable and wireless. Wireless systems include radio telemetry or data loggers. In a cable system, power is obtained through a cable and the signals are free from any radio frequency interference or dropout.

To obtain phasic activity, further signal processing is performed on the raw signal. EMG signals are usually full-wave rectified, smoothed and enveloped. The most common evaluation is the phasic activity of muscle in a gait-analysis context. A correlation between the EMG signal output and muscle force magnitude can be evaluated, but it is only possible to derive the relationship for isometric muscle contraction [106].

8.5.2 Energy Consumption

Energy consumption during gait and body movement is used to quantify activity. The standard and the most reliable method measures the amount of oxygen consumed by the body. The basis for this convention is the physiological production of metabolic energy from the combination of oxygen and carbon to form carbon dioxide. The oxygen consumption is determined by measuring the oxygen in expired air during steady-state walking. Conversion to more convenient energy units is accomplished using the equation

$$E = \frac{4.92 V_E}{100}(20.93 - F_{O_2})$$

where V_E is the volume of expired air in L/min, E is the energy in kilocalories/minute, and F_{O2} is the percentage of oxygen in the expired air.

Energy consumption measurements can be tedious and difficult to accomplish. It would be convenient to have a method of measuring energy consumption instantaneously over a short period of time. One such method is ambulatory monitoring. Energy consumption based on the measurement of work done on the body center of mass is highly correlated with the work done on body segments. Thus, the mechanical work done by body segments can be estimated from their motion. In addition, the average acceleration is correlated with energy consumption [107]. This assumption can be used to estimate physical and habitual activity.

Quantitative assessment of physical and habitual activity is difficult, but reliable and valid measures are available. Questionnaire assessments are the most widely used method for measuring physical activity in epidemiological studies. Physical activity monitors based on accelerometry may be suitable for some studies, and several studies have used such devices. The gold standard for physical activity measurement is the doubly labeled water method. This technique provides a reliable estimate of total energy expenditure, but cost and complexity limit its usefulness. Physical activity can be measured as the integration of acceleration per unit time [107]. Several studies based on accelerometry have been attempted [108–111]. Bouten et al. assumed that energy expenditure during activity equaled total energy expenditure less metabolism during sleep. There is a linear relationship between energy expenditure during activity and the sum of the integral of the absolute value of the accelerometer in three directions (vertical, lateral, anteroposterior) with relatively high correlation [108]. The validity and usefulness of such devices have been analyzed precisely [109, 110].

Simple acceleration pickups making use of the piezoelectric principle have been widely used. For example, an acceleration pick-up attached to the wrist has been used to monitor body motion during sleep [112–115].

8.6 References

1 Christensen, H.W. Precision and accuracy of an electrogoniometer. *J. Manip. Physiol. Therap.* **1999**, *22*, 10–14.

2 Chao, E.Y.S. Justification of triaxial goniometer for the measurement of joint rotation. *J. Biomech.* **1980**, *13*, 989–1006.

3 Ketterkamp, D.B., Johnson, R.J., Smidt, G.L., Chao, E.Y.S., and Walker, M. An electrogoniometric study of knee motion in normal gait. *J. Bone Jt Surg. A* **1970**, *52*, 775–790.

4 Johnson, R.C. and Smidt, G.L. Measurement of hip-joint motion during walking. *J. Bone Jt Surg. A* **1969**, *51*, 1083–1094.

5 Rehabilitation Engineering Centre. www.lwc-eirec.go.jp/seika/seika-flexgonioE.html

6 Ball, P. and Johnson, G.R. Reliability of hindfoot goniometry when using a flexible electrogoniometer. *Clin. Biomech.* **1993**, *8*, 13–19.

7 Tesio, L., Monzani, Gatti, R. Flexible electrogoniometers: kinesiological advantages with respect to potentiometric goniometers. *Clin. Biomech.* **1995**, *10*, 275–277.

8 CyberGlove. www.immersion.com/products/overview.shtml

9 Rudolf, F., Jornod, A., Bergqvist, J., and Leuthold, H. Precision accelerometer with µg resolution. *Sensors Actuators A* **1990**, *21–23*, 297–302.

10 Suzuki, S., Tuchitani, S., Sato, K., Ueno, S., Yokota, Y., Sato, M., and Esashi, M. Semiconductor capacitive-type accelerometer with PWM electrostatic servotechniques. *Sensors Actuators A* **1990**, *21–23*, 316–322.

11 Morris, J.W. Accelerometry – a technique for the measurement of human body movements. *J. Biomech.* **1973**, *6*, 729–736.

12 Willemsen, A.T.H.M., Frigo C., and Boom, H.B.K. Lower extremity angle measurement with accelerometers – error and sensitivity analysis. *IEEE Trans. Biomed. Eng.* **1991**, *38*, 1186–1193.

13 Evans, A., Duncan, G., and Gilchirst, W. Recording acceleration in body movements. *Med. Biol. Eng. Comput.* **1991**, *29*, 102–104.

14 Smidt, G.L., Arora J.S., and Johnston, R.C. Accelerographic analysis of several types of walking. *Am. J. Phys. Med.* **1971**, *50*, 285–300.

15 Aminian K., Rezakhanlou, A.E., Fritsch, C., Leyvraz P.F., and Robert, P. Temporal feature estimation during walking using miniature accelerometers: an analysis of gait improvement after hip arthroplasty. *Med. Biol. Eng. Comput.* **1999**, *37*, 686–691.

16 Breniere, Y. and Dietrich, G. Heel-off perturbation during gait initiation: biomechanical analysis using triaxial accelerometry and a force plate, *J. Biomech.* **1992**, *25*, 121–127.

17 Sekine, M., Tamua, T., Togawa T., and Fukui, Y. Classification of waist-acceleration signals in a continuous walking record. *Med. Eng. Phys.* **2000**, *22*, 285–291.

18 Aminian, K., Robert, P., Jequiuer, E., and Schuts, Y. Estimation of speed and incline of walking using neural network. *IEEE Trans. Instrum. Meas.* **1995**, *44*, 743–746.

19 Williamson, R. and Andrews, B.J. Gait event detection for FES using accelerometers and supervised machine learning. *IEEE Trans. Rehab. Eng.* **2000**, *8*, 312–319.

20 Hansson G.Å., Asterland, P., Holmer, N.G., and Skerfving S. Validity and reliability of triaxial accelerometers for inclinometry in posture analysis. *Med. Biol. Eng. Comput.* **2001**, *39*, 405–413.

21 Sabelman, E.E. The Accelerometric Motion Analysis for Mobility Diagnosis and Therapy: A Real-World Application of Wearable Computer. In *Proc. ASME March Dinner Meeting*, Sunnyvale, CA, 16 March **2000**.

22 Watson Industries. www.watson-gyro.com/

23 Gyrosensor Murata. www.murata.com/sensor/index.html

24 Silicon VSG Vibrating Structure Gyroscope. www.spp.co.jp/English/jigyou/jigyou-e.html

25 Clark, W.A., Howe, R.T., and Horowitz, R. Surface Micromachined Z-Axis

Vibratory Rate Gyroscope. In *Proc. Solid State Sensor and Actuator Workshop*, Hilton Head, SC, June **1996**.

26 Miyazaki, S. Long-term unrestrained measurement of stride length and walking velocity utilizing a piezoelectric gyroscope. *IEEE Trans. Biomed. Eng.* **1997**, *44*, 753–759.

27 Williamson, R. and Andrews, B. J. Detecting absolute human knee angle and angular velocity using accelerometers and rate gyroscopes. *Med. Biol. Eng. Comput.* **2001**, *39*, 1–9.

28 Tong, K. and Granat, M. H. A practical gait analysis system using gyroscopes. *Med. Eng. Phys.* **1999**, *21*, 87–94.

29 Williamson, R. and Andrews, B.J. Sensor systems for lower limb functional electrical stimulation (FES) control. *Med. Eng. Phys.* **2000**, *22*, 313–325.

30 Mead, J., Peterson, N., Grimby, G., and Mead, J. Pulmonary ventilation measured from body surface movements. *Science* **1967**, *156*, 1383.

31 Rolfe, P. A magnetometer respiration monitor for use with premature babies. *J. Biomed. Eng.* **1971**, *6*, 402.

32 Ashutosh, K., Gilbert, R., Auchincloss, J. H., Erlebacher, J., and Peppi, D. Impedance pneumograph and magnetometer methods for monitoring tidal volume. *J. Appl. Physiol.* **1974**, *37*, 964.

33 Stagg, D., Goldman, M., and Davis, J. N. Computer-aided measurement of breath volume and time components using magnetometers. *J. Appl. Physiol.: Respir. Environ. Exercise Physiol.* **1978**, *44*, 623.

34 Saunders, N. A., Kreitzer, S. M., and Ingram, R. H. Rib cage deformation during static inspiratory efforts. *J. Appl. Physiol.* **1979**, *46*, 1071.

35 Robertson, C. H., Jr., Bradley, M. E., and Homer, L. D. Comparison of two- and four-magnetometer methods of measuring ventilation. *J. Appl. Physiol.: Respir. Environ. Exercise Physiol.* **1980**, *49*, 355.

36 Raab, F. H., Blood, E. B., Steiner, T. O., and Jones, H. R. Magnetic position and orientation tracking system. *IEEE Trans. Aerospace Electron. Syst.* **1979**, *15*, 709.

37 Polhemus sensor. www.polhemus.com

38 Day, J. S., Murdoch, D. J., and Dumas, G.A. Calibration of position and angular data from a magnetic tracking device. *J. Biomech.* **2000**, *33*, 1039–1045.

39 Logan, S. E. and Groszevski, P. Dynamic wrist motion analysis using six degrees of freedom sensors. *Biomed. Sci. Instrum.* **1989**, *25*, 213.

40 Russell, P., Weld, A., Pearcy, M. J., Hogg, R., and Unsworth, A. Variation in lumbar spine mobility measured over a 24-hour period. *Br. J. Rheumatol.* **1992**, *31*, 329.

41 Pearcy, M. J. Twisting mobility of the human back in flexed postures. *Spine* **1993**, *18*, 114.

42 Biryukova, E. V., Roby-Brami, A., Frolov, A. A., and Mokhtari, M. Kinematics of human arm reconstructed from spatial tracking system recordings. *J. Biomech.* **2000**, *33*, 985–995.

43 Spyers-Ashby, J. M., Stokes, M. J., Bain, P. G., and Roberts, S. J. Classification of normal and pathological tremors using a multidimensional electromagnetic system. *Med. Eng. Phys.* **1999**, *21*, 713–723.

44 Spectron. www.spectronsensors.com

45 Clino Ltd. www.clino.co.uk/products.htm

46 The Fredericks Company. www.frederickscom.com/products/tilt_overview.html

47 Shaevitz. www.msiusa.com/shaevitz/product_tour.asp

48 Terjesen, T., Anda, S., and Svenningsen, S. Femoral anteversion in adolescents and adults measured by ultrasound. *Clin. Orthop.* **1990**, *256*, 274–279.

49 Tanaka, S., Yamakoshi, K., and Rolfe, P. New portable instrument for long-term ambulatory monitoring of posture change using miniature electro-magnetic inclinometers. *Med. Biol. Eng. Comput.* **1994**, *32*, 357–360.

50 Kemp, B., Jansen, A. and van der Kamp J. M. W. Body position can be monitored in 3D using miniature accelerometers and earth-magnetic sensors. *EEG and Clin. Neurophysiol.* **1998**, *109*, 484–488.

51 Weir, R. F. and Childress, D. S. A portable, real-time, clinical gait velocity analysis system. *IEEE Trans. Rehab. Eng.* **1997**, *5*, 310–321.

52 B & L Engineering. www.bleng.com/sa.htm

53 BTS. www.bts.it/

54 Motion Lab Systems. www.emgsrus.com

55 Hill, K. D., Goldie, P. A., Baker, P. A., and Greenwood, K. M. Retest reliability of the temporal and distance characteristics of hemiplegic gait using a footswitch system. *Arch. Phys. Med. Rehabil.* **1994**, *75*, 577–583.
56 Muybridge, E. *The Human Figure in Motion.* Dover Publications, **1955**.
57 VICON. www.vicon.com
58 Woltring, H. J. New Possibilities for human motion studies by real-time light spot position measurement. *Biotelemetry,* **1974**, *1*, 132.
59 Woltring, H. J. New possibilities for human motion studies by real-time light spot position measurement. *Biotelemetry* **1974**, *3*, 132.
60 Macellari, V. CoStel, a computer peripheral remote sensing device for 3-dimensional monitoring of human motion. *Med. Biol. Eng. Comput.* **1983**, *21*, 311.
61 CODAmpx3. www.charndyn.com/
62 Optrak. www.ndigital.com/certus.html
63 Motion Analysis. www.motionanalysis.com/index.html
64 Adams, L. P. and Spirakis, A. Stereo photogrammetry. Orr, J. F. and Shelton, J. C. eds. *Optical Measurement Methods in Biomechanics.* Chapman & Hall, London **1997**, 17–38.
65 Dapena, J., Harman, E. A., and Miller, J. A. Three-dimensional cinematography with control object of unknown shape. *J. Biomech.* **1982**, *15*, 11–19.
66 Harze, H. High precision three-dimensional photogrammetric calibration and object-space reconstruction using a modified DLT-approach. *J. Biomech.* **1988**, *21*, 533–538.
67 Sati, M., de Guise, J. A., Larouche, S., and Druin, G. Quantitative assessment of skin–bone movement at the knee. *The Knee* **1996**, *3*, 121–138.
68 Sati, M., de Guise, J. A., Larouche, S., and Drouin, G. Improving *in vivo* knee kinematic measurements: Application to prosthetic ligament analysis. *The Knee* **1996**, *3*, 179–190.
69 Ball, K. A. and Pierrynowski, M. R. Estimation of six degree of freedom rigid body segment motion from two-dimensional image data. *Human Movement Science* **1995**, *14*, 139–154.
70 Cappozzo, A., Cantani, F., Croce, U. D., and Leardini, A. Position and orientation in space of bones during movement: anatomical frame definition and determination. *Clin. Biomech.* **1995**, *10*, 171.
71 AMTI. www.amtiweb.com
72 Bertec. www.bertec.com/products/index.html
73 Kistler. www.kistler.ch/web/portal/.nst/?open
74 Cavagna, G. A. Force platforms as ergometers. *J. Appl. Physiol.* **1975**, *39*, 174.
75 Murray, M. P., Seireg, A., and Scholz, R. C. Center of gravity, center of pressure, and supportive force during human activities. *J. Appl. Physiol.* **1967**, *23*, 831.
76 Ishida, A. and Imai, S. Responses of the posture-control system to pseudo-random acceleration disturbances. *Med. Biol. Eng. Comput.* **1980**, *18*, 433.
77 Spolek, G. A. and Lippert, F. G. An instrumented shoe – A portable force measuring device. *J. Biomech.* **1976**, *9*, 779.
78 Kljajic, M. and Krajnik, J. The use of ground reaction measuring shoe in gait evaluation. *Clin. Phys. Physiol. Meas.* **1987**, *8*, 133.
79 Miyazaki, S. and Iwakua, H. Foot-force measuring device for clinical assessment of pathological gait. *Med. Biol. Eng. Comput.* **1978**, *16*, 429.
80 Miyazaki, S. and Ishida, A. Capacitive transducer for continuous measurement of vertical foot force. *Med. Biol. Eng. Comput.* **1984**, *22*, 309.
81 Wertsch, J. J., Webster, J. G., Tomkins, W. J. A portable insole plantar pressure measurement system. *J. Rehabil. Res. Dev.* **1992**, *29*, 13–18.
82 PEDAR. www.novelelectronics.com
83 F-scan. www.tekscan.com
84 Insole Force Sensor. lifesci.arc.nasa.gov/~rwhalen/insole.html
85 Nitta. www.nitta.co.jp/english/index-e.htm
86 Cavanagh, P. R., Bewitt, F. G., and Perry J. E. In-shoe plantar pressure measurement: a review. *Foot* **1992**, *2*, 185–194.
87 Randolph, A. L., Nelson, M., Akkapeddi, S., Levin, A., and Alexandrescu, R. Reliability of measurement of pressure applied on the foot during walking by a

computerized insole sensor system. *Arch. Phys. Med. Rehabil.* **2000**, *81*, 573–578.

88 McPoil, T. G., Cornwall, M. W., and Yamada, W. A comparison of two in-shoe plantar pressure measurement systems. *The Lower Extremity* **1995**, *2*, 95–103.

89 Lord, M. Foot pressure measurement: A review of methodology. *J. Biomed. Eng.* **1981**, *3*, 91.

90 Aritomi, H., Morita, M. and Yonemoto, K. A simple method of measuring the footstool pressure of normal subjects using prescale pressure-detecting sheets. *J. Biomech.* **1983**, *16*, 157.

91 Prescale. www.prescale.com

92 Betts, R. P., Franks, C. I., Duckworth, T., and Burke, J. Static and dynamic foot-pressure measurements in clinical orthopaedics. *Med. Biol. Eng. Comput.* **1980**, *18*, 674.

93 Arcan, M. and Brull, M. A. A fundamental characteristic of the human body and foot, the foot-ground pressure pattern. *J. Biomech.* **1976**, *9*, 453.

94 Cavanagh, P. R. and Ae, M. A technique for the display of pressure distribution beneath the foot. *J. Biomech.* **1980**, *13*, 69.

95 Foot Force Imprinter. www.acor.com/Sizing%20Tools.htm

96 Grieve, D. W. and Rashdi, T. Pressure under normal feet in standing and walking as measured by foil pedobarography. *Ann. Rheum. Dis.* **1984**, *43*, 816–818.

97 Dhanendran, M., Hutton, W. C., and Parker, Y. The distribution of force under the human foot – an on-line measuring system. *Meas. Control* **1978**, *11*, 261.

98 Soames, R.W., Blake, C.D., Scott, J. R. R., Goodbody, A., and Brewerton, D. A. Measurement of pressure under the foot during function. *Med. Biol. Eng. Comput.* **1982**, *20*, 489–495.

99 Nicol, K. A new capacitive transducer system for measuring force distribution statically and dynamically. *Proc. Transducer Tempcon.* **1981**, *81*, 1.

100 Footmaxx. system force plate. www.footmaxx.com

101 The PVDF sensor. www.ktech.com/research_development/applied_physics/piezoelectric.cfm

102 GaitRite. www.gaitrite.com

103 Gait Mat. www.gaitmat.com

104 EMED. www.novelelectronics.com/nav%20emed.htm

105 Ohyama, M., Tomita, Y., Honda, S., and Matsuo, N. Active wireless electrodes for multichannel surface electromyography. *Trans. IEE Japan* **1999**, *E119*, 279–284.

106 Gottlieb, G. L. and Agarwal, G. L. Dynamic relationship between isometric muscle tension and electromyogram in man. *J. Appl. Physiol.* **1971**, *30*, 345–351.

107 Wong, T. C., Webster, J. G., Montoye, H. J., and Washburn, R. Portable accelerometer device for measuring human energy expenditure. *IEEE Trans. Biomed. Eng.* **1981**, *28*, 467–471.

108 Bouten, C. V., Westerterp, K. R., Verduin M., and Janssen, J. D. Assessment of energy expenditure for physical activity using a triaxial accelerometer. *Med. Sci. Sports Exer.* **1994**, *26*, 1516–1523.

109 Bouten, C. V., Koekkoek, K. T., Verduin, M., Kodde, R., and Janssen, J. D. A triaxial accelerometer and portable data processing unit for the assessment of daily physical activity. *IEEE Trans. Biomed. Eng.* **1997**, *44*, 136–147.

110 Aminian, K., Robert, P., Buchser, E. E., Rutschmann, B., Hayoz, D., and Depairon, M. Physical activity monitoring based on accelerometry: validation and comparison with video observation. *Med. Biol. Eng. Comput.* **1999**, *37*, 304–308.

111 Bouten, C. V., Sauren, A. A., Verduin, M., and Janssen, J. D. Effects of placement and orientation of body-fixed accelerometers on the assessment of energy expenditure during walking. *Med. Biol. Eng. Comput.* **1997**, *35*, 50–56.

112 Krippke, D. F., Mullaney, D. J., Messin, S., and Wyborney, V. G. Wrist actigraphic measures of sleep and rhythms. *EEG and Clin. Neurophysiol.* **1978**, *44*, 674.

113 Webster, J. B., Messin, S., Mullaney, D. J., and Kripke, D. F. Transducer design and placement for activity recording.\kep *Med. Biol. Eng. Comput.* **1982**, *20*, 741.

114 Sadeh, A., Hauri, P. J., Kripke, D. F., and Lavie, P. The role of actigraphy in the evaluation of sleep disorders. *Sleep* **1995**, *18*, 288–302.

115 Actigraphy. www.ambulatory-monitoring.com

9
Cardiac Pacemakers

B. Hensel, G. Czygan, I. Weiss, and T. Nappholz

This contribution is dedicated to Prof. Dr. Max Schaldach who died tragically in an aircraft accident on 5 May 2001. He agreed to contribute to this book and review the history and current state of the art in the field of sensors for cardiac pacemakers. Sadly he could not prepare the article before his death. Representing all his many coworkers, we have finished the work and dedicate it to his memory.

9.1
Introduction

Modern cardiac pacemakers are much more than simple clock generators that periodically deliver electrical impulses to the heart; they depend to a high degree on the data that they continuously acquire from the cardiovascular system. The pacemaker electrodes serve not only as actuators but also as the primary sensors in the control loop that adapts the pacing rate to the physiological demand. The reliable detection of intrinsic electrical excitations, the monitoring of stimulated events, and the measurement of intracardiac impedance to determine minute ventilation and cardiac contractility depend strongly on the electrode geometry and the properties of the interface with the biological tissue.

The focus of this chapter is on the electrode as the primary sensor for the pacemaker. Systems for rate adaptation that utilize the electrodes to measure intracardiac impedance are the key issue, while pacemakers that require additional devices such as acceleration sensors are not treated here [1].

The role of the electrodes as sensors will gain even more importance in future generations of pacemakers that will be able to monitor the state of the cardiovascular system by dedicated signal processing and by sophisticated analyses of long time series of the heart beat and thus provide the basis for novel forms of diagnosis and therapy.

Sensors in Medicine and Health Care
Edited by P. Å. Öberg, T. Togawa, and F. A. Spelman

ISBN: 3-527-29556-9

9.2
The Pacemaker Electrode as the Primary Sensor in the Cardiac Control Loop

As the interface between the pacemaker and the myocardium, the electrode fulfills two fundamental purposes. First, it transmits the electrical stimulus generated by the pacemaker to the heart. Second, it picks up an electrical signal and transfers it to the implant, where it is evaluated for control of pacemaker timing. While much effort has been invested in electrode development to optimize stimulation requirements, research has more recently focused on the electrode-sensing properties.

9.2.1
The Interface Between the Electrode Surface and the Myocardium

Pacing and sensing activities of the pacemaker electrode are determined by the electrical transmission properties of the boundary between the electrode and the myocardium. This interface includes an electrochemical phase boundary between a metal and an electrolyte solution [2, 3].

The special phase boundary properties are largely based on two effects. First, the metal and the electrolyte contain different electrical conduction systems. In a metal it is the electrons, and in an electrolyte solution it is ions, that are the charge carriers. The ions are solvated by water molecules due to electrostatic attraction. Second, each electrode surface is covered with a layer of adsorbed water molecules. The potential difference that exists between the electrochemical potential of the electrode and that of the solvated ions creates a molecular monolayer through electrostatic interaction between the charged metal surface and the polar water molecules. A phase boundary results, the so-called Helmholtz double layer: the electrode surface and ions are separated by a double layer of water molecules (Fig. 9.1). Thus, the ions of the electrolyte do not come into contact with the electrode surface, and direct electron exchange is extremely rare at equilibrium, which provides insulation.

The Helmholtz double layer largely influences the current conduction through the phase boundary via three main charge transfer mechanisms.

- The phase boundary resembles a plate capacitor in which the metallic electrodes and the electrolyte represent both plates, while the molecular double layer is the dielectric. When a charge is transmitted, a capacitor-like shift of the metallic electrons and the electrolyte ions occurs. This shifting current charges the phase boundary.
- Electrons penetrate the Helmholtz double layer directly and react with the electrolyte constituents. These Faraday current flows cause electrochemical changes of the electrolytic contents (modifications of carbohydrates and proteins, or electrolysis of water). Since activation energy must be exceeded for most chemical reactions, the Faraday current flows depend extensively on the potential ratios at the phase boundary and occur only at higher voltages. Owing to the diffusion

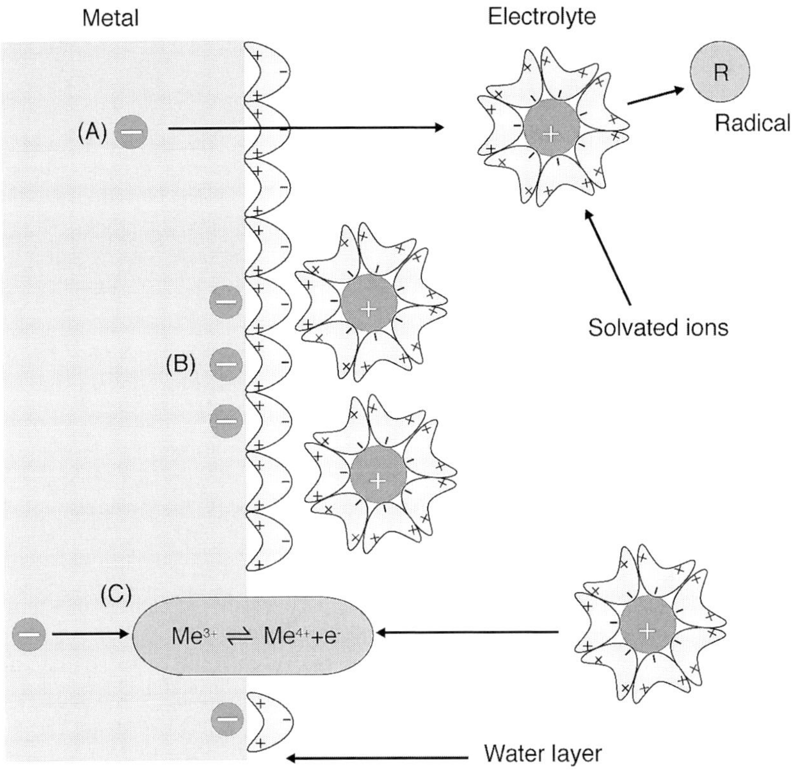

Fig. 9.1 Structure of the phase boundary between metal and electrolyte and the current-transport mechanism through the contact surface. A) Irreversible direct electron transfer; B) capacitive indirect charge transfer; C) reversible surface redox (reduction–oxidation) reaction.

of the reaction product away from the phase boundary, these reactions are irreversible.

- The third transport mechanism is based on reversible electrochemical redox reactions on the electrode surface, which activate the adsorption or desorption of solution constituents and change the oxidation states of metal atoms at the electrode surface.

For applications in pacemaker therapy, the nonlinear, but completely reversible capacitive current through the phase boundary is the most significant. In a simple equivalent circuit diagram, the contact surface is described by the parallel circuit of the capacitor, C_H and the Faraday resistance, R_F (Fig. 9.2). The circuit is completed by the frequency-independent lead resistance R_L. The potential dependence of the electrochemical reactions is not taken into account here. The specific Helmholtz capacity (C_H) has a value of about 10 μF/cm^2 for smooth metallic electrode surfaces. However, the electrode capacity can be increased by three orders of magnitude (up to 50 mF/cm^2) by coating processes described below (fractal coating).

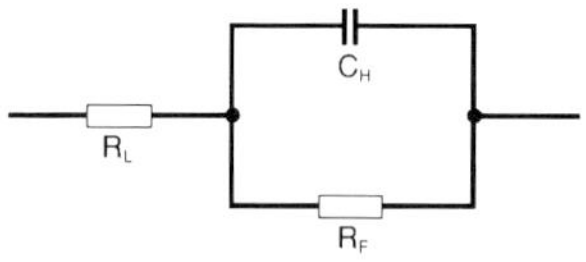

Fig. 9.2 Equivalent circuit diagram for the electrical characteristics of the interface: C_H, Helmholtz capacity; R_F, Faraday resistance; R_L, lead resistance.

Faraday currents are largely negligible with these coated pacemaker electrodes as there is only a slight voltage drop over the phase boundary.

The interface must be viewed in connection with the pacemaker system and the properties of the biological surroundings [4]. The equivalent circuit diagram in Fig. 9.3 illustrates the measurement of cardiac signals. The pacemaker electrodes detect the potential drop of currents in the extracellular space that are generated by the summed transmembrane potentials of many individual cells of the myocardium. This source is represented in the equivalent circuit diagram by the source voltage, $U_S(t)$, with a source resistance, R_S. A parallel circuit of capacitors and resistors denotes electrodes that sense the signals. The leads are connected to the pacemaker input channel with an input resistance, R_I. The pre-amplified signals are then filtered to suppress interference signals, as well as steady components foreign to the signal.

Since the intracardially measured potentials have low signal amplitudes, occasionally only a few millivolts, the susceptibility to interference by external and system-caused disturbances should be minimized for a high signal-to-noise ratio and good signal quality. Each amplifier creates a noise voltage, U_N, and noise current, I_N, which interfere with the measurement. While the degree of noise voltage is

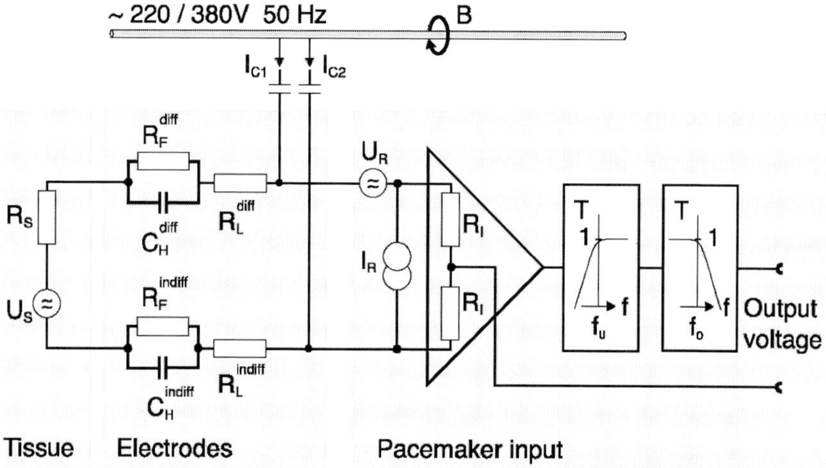

Fig. 9.3 Equivalent circuit diagram for the input channel of the pacemaker with the signal source (biological tissue), the electrodes, the pacemaker input consisting of the amplifier, and the subsequent filters.

largely determined by the amplifier technology, the influence of the noise current depends on the impedance connected to the pacemaker input.

External interference is caused when biosignals are recorded around electrical devices that generate electrical and magnetic stray fields. The fields capacitively couple into the leads or the body, causing a disturbance within the conduction current, I_{C1} and I_{C2}. By flowing through the electrodes and the body to the ground, these currents create disturbances in the measured signals, depending on the system impedance. At frequencies <6 Hz, noise due to electrochemical reactions at the electrodes contributes an additional disturbing influence. The noise decreases, as the active electrode surface area is enlarged [5].

9.2.2 Electrode Requirements

From a clinical standpoint, electrode development must meet the following demands.

- Minimize the charge necessary for pacing, supply high signal amplitudes for a good signal-to-noise ratio, eliminate afterpotentials for measuring and analyzing cardiac signals, and provide materials ensuring electrode biocompatibility and long-term stability.
- The pacing and sensing properties are decisively influenced by the impedance of the phase boundary, Z_{PB}. The latter is described by the following equation

$$|Z_{PB}| = \left(\frac{R_F^2 + 2R_F R_L}{1 + R_F^2 \omega^2 C_H^2} + R_L^2\right)^{1/2}$$

where ω is the frequency, R_F the Faraday resistance, R_L the lead resistance, and C_H the Helmholtz capacity. The circuit is high-pass and the frequency dependence agrees well with that observed experimentally for the interface between a smooth metal electrode and a physiological saline solution (Fig. 9.4). As a result of the strong increase of Z_{PB} below 10 kHz, low-frequency sections of electrical signals are attenuated during transmission. This filtering characteristic has an effect on the current conduction during both pacing and sensing.

The stimulation pulses emitted from the pacemaker have high signal amplitudes in their low-frequency components and the strong attenuation of this frequency band leads to losses, which can only be reduced by diminishing the interface impedance at low frequencies. Since the Faraday resistance is determined by the electrolyte composition, the only option is to enlarge the interface capacity according to the above equation. As the dashed curve in Fig. 9.4 shows, an increase in the capacity of the contact surface by three orders of magnitude leads to a low impedance over a wide range of frequencies. This has an additional advantage in that the voltage from the pacemaker no longer drops at the low-impedance interface, but rather at the other components of the pacing system. Based on the nonlinear character of the voltage-dependent Faraday resistance, the voltage necessary

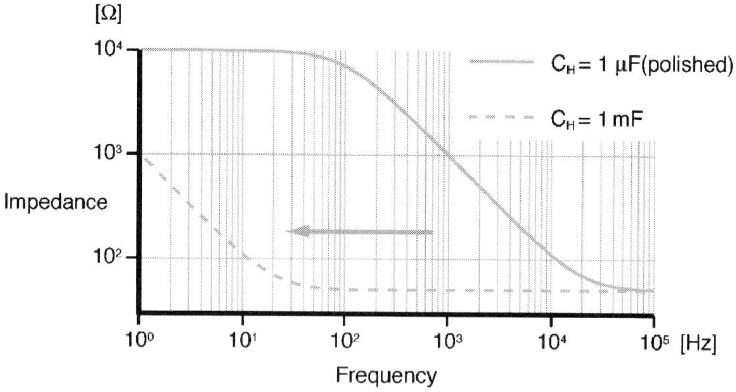

Fig. 9.4 Impedance of electrode–electrolyte system. The values of the solid curve correspond to those of a polished metal surface. The dashed curve results from an interface capacity increased by three orders of magnitude ($R_F = 10$ kΩ, $R_L = 50$ Ω): C_H, Helmholtz capacitance.

for the Faraday reactions at the contact surface is no longer attained. Thus, the irreversible reactions leading to inflammation caused by released radicals are prevented. Electrodes with small geometric surfaces, yet very large interface capacities, are considered optimal from the point of view of energy-conserving pacing.

A closer observation of the measuring systems also shows a need for a large interface capacity. It is true that in many modern pacemakers the input resistance is large enough to avoid a filtering response as a result of capacitive interface properties. But the input resistance only has a small effect on the signal-to-noise ratio and the influence of externally coupled disturbances. The larger the electrode impedance, the stronger is the effect on the measurement by disturbing currents in the system. A frequency analysis of a ventricular evoked response (VER) (Fig. 9.5) also shows that most of the spectral parts of the signals lie below 100 Hz, where the interface impedance of a smooth metal surface is very high. Analogous to the requirements for pacing properties, the interface capacity must also be as large as possible for optimized electrode-sensing characteristics.

Another aspect results from the interaction of the pacing and sensing system. During pacing the capacitive system components are charged. When measuring cardiac signals directly after the stimulus, the remaining charge at the interface is disturbingly noticeable. The resulting afterpotential is superimposed on the measured signals, thereby making their exact evaluation nearly impossible. Attempts quickly to de-charge the interface after the stimulus through charge-controlled counter-pulses have been impractical due to the energy consumption that they require. Similarly, implanting additional electrodes for signal measuring purposes is, from a clinical standpoint, unreasonable due to the increased risk and effort [6–8]. Therefore, the only solution is to develop electrodes that guarantee a negligible afterpotential, thus enabling pacing and sensing with the same pair of electrodes. Further quantitative

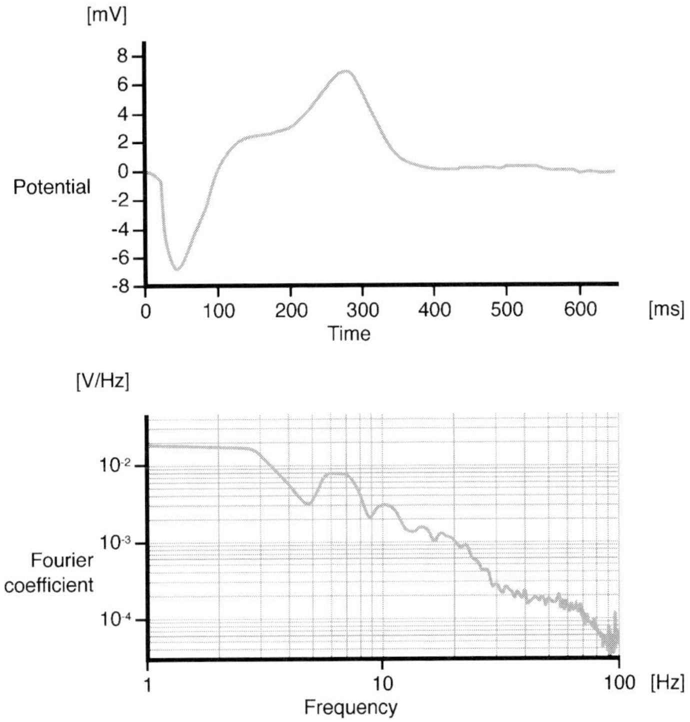

Fig. 9.5 Ventricular evoked response (VER) and its frequency analysis.

considerations lead to the requirement that the capacity must exceed 10 mF/cm^2 to reduce afterpotentials to a negligible value below 1 mV.

9.2.3
Design Realization

One approach to increasing the interface capacity is to enlarge the electrochemically active electrode surface, since the interface capacity (as in a plate capacitor) is proportional to the area and the dielectric constant of the insulating layer. To fulfill the above mentioned electrode requirements, the active electrode surface must be increased by at least three orders of magnitude. Since the geometric electrode surface for energy-saving pacing must be small, an increase in the active surface area can only be accomplished by restructuring the electrode tip. Sintering or electrochemical etching methods have been implemented. However, these approaches resulted in insufficient enlargement by factors of only five to ten. With the advent of fractal electrodes, this problem has been resolved.

The fractal principle is demonstrated in Fig. 9.6. The hemispherical electrode tip is covered with smaller hemispheres of identical shape, consequently nearly doubling the electrode surface. This approach is repeated according to the fractal

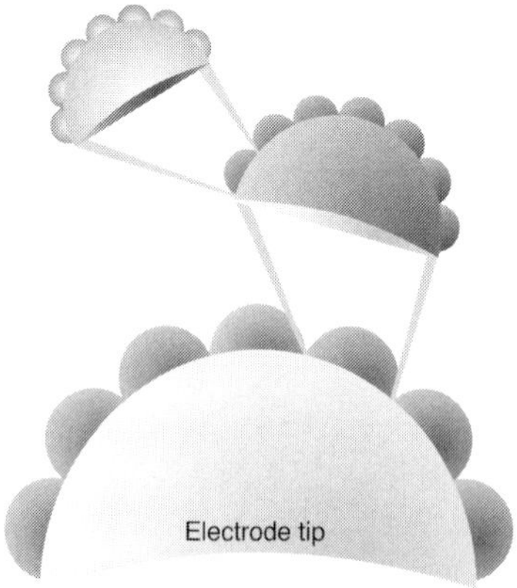

Fig. 9.6 Fractal coating principle: a hemispherical electrode tip is repeatedly covered with ever-smaller hemispheres.

Fig. 9.7 Fractal iridium coating at 10 µm (above, scanning electron microscope) and 100 nm (below, scanning tunneling microscope) magnification.

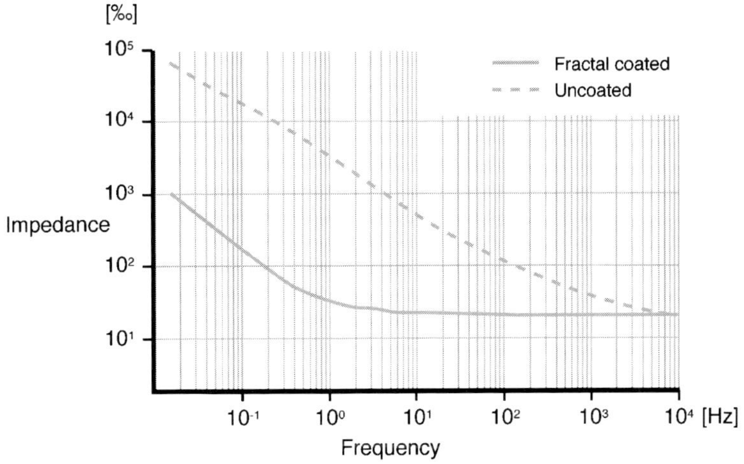

Fig. 9.8 Impedance spectrum of a fractal coated (iridium) versus an uncoated electrode.

principle with ever smaller structures. After n steps, the electrochemically active electrode surface is enlarged by a factor of 2^n. Given this relationship, a large active surface can be achieved, even on small electrode tips.

The strategy for technologically realizing the fractal coating principle is the application of fractal metal layers on the electrode substrate by physical vapor deposition (PVD). The process allows a growth of fractal layers. Iridium has been found to be an ideal coating material, superior to others such as titanium nitride [9]. Known for its biocompatibility, this precious metal is inert under electrical loads during pacemaker therapy and defibrillator use.

Fig. 9.7 shows images of fractal iridium coating taken by a scanning electron microscope and scanning tunneling microscope. Although these images are at completely different levels of magnification, fractal structures are present and the existence of repetitive self-similar structures is shown.

Fractal surface coating changes the electrical characteristic of the electrode/electrolyte interface significantly. Fig. 9.8 shows the results of potentiostatic measurements of the interface impedance of a fractal coated and uncoated electrode. With fractal coating, the cut-off frequency below which impedance and attenuation increase decisively is lower than that of uncoated electrodes by more than three orders of magnitude from >1000 Hz to <1 Hz. The resulting interface capacity can be increased 5000-fold, with values of up to 50 mF/cm^2 for the fractal coated substrate, compared with 10 $\mu F/cm^2$ for the uncoated substrate. Since the electrode properties rely on the fractal structure of the surface and iridium is inert in a biological environment, the electrical response retains long-term stability.

9.3 Rate Adaptation by Minute Ventilation

In the past 40 years cardiac pacing has evolved from a simple fixed rate oscillator, generating pacing pulses at 72 pacing pulses per minute (PPM), to a sophisticated system operating in multiple chambers and controlled by a variety of sensors in addition to the sinus node. Rate control of a pacemaker by a sensor has been a reality for the latter half of this long history. First implemented with an accelerometer, it then gradually expanded to a very small number of truly practical 'physiological' sensors. Considered 'practical' as they gave a reasonable indication of the metabolic state of the majority of patients implanted with the device. The 'surviving few' are: cardiac repolarization (QT), minute ventilation (MV), cardiac contractility with impedance measurement, and accelerometer in the lead.

9.3.1 Brief History and Implementation

Respiration rate, which is one of the components of MV, was the first candidate identified by Krasner [10, 11] in 1966 for rate control of pacemakers. Krasner, like Funke [12] some years later, did not realize a working system. Rossi [13–15] in collaboration with Biotec SPA (Bologna, Italy) achieved a working system in 1983. Unfortunately, this system only measured respiration rate and did this with a special lead that had to be implanted beneath the skin and across the sternum. This approach suffered both lead problems and artifact problems due to lead erosion and continuous lead movement. To avoid the problems experienced by Rossi, Nappholz [16, 17] and coworkers succeeded in monitoring true MV from a standard bipolar lead in the ventricle. This solution addressed a number of critical issues: it eliminated the need for a subcutaneous lead, measured impedance in a 'tri-polar' configuration, and measured MV instead of rate. The need to eliminate the subcutaneous lead was obvious and so was the requirement to measure MV instead of respiration rate. The later was made clear by the extensive work of Pierce [18, 19] and Whipp [20], showing considerable improvement in the correlation of MV with exercise compared with just respiration rate. The true advantage of MV over other commonly considered parameters such as mixed venous temperature, pH, and venous oxygen saturation was demonstrated by the extensive work of McElroy [21, 22]. The utility of MV measurement in patients with congestive heart failure was demonstrated by the work of Weber [23]. The tri-polar measurement configuration is shown in Fig. 9.9, in which the fundamentals of the respiration measurement are laid out. A constant current (I_v or I_a) is injected between the pacemaker case and the selected lead. It can be seen that as we measure the 'thoracic impedance' (r_{res}) we can use either the ventricular lead or the atrial lead. In either case it will pass through the same impedance. The current pulse shown has an amplitude of the order of 1 mA and a duration of 15 μs. It is preferably bi-phasic (shown) to minimize interference on external monitoring devices, such as ECG monitors. The current pulse is delivered normally at about 20

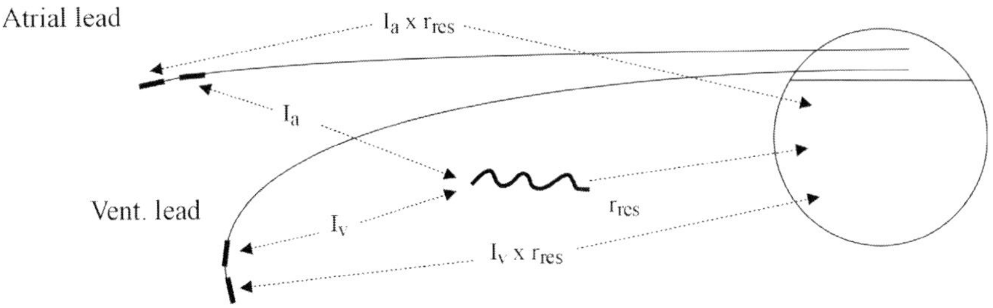

Fig. 9.9 Tri-polar measurement configuration for respiration measurement.

times a second. Injecting the current into the ring increases the safety margin for the patient to about 500% [24] and also allows a tri-polar measurement. As the pacemaker case is geometrically a large surface the tri-polar implementation approaches a quadrupolar measurement, which is the recommended approach to minimization of unpredictable artifacts generated by the use of the same electrodes for both the generation of the measurement current and the monitoring of the impedance. Although the use of a current pulse, as described, has the benefit of safety and minimum current drain from the implant, it does leave the measurement susceptible to certain types of external interference that occur at multiples of the sampling frequency. Certain more commonly utilized monitoring equipment has been known to inappropriately modify the pacing rate of the implanted device [25]. Further details of the setup for measuring the cardiac impedance are discussed below.

9.3.2 The Measurement of MV

The standard approach to measuring MV is to filter between 0.1 and 1 Hz the reconstructed respiration signal, obtain its rate and its average value (tidal volume). This approach is most suitable as the respiration signals are extremely small and distorted by various sources of interference. The most prominent interference is modulation due to stroke volume. A sample waveform is shown in Fig. 9.10.

Even after filtering, the signal still has a distortion due mostly to stroke volume modifications. To ensure an accurate measurement of rate, the polarity of the signal (related to the long-term average reference, shown as a line) is continuously determined in the form of a majority count. If most of the counts are positive then the signal is positive. This allows a good assessment of the respiration rate even in the presence of noise. To minimize the effect of the interference on the tidal volume measurement the signal is rectified and the average is then used as the tidal volume. The product of these two gives the MV value. The use of this measured MV to control the pacing rate is shown in Fig. 9.11 where the relationship between heart rate and MV is plotted. In the initial implementation of MV control in pacemakers the relationship between MV and heart rate was assumed to be linear. This gave rise to the conversion constant called rate response factor

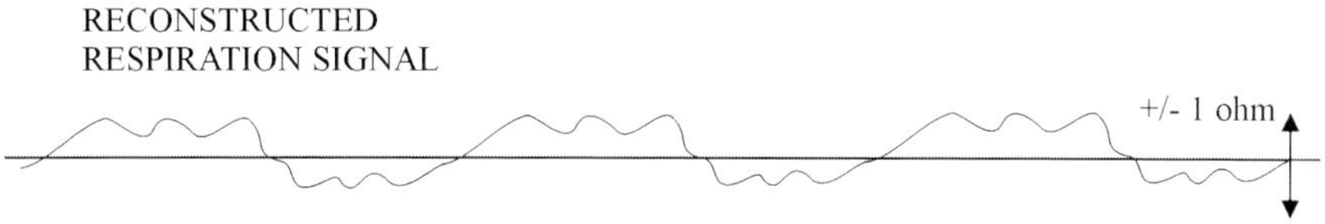

Fig. 9.10 Sample waveform of the respiration signal.

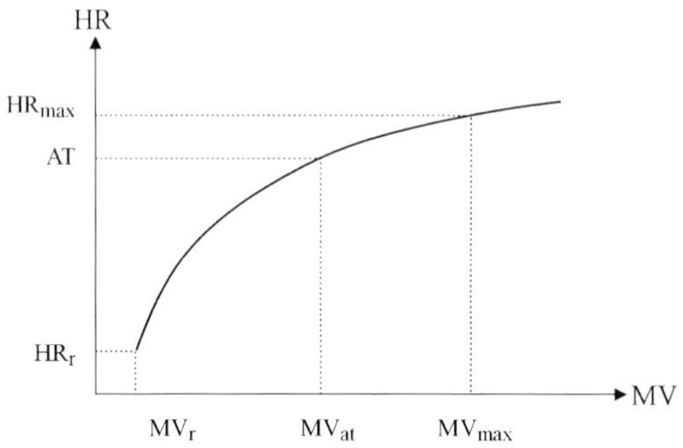

Fig. 9.11 Relationship between heart rate and minute ventilation (MV).

(RRF). With the patient at rest the average MV gave the value of MV_r. During exercise the value of MV_{max} was measured. The RRF would be given by

$$RRF = \frac{MV_{max} - MV_r}{HR_{max} - HR_r} .$$

The resting value, MV_r, is the long-term average over a number of hours. The distribution of the relationship between HR and MV has been characterized [22, 26, 27] and found to be in the range 1.4–2.2 bpm/L. This closely corresponds to the RRF values obtained in practice. The RRF is obtained in practice in one of three ways: by exercising the patient to obtain MV_{max} and then automatically computing RRF; by having the patient walk down the hall, which is equivalent to the 'low-intensity treadmill exercise' (LITE) [28]; or by automatic determination by matching sensor response to a 'target rate histogram' [29, 30]. Because of the specificity of the MV measurement, the trend with the manufacturers is to offer the automatic adaptation option with this parameter and when in combination with another sensor [30]. The problems associated with RRF being just a single slope are also now generally accepted [31].

9.3.3
Clinical Utility of MV for Controlling Pacing Rate

For any sensor to be accepted for incorporation into a critical device it must meet stringent requirements in sensitivity, specificity, and repeatability. There is no doubt about MV being a reliable parameter for controlling heart rate. From numerous studies cited above it is clear that respiration not only supplies the essential metabolites to the circulatory system but it also monitors the effectiveness of the pump in purging carbon dioxide. So when we talk about the essential characteristics of a sensor we are talking about the 'conversion' of MV into the electrical signal for rate control: the transduction of the respiratory signals into electrical information.

9.3.3.1 **Sensitivity of the MV Sensor**

As intimated before, the actual electrical signal obtained from the bipolar leads is very small, in the order of a few ohms. From clinical work we know that MV changes from rest to exercise in excess of 400%. From the early work on the Meta pacemaker [32] we see that the RRF values obtained had a very small standard deviation, shown in Fig. 9.12. This would be an indication of reliable transduction of MV. In the figure we see that the RRF values fall on an orderly 'normal' distribution. The exceptions are the two RRF values of 20 and 10 and it was concluded that they tended to be programmed because of convenience where the physician decided to bypass the treadmill calibration. There are a multitude of studies indicating the close correlation of the MV sensor and exercise (or oxygen consumption). A more recent study by Bonnet [33] best sums up the data, as this study evaluated four electrode configurations including a floating lead in the atrium on a total of 51 patients and 57 exercise studies. The correlation between actual MV measurement and that measured by the sensor was in excess of 0.86 for all the lead configurations. Other studies looked specifically for correlation at the 'onset and cessation' of exercise [34]. In all these studies it was found that even with the filter delays the correlation was not significantly effected in these transient conditions. With regards to the influence of pulmonary disease, many studies have been carried in patients suffering from various pulmonary disorders [32, 35] and apart from a slightly higher RRF value no other disadvantage was seen.

9.3.3.2 **Specificity of the MV Sensor**

Because of the nature of the application of the sensor, 'specificity' is a critical consideration. It is highly desirable that the sensor responds to MV and to as few unrelated influences as possible. Early in the clinical evaluation of the Meta MV24 it was established that regular arm motion involving the pectoral musculature (in

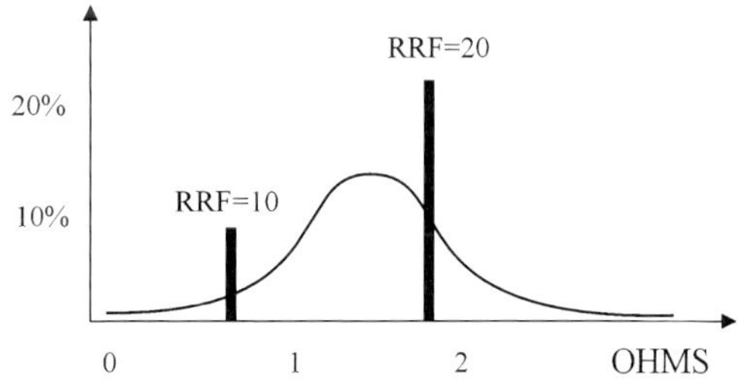

Fig. 9.12 Distribution of the rate response factor (RRF).

which the pacemaker was implanted) contributed to a rate increase. Obviously vigorous arm motion itself is metabolically demanding hence the artifact component was of lesser relevance. Stroke volume, which was recognized early in the development as a possible source of interference, became much less of a problem [36] after appropriate care in filtering was applied. Posture during exercise was found to be an influence. In a study carried out by Dura [37] on 19 patients implanted with the Kappa 400 (Medtronic, Minneapolis MN), it was found that posture influenced the ratio of the measured impedance signal to the actual MV value. This could be due to the modification to the static and dynamic geometry of the thoracic viscera. The biggest problem with specificity still remains the potential of interference from external electrical monitors [25] and electronic devices [38]. This problem could be eliminated by appropriate design, but this has not yet been done.

9.3.3.3 Repeatability of the MV Sensor

For a sensor to be useful it must be repeatable. This requires the MV sensor to have the same RRF value after a certain time. In the case of a non-automatic system it should maintain its calibration over a period of years. Early work by Li [39] indicated that once the device had been calibrated appropriately it had no problems with maintaining calibration.

9.4 Rate Adaptation Based on Cardiac Contractility

9.4.1 General Conception

9.4.1.1 Open-Loop vs. Closed-Loop Control

Different sensor concepts to detect a physical quantity by which the metabolic demand can be assessed were proposed and implemented [1]. These approaches result in an open-loop or closed-loop control of the heart rate, depending on whether the stimulation rate has a physiological feedback effect on the sensor signal or not [1, 40]. If the sensor signal results from an external source, a quantity that is not controlled by the physiological controller of the circulatory system (which is located in the medulla oblongata), an open-loop control system is implemented. One example is an acceleration sensor that detects body movements, currently the most widely used sensor principle for rate-adaptation. Physical exertion results in acceleration forces on the sensor integrated into the pacemaker. From the acceleration signal the pacing rate is adjusted. However, the pacing rate has no effect on the body movements. Another example is the MV that has been discussed above.

In contrast, in a closed-loop system the sensor signal is influenced by the stimulation rate. If, for example, the stimulation rate is increased inadequately, this will in turn induce a change of the control signal in the opposite direction: there is a negative feedback. Hence the pacemaker will decrease the stimulation rate again. As the physiological controller is integrated into the loop, a stable setpoint can be found for each metabolic state.

9.4.1.2 Cardiovascular Regulation, Sensor Concept

A sensor application for rate-adaptive pacing is presented, which allows physiological closed-loop control of the cardiovascular system based on detection of the cardiac contraction dynamics. Fig. 9.13 depicts the basic concept.

The main task of the cardiovascular system is to provide adequate blood perfusion of the body organs and muscles. To ensure sufficient perfusion the mean arterial blood pressure (MABP) needs to be maintained and controlled. The blood pressure results from the cardiac output (CO) and the total peripheral resistance (TPR). The TPR is adjusted by local regulative mechanisms and by the autonomic nervous system due to the current metabolic demand. For example, it is decreased by vasodilatation of the muscle arterioles upon physical exercise. The blood pressure is sensed by the baroreceptors and this information is conducted to the medullary cardiovascular control center via afferent nerve pathways. To maintain a certain MABP the CO

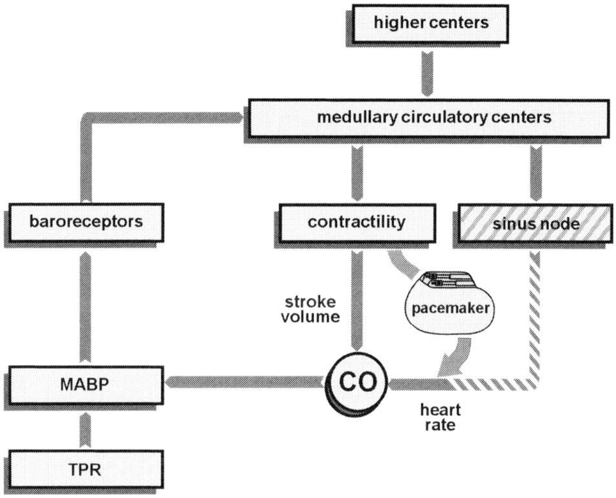

Fig. 9.13 Circulatory control loop with impaired sinus node and pacemaker. The cardiac output (CO) results from the stroke volume (SV) and the heart rate (HR): $CO = SV \times HR$. The total peripheral resistance (TPR) and the CO determine the mean arterial blood pressure (MABP). The TPR is adjusted to the current metabolic demand. From the viewpoint of control theory it can be considered as a disturbance of the control loop. The blood pressure is sensed by the baroreceptors and is fed back to the control centers located in the medulla oblongata. From the physiological controller, via the autonomic nervous system, the contractility and the sinus node are regulated. In a chronotropic incompetent patient the sinus node is impaired. A rate-adaptive pacemaker, sensing the inotropic state, replaces the function of the sinus node.

(the blood volume) that is pumped into the circulation within a certain time, needs to be adjusted. The CO is modulated by the heart rate (chronotropic state) and the contractility (inotropic state). The contractility is a measure for the force and the dynamics of the myocardial contraction. Both inotropy and chronotropy are controlled by the autonomic nervous system via the sympathetic and parasympathetic efferent nerves and the related hormones. In chronotropic incompetent patients the sinus node, the natural pacemaker of the heart, is impaired resulting in an inadequate heart rate increase on physical or mental exertion.

For physiological rate-adaptive pacing with an artificial pacemaker it is desirable to detect and evaluate the tone of the autonomic nervous system. Even in chronotropic incompetent patients this information is included in the inotropic state: in the contractility of the myocardium. With a sensor that is able to detect the contractile state, and an algorithm that transfers this signal into an appropriate stimulation rate, the physiological control loop of the cardiovascular system can be reestablished. The medullary circulation center is in charge of the regulation. The pacemaker only 'bridges' the sick sinus node: chronotropic insufficiency.

9.4.2 Impedance Sensor

9.4.2.1 Sensor Implementation

Changes in cardiac contractility can be assessed by the electrical intracardiac impedance signal. To measure the impedance with an implanted pacemaker small subthreshold current pulses with a constant amplitude of 200 μA are injected between the tip electrode of the ventricular lead and the pacemaker housing. To reduce polarization artifacts and to balance the charges on the electrode the current pulses have a biphasic shape with alternating polarity. The dominant frequency component of the injection current is approximately 16 kHz, as shown in Fig. 9.14. For other implementations sinusoidal currents were used in the past [41]. An advantage of the pulsed current is that it keeps the energy consumption low and thus saves the life of the pacemaker battery. The resulting voltage is measured between the two points shown in Fig. 9.14.

For evaluation of the impedance signal by the pacemaker processor the voltage must be synchronously demodulated, sampled, filtered, amplified, and converted to a digital value. Fig. 9.15 depicts the measurement circuit for the complete signal path.

The impedance signal is composed of an offset of about 400–1500 Ω, depending on the electrode quality and position, and a superimposed alternating component with an amplitude of approximately 5–10% of the offset. As only the alternating impedance component is evaluated, the offset is removed by the bandpass filter with a bandwidth of 0.4–40 Hz before the signal is amplified. Also the signal modulation by respiration is diminished by the filter. The gain factor of the amplifier is programmable via software.

Fig. 9.14 also displays the potential distribution resulting from the injected current. For simplification a small spherical tip electrode and a sphere as counter

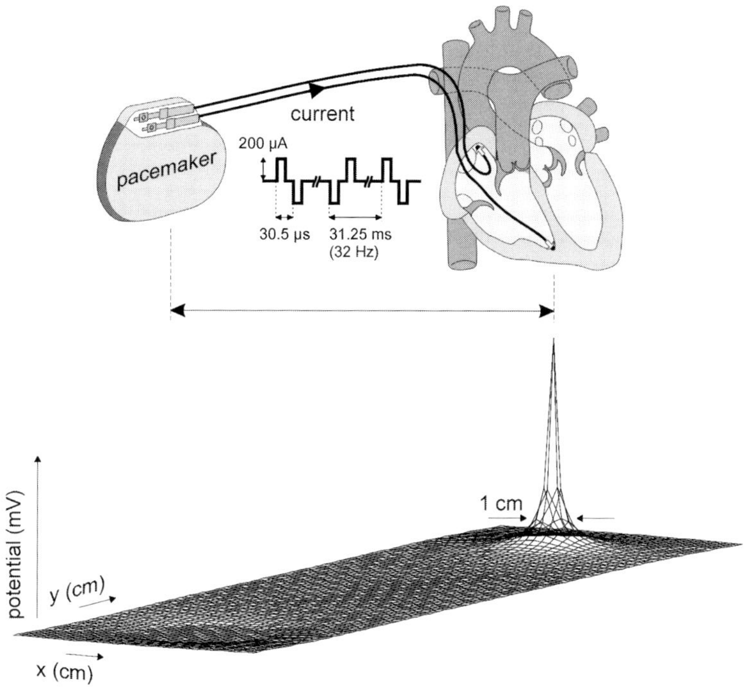

Fig. 9.14 Principle of unipolar (measurement between tip electrode and pacemaker case) intracardiac impedance measurement: resulting potential distribution.

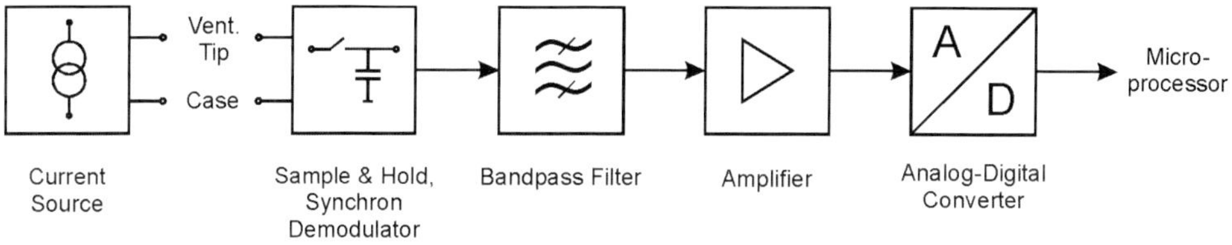

Fig. 9.15 Block diagram of the impedance measurement circuit integrated into the pacemaker electronics.

electrode were assumed for this computation [42, 43]. Field calculations indicate that the major part of the potential difference (the voltage) drops within a small volume around the tip electrode, where the current density is high. It follows that with this configuration basically the impedance of the right-ventricular apex is monitored. As shown below, the course of the impedance mirrors the apical contraction dynamics. The influence of respiration and ventricular volume changes on the signal is minimal.

9.4.2.2 **Origin of the Impedance Signal, Tissue Properties**

The varying proportion of blood and myocardium during a contraction cycle in the vicinity of the ventricular tip electrode cause characteristic changes in the impedance signal. In Fig. 9.16 the two extreme states of the right-ventricular apex, maximum relaxation at end-diastole and maximum contraction at end-systole, are depicted. The impedance is altered as blood and myocardial tissue possess different conductivities.

The myocardium is electrically anisotropic. The conductivity parallel to the muscle fibers is about twice of the transversal conductivity and the values in Tab. 9.1 were determined.

The conductivity of blood was measured to be approximately 6 mS/cm (6.17 mS/cm at low frequencies [44] 5.7–6.8 mS/cm at 1 kHz [47]). The conductivity also depends on the hematocrit and on the flow velocity as described by Sakamoto et al. [48]. As the hematocrit is constant for an individual, or at least does not change very quickly, its contribution is not relevant here. For laminar flow the erythrocytes accumulate near the axis of the vessel and orientate in the flow direction. The blood becomes electrically anisotropic. Based on the data from Sakamoto et al. the conductivity change of laminar flowing blood in the right ventricle during a contraction cycle was estimated to be approximately 1% in the flow direction [49]. For the described sensor concept the impedance measurement is restricted to the right-ventricular apex. As the flow is not laminar in this region, only very small conductivity changes are expected. Therefore the influence of blood velocity on the conductivity is neglected here.

Although the conductivities of myocardium and blood generally are frequency dependent, the inherent reactive component is small in the frequency range below 100 kHz [50]. The phase shift between voltage and injected current is negligible for blood [51] and smaller than 5° for myocardium [45] in the range up to 100 kHz.

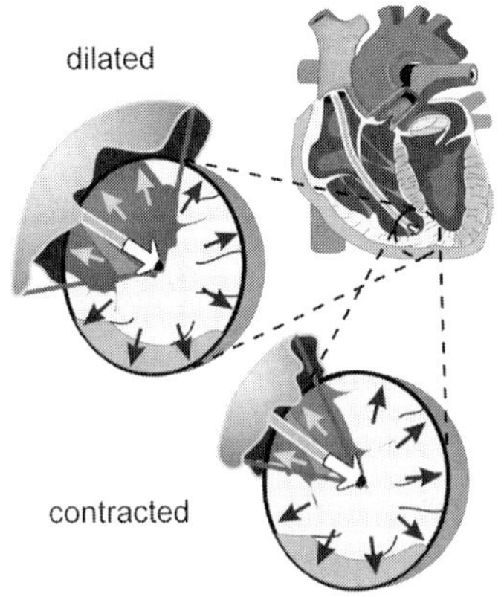

Fig. 9.16 Local geometry changes in the vicinity of the ventricular tip electrode in the two extreme states of the contraction cycle. At end-diastole, at maximum ventricular filling, the measurement volume contains a high portion of blood, whereas in the end-systolic state the myocardium dominates.

Tab. 9.1 Longitudinal and transversal myocardial conductivity. In some references the anisotropy was not determined: LF, low frequency.

<table>
<tr><th>Species</th><th>Frequency (kHz)</th><th>Longitudinal conductivity (mS/cm)</th><th>Transversale conductivity (mS/cm)</th><th>Ref.</th></tr>
<tr><td>Dog</td><td>LF</td><td>4.0</td><td>1.8</td><td>[44]</td></tr>
<tr><td>Dog</td><td>10</td><td colspan="2">1.7</td><td>[45]</td></tr>
<tr><td>Human</td><td>10</td><td colspan="2">4.6</td><td>[45]</td></tr>
<tr><td>Human</td><td>100</td><td colspan="2">4.8</td><td>[45]</td></tr>
<tr><td>Dog</td><td>30</td><td>4.0</td><td>2.4</td><td>[46]</td></tr>
</table>

9.4.3 Rate-adaptive Pacing Based on Intracardiac Impedance

9.4.3.1 Unipolar Impedance Signal

It has been shown how cardiac contraction modulates the course of the unipolar impedance signal. For a constant inotropic state, such as the resting state, this results in a typical impedance signal. If the myocardial contractility increases, the impedance course also changes. Fig. 9.17 demonstrates the change in the impedance signal provoked by application of a positive inotropic drug [52]. The maximum derivative of the invasively measured ventricular blood pressure (dp/dt_{max}) and a ventricular inotropic parameter (VIP) derived from the impedance signal were correlated for 12 patients. The dp/dt_{max} is a well-accepted measure of ventricular inotropy [53]. For VIP determination the slope (the difference between two samples) in a predefined region of interest was computed. This region was defined by the biggest difference in slope between rest and the maximum dobutamine dose during three consecutive sampling intervals of 7.8 ms duration each. A high correlation was found between the VIP and dp/dt_{max} as shown in Fig. 9.18.

Several other studies were performed to investigate the influence of the cardiac sympathetic innervation on the unipolar impedance signal. Bingelli et al. correlated the impedance with the muscle sympathetic activity (MSA), using microneurography of the peroneal nerve, during a cold pressor test. The inotropic state computed from the impedance signal and the MSA were well correlated in eight out of nine patients ($R^2=0.86$–0.99, $P<0.0001$) [54]. Witte et al. induced a regional ischemia in different regions of the myocardium during percutaneous coronary angioplasty (PTCA) in seven patients [55]. As a result they found the impedance indicating an increased contractile state when the ischemic region was in the left ventricular myocardium, but a decreased state when it was in the right ventricular apex. The increased contractility is explained by an increased sympathetic activity to compensate the ischemia resulting in a higher contractility of the non-ischemic regions. As the impedance determines the local contractility, a decrease is detected when a coronary artery is occluded that supplies the region where the impedance measurement is performed.

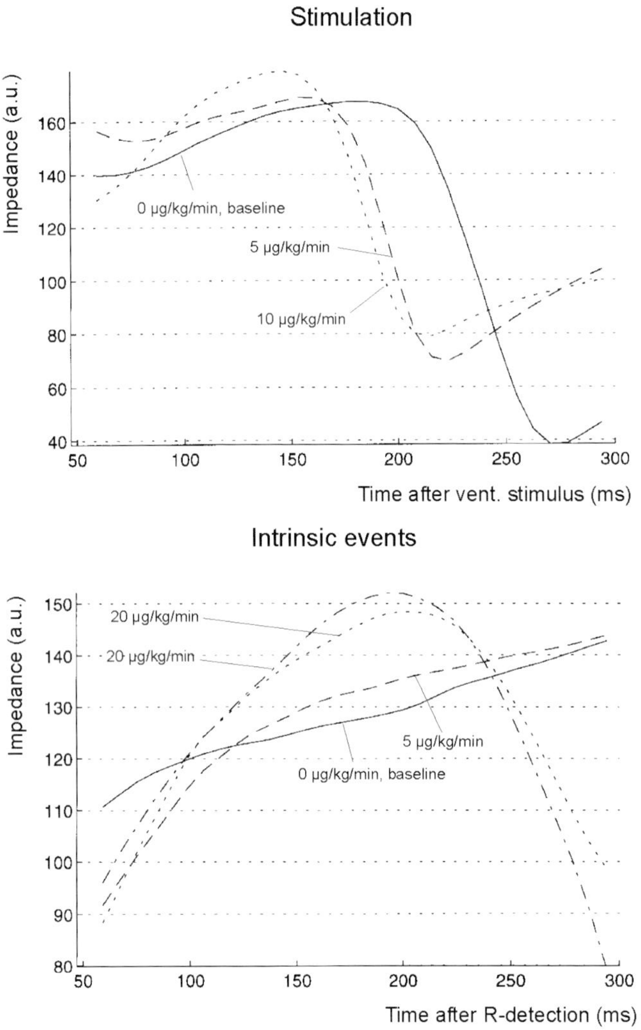

Fig. 9.17 Right-ventricular impedance at different dobutamine doses for intrinsic and paced ventricular events within a certain period after the ventricular event for the same patient. From [52].

9.4.3.2 **Signal Evaluation**

Recently a more general approach to evaluating the unipolar impedance signal was introduced that requires no manual sensor calibration to extract a ventricular inotropic parameter. With this method the difference in morphology between the currently measured impedance curve, for example, influenced by an increased sympathetic tone, and a continuously updated reference curve that reflects the resting state, is evaluated. The adaptive rate is computed from a 'differential area', that is, the absolute area between the reference curve and the current curve,

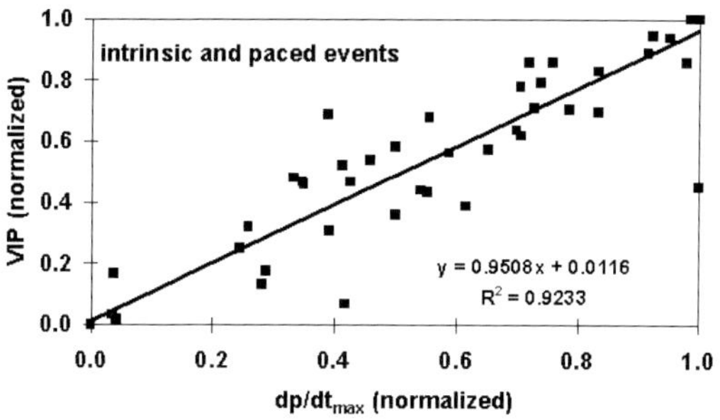

Fig. 9.18 Correlation of the ventricular inotropic parameter (VIP) and dp/dt_{max} for 12 patients with sick sinus syndrome. Each sample marks a certain dobutamine dose for one patient. Paced and sensed events were pooled. From [52].

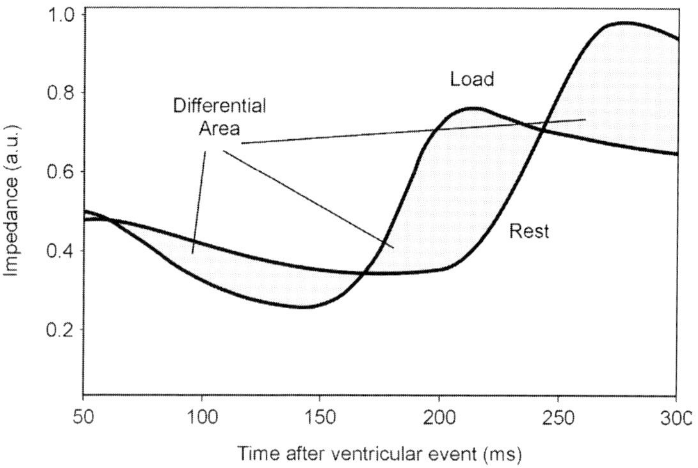

Fig. 9.19 The impedance curve is evaluated for the period 50–300 ms after a ventricular event, which involves isovolumetric contraction and the beginning of the ejection phase. The differential area between the rest and load curves (shown in gray) is multiplied by a self-adjusted response factor to determine the pacing rate.

weighted with a response factor (Fig. 9.19). As the curve morphologies might vary for different patients, the response factor is automatically adapted for each patient individually to obtain an optimal rate response. Additionally the measurement gain is adjusted by the algorithm, so that the sensor signal always is recorded with the optimum resolution. By the automatic mechanisms the system adapts to long-term changes that might alter the impedance signal, for example, electrode ingrowth, progredient heart disease, or medication affecting the inotropy.

9.4.3.3 **Clinical Results**

The results from a clinical multicenter study (Figs. 9.20 to 9.22) demonstrate the effectiveness of the impedance sensor concept for contractility-based rate-adapta-

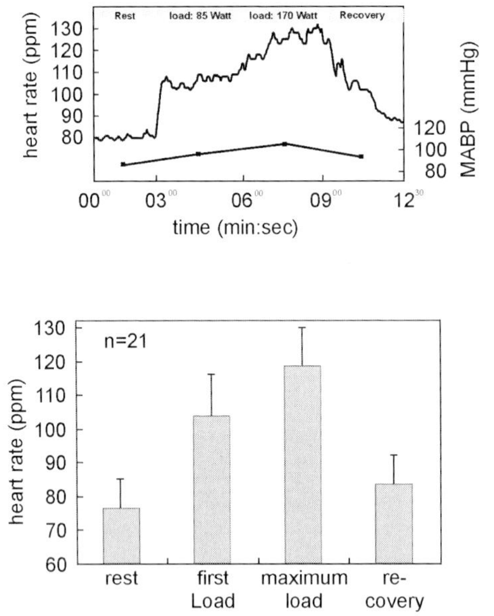

Fig. 9.20 Top: Pacing rate and mean arterial blood pressure (MABP) of a young chronotropically incompetent patient undergoing ergometry with two load steps. Bottom: averaged heart rate of 21 patients during ergometry (rest – mean values, first and maximum load: maximum value, recovery: final value). Both from [56].

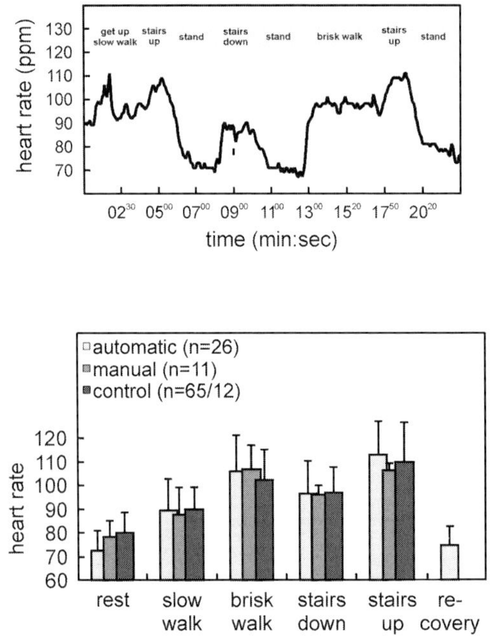

Fig. 9.21 Top: Pacing rate of a chronotropically incompetent patient during an ambulatory test simulating normal daily activity. Bottom: averaged heart rates of 26 patients during an ambulatory exercise test. The results are compared with those from 11 patients with a manually-calibrated impedance-based rate-adaptive pacemaker and a healthy control population (rest, slow walk, brisk walk: $n = 65$, stairs down, stairs up: $n = 12$). Both from [56].

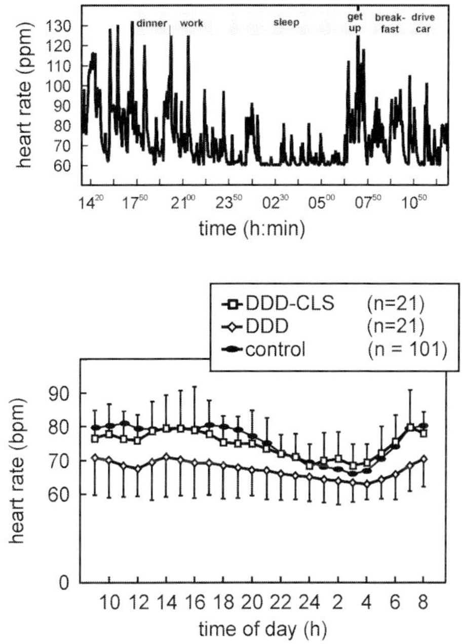

Fig. 9.22 Top: 24 h Pacing rate trend of a chronotropically incompetent patient during normal daily activities. Bottom: hourly averaged heart rates in DDD (dual-chamber pacing without rate-adaptation) and DDD-CLS (rate-adaptation by contractility) pacing modes of 21 patients during normal daily activities. The results are compared with those of healthy individuals. Both from [56].

tion using the above described evaluation method [56]. To investigate the sensitivity of the sensor the results of the exercise tests and the averaged 24 h heart rate course were compared with a control group of healthy individuals in the same age range. A close correlation was found. As the sensor detects changes of the sympathetic tone, it not only reacts on physical but also on psychological stress conditions. It has been shown that an impedance sensor-based pacemaker increases the stimulation rate properly during a color-word test [57]: a standardized mental stress test [58]. Conversely, sensor principles based on body acceleration are not able to react in a proper way under mental stress conditions when the body is not moved [59]. From the ambulatory test results shown in Fig. 9.21 the specificity of the impedance sensor can be observed. The adaptive pacing rate is graded depending on the kind of exertion performed. So the rate increase is higher when the patients climb up stairs compared with descending. Also in this case an accelerometer-driven pacemaker might deliver non-physiological results, as in most individuals the body movements will be stronger for descending than for climbing [60].

The contractility sensor was compared with other sensors for rate-adaptive pacing [59]. The principles evaluated in this study, besides the contractility sensor, were based on acceleration, minute ventilation, minute ventilation combined with acceleration, QT-interval and QT-interval combined with acceleration. The maximum rate reached during different physical and psychical stress tests was compared for 27 pacemaker patients with a healthy control group. The results for the contractility sensor were comparable to those of the control group for all challenges. The minute ventilation-based pacemakers showed good results as well, except for mental stress where it did not react properly. The other systems had weaknesses for more than one stress challenge.

9.5 References

1 Lau, C.-P. The Range of Sensors and Algorithms Used in Rate Adaptive Cardiac Pacing. *PACE* **1992**, *15*, 1177–1211.

2 Bolz, A., Schaldach, M., Hubmann, M. Interface Aspects of Stimulating Electrodes. in *Proc. Ann. Int. Conf. IEEE Eng. Med. Biol. Soc.* **1991**, *13*, 714–716.

3 Hamann, C. H., Vielstich, W. *Elektrochemie I. Leitfähigkeit, Potentiale, Phasengrenzen*. VCH, Weinheim, **1985**.

4 Bolz, A. Thin-Film Technology and Hybrid Design in Biomedical Engineering. *Progress in Biomedical Research, Collection of Reprints* **1989–1996**, *1*, 315–346.

5 Falterstein, A. H. A General Analysis of Voltage Fluctuations of Metal Electrolyte Interfaces. *Med. Biol. Eng. Comput.* **1966**, *4*, 589–594.

6 Brouwer, J., Nagelkerke, D., de Jongste, M. J. L., Boute, W., den Heijer, P., Lie, K. I. Analysis of the Morphology of the Unipolar Endocardial Paced Evoked Response. *PACE* **1990**, *13*, 302–313.

7 Curtis, A. B., Maas, S. M., Domijan, A. D., Keim, S. G., Duran, A. A Method for Analysis of the Local Atrial Evoked Response for Determination of Atrial Capture in Permanent Pacing Systems. *PACE* **1991**, *14*, 1576–1581.

8 Livingston, A. R., Callaghan, F. J., Byrd, C. L., Heemels, J. P., Hollander, P., van Mechelen, R., Chappin, J. J. Atrial Capture Detection with Endocardial Electrodes. *PACE* **1988**, *11*, 1770–1776.

9 Bolz, A. Die Bedeutung der Phasengrenze zwischen alloplastischen Festkörpern und biologischen Geweben für die Elektrostimulation. Schiele & Schön, Berlin, **1995**, 76–81.

10 Krasner, J. L., Voukydis, P. C. A Physiologically Controlled Cardiac Pacemaker. *J. Assoc. Adv. Med. Instrum.* **1966**, *1*, 14.

11 Krasner, J. L., Nardella, P. C. US Patent No. 3593718; **1967**.

12 Funke, H. D. Ein Herzschrittmacher mit belastungsabhängiger Frequenzregulation. *Biomed. Techn.* **1975**, *20*, 225.

13 Rossi, P., Plicchi, G., Canducci, G., Rognoni, G., Aina, F. Respiration Rate as a Determinant of Optimal Pacing Rate. *PACE* **1983**, *6*, 502.

14 Rossi, P., Plichi, G., Canducci, G., Rognoni, G., Aina, F. Respiration as a Reliable Physiological Sensor for the Control of Cardiac Pacing Rate. *Br. Heart J.* **1984**, *51*, 7.

15 Rossi, P., Rognoni, G., Ochetta, E., Aina, F., Prando, M. D., Plicchi, G., Minella, M. Respiration Dependent Ventricular Pacing Compared with Ventricular and Atrio-Ventricular Synchronous Pacing. Aerobic and hemodynamic variables. *J. Am. Coll. Cardiol.* **1985**, *6*, 646.

16 Nappholz, T. A., Lubin, M. Measuring Minute Ventilation with a Pacing Catheter. *Proc. Third Asian Symp. Cardiac Pacing and Electrophysiology*, Melbourne, Australia, **1985**.

17 Nappholz, T. A., Valenta, H. Electrode Configurations for a Respiratory Impedance Measurement Suitable for Rate-Responsive Pacing. *PACE* **1986**, *9*, 960.

18 Pearce, D. H., Milhorn, H. T., Holloman, G. H., Reynolds, W. J. Computer Based System for the Analysis of Respiratory Response to Exercise. *J. Appl. Physiol.* **1977**, *42*, 968.

19 Pearce, D. H., Milhorn, H. T. Dynamic and Steady State Respiratory Responses to Bicycle Exercise. *J. Appl. Physiol.* **1976**, *42*, 959.

20 Whipp, B. J., Wasserman, K. Oxygen Uptake Kinetics for Various Intensities of Constant-Load Work. *J. Appl. Physiol.* **1972**, *33*, 351.

21 McElroy, P., Weber, K.T., Nappholz, T.A. Heart Rate, Ventilation, Mixed Venous Temperature, pH and Oxygen Saturation during Incremental Upright Exercise. *Proc. Third Asian Pacific Symp. Cardiac Pacing and Electrophysiology*, Melbourne, Australia, **1985**.

22 McElroy, P., Janicki, J.S., Weber, K.T. Physiologic Correlates of the Heart Rate Response to Upright Isotonic Exercise: Relevance to Rate-Responsive Pacemakers. *J. Am. Coll. Cardiol.* **1988**, *11*, 94.

23 Weber, K.T., Kinsevawitc, G.T., Janicki, J.S. Oxygen Utilization and Ventilation during Exercise in Patients with Chronic Cardiac Failure. *Circulation* **1982**, *65*, 1213.

24 Seger, J. Safe Amplitude and Pulse Width for Minute Ventilation Pulses. *PACE* **1997**, *20*, 1173.

25 Chew, E.W., Troughear, R.H., Kuchar, D.L., Thorburn, C.W. Inappropriate Rate Change in a Minute Ventilation Rate Responsive Pacemaker due to Interference by Cardiac Monitors. *PACE* **1997**, *20*, 276.

26 Soucie, L.P., Carey, C., Woodend, A.K., Tang, A.S. Correlation of the Heart Rate-Minute Ventilation Relationship with Clinical Data. *PACE* **1997**, *20*, 1913.

27 Rickli, H., MacCarter, D.J., Maire, R., Amann, F.W., Candinas, R. Age and Sex Related Changes in Heart Rate in Ventilation Coupling. *PACE* **1997**, *20*, 105.

28 Lewalter, T., MacCarter, D., Jung, W., Bauer, T., Schimpf, R., Manz, M., Lüderitz, B. The 'Low Intensity Treadmill Exercise' Protocol for Appropriate Rate Adaptive Programming of Minute Ventilation Controlled Pacemakers. *Pacing Clin. Electrophysiol.* **1995**, *18*, 1374–1387.

29 Ritter, P. Clinical Evaluation of an Automatic Slope Calibration Function in a Minute Ventilation Controlled DDR Pacemaker. *PACE* **1997**, *20*, 1522.

30 Leung, S.K., Lau, C.P., Tang, M.O., Leung, Z., Yakimov, K. An Integrated Dual Sensor System Automatically Optimized by Target Rate Histogram. *Pacing Clin. Electrophysiol.* **1998**, *21*, 1559–1566.

31 Brachmann, J. Minimal Algorithmic Changes in the Rate Adaptive Pacemaker Slope cause Significant Differences in the Patient Exercise Capacity. *ACC Abstracts*, March **1991**.

32 Lau, C.P., Antoniou, A., Ward, D.E., Camm, A.J. Reliability of Minute Ventilation as a Parameter for Rate Responsive Pacing. *PACE* **1989**, *12*, 321.

33 Bonnet, J.L., Ritter, P., Pioger, G. Measurement of Minute Ventilation with Different DDR Pacemaker Electrode Configurations. Investigators of Multicenter Study Evaluating the Chorus, R.M. and Opus, R.M. Pacemakers. *Pacing Clin. Electrophysiol.* **1998**, *21*, 4–10.

34 Mond, H.G. *Rate-Responsive Cardiac Pacing*. Telectronics Pacing Systems, Englewood, CO, **1990**.

35 Kay, G.N., Bubien, R.S., Epstein, A.E., Plumb, V.J. Rate-Modulated Pacing Based on Transthoracic Impedance Measurements of Minute Ventilation: Correlation with Exercise Gas Exchange. *J. Am. Coll. Cardiol.* **1989**, *15*, 1283.

36 Kay, G.N. Transthoracic Impedance Measurement with Minute Ventilation Sensing Pacemakers: Discrimination of Respiratory and Stroke Volume Components. *PACE* **1989**, *12*, 671.

37 Dura, F. Influence of Posture, Breathing Pattern, and Type of Exercise on Minute Ventilation Estimation by a Pacemaker Transthoracic Impedance Sensor. *Pacing Clin. Electrophysiol.* **2000**, *23*, 1767–1771.

38 Daryl, D. Inappropriate Rate Rise in a Meta 1256 Pulse Generator due to Proximity to Anti-Theft Device. Telectronics internal document, Englewood, CO, **1989**.

39 Li, H., Neubauer, S.A., Hayes, D.L. Follow Up of Minute Ventilation Rate Adaptive Pacemaker. *PACE* **1992**, *15*, 1826.

40 Schaldach, M. Control Aspects of Cardiac Output Adjustment. In: Schaldach, M., *Electrotherapy of the Heart*. Springer, Berlin, **1992**, 73–86.

41 Hochgraf, C. Rate Adaptation by Electrogram and Intracardiac Impedance. In: Webster J.G., *Design of Cardiac Pacemakers*. IEEE Press, New York, **1995**; 369–396.

42 Schaldach, M., Urbaszek, A. Unipolar Intracardiac Impedance – An Optimal Biosensor For Closed-Loop Rate-Responsive Pacing. *Proc. IX Int. Conf. Electrical Bioimpedance.* Heidelberg **1995**, 137–140.

43 Urbaszek, A. *Konzeption und technische Lösungen zur Optimierung der frequenzadaptiven Elektrostimulation des Herzens.* Schiele & Schön, Berlin, **1996**.

44 Rush, S., Abildskov, J. A., McFee, R. Resistivity of Body Tissues at Low Frequencies. *Circul. Res.* **1963**, *12*, 40–50.

45 Duck, F. A. *Physical Properties of Tissue.* Academic Press, London San Diego, **1990**.

46 Steendijk, P., van der Velde, E. T., Baan, J. Dependence of Anisotropic Myocardial Electrical Resistivity on Cardiac Phase and Excitation Frequency. *Basic Res. Cardiol.* **1994**, *89*, 411–426.

47 Geddes, L. A., Baker, L. E. The Specific Resistance of Biological Material – A Compendium of Data for the Biomedical Engineer and Physiologist. *Med. Biol. Eng.* **1967**, *5*, 271–293.

48 Sakamoto, K., Kanai, H. Electrical Characteristics of Flowing Blood. *IEEE Trans. Biomed. Eng.* **1979**, *26*, 686–695.

49 Boheim, G. Intrakardiale Impedanzmessung zur frequenzadaptiven Stimulation des Herzens. PhD Thesis, Friedrich-Alexander Universität Erlangen-Nürnberg, Erlangen, **1988**.

50 Baker, L. E. Principles of the Impedance Technique. *IEEE Eng. Med. Biol.* **1989**; 11–15.

51 Schwan, H. P. Electrical Properties of Blood and its Constituents: Alternating Current Spectroscopy. *Blut* **1983**, *64*, 185–197.

52 Osswald, S., Cron, T., Grädel, C., Hilti, P., Lippert, M., Ströbel, J., Schaldach, M., Buser, P., Pfisterer, M. Closed-Loop Stimulation Using Intracardiac Impedance as a Sensor Principle: Correlation of Right Ventricular dP/dtmax and Intracardiac Impedance During Dobutamine Stress Test. *PACE* **2000**, *23*, 1502–1508.

53 Opie, L. H. Ventricular Function. In: Opie, L. H., *The Heart – Physiology, from Cell to Bedside,* 3rd edn., Lippincott-Raven, Philadelphia: **1998**

54 Bingelli, C., Duru, F., Corti. R., Sudano, I., Spieker, L. E., Turina, A., Westermann, P., Ostermeier, M., Rahn, M., Luscher, T. F., Noll, G., Candinas, R. Autonomic Nervous System-Controlled Cardiac Pacing: A Comparison Between Intracardiac Impedance Signal and Muscle Sympathetic Nerve Activation. *PACE* **2000**, *23*, 1632–1637.

55 Witte, J., Reibis, R., Ströbel, J., Schaldach, M. The Response of the Intracardiac Impedance Signal to a Provoked Regional Ischemia during PTCA. *Proc. 7th Eur. Symp. Cardiac Pacing,* Istanbul **1995**; 555–559.

56 Novak, M., Hoffmann, G., Schaldach, M. Multi-Center Investigations with Automatically Initialized Closed Loop Stimulation – Rate Response during Daily Life and Physical Exercise Tests. *Prog. Biomed. Res.* **1998**, *3*, 147–151.

57 Witte, J., Reibis, R., Pichlmaier, A. M., Ebner, E., Malinowski, K., Rödiger, W., Niederlag, W., Wunderlich, E., van Woersem, R. J., Res, J. C. J., Merkely, B., Lucchese, F. A., Halperin, C., Greco, O. T., Beljaev, O. V., Vaskelyte, J., Schaldach, M. ANS-Controlled Rate-Adaptive Pacing – A Clinical Evaluation. *Eur. JCPE* **1996**, *6*, 53–59.

58 Steptoe, A., Vögele, C. Methodology of Mental Stress Testing in Cardiovascular Research. *Circulation* **1991**, *83*, II.14–II.24.

59 Malinowski, K. Interindividual Comparison of Different Sensor Principles for Rate Adaptive Pacing. *PACE* **1998**, *21*, 2209–2213.

60 Lau, C.-P. The Sensing of Body Movement and Acceleration Forces. In: Lau, C.-P. *Rate Adaptive Cardiac Pacing: Single and Dual Chamber.* Futura Publishing, Mount Kisco, NY, **1993**, 73–99.

10
Cochlear Implants

F. A. Spelman

10.1
Introduction

Cochlear implants are designed to help the deaf contact the auditory environment. The problem of sensorineural deafness is immense. The number of potential users has been estimated by some as 900000 in the USA alone [1]. Other estimates range as high as 2000000 sensorineural deaf in the USA. The preferred treatment for sensorineural deaf patients is the cochlear implant. The reason for that preference is that sensorineural deaf subjects cannot perceive sound without extraordinary aid. Acoustic hearing aids cannot help because the basic transduction system of the inner ear is no longer operative because either the cochlea or the auditory nerve is damaged [1].

This chapter introduces the reader to cochlear prostheses and some of the materials problems that must be faced by their designers. The concentration of the chapter is on cochlear implants rather than on implants in the cochlear nucleus, brainstem implants. The reason for this is based on use: more than 40000 people use cochlear implants, while less than 100 have brainstem implants [2]. At present, cochlear implants are in wide and rapidly increasing clinical use, while brainstem implants are still in the experimental stage of development. There is no description of tactile prostheses [3]. In this chapter, I provide a brief introduction to the physiology of the auditory system, a description of cochlear implants, their design and use, and finally to suggest some issues that implant designers must face in the future. The descriptions are introductory, but provide a framework for further study in depth.

10.2
The Auditory System

Fig. 10.1 shows the auditory system in a simplified form. Sound impinges on the external ear and is guided by way of the ear canal to the tympanic membrane (eardrum). The tympanic membrane vibrates with a relatively large displacement and low pressure. The ossicles of the middle ear act as an acoustic impedance

Sensors in Medicine and Health Care
Edited by P. Å. Öberg, T. Togawa, and F. A. Spelman

ISBN: 3-527-29556-9

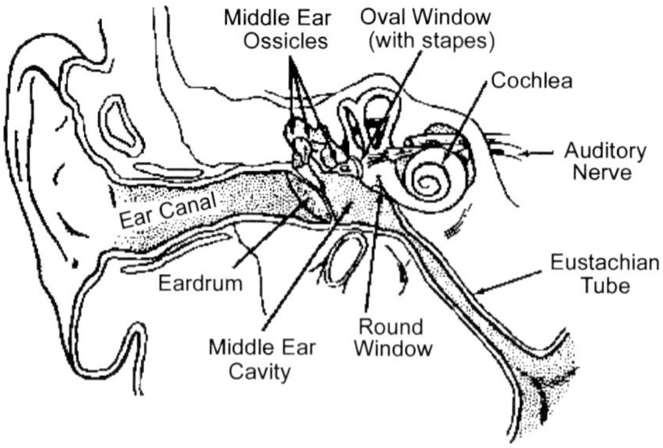

Fig. 10.1 Sketch of the peripheral auditory system, showing the anatomy of the external ear, ear canal, eardrum, middle ear, cochlea, and auditory nerve [1]. Reproduced with permission. © 1980 IEEE.

transformer to change the vibration to relatively small displacement and high pressure at the oval window. The cochlea, the spiral-shaped organ of the inner ear, contains the cells that convert mechanical motion into the electrochemical signals that are recognized by the nervous system [4, 5]. Several sites on the Web provide animations of the operations of the components of the auditory system. One such site may be found at http://www.neurophys.wisc.edu/animations. Other animations and data are maintained in a 'virtual library' that has been assembled by the Association for Research in Otolaryngology at its web site http://www.aro.org. Geisler refers to both sites in his work, *From Sound to Synapse* [5].

10.2.1 The Auditory Periphery

The cochlea is spiral shaped, with its broad base pointing medially and posteriorly, and with its pointed apex directed anteriorly and laterally (Fig. 10.1). Inside the cochlea are three chambers, the scala tympani, scala vestibuli, and scala media. The scala media is bounded by the basilar membrane and Reissner's membrane. The basilar membrane and the fluids of the three scalae form a traveling wave structure that responds to mechanical vibration such that high frequencies produce maximum excursions at the basal turn of the cochlea and low-frequency sounds produce maximum excursions at the apex of the cochlea [4].

The hair cells, one row of inner hair cells nearer the center of the cochlea and three rows of outer hair cells, comprise one part of the organ of Corti. It is the inner hair cells that convert the mechanical motion of the basilar membrane into electrochemical signals that are interpreted by the central nervous system as sound. When the cilia of the hair cells bend, the cells hyperpolarize and depolarize, sending neural transmitter substances to the peripheral processes of the auditory neurons during depolarization [4].

The mechanical properties of the basilar membrane combine with active micromechanical properties of the outer hair cells to provide tuning along its length. Hence, the apical inner hair cells are stimulated by low frequencies, while the basal cells are excited by high frequencies. The distribution of frequencies is approximately logarithmic along the basilar membrane. In the human, the auditory system responds to frequencies from 20 Hz to 20 kHz [4, 5].

The inner hair cells connect to the vast majority of more than 30000 afferent auditory neurons, and are thought to be the sensory cells of the system [6]. The outer hair cells connect to efferent auditory neurons and actively adjust the damping, and thus some of the fine tuning of the system. However the behavior of the outer hair cells is under active investigation and some debate at this time [4, 5].

10.2.2 The Central Auditory System

The tonotopic, that is, frequency-based, organization of the auditory system is preserved throughout the auditory system. Binaural signals travel from the cochlea to the cochlear nucleus, where the neurons send data to the contralateral trapezoid body, the superior olivary complex, the inferior colliculus, the medial geniculate nucleus, and thence to the auditory cortex [7]. The fastest and most direct path to the auditory cortex occurs for the obligatory synapses of the cochlear nucleus, inferior colliculus, medial geniculate nucleus, and the cortex.

10.2.3 Damage to the Auditory System

This discussion will be limited to damage produced in the auditory periphery, indeed, to damage that is found in the cochlea itself. Deafness can be the result of illness, traumatic injury, congenital abnormality, or the application of ototoxic drugs [8]. In all cases, either the hair cells are damaged or there is damage to the auditory neurons. In both cases, the auditory neurons can degenerate after loss of stimulation. However, the correlation between the etiology and the population of neurons in the cochlear nucleus is not significant [9]. While Hinojosa and Marion's finding is puzzling, it is reassuring since it suggests that sufficient cells can survive to make cochlear prostheses useful [9]. However, the function of the auditory system deteriorates after the onset of deafness, suggesting that degeneration of the central nervous system (CNS) takes place [10, 11]. In a consensus meeting at the NIH and devoted to the cochlear implant, the participants concluded that there was slight correlation between etiology and success with the cochlear implant, with the exception of the duration of deafness [10].

10.2.4 Neural Plasticity and the Implantation of Children

The clinical and research communities were initially cautious about implanting children, despite the long understanding that the auditory system responds to stimulation, exhibiting both structural and functional change in response to normal hearing [12]. The safety and efficacy of cochlear implants has increased their use in children. Deaf children are being implanted at the age of 18 months to 2 years [13–15], and in some cases 12 months [16]. The children who receive implants progress as well or better than do their peers with acoustic hearing aids [13, 14, 17].

10.3 Cochlear Implants

Cochlear implants have a lineage that extends from the time of Alessandro Volta. Indeed, Simmons discovered an account of an experiment performed by Volta who wondered whether electrical excitation could elicit an auditory response [18]. He connected a series stack of electrochemical cells to electrodes that he placed in each ear, and reported hearing '...a sound like a boiling viscid fluid.' He performed the experiment only once. The first modern account of the use of a cochlear implant in a human subject was reported by Djourno and Eyries, whose subject could perceive the rhythm of speech, but could not identify spoken words [19].

Cochlear implants have been used clinically since the 1960s [18, 20–24]. Both single-channel and multi-channel devices have been implanted. At this time, multi-channel devices are preferred because they provide patients with more information than do single-channel prostheses.

From the beginning of clinical implantation of cochlear prostheses, there have been reports of success with the devices. However, it is over the past 15 years that the cochlear implant has achieved startling clinical success. As of the time of this writing, more than 40 000 people use cochlear implants, with recipients ranging in age from 12 months to 80 years. The vast majority are able to converse normally with the aid of lip-reading; even those who cannot converse, can detect critical auditory cures like approaching vehicles, telephone rings, and fire alarms [25].

10.3.1 Block Diagram of a Cochlear Implant

Fig. 10.2 is a block diagram of a multichannel cochlear implant. While there are differences in the specific designs of the various implants that are available, the fundamental design makes use of the same components.

Auditory signals are transduced with a microphone that is usually placed behind the ear of the user. In most of today's implants, amplification and signal processing are done with electronics located in a body-worn processor, although behind the ear processors have been used. All manufacturers with extensive clinical experience now offer behind-the-ear (BTE) processors. Patients prefer the BTE processors to the larger, belt-worn units.

The amplifier may be followed by a signal compression block to mimic the logarithmic amplitude processing that is part of the normal auditory system. In most implants, however, compression is a part of the multichannel processor that

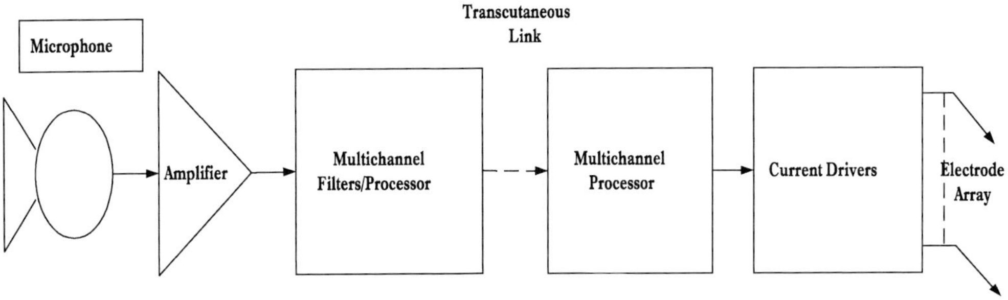

Fig. 10.2 Block diagram of a multichannel cochlear implant. The external components are to the left of the dotted arrow, while the implanted components are to the right of the arrow. The dotted arrow represents a transcutaneous signal link to send data from the external components to the internal components [26].

follows the amplifier. In that processor, signals are filtered, possibly compressed, and converted into a serial code that is transmitted across a transcutaneous link to an internal processor, implanted in the temporal bone of the patient. The internal processor converts the serially coded data into a stream of data that can be delivered to the array of current drivers that deliver pulses of controlled current to the electrodes of the implant's array. The number of electrodes that is implanted varies from 16 to 24. Two manufacturers use the transcutaneous data link to transfer data both to and from the implanted unit.

10.3.2 External and Internal Components

The cochlear implant is comprised of components that are both internal and external. The greater part of signal processing is accomplished with external components. Most processors employ digital signal processing to filter and compress the auditory data; the processors consume too much power to permit them to be placed inside the temporal bone of a patient. Batteries must be recharged or replaced daily. That need means that recharging will probably be required daily if the processors are reduced sufficiently in size to be implanted subcutaneously.

The transcutaneous links transfer power and data from the external to the internal processor. Power is transferred inductively, with external and internal coils held in proximity by internal and external magnets. Data may be transferred via inductive links. Inductive links are used to send data bidirectionally. For example, it is possible to record neural responses to stimuli or the electrical impedances of the stimulating electrode and to measure electrode impedances and potentials via back telemetry [27].

An earlier implant was produced with a percutaneous connector (Ineraid, Symbion, Utah). That implant is presently in use by several patients, but is no longer manufactured. A few implants have used percutaneous connectors for laboratory testing, but the connectors are temporary and later replaced with the standard inductive link. The connectors have been relatively free of trouble, but, because they may extend as much as a centimeter above the surface of the skin, are susceptible to mechanical damage and have been abandoned for that reason.

A new commercial firm, Advanced Cochlear Systems, is developing a transcutaneous optical link to provide the high-speed, bidirectional transfer of data between the external and internal processors. Power transfer will remain via electromagnetic links, but, as the data rates increase because of increased numbers of electrodes and a concomitant need for more information transfer, the optical link becomes a necessary component of the cochlear implant. The manufacturer reports being able to transfer more than 10 Mbits/s across the skin [28].

The internal processor parses the serial data code into signals that can be delivered via current drivers to the electrodes implanted in the scala tympani. The auditory signal can be encoded into controlled electrical current in a number of ways, ranging from compressed analog signals to pulsatile signals. In the former method, signals are filtered, compressed and delivered directly to the appropriate electrodes in the scala tympani (low-frequency signals to the apical electrodes and high-frequency signals to the basal electrodes). Other techniques employ pulsatile currents that are delivered to the electrodes in the scala tympani. Frequency is encoded by place (electrode location) and signal magnitude is encoded by varying the amplitude or pulse width of the signal. Pulse repetition frequency is most often a function of the rate at which the auditory signals are sampled and the rate at which electrodes can be driven, although high data rates have been shown recently to help the artificial system mimic the stochastic neural discharge behavior of the normal system [29, 30].

Time multiplexing is employed in today's implants because of the field interactions that exist between implanted electrodes. Those interactions have been observed experimentally [31] and in models of the implanted cochlea [32]. Two approaches to solve the problem are to bring the electrode array into apposition with the neurons of the cochlea, in the central (modiolar) wall, and to apply currents in such a way as to provide limited potential fields [26, 32, 33].

The electrode arrays are placed in the scala tympani of the cochlea. Ideally, the electrodes should be located as near to the cochlear neurons as is possible [33, 34]. Apposition to the neurons is beneficial for two reasons:

- thresholds of excitation for the neurons can be smaller, an effect that usually increases the dynamic range of stimulation;
- cross-talk between electrodes is reduced, since the fields produced by individual electrodes are more limited spatially when the electrodes are close to the neurons.

Thus, cochlear implant manufacturers are developing techniques to bring electrode arrays into apposition with the modiolar wall of the scala tympani, the wall that separates the scala from the cell bodies of the neurons.

A simple example problem illustrates the effect of distance on the potential field produced by a monopole, a single electrode referenced to a second contact at a distance. Consider a point source electrode placed at the origin of a Cartesian coordinate system. Let the electrode be placed on an insulator that is planar and that occupies the half space defined by $z<0$. The potential field that is produced in the space $z>0$ is

$$V(x, y, z) = \rho \frac{I}{2\pi\sqrt{x^2 + y^2 + z^2}}$$

where $V(x,y,z)$ is the potential found in the half space $z>0$; x, y, z are the spatial coordinates in the system, ρ is the resistivity of the conductive half space in Ωm. Consider a conductive medium (0.9% saline) filling the half space and whose resistivity is 0.7 Ωm. Current is 10 μA. If $y=0$ m and z has values of 20 μm and 100 μm, the half-amplitude width of the potential field taken along the x dimension is 70 μm for the nearer calculation and 352 μm for the further calculation. The peak potentials are 56 and 11 mV respectively for the two calculations (Fig. 10.3).

In summary, the designers of cochlear implants mimic the operation of the normal ear. The cochlear implant employs an input transducer to convert the acoustic signal into an electrical signal, filters to decompose the acoustic signal into data that can be processed by the central nervous system, and electrodes to deliver electrical signals to the neurons that remain in the inner ear.

Models have suggested several ways to limit the extent of the fields in the inner ear and, thus, the likely responses of auditory neurons [35–38]. The models range

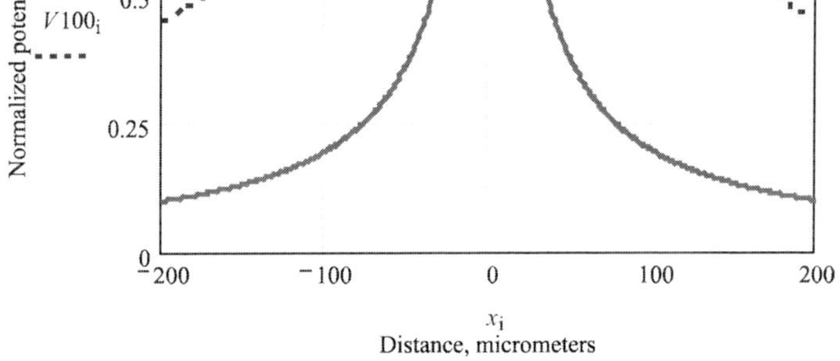

Fig. 10.3 Relative potential against distance for a monopole 20 and 100 μm from the plane of the monopole. V_{20i} is the potential at 20 μm from the plane, and V_{100i} is the potential at 100 μm from the plane. Both potentials are normalized to their maxima. Source current is 10 μA and the half-amplitude distance is smaller for the nearer than the further distance.

from simple electromagnetic models, like the one shown here, to Rubinstein's Fourier analytical model [38], to Frijns' detailed models that combine the boundary element method of analysis with neural modeling [32]. The models suggest several approaches to the design of cochlear implants.

First, electrode arrays must be placed near the cells that they will excite. This reduces the threshold of excitation of the cells as well as the extent of the fields in proximity to the excitable neurons. Second, if electrode arrays have large numbers of closely spaced electrodes, it is possible to drive several electrodes simultaneously to limit the spread of potentials in the inner ear [33, 36, 39, 40]. The use of quadrupolar excitation (called tripolar by Middlebrooks) produces limited neural responses as well as limited field spread [39, 41].

The quadrupole source is a special case of drive to a triad of electrodes. The central electrode carries current I, and the two equally spaced, adjacent electrodes each carry current $-I/2$. The quadrupole name comes from the coalescence of two dipoles, each driven with $\pm I/2$, merging the two positive current sources. Of course, a quadrupolar electrode configuration can carry either positive or negative current at its maximum location.

Models provide both qualitative and quantitative insight into the behavior of cochlear implants. The elegant models of Rubinstein and Frijns show the clear relationships between the relative positions of electrodes and neurons in simple geometries [38, 40] and in complex, realistic anatomy [32, 35, 42].

The models have suggested several experiments, one of the most exciting of which has been Middlebrooks and Bierer's assessment of the effects of driving combinations of electrodes to reduce the effects of potential spread [41]. Responses of cells in the auditory cortex were limited. The tonotopic extent of the responses decreased monotonically and significantly as the stimulus configuration changed from monopolar to bipolar to quadripolar (tripolar) [41]. Modeling had predicted that that would be the case [33, 40].

More recently, Frijns questioned the validity of using arrays that were designed to be perimodiolar from the base to the apex of the cochlea [32]. Frijns developed a three-dimensional, spiral model of the cochlea. He placed a putative Clarion Hi-Focus electrode array, modeling it both with and without its positioner.

He compared his results to those obtained for a putative guinea pig cochlea. He found that the model produced decreased thresholds of excitation and high spatial selectivity. His anatomically faithful model showed that the size and geometry of the apical turns of the cochlea were likely to make a perimodiolar electrode beneficial in the upper turns of the organ [32]. However, some of that benefit may be reduced by cross-talk occurring between turns in the apical region: specifically, dynamic range may be decreased [32].

10.3.3
The Principle of Operation of Some Implants

Cochlear implants are used by between 30 000 and 40 000 patients throughout the world. The three implants that are most often used are produced by Cochlear Pty.

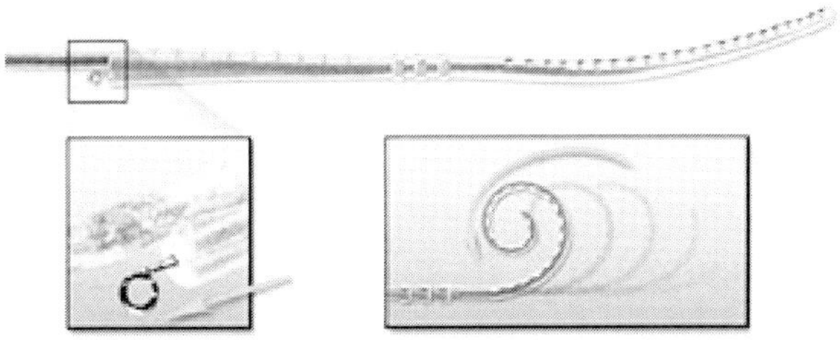

Fig. 10.4 The Cochlear Contour electrode array (see text). The array is pre-shaped to fit the cochlea and is straightened by a metal stilette that is withdrawn during implantation. From www.cochlear.com.

of Sydney, Australia, Advanced Bionics of Sylmar, CA, and Med-El of Innsbruck, Austria. This section outlines the strategies used by each manufacturer to deliver signals to the users of its implants.

10.3.3.1 The Nucleus Implant

Cochlear Pty., Limited has sold more than 33 000 implants at the time of writing. For up-to-date information on the Nucleus implant, visit www.cochlear.com. Those implants make use of either 22 or 24 electrodes. In both cases, 22 electrodes are placed in the scala tympani of the inner ear. The 24-electrode array uses two external electrodes as external reference electrodes. The Nucleus Contour electrode array is shown in Fig. 10.4. A discussion of the Contour array, its insertion and its benefits is given below. The external processor extracts frequency information from 20 external filters, sampling the auditory signal to determine which frequency bands contain the maximum energy (SPEAK [25, 43]). The envelopes of the signals taken from the filters whose energy is greatest are sampled and the data are distributed to electrodes within the array, driving between five and ten electrodes within a given sequence. The electrodes are driven with biphasic rectangular pulses sequenced rapidly in time. The auditory signal data are updated at rates that may be more than 1600 pulses per second per electrode [43].

The electrode array is slender, ranging in diameter from 0.4 mm at the apical end to 0.6 mm at the basal location. The array is composed of platinum rings, each ring is approximately 0.35 mm in width, and the edge-to-edge separation is about 0.4 mm. The insulated platinum/iridium wires that connect to each electrode are led back through the silicone rubber carrier of the electrodes [44]. The electrodes are inserted into the scala tympani via a small hole that is drilled through the temporal bone. In many cases, the electrode array resides near the lateral wall of the scala tympani when it is implanted in the ear of a human subject [45]. Recent reports of a new array describe a curved array, the Contour Array. It is curved to hug the modiolar wall of the scala tympani. A metal stilette can be inserted to straighten the array during insertion into the cochlea [46] (www.cochlear.com).

10.3.3.2 The Clarion

The Clarion cochlear implant is the commercial development of the Advanced Bionics Corporation (Sylmar, CA). For up-to-date information about the Clarion implant, visit www.cochlearimplant.com.

The Clarion Multi-Strategy implant offers several different strategies to aid its users [47]. Its processor can be adjusted to operate with several different forms of excitation. They include Simultaneous Analog Stimulation (SAS), the Paired Pulsatile Sampler (PPS) and the Continuous Interleaved Sampler (CIS) [48]. Referring to Fig. 10.1, the microphone and amplifier are followed by an analog/digital converter and by software that compresses the auditory signal, usually logarithmically. The compressed signal is decomposed into a set of digitally filtered signals. The digitized and filtered outputs are transmitted across the r.f. link to an internal processor, employing a 49 MHz AM signal [47]. The data are demodulated, demultiplexed and transmitted to current sources that drive electrodes directly with the compressed and filtered analog signals [47]. The sampling rate is 13000 samples per second per channel, with an aggregate sampling rate of 104000 samples per second. The system operates similarly to the Compressed Analog (CA) systems that have been used earlier [26, 49]. The SAS system drives several electrodes simultaneously, but cannot overcome the interference that occurs as a result of field interactions among electrodes.

Alternatively, the Clarion offers Continuous Interleaved Stimulation (CIS). Wilson and his colleagues originally developed CIS to overcome the field interactions between channels in the Ineraid implant [31]. CIS is a temporal multiplexing scheme that rapidly delivers pulsatile signals to individual electrodes in an electrode array. Fig. 10.5 illustrates the operation of the CIS processor and drivers. The acoustic signal is amplified and filtered. The filtered signals are rectified and low-pass filtered. With this processing technique, the envelope of each signal is detected. The envelope information is compressed and then used to modulate the amplitude of a biphasic rectangular pulse pair, while pulse width is constant. A specific pulse pair is delivered to a specific electrode. Pulses are updated 833 times per second in the Clarion 1.0 processor [50]. In the case of the Clarion implant, eight electrodes may be driven, although eight pairs of electrodes are provided via the Clarion electrode array.

Advanced Bionics allows the use of paired pulsatile stimulation (PPS) [47, 48]. PPS drives two electrodes simultaneously, choosing pairs of electrodes with a maximum physical separation between them. The physical separation avoids the interference between electrodes that is caused by field interactions. The pulses that are delivered are biphasic pulses, like those shown in Fig. 10.5. PPS has not been used as extensively as have SAS and CIS.

Osberger and Fisher reported that 71 patients who were given the Clarion Multi-Strategy implant improved hearing performance when they were allowed to use a preferred mode of stimulation. Interestingly, the patients who preferred the SAS mode had been deaf for shorter times than those who preferred CIS [51]. The preference may be the result of the better isolated electric fields that are produced by the HiFocus Array.

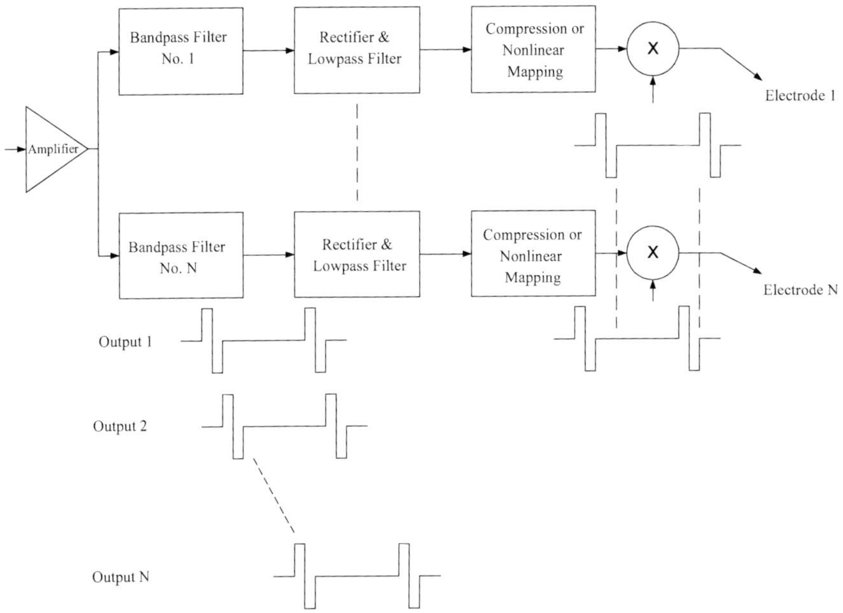

Fig. 10.5 The Continuous Interleaved Sampling technique developed by Wilson et al. The technique uses an envelope detector to amplitude modulate pulses that are multiplexed among the electrodes of the array. The technique has been successful in overcoming the electric field interactions among electrodes [26].

Like Cochlear Pty., Advanced Bionics offers a method to appose the electrodes of their array to the cells in Rosenthal's canal, behind the modiolar wall of the scala tympani. Their technique employs a 16 contact electrode array that is inserted into the scala tympani and then moved centrally to the modiolus by means of a polymeric die that is slid between the lateral wall and the electrode array (Fig. 10.6). There are reports of success with the technique in which speech comprehension is improved when patients receive the new array [52]. The HiFocus Array is discussed below.

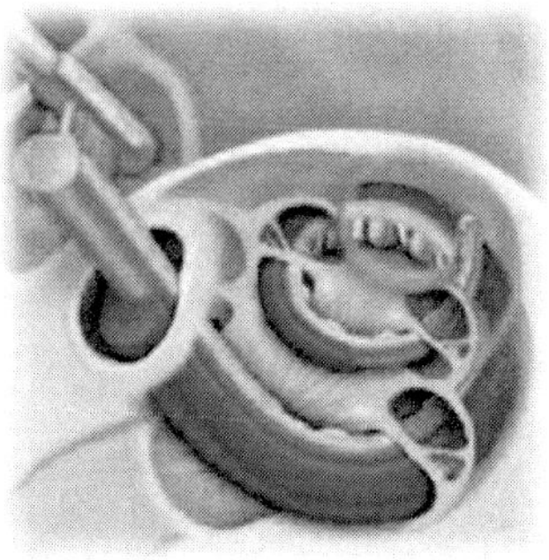

Fig. 10.6 The Advanced Bionics electrode array and apposition device. The array is light blue, and the device is dark blue, from www.cochlearimplant.com.

10.3.3.3 The Med-El Implant

Med-El Corporation (Innsbruck, Austria, www.medel.com) produces the Combi 40 and the Combi 40+ implants. Both employ a slim electrode array with 24 contacts, all which can be placed within the scala tympani of a human patient to a length of 31 mm, according to the data provided by Med-El. That places the array fully into the apical part of the cochlea, providing the opportunity to excite low-frequency neurons. Med-El has attempted to optimize their electrode array for monopolar stimulation.

Their primary signal processing method employs Continuous Interleaved Stimulation. At this time they report the highest multiplexing rate: 18000 pulses per second. For the 12-electrode arrays employed by the Combi 40, the data can be updated at a rate of 1500 samples per second. Med-El employs an alternative technique, a technique like that of Cochlear Pty., in which they choose the frequencies in the auditory spectrum at which energy is maximum, and then drive the electrodes that are matched to those frequencies in the ear of a given patient. The data rates are higher for the Med-El '*n* of *m*' strategy, which updates information at a rate of about 2000 pulses per channel [53]. The Combi 40 and 40+ series of implants are in use in Europe, but at this time are undergoing clinical trials in the USA.

Jolly et al. suggested a novel solution to the placement of electrode contacts near the modiolar wall of the cochlea [54]. The array, discussed below, employs a shape-memory wire in compression to force the contacts into proximity to the medial wall of the cochlea.

10.4 Arrays for Specific Cochlear Implants

10.4.1 Electrode Arrays

The cochlear electrode array is the transducer that permits the cochlear prosthesis to communicate with the remaining cells of the auditory nerve. Several issues are intimately related to the materials with which the array is made. How many contacts are optimal? Of what metal should the contacts be made? How can stimulating fields be focused on target neurons? How close can the electrodes be placed to the target neurons? If a sheath of tissue grows around the electrode array, what is the likely result? What problems can arise with the electrode arrays that are used?

10.4.1.1 Numbers of Contacts

There are about 31400 cochlear nerve fibers and between 25000 and 30000 cell bodies in the spiral ganglion [6]. The nerve fibers are organized into bundles or fascicles, which are roughly 200 μm wide and are spaced an equal distance apart

[55]. Thus, an attempt to bring thousands of electrodes to the nerves will be anatomically as well as technically futile.

The largest number of contacts placed in the scala tympani is 22, in the Contour Array of Cochlear Corporation. Are 22 contacts sufficient to provide speech information to implant recipients? Perhaps they are. Shannon and his coworkers suggest that only a few channels are necessary, perhaps as few as three or four [56]. Shannon decomposed speech signals into frequency bands, used the envelopes of the frequencies to amplitude modulate white noise bands of center frequencies and bandwidths that were equivalent to those of the bandpass filters. However, their studies were done in noise-free environments. When noise is added to acoustic information, more channels of acoustic data are required [57]. The answer seems to be more than three, less than 100, under ideal circumstances, that is, conditions in which the neurons of the auditory nerve are present in normal numbers.

However, two problems arise. First, neurons may not be present in large numbers [9]. Second, auditory neurons occur as bundles or fascicles in the inner ear. The fibers do not occur as neural sheets in the ear [55]. Third, the variable spacing of fascicles may mean that electrodes lie in the spaces between nerve bundles rather than directly near the bundles themselves, and that some contacts will be more effective than will others. Thus, having as many as 72 electrode contacts in an array (Fig. 10.7) may improve the operation of the cochlear prosthesis [58].

The development of high-density electrode arrays depends as much on materials issues as on design that can be placed in the scala tympani near the target neurons. Delays in the development of the wire-based MicroHelix electrode design have been caused primarily by failures in the insulating materials that are used to cover the array and the wires that comprise it. The array consists of layers of 25 μm diameter wires that are served around a central core. The wires have been individually insulated with seven-micron-thick polyester insulation, and covered with silicone rubber. Electrode sites are made by ablating the insulation at one location per wire with an excimer laser (Fig. 10.7). Cross-talk has been measured between the wires; the cross-talk can be reduced by hand-selecting the wires

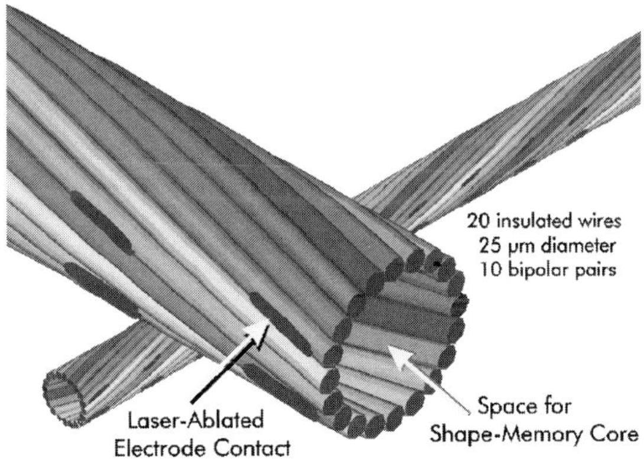

Fig. 10.7 Sketch of the array developed by PI Medical and the University of Washington. PI Medical is now called MicroHelix. This technology is the basis of the MicroHelix electrode array.

that are used for the array to ensure that they are pinhole free and that their insulating polymers adhere to the metals that are used as conductors in the array.

Electrode arrays have been conceived as monolithic circuits; at this time no monolithic arrays have been implanted in the inner ears of human subjects. The present electrode arrays are built semi-automatically and require substantial effort to achieve their successful completion. MicroHelix's wire bundle array offers an approach that lends itself to automatic construction. However, the array suffered from insulation failures when it was investigated by the author and his students. Later versions may be more reliable.

Other techniques, either based on the techniques of microcircuit manufacture or on the methods introduced by MicroHelix [58] are necessary to increase substantially the number of electrodes that can be inserted into the cochlea. Several manufacturers are rumored to be pursuing high-density array technology, but nothing has been reported publicly at this time.

10.4.1.2 **Contact Metals**

Cochlear electrode arrays employ platinum-iridium electrodes (Pt 90%, Ir 10%). The arrays are long-lived and safe, as long as safe limits of charge are delivered through the individual contacts. The safe limits were established in the late 1970s when Brummer and Turner performed landmark experiments on the charge-carrying capacity of Pt electrodes [59, 60]. To avoid critical electrode potentials, the electrodes were reduced from their maxima by a factor of four [61]. Tykocinski and coworkers used an electrochemical means to increase the surface area of platinum electrodes, something like the use of platinum black. The successfully implanted and used the arrays in the scalae tympani of cats, finding no evidence of damage from the use of their arrays [62].

Recent work has led to the consideration of oxides of iridium (Ir:O) for electrode arrays [61]. The oxides of iridium demonstrate markedly increased charge-carrying capacity [61, 63]. Recent work has shown that iridium oxide can be deposited electrochemically on the surfaces of Pt, Au, PtIr, and 316LVM stainless steel, as well as plated silicon substrates, and that the deposited surfaces can carry the equivalent charge densities to activated iridium surfaces [63]. That work holds great promise for the development of small, high-density electrode arrays, since the Ir:O surfaces can be applied to either wire bundle arrays or monolithic arrays, as the latter are developed.

McCreery and coworkers investigated the effects of driving activated iridium electrodes, producing an oxide covering over the surfaces of the microelectrodes. The microelectrodes had geometric surface areas ranging from 100 to 500 μm^2. The microelectrodes were implanted in the cochlear nuclei of cats. When the investigators drove currents whose charges ranged from 1.8–3.6 nC/phase, they found no apparent histological damage, although they reported an increased number of macrophages in the regions near the electrode sites [64]. The implication is that oxides of iridium can be used to drive neural tissues safely.

The safety of Ir:O and the recent demonstration of its ease of manufacture suggest its use for cochlear implants. However, before it earns approval by regulatory agencies, more tests will be required. A high-density electrode array must position more than 22 electrodes in the 30 mm length of the human cochlea. If the number of electrodes is doubled or tripled, the electrodes must be small. Hence, large charge densities will be the order of the day.

10.4.1.3 Focusing Fields

The present signal-processing techniques that are applied to cochlear implants often are designed to overcome the field interactions that occur in cochlear implants. Wilson's continuous interleaved stimulation strategy was developed for that purpose [31]. Can field interactions be reduced substantially? A number of studies suggest that it should be possible to drive the electrodes in a cochlear implant in a way that will limit the half-maximum potential fields to the spacing between electrodes in the array [33, 36, 37, 39]. Suesserman and Spelman suggested focusing as a result of modeling results [37] and Jolly et al. confirmed the modeling results in vivo, both as passive potential distributions in the cochlea [36] and as narrowed neural responses in the auditory nerve [39]. More recently, Middlebrooks and Bierer reported that they observed focusing of excitation in the auditory cortex when tripolar (quadrupolar) stimulation was used [41].

These studies offer promise for the use of focused fields to avoid the interactions among the fields that are produced by the simultaneous drive of current into stimulating electrode arrays. However, while there is evidence that neural interactions can be reduced [39, 41], there are no behavioral studies that demonstrate conclusively that focused fields reduce interactions. Middlebrooks' and Bierer's data suggest that field interactions can be reduced by using quadrupolar stimulation. That may not necessarily lead to improved speech comprehension [65], although theoretical understanding of hearing would suggest that it should.

10.4.1.4 Proximity to Neurons

Fig. 10.3 shows clearly that an electrode that is near its target cells can produce a more limited potential field than does a more distant source. Understanding that proximity pays has been understood since the times of early investigations of the cochlear implant [34]. Recently, designers of electrode arrays have developed several approaches to bring contacts near to the neurons that are found in the modiolar (medial) wall of the cochlea. Present arrays are designed to go further into the cochlea than did their predecessors, in order to be closer to the low-frequency neurons that lie in the apical turns of the organ and are necessary to encode speech.

Four techniques have been developed to bring the electrode arrays into contact with the modiolar wall of the cochlea. Each will be discussed in the following paragraphs.

The Advanced Bionics Array

Advanced Bionics uses a polymeric die, electrode positioning system (EPS), to force its array from the lateral wall to the medial wall of the scala tympani (Fig. 10.6). In addition to gaining apposition to the target neurons, the array increases the focus of its fields by means of polymeric separations between each of the electrodes in the array. Patients who use the HiFocus array have increased word comprehension and sentence comprehension when they are compared to patients who use the earlier designs of the same manufacturer [52]. The HiFocus array does not move as far into the cochlear spiral as did its predecessor, the Spiral [46].

The HiFocus array is designed to bring the electrode arrays closer to target neurons, and to reduce field spreading. Both goals are achieved in part by use of the EPS. The closer an electrode is to the location at which its potential field is measured, the narrower will be the spread of that field. Investigation of the simple field equation that was given earlier shows that as the distance in the y and z dimensions goes to zero, that the x variation of the field behaves like the Dirac delta function, $\delta(x)$. In the limit, the width of the potential field becomes infinitesimal. Of course, electrodes of finite dimension cannot produce fields that are infinitesimally wide. However, the limiting dimension of the field is minimized as the distance from the electrode is reduced.

The HiFocus array's field is reduced additionally by its physical layout. Each electrode in the array is separated from adjacent contacts by means of a silicone barrier. In combination with the EPS, the barriers that bound each electrode are placed in contact with the modiolar wall, effectively isolating one electrode from the others in the array. The contact cannot form a perfect barrier, however, and the modiolar bone is not a perfect insulator [66]. Some spread of the fields will result with this design, but it should be reduced.

The EPS achieves the goal of placing the HiFocus array in apposition with the medial wall of the scala tympani. However, it must do that at the expense of rubbing against the lateral wall of the scala. Tests performed in cadaveric temporal bones indicate that there is no more damage with the new array than with its predecessors [46, 67]. The tests have not included histology after long-term implantation. Does damage occur in the periosteum, for example? At the time when this chapter was written there was no evidence of such damage; in the summer of 2002 the HiFocus and EPS were withdrawn after evidence of damage was found (see the website of the US FDA). A new array is being used without the EPS.

Cochlear Corporation

Cochlear Corporation's Contour electrode array follows a different philosophy of design. The array is molded to match the shape of the cochlear spiral at the modiolar wall. When the array is located inside the scala tympani, it moves into apposition with the medial wall. The electrode carries electrodes that are band segments that do not encircle the carrier. Previous designs from this organization employed banded electrodes that circled around the carrier. The contacts of the

Contour array are arranged on the side of the array that contacts the medial wall. The Contour array can be inserted further into the scala tympani than can the banded array [46].

The Contour array has a central chamber into which a metal stilette can be inserted. The stilette is straight. When it is inserted into the array, the array straightens for insertion into the cochlear spiral. As the surgeon implants the array in the scala tympani, she can remove the stilette during insertion, allowing the electrodes to meet the modiolar wall. The spiral-shaped array contacts the modiolar wall, with the electrodes oriented appropriately close to the auditory neurons.

The Contour array has characteristics that provide more focused fields than did the banded array. First, its spiral shape brings it closer to the target neurons. Second, the electrode design, which is a partially banded design, limits the potential fields that are produced by the array. The banded array produces current density vectors that point in all directions around the longitudinal axis of the structure. The Contour array, on the other hand, produces current density vectors whose directions are limited by the circumferential extent of the contacts. Hence, its fields are more limited.

Those limited fields have been gained at the expense of a preferred direction of insertion. The banded geometry of the earlier array was not sensitive to rotational direction. The Contour array requires that it be inserted such that its contacts lie on the medial side of the device. That is a small price to pay for the improved performance that comes from proximity to the target neurons.

Med-El

Med-El has developed a perimodiolar electrode array that is based on their earlier electrode design [54]. The array is constructed with a small groove molded into its lateral side. A 100 μm diameter Nitinol wire is inserted into the lateral side of the array, allowed to exit a tubular housing and then lies within the groove. The electrode array is inserted into the scala tympani and the wire is held in place while the electrode array is gently retracted by the surgeon [54]. The process leaves the wire near the lateral wall while the electrode array is placed in apposition to the medial wall [54]. The placement of the array has been tested in fresh human temporal bones. It has not been tested in human subjects.

MicroHelix Array

The array design that is shown in Fig. 10.7 is designed to be placed in apposition with the modiolar wall of the cochlea. The array is designed as a spiral of wires that are served around a central core. The core is removed and replaced by a Nitinol wire.

Nitinol is an alloy of nickel and titanium that has shape memory. If the wire is wrapped around a mandrel with a shape that matches the cochlear spiral, it can be heated to a temperature of 700–800 degrees Fahrenheit and then cooled. It will hold the shape of the mandrel. If the wire is then straightened, it will remain straight until it is heated to a temperature that either exceeds or falls below a design 'transition temperature'. The choice of whether the Nitinol must exceed or fall below the transition temperature is a function of the composition of the alloy. For further information, contact Shape Memory Applications, Inc. or visit their web site at www.sma-inc.com [68].

In the case of the MicroHelix array, the shape memory wire core has a 30 µm groove machined on one side. An insulated gold wire whose diameter is 25 µm is placed in the groove and welded to the Nitinol wire at its apical end. If a current source is applied to the basal ends of the Nitinol and gold wires, the Nitinol heats from the electric power that is dissipated along its length. It curls as it heats, finally assuming the shape of the medial wall of the scala tympani [69]. This system has been tested *in vitro*, but has never been tried *in vivo*.

Four techniques have been employed to bring cochlear electrode arrays into apposition with the modiolar wall of the scala tympani. Two have been tested in human subjects and appear to work well.

10.4.1.5 The Effect of a Sheath

The foreign body responses of animals isolate artificial materials from the animal tissues. Cochlear implants are no exception. In animals, sheaths of connective tissue surround the electrode arrays that are implanted in the scala tympani [70]. The same phenomenon has been reported by surgeons to occur in human patients [71].

10.4.1.6 Problems

There are several undesirable effects of such sheaths. First, the sheaths are about 40 µm thick. They remove the electrodes from the excitable cells. That increases the amount of current necessary to excite the cells. Second, the sheaths have higher resistivities than the scalar fluids, by factors of three or more. Attenuation will occur as a result of having a sheath interposed between the electrodes and their target cells. Third, tests in our laboratory indicate that the sheaths are anisotropic, potentially directing current density vectors inappropriately (Fig. 10.8).

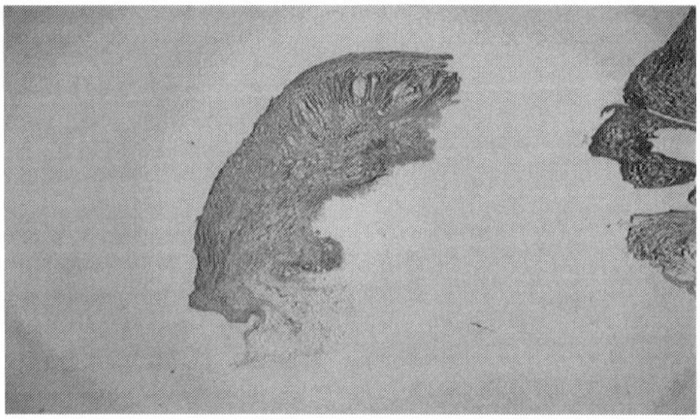

Fig. 10.8 Human connective tissue sheath taken during explant of cochlear electrode array. The tissue is stained for collagen. The fibers are in several different directions, producing anisotropic resistivity.

Foreign body reactions are undesirable. Designers of cochlear implants await developments from the biomaterials community to develop surface treatments that will reduce or eliminate the growth of connective tissue around the electrode arrays.

10.4.2 The Internal Processor

Fig. 10.2 outlines the subsystems of the cochlear prosthesis. The internal processor takes its power and data from the transcutaneous link, using both to assign drive currents to the contacts of the electrode array and to distribute the signals to each of the electrodes. Power and data are transmitted via an HF link. Patrick mentions using a carrier frequency of 2.5 MHz [44]; Advanced Bionics transmits data to its processor via an AM system whose carrier is at 49 MHz [47]. Advanced Bionics returns data from the implanted array via an FM link centered at 10.7 MHz [47]. The producers of cochlear implants have used percutaneous connectors [31], but they are not used clinically, but percutaneous connectors have been and are sometimes used for research studies in human subjects.

10.4.2.1 The Effects of Sampling and Data Rates

Auditory signals range from 20 Hz to 20 kHz for human hearing and from 300 to 3000 Hz for the communication of speech [5]. Following classical sampling theory, that suggests sample rates of more than 6000 samples/s for speech communications and at least 40 000 samples/s for the full human range of hearing.

The data that must be transferred to the internal processor include more than the stream of, for example, eight- or nine-bit samples of acoustic data. Data include information about the region of the acoustic world that is represented, that is, the outputs of filters that parse the signals into their component frequencies; directions to individual electrodes within the array, so that the signals are sent to the electrodes that are optimum for a specific sound token; time and order of excitation for techniques like CIS, SPEAK, or PPS; magnitudes of current to be mapped to individual electrodes in the array. Hence, there is a requirement for bandwidth in excess of that required by the acoustic signal alone.

10.4.2.2 Processing

A block diagram of a hypothetical internal processor is given in Fig. 10.9. The processor is shown in five blocks, all of which rely on radio-frequency signals for power. The r.f. signals provide both data and power to the circuit. The power system develops internal d.c. from its carrier signal. The three major manufacturers of cochlear implants employ this technique. Data rates can be substantial, with signal transfers at as much as 104 000 samples/s in the Advanced Bionics Clarion [47].

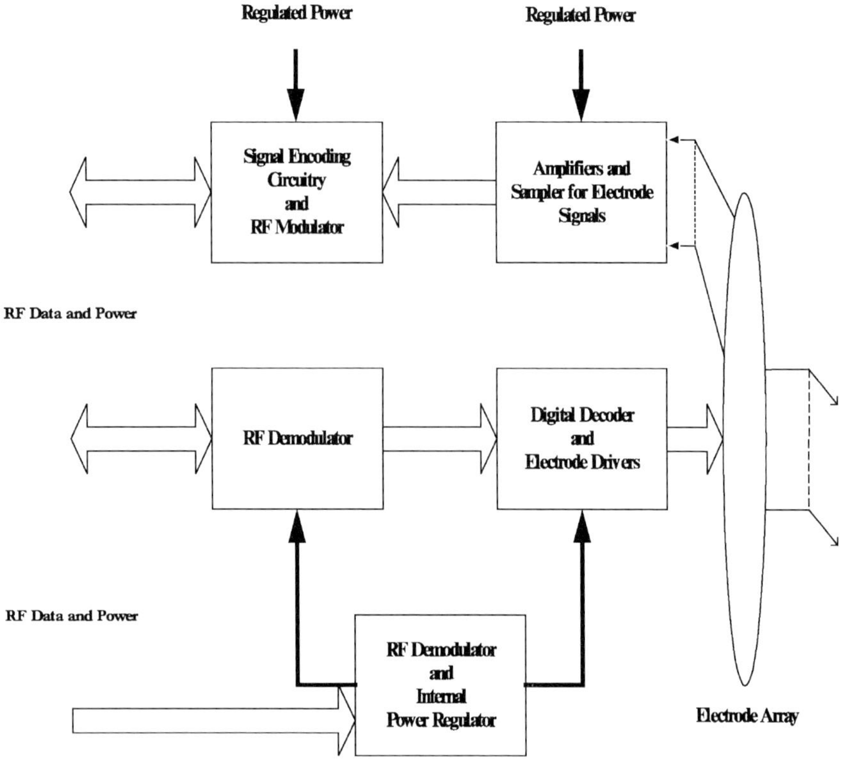

Fig. 10.9 Hypothetical block diagram for the internal processor of a cochlear implant. This processor receives and transmits its data and power from an rf link. The bold lines show power distribution; other arrows show data paths.

10.4.2.3 Phase Information

Present implants process auditory information by computing either envelope data from speech sounds [50]. The instantaneous phase of speech is lost in the data acquisition and coding methods that are used. Phase data are important, however, for the understanding of speech and music, or for the understanding of speech in noise, or for the understanding of music. If phase data are reversed by as little as 10–20 μs in a sequence of running speech, comprehension is reduced severely [72].

That anecdotal result is in interesting agreement with other psychophysical results that demonstrate that the threshold of detection of an interaural time difference is 6 μs [73]. The latter figure illustrates the capability of the normal system, a capability that is increasingly important for the location of sound sources in space. Thus, the processing of the fine temporal structure of auditory signals will likely become more important as binaural implants are used more often in the future. The encoding schemes used in today's implants ignore temporal fine structure.

10.4.2.4 **Packaging**

Cochlear implants are offered in two configurations at the time that this chapter is being written. Implants are preponderantly packaged as body-worn processors that are wired to head-mounted microphones and inductive coils. The microphones acquire acoustic data from the vicinity of the subject's ears, and the inductive coils provide power and encoded data to the internal processors that are implanted in the temporal bones of the users. Less common are external processors that are worn behind the ear (BTE processors). Valid technical reasons underlie this package design. The complex processing that is done by Cochlear Corporation and Advanced Bionics requires substantial circuitry and substantial battery power to allow it to run. They must sacrifice either processing power or battery life in their BTE processors. The simpler signal processing that is employed by Med-El requires less electronics and consumes less power. The Med-El processor is offered as a BTE device. The need for large batteries makes the BTE processor larger than that used in an acoustic hearing aid that is worn behind the ear. The Med-El processor is comparable in size to an acoustic aid.

The users of implants generally express delight over the BTE processor. When the ESPrit 3G (Cochlear Corporation) is used, users have demonstrated improved speech processing capabilities when they used it (www.cochlear.com). While the results are in a limited group ($n=37$), 70% of the users preferred the new processor to the belt-worn device, while doing as well or better in tests of speech comprehension.

Advanced Bionics has maintained the full capabilities of their new Platinum Processor in the BTE processor. However, they have sacrificed battery life in order to do so. Patients must change batteries more than twice daily if they use the Advanced Bionics BTE processor over a full period of wakefulness. Advanced Bionics offers several convenient battery options to solve that problem, however (www.bionicear.com).

The Med-El Combi 40+ BTE processor provides the full capabilities of the CIS system in a small package (www.medel.com). Med-El makes clear that its BTE is available to all users, while its competitor processors are limited to specific groups of users. The advantage is also a limitation: the Combi series of Med-El processors employ the CIS strategy alone. It remains to be seen whether that limitation will be serious or not.

The external processors that are used by all manufacturers employ conventional electronic packaging techniques. The three units are similar in their packaging strategies. All of the internal processors are implanted in the temporal bone of the skull. The processors must have minimal thickness when they are planted in the skull of a very young child, since the temporal bone is less than a centimeter thick in the case of the infant or toddler skull. The internal processors vary among the three in the materials that are used in the packaging. The Clarion processor is packaged in ceramic material and has dimensions of $3.7\times2.5\times0.6$ cm. The receiver coil is sealed in the ceramic case. The feedthrough connections for the electrode array are hermetic [47].

The internal processor of Cochlear Corporation is housed in a titanium case that is encapsulated in silicone elastomer [74]. The use of the combination has resulted in a small but significant increase in reliability (vide infra) [74]. The use of titanium requires that the receiving coil be placed externally to the case. Feedthroughs are ceramic. Dimensional data are not available, although the pictures of 10-year-old internal processors appear to have somewhat smaller dimensions than those of the Clarion unit [44].

10.4.2.5 Magnets

All of the implants that are produced today use rare-earth magnets to maintain the position of the coils that transfer data into and out of the internal processor. The technology is reliable; other alternatives have not been explored for systems that employ transcutaneous transmission of data and power.

10.4.2.6 Reverse Telemetry

Data are transferred out of the processor to provide a wealth of information to the clinicians who provide implants to patients and to researchers and clinicians who test the implants. It is possible to record data from both used and unused electrode sites on the arrays that are implanted in patients [27, 75, 76]. Abbas and his colleagues have developed several elegant techniques that enable them to acquire responses from the auditory neurons of cochlear implant users by employing the back telemetry features of the internal processor, and by eliminating stimulus artifact computationally [27].

Other investigators have measured the electric fields produced by cochlear implants. They drive the implant electrodes as monopoles and record the potentials induced at the sites of the unused electrodes. The data can be fitted to models of the electroanatomy of the implanted ear to provide insight into the operation of the device [76].

Both Cochlear and Advanced Bionics provide telemetry to monitor the state of the electrode array and the processors. The Clarion does not permit an investigator to monitor evoked neural potentials, but it is used for monitoring for the lifetime of the implant [47].

Reverse telemetry is provided by the Nucleus and Advanced Bionics implants; it is not presently a part of the Med-El Combi 40+.

The development of reverse telemetry is a boon to the developers and users of cochlear implants. At the simplest level, electrode impedances can be measured and recorded during the life of the implant, as can field potentials in the inner ear. During assessment and fitting, investigators can study the threshold and growth functions of auditory neurons under stimulation. Reverse telemetry has progressed from an interesting feature to a necessity in modern cochlear implants.

10.4.2.7 Safety and Reliability

Cochlear implants have been used clinically for more than two decades. Part of that acceptance is the result of the excellent safety record that has been compiled for the devices. The cochlear implant is a remarkably reliable device, and has been for at least a decade. Indeed, its cumulative survival percentage ranged from 94.9% (12 year) to 99.2% (1 year) in more than 16427 users of the Cochlear CI22 implant [74].

Pfingst comments on the low levels of injury that accompany cochlear implants over the long term, with injuries confined largely to minor disturbances of the skin [25]. He suggests that two concerns of electrical stimulation could be the result of high rates of stimulation that can affect the threshold responses of neurons and the potential for residual accumulation of charge on electrodes [25]. In the former case, animal studies must be continued to determine the effects of high rates of stimulation (>2 kHz) on neurons, and care in design must be exerted to guarantee that charge accumulation is negligible.

Lehnhardt and coworkers reported on the reliability of Nucleus CI22 and CI124 implants [74]. They computed cumulative survival percentages over several years and thousands of implants. Their results over 4 years of implantation of the CI22, the older implant of the two used, was 99% in adults and 98.8% in children ($n=8145$). The newer device had a one year cumulative survival of 99.2% in both adults and children ($n=1650$) [74].

Failures cited were due to damage to the receiver antenna or to the electrode array, and damage to internal components from electrostatic discharge. The authors do not cite the number of failures produced by each of the modes mentioned above [74]. They do cite an improvement in reliability that resulted from modifications of the design of their internal processor. The use of titanium encapsulated in silicone elastomer improved the 4 year cumulative survival from 97.1% with the earlier design to 99% in adults and 98.8% in children when the new technique was used [74].

10.4.3 Cost

The cost of a cochlear implant is about $30000 (US) [77]. In a careful study of cost-utility of the multichannel cochlear implant, Palmer and coworkers assessed the quality-adjusted life year of the multichannel implant, and, assuming a 22 year life expectancy, computed a marginal cost per quality-adjusted life year [77]. The marginal cost of $14670 for cochlear implants compares favorably with marginal costs of $21964 for severe hypertension; $63143 for estrogen therapy for post-menopausal women; and $83688 for coronary artery bypass graft for single-vessel disease with moderate to severe angina [77]. The cochlear implant appears to be competitive in cost compared to other treatments.

10.5 Directions for the Future

Cochlear implants are a major bioengineering success story. Their spectacular clinical acceptance is likely the result of the logical approach that has been taken to produce them. Pfingst cites four prominent reasons for their clinical attainment [25]:

- a foundation in strong basic science,
- a keen desire to match technology to deaf patients,
- government support and active technology transfer, and
- redundancy in the nature of acoustic speech signals.

The first three reasons cited above form the definition of bioengineering research and development. Engineering product design moves from understanding developed in basic research through strong marketing and astute planning, through a combined technological and financial base.

While present implants are accepted enthusiastically, there is room for substantial improvement. Patients recognize words at a level of about 40% correct, while they do much better when they have the contextual cues that come with sentences (www.cochlear.com). Part of the limits of recognition may be a function of the mode of stimulation, imposed by interference between and among electrodes in the array and of the cochlear anatomy [32, 78].

The directions that will be taken by future cochlear implants will probably be directed by the need for independent channels and the necessity to guarantee perimodiolar placement of cochlear electrode arrays.

10.5.1 The Need for Independent Channels

Information transfer from the auditory environment is limited severely by the limits of electrode arrays. Present-day arrays allow designers to drive one or two electrodes at a time. The limitations are imposed by the spread of fields within the cochlea. The auditory neurons are excited at several locations simultaneously in the normal ear [5]. Modeling and experimental testing suggest that by increasing the number of electrode sites on an array and driving them as multipoles, that field spread could be reduced and aggregate data rates increased [33, 36, 37, 41]. Future electrode arrays are likely to have at least twice the maximum number of electrode sites that are used today.

If the electrode sites are driven simultaneously to produce shaped fields, then independent channels will be generated by that field shaping. That could allow the simultaneous and independent excitation of groups of neurons within the auditory nerve.

10.5.2
Perimodiolar Location

Moving the cochlear electrode array nearer to the auditory neurons decreases the threshold currents necessary to excite cells. It limits the breadth of potential fields in the cochlea as well. Results in animals and today's patients support that assertion [79, 80]. Both Cochlear Corporation and Advanced Bionics are producing perimodiolar arrays; Med-El is developing such a device [54].

Future cochlear electrode arrays will be perimodiolar arrays, although the approach to perimodiolar placement is likely to be different from the methods employed today. The benefits of placing electrodes near to excitable cells far outweigh the difficulties. The issues raised by Frijns in his concern about cross-talk in the apical turns of the implanted cochlea [32] may well be overcome by using multi-electrode drives.

10.5.3
Tissue Growth

At present, no one has addressed a means to reduce or eliminate the growth of connective tissue around electrode arrays. Concerns about potential increases in threshold and decreases in spatial selectivity argue that tissue growth should be reduced or eliminated. Developments in biomaterials or in biological coatings for electrode sites and electrode arrays may permit the designers of cochlear implants to produce arrays that have negligible tissue growth.

10.5.4
Power Consumption

The improvements proposed above can provide potential benefits, but also present challenges to the designers of cochlear implants. The greatest obstacle faced when electrode numbers increase and are placed close to the modiolar wall is the increase in power consumption that accompanies the design changes.

If electrodes are driven as multiples, power consumption must increase. Either bipolar or quadrupolar configurations consume twice the power of monopolar drives. Jolly's pseudo-quadrupolar drive does not [36]. Quadrupole (tripole) drives may require more power to achieve threshold. That is a consequence of a reduction in peak potential by a multi-polar array when it is compared to a monopole [33].

If the thresholds are constant across an array of independent channels, the simultaneous drive of multiple channels increases linearly with the number of channels driven. Thus, improved speech perception may be purchased at the cost of increased power consumption and decreased battery life.

If electrode numbers are increased, the surface areas of the electrodes must decrease. That will increase electrode impedance [63], in turn increasing power consumption if the driven current remains the same. When electrodes are placed near to the bony modiolar wall, electrode impedance increases [79]. The likely cause of the increase is the larger resistivity of the bony tissues compared to that of the fluids of the scala tympani [78]. The closer the electrodes of an array are to bony tissue, the greater will be the magnitude of their impedance, since impedance is proportional to the resistivity of the tissue in proximity to the electrode site.

Hence, perimodiolar electrode arrays will likely have larger electrode impedance and will consume more power than arrays that are further from bony tissues. Arrays with more contacts and perimodiolar placement will produce independent channels, but probably at a substantial increase in power consumption.

The complexity of signal processing in both the internal and external processor must increase as the number of channels increases. Increased complexity is likely to increase the power consumed by both the external and internal processors.

Of course that increase in power requirements will impose several design problems: the development of small, high-capacity batteries; a means to handle the likely increase in heat within the drive circuitry; a potential increase in the size and weight of both the internal processor and its battery.

10.5.5 Binaural Implants

Cochlear Corporation has implanted some patients with bipolar prostheses. The results are preliminary; it is probable that binaural implants will be used in the future. It may be that some patients will have a cochlear implant in one ear and a hearing aid in the other. Both approaches are likely to provide benefit, and are likely to be offered.

10.6 Conclusions

Cochlear implants are a marvelous success that has been provided by teams of bioengineers, physiologists, psychophysicists, audiologists, and physicians. The best implants come from the collaborations inherent in interdisciplinary teams. The successful cochlear implants still have limitations. They provide an imperfect contact with the auditory environment. Substantial improvements in the behavior of cochlear implants may well come from high-density, perimodiolar arrays, used in implants with processors that are optimized to take advantage of independent channels. The development of new electrodes and processors will be a challenge, as will the development of power sources to drive the new devices. One of the exciting achievements of bioengineering provides interesting challenges for the future.

10.7 Acknowledgments

The author acknowledges the careful review of the manuscript by Ben Clopton. Further, he deeply appreciates the ideas and interesting discussions provided by Ben Clopton and Scott Corbett, as well as the work done by Sally Jandrall to compile statistics about the implants that are currently in clinical use, and to present those statistics concisely and cogently.

10.8 References

1 Levitt, H. and Nye, P.W., Sensory training aids for the hearing impaired (prologue), in *Sensory Aids for the Hearing Impaired*, H. Levitt, J.M. Pickett, and R.A. Houde, Editors. IEEE Press, NY **1980**, 3–28.

2 Brackmann, D.E., et al., Auditory brainstem implant: I. Issues in surgical implantation. *Otolaryngol. Head Neck Surg.* **1993**, *108*, 624–633.

3 Martin, M.C., Alternatives to cochlear implants, in *Cochlear Implants*, R.A. Schindler and M.M. Merzenich, Editors. Raven, NY **1985**.

4 Dallos, P., Popper, A.N., Fay, R.R., Editors. *The Cochlea*, 1st edn. Springer Handbook of Auditory Research, Vol. 8. Springer, NY **1996**, 551.

5 Geisler, C.D., *From Sound to Synapse: Physiology of the Mammalian Ear*, 1st edn. Oxford University Press, NY **1998**, 381.

6 Ryugo, D.K., The auditory nerve: peripheral innervation, cell body morphology and central projections, in *The Mammalian Auditory Pathway: Neuroanatomy*, D.B. Webster, A.N. Popper, and R.R. Fay, Editors. Springer, NY **1992**, 23–66.

7 Rubel, E.W. and Dobie, R.A., The auditory system: central auditory pathways, in *Textbook of Physiology: Excitable Cells and Neurophysiology*, H.D. Patton, et al., Editors. Saunders, Philadelphia **1989**, 386–411.

8 Pyman, B.C., et al., Preoperative evaluation and selection of adults, in *Cochlear Prostheses*, G.M. Clark, Y.C. Tong, and J.F. Patrick, Editors. Churchill Livingstone, Melbourne **1990**, 125–135.

9 Hinojosa, R. and Marion, M., Histopathology of profound sensorineural deafness. *Ann. NY Acad. Sci.* **1983**, *405*, 458–484.

10 NIH, *NIH Consensus Statement: cochlear Implants in Adults and Children*. National Institutes of Health, Bethesda, MD **1995**.

11 Battmer, R.D., Reid, J.M. and Lenarz, T., Performance in quiet and in noise with the Nucleus Spectra 22 and the Clarion CIS/CA cochlear implant devices. *Scand. Audiol.* **1997**, *26*, 240–246.

12 Brugge, J.F., An overview of central auditory processing, in *The Mammalian Auditory Pathway: Neurophysiology*, A.N. Popper and R.R. Fay, Editors. Springer, NY **1991**, 1–34.

13 Anonymous, Clarion pediatric results, in *Clarion Promotional Brochure*, Advanced Bionics, Inc., Sylmar, CA, USA, **1997**.

14 Clark, G.M., Electrical stimulation of the auditory nerve: the coding of frequency, the perception of pitch, and the development of speech processing strategies for profoundly deaf people. *Clin. Exp. Pharmacol. Physiol.* **1996**, *23*, 766–776.

15 Meklenburg, D.J., et al., Auditory (re)habilitation for implanted deaf children and teenagers, in *Cochlear Prostheses*, G.M. Clark, Y.C. Tong, and J.F. Patrick, Editors. Churchill Livingstone, Melbourne **1990**, 207–221.

16 Skinner, M.W., in *Conf. Implantable Auditory Prostheses*, Pacific Grove, CA **2001**.

17 Osberger, M.J., Current issues in cochlear implants in children. *Hearing Rev.* **1997**, 4(October), 28–31.

18 SIMMONS, F.B., Electrical stimulation of the auditory nerve in man. *Arch. Otolaryngol.* **1966**, *84*(July), 24–76.

19 DJOURNO, A. and EYRIES, C. Prothese autitive par excitation electrique a distance du nerf sensoriel a l'aide d'un bobinage inclus a demcure. *Presse Med.* **1957**, *35*, 14–17.

20 HOUSE, W.F., BERLINER, K.I., Cochlear implants: from idea to clinical practice, in *Cochlear Implants: A Practical Guide*, H. COOPER, Editor. Singular Publishing, San Diego, CA **1991**, 9–33.

21 CLARK, G.M., et al., A multiple-electrode hearing prosthesis for cochlear implantation in deaf patients. *Med. Prog. Technol.* **1977**, *5*, 127–140.

22 HOCHMAIR-DESOYER, I.J., et al., Four years of experience with cochlear prostheses. *Med. Progr. Technol.* **1981**, *8*, 107–119.

23 HOCHMAIR, E.S., HOCHMAIR-DESOYER, I.J., and BURIAN, K. Experience with implanted auditory nerve stimulator. *Trans. ASAIO* **1979**, *XXV*, 357–360.

24 MERZENICH, M.M., et al., Neural encoding of sound sensation evoked by electrical stimulation of the acoustic nerve. *Ann. Otol. Rhinol. Laryngol.* **1973**, *82*, 486–504.

25 PFINGST, B.E., Auditory prostheses, in *Neural Prostheses for Restoration of Sensory and Motor Function*, J.K. CHAPIN and K.A. MOXON, Editors. CRC, Boca Raton, FL **2001**, 296.

26 SPELMAN, F.A., The past, present and future of cochlear prostheses. *IEEE Eng. Med. Biol.* **1999**, *18*(3), 27–33.

27 ABBAS, P.J., et al., Electrically evoked compound action potentials recorded from subjects who use the Nucleus CI24M device. *Ann. Otol. Rhinol. Laryngol.* **2000**, *185*(12), 6–9.

28 Anonymous, *Advanced Cochlear Systems Web Site*, www.advcoch.com, Snoqualmie, WA **2001**.

29 RUBINSTEIN, J.T., et al. Stochastic resonance: can it be exploited by speech processors? in *Conf. Implantable Auditory Prostheses*, Pacific Grove, CA **1997**.

30 BRUCE, I.C., et al., A stochastic model of the electrically stimulated auditory nerve: single-pulse response. *IEEE Trans. Biomed. Eng.* **1999**, *46*, 617–629.

31 WILSON, B.S., et al., Better speech recognition with cochlear implants. *Nature* **1991**, *352*(July 18), 236–238.

32 FRIJNS, J.H.M., BRIAIRE, J.J. and GROTE, J.J., The importance of human cochlear anatomy for the results of modiolus-hugging multichannel cochlear implants. *Otol. Neurol.* **2001**, *22*, 340–349.

33 SPELMAN, F.A., et al., The effects of electrode configuration on potential fields in the electrically-stimulated cochlea: models and measurements. *Ann. Otol. Rhinol. Laryngol.* **1995**, *104*(Suppl. 166), 131–136.

34 MERZENICH, M.M. and WHITE, M.W., Cochlear implants. The interface problem, in *Functional Electrical Stimulation: Applications in Neural Prostheses*, F.T. HAMBRECHT and J.B. RESWICK, Editors. Marcel Dekker, NY **1977**, 321–340.

35 FRIJNS, J.H.M., DE SNOO, S.L. and TEN KATE, J.H., Spatial selectivity in a rotationally symmetric model of the electrically stimulated cochlea. *Hearing Res.* **1996**, *95*, 33–48.

36 JOLLY, C.N., SPELMAN, F.A. and CLOPTON, B.M., Quadrupolar stimulation for cochlear prostheses: Modeling and experimental data. *IEEE Trans. Biomed. Eng.* **1996**, *43*, 857–865.

37 SUESSERMAN, M.F. and SPELMAN, F.A., Lumped-parameter model for in vivo cochlear stimulation. *IEEE Trans. Biomed. Eng.* **1993**, *40*, 234–235.

38 RUBINSTEIN, J.T. and SPELMAN, F.A., An analytical model for extracellular stimulation of nerve: I, Passive unmyelinated fiber. *Biophys. J.* **1988**, *54*, 975–981.

39 JOLLY, C.N., et al., Guinea pig auditory nerve response triggered by a high-density electrode array. *Med. Prog. Technol.* **1997**, *21*(Suppl.), 13–23.

40 CLOPTON, B.M., et al., Electrode configuration and spread excitation: Compartmental models of spiral ganglion cells. *Ann. Otol. Rhinol. Laryngol.* **1995**, *104*(Suppl. 166), 115–117.

41 MIDDLEBROOKS, J.C. and BIERER, J.A., Cortical images of cochlear implant stimuli, in *Proc. Conf. Implantable Cochlear Prostheses*, Pacific Grove, CA **2001**.

42 Frijns, J. H. M., de Snoo, S. L. and Schoonhoven, R., Potential distributions and neural excitation patterns in a rotationally symmetric model of the electrically stimulated cochlea. *Hearing Res.* **1995**, *87*, 170–186.

43 Vandali, A. E., et al., Speech perception as a function of electrical stimulation rate: Using the Nucleus 24 cochlear implant system. *Ear Hearing* **2000**, *21*, 608–624.

44 Patrick, J. F., et al., Engineering, in *Cochlear Prostheses*, G. M. Clark, Y. C. Tong, and J. F. Patrick, Editors. Churchill Livingstone, Edinburgh **1990**, 99–125.

45 Skinner, M. W., et al., Determination of the position of Nucleus cochlear implant electrodes in the inner ear. *Am. J. Otol.* **1994**, *15*, 644–651.

46 Rebscher, S. J., et al. Insertion trauma, mechanical performance and optimum dimensions: refining a second generation of cochlear implant electrodes, in *Proc. Conf. Implantable Auditory Prostheses*, Pacific Grove, CA **2001**.

47 Kessler, D. K., The CLARION Multi-Strategy Cochlear Implant. *Ann. Otol. Rhinol. Laryngol.* **1999**, *177*(Suppl.), 8–16.

48 Zimmerman-Phillips, S. and Murad, C., Programming features of the CLARION Multi-Strategy Cochlear Implant. *Ann. Otol. Rhinol. Laryngol.* **1999**, *177*(Suppl.), 17–21.

49 Eddington, D. K., Speech discrimination in deaf subjects with cochlear implants. *J. Acoust. Soc. Am.* **1980**, *68*, 885–891.

50 Loizou, P. C., Signal-processing techniques for cochlear implants. *IEEE Eng. Med. Biol.* **1999**, *18*(3), 34–46.

51 Osberger, M. J. and Fisher, L., SAS-CIS preference study in postlingually deafened adults implanted with the Clarion cochlear implant. *Ann. Otol. Rhinol. Laryngol.* **1999**, *108*(4), 74–78.

52 Koch, D. B. and Osberger, M. J., in *Proc. Conf. Implantable Auditory Prostheses*, Pacific Grove, CA **2001**.

53 Anonymous, *www.medel.com* **1999**, Med-El, Innsbruck, Austria.

54 Jolly, C. N., et al., Principles and outcome in perimodiolar positioning. *Ann. Otol. Rhinol. Laryngol.* **2000**, *185*(Suppl.), 20–23.

55 Spelman, F. A. and Voie, A. H., Fascicles of the auditory nerve in the human cochlea: measurements in the region between the spiral ganglion and the osseous spiral lamina, in *Proc. 19th Ann. Midwinter Meeting of the Association for Research in Otolaryngology*, St. Petersburg, FL **1996**.

56 Shannon, R. V., et al., Speech recognition with primarily temporal cues. *Science* **1995**, *270*(13 October), 303–304.

57 Rubinstein, J. T. and Miller, C. A., How do cochlear prostheses work? *Curr. Opin. Neurophysiol.* **1999**, *9*, 399–404.

58 Corbett, S. S., et al., High density cochlear implant and method of manufacturing the same. US Patent no. 5630839, **1997**.

59 Brummer, S. B. and Turner, M. J., Electrical stimulation with Pt electrodes II: estimation of maximum surface redox (theoretical non-gassing) limits. *IEEE Trans. Biomed. Eng.* **1977**, *24*, 440–444.

60 Brummer, S. B. and Turner, M. J., Electrical stimulation with Pt electrodes I: a method for determination of real electrode areas. *IEEE Trans. Biomed. Eng.* **1977**, *24*, 436–440.

61 Robblee, L. S. and Rose, T. L., Electrochemical guidelines for selection of protocols and electrode materials for neural stimulation, in *Neural Prostheses: Fundamental Studies*, W. F. Agnew and D. B. McCreery, Editors. Prentice Hall, Englewood Cliffs, NJ **1990**, 25–66.

62 Tykocinski, M., et al., Chronic electrical stimulation of the auditory nerve using high surface area (HiQ) platinum electrodes. *Hearing Res.* **2001**, *159*, 53–68.

63 Meyer, R. D., et al., Electrodeposited iridium oxide for neural stimulation and recording electrodes. *IEEE Trans. Neural Sys. Rehab. Eng.* **2001**, *9*, 2–11.

64 McCreery, D. B., et al., Stimulation with chronically implanted microelectrodes in the cochlear nucleus of the cat: histologic and physiologic effects. *Hearing Res.* **1992**, *62*(1), 42–56.

65 Pfingst, B. E., Zwolan, T. A. and Holloway, L. A., Effects of stimulus configuration on psychophysical operating levels and on speech recognition with cochlear

implants. *Hearing Res.* **1997**, *112* (October), 247–260.

66 Suesserman, M.F. and Spelman, F.A., Quantitative, in vivo measurements of inner ear tissue resistivities I. In vitro characterization. *IEEE Trans. Biomed. Eng.* **1993**, *40*, 337–343.

67 Rebscher, S.J., et al., Strategies to improve electrode positioning and safety in cochlear implants. *IEEE Trans. Biomed. Eng.* **1999**, *46*(3), 340–352.

68 Anonymous, *Shape Memory Applications, Inc.*, **2001**.

69 Spelman, F.A., et al., *Cochlear Implant with Shape Memory Material and Method for Implanting the Same.* MicroHelix, USA **1998**.

70 Leake, P.A., Kessler, D.K. and Merzenich, M.M., Application and Safety of Cochlear Prostheses, in *Neural Prostheses: Fundamental Studies*, W.F. Agnew and D.B. McCreery, Editors. Prentice-Hall, Englewood Cliffs, NJ **1990**, 253–296.

71 Gates, G.A., *Connective tissue surrounding cochlear implants*, personal communication, **2000**.

72 Clopton, B. and Lineaweaver, S., *The importance of phase information for the encoding of speech*, Internal Report, Advanced Cochlear Systems, Snoqualmie, WA, USA, **2001**.

73 Irvine, D.R.F., Physiology of the Auditory Brainstem, in *The Mammalian Auditory Pathway: Neurophysiology*, A.N. Popper and R.R. Fay, Editors. Springer, NY **1992**, 153–231.

74 Lehnhardt, M., von Wallenberg, E. and Brinch, J.M., Reliability of the Nucleus CI22 and C124M Cochlear Implants. *Ann. Otol. Rhinol. Laryngol.* **2000**, *185*(Suppl.), 14–16.

75 Zilberman, Y. and Santogrossi, T., Back-telemetry and the Clarion cochlear prosthesis. *Ann. Otol. Rhinol. Laryngol.* **1995**, *166*(Suppl.), 146–147.

76 Faltys, M., et al. Electric field imaging methodology, in *Proc. Conf. Implantable Auditory Prostheses*, Pacific Grove, CA **2001**.

77 Palmer, C.S., et al., A prospective study of the cost-utility of the multichannel cochlear implant. *Arch. Otolaryngol. Head Neck Surg.* **1999**, *125*, 1221–1228.

78 Spelman, F.A. Measurements of the resistivity of bony tissues of the cochlea, in *Proc. 9th Ann. Conf. IEEE/EMBS*, Boston, MA **1987**, IEEE Press.

79 Watts, S.J., Lindsen, P., and Hawthorne, M., Medialization of electrode array in cochlear implantation. *J. Laryngol. Otol.* **2000**, *114*(10), 746–749.

80 Lenarz, T. Channel interactions with different electrode arrays: Results from animal studies and psychophysics and relation to possible damage, in *Proc. Conf. Implantable Auditory Prostheses*, Pacific Grove, CA **2001**.

11
Sensors for Catheter Applications

P. J. French, D. Tanase, and J. F. L. Goosen

11.1
Introduction

Since the beginnings of medicine, sensors have been used for diagnostics. In many cases this has been simply the senses of the doctor or surgeon, but also simple aids to give more precise information on the condition of the patient have been employed. A simple example is the thermometer for diagnosing a fever. In this case the doctor can easily estimate the fever without an aid, but the thermometer gives a more precise measurement and also makes monitoring over a period easier. In the operating theatre sensors are playing an ever-increasing role. In particular, the use of minimally invasive surgery has deprived the surgeon of using his or her own sensors and therefore micro-sensors have to be employed.

Sensors used in medicine can be divided into two main groups, external and internal, and both of these can be further subdivided as listed below:

- External
  - *In situ:* where the measurements are made directly on the patient
  - *In vitro:* analysis where a sample is taken and measurement is made in the laboratory.
- Internal
  - Short term: often on catheters, where measurements are made over a period of less than a few hours
  - Medium term: this can be for post-operative control where the patient is monitored for up to several months
  - Long term: the ultimate test for biocompatibility; in this case the sensor must operate for a number of years in, for example, pacemakers.

For each of these applications there are demands on the sensor in terms of sterility, biocompatibility, reliability/stability and packaging. Demand for sterility implies the use of disposables or the application of a re-sterilization technique. The use of disposables is limited by the price of the (dedicated) catheter. On the other hand, many sterilization techniques damage the probes.

The use of catheters for medical applications can be dated back to 3000 bc when the Egyptians used gold and silver pipes inserted into the bladder [1]. The earliest known

Sensors in Medicine and Health Care
Edited by P. Å. Öberg, T. Togawa, and F. A. Spelman

ISBN: 3-527-29556-9

cardiac catheterizations were performed by Hales in 1711 using brass pipes to enter the ventricles of a horse. He inserted these pipes into the venous and arterial system, using as insertion site the jugular vein and the carotid artery. Two centuries later, in 1929, in a small hospital in Germany, Werner Forssmann (1904–79), a young surgical resident, inserted a catheter in his antecubital vein and went to an x-ray room to document the position of the catheter that he pushed into his right atrium [2]. In this way, he proved that a catheter could be inserted safely into a human heart. However, at that time it was believed that any entry into the heart would be fatal, so Forssmann was immediately fired, despite the importance of his discovery. He never returned to research in cardiology, but was awarded the Nobel Prize in 1956, along with cardiology innovators Cournand and Richards, for his pioneering efforts.

In 1953, Sven-Ivar Seldinger (1921–99) published the description of a percutaneous technique to enter an artery with a catheter. Through this technique, access is gained by means of a catheter to any part of the body via the cardiovascular system, using x-ray fluoroscopy as a guide [3]. Ten years later, in January 1964, Charles T. Dotter (1920–85) and his trainee, Melvis Judkins (1922–85), first used a catheter for percutaneous transluminal angioplasty (PTA) in an 82-year-old female patient, suffering from a non-healing left-foot ulcer and gangrenous toes [4]. Eight months after the procedure, the patient was ambulatory, the ulceration was gone and the gangrenous toes were healed. The patient died 3 years later of congestive heart failure, but at that time she was 'still walking on her own feet'. Throughout his medical career, Dotter 'dedicated his considerable energy to the belief that there is always a better way to treat disease'. Driven by his own motto, 'if a plumber can do it to pipes, we can do it to blood vessels', Dotter was convinced that a patient could be treated without the scalpel.

During one of his classes at the University of Oregon Medical School, he explained to his students, 'I've been standing here and talking to you for about 20 minutes and all this time I have had a catheter in my heart. Now, I'll show you what a normal heart reading looks like'. So, he went and he plugged himself into the machine – a man standing there with a catheter in his heart – and he moved it among the chambers of the heart as he stood there and he explained what the graphs represented. It was an absolutely horrifying example, but it was the kind of thing he did, to say, 'it's perfectly safe, it can be done, it isn't dangerous' [5].

After employing Dotter's techniques of transluminal angioplasty, Andreas Grüntzig (1939–85), a young German physician working at the University Hospital in Zürich, began playing with the idea of adding a balloon to the existing catheters. He started by making prototypes and eventually developed a double-lumen catheter fitted with a poly(vinyl chloride) balloon. One lumen was used for balloon inflation and the other to pass a guide wire through it. Using this catheter first in animal studies, he succeeded, in 1977, in performing the first coronary angioplasty on a conscious human [6]. With this, a new revolution in medicine began, which is still continuing nowadays. 'It was these scientists, often working alone or in small groups and usually without grant support, who overcame major technical obstacles, suffered the vagaries of clinical research, endured the intransigence of their colleagues as well as their own self-doubt and achieved great leaps of knowledge, thereby advancing medicine by extraordinary bounds' [7].

Even though the list of inventors and contributors to interventional radiology is long, owing to his extensive contributions to medicine, Dotter has been credited with developing a new medical speciality, nowadays called interventional radiology.

This chapter considers three aspects of catheters and the use thereof. The first is the medical background, which shows the benefits to both patient and hospitals. The second aspect is the navigation and localization of the catheter. Since the surgeon has no direct visual contact with the catheter, a number of aids to navigation have been developed. Finally, we discuss the sensors themselves, where each signal domain is taken in turn. In some cases examples are given of devices already applied to catheters and in others only potential sensors for a certain application.

11.2 Medical Background

Catheters are used in a variety of medical procedures where access is limited. Generally, they are used to reach areas inside the body through natural pathways, without the need for open surgery. This minimizes damage to healthy tissue, reducing the recovery time, chances of infection and discomfort to the patient. As these procedures require only a local anesthetic, the risk associated with general anesthesia is avoided. Catheters are also used to monitor the patient after surgery by leaving a catheter in the wound and draining. This is common after head trauma to monitor intercranial pressure and draining of liquor. All the benefits notwithstanding, the limited access makes the procedures more demanding for the physician and limits the possible procedures that can be performed using catheters. Catheters are mainly used in two areas: the circulatory system (blood vessels and heart) and the urinary tract and bladder. For further reading on the medical background, see [8–10].

11.2.1 Circulatory System

The circulatory system (Fig. 11.1) serves to move blood around the body to bring oxygen and nutrients to the tissues of the body and also maintain correct body temperature. As oxygen and other chemicals diffuse out of the blood and into the fluid surrounding, the cells of the body's tissues, waste products, diffuse into the blood to be carried away. Blood circulates through organs such as the liver and kidneys, where wastes are removed, and back to the lungs where carbon dioxide is dumped and a fresh dose of oxygen taken in. Then the process repeats itself. This process of circulation is necessary for continued life of the cells, tissues and eventually of the whole body. The heart serves the vital function of pumping the blood around to keep this transport going.

The circulatory system can be divided into several parts, which are the heart, the arteries and the veins. The arteries and arterioles carry blood from the heart to the tissue and organs, veins and venules carry blood back to the heart, and the

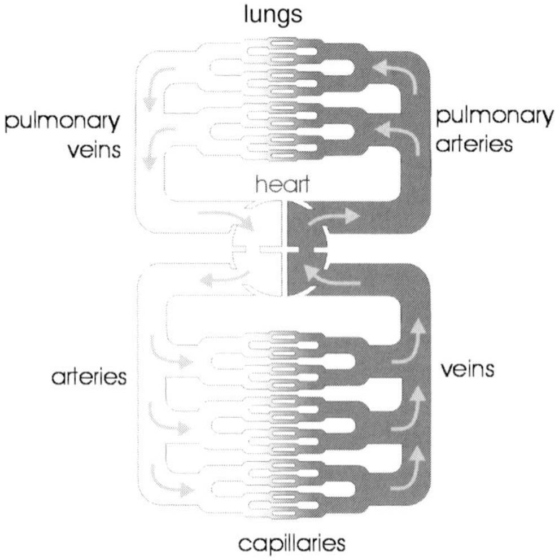

Fig. 11.1 Schematic representation of the circulatory system.

capillaries, which are the very small vessels, ensure the exchange of chemicals between the blood and the tissue. Overall, a human has approximately 10 000 km of blood vessel, ranging in diameter from the vena cava (largest vein) of 3 cm and aorta of 2.5 cm to the capillaries of 8 μm. This system of blood vessels contains on average 5 L of blood.

A problem with the circulation of blood somewhere in the body can pose a serious health risk. These cardiovascular diseases cause almost half the deaths in the Western world.

11.2.1.1 Circulatory Problems

There are many causes of circulatory problems, which include disorders that patients are born with, such as arteriovenous malformation (a local, complex, tangled web of arteries and veins), vascular diseases such as arteriosclerosis and problems resulting from trauma. These afflictions result in an obstruction to the flow of blood, internal bleeding or undesired circulation.

In the case of an obstruction, little or no blood, and therefore oxygen and nutrients, will reach the tissue further on in the circulation, which results in ischemia

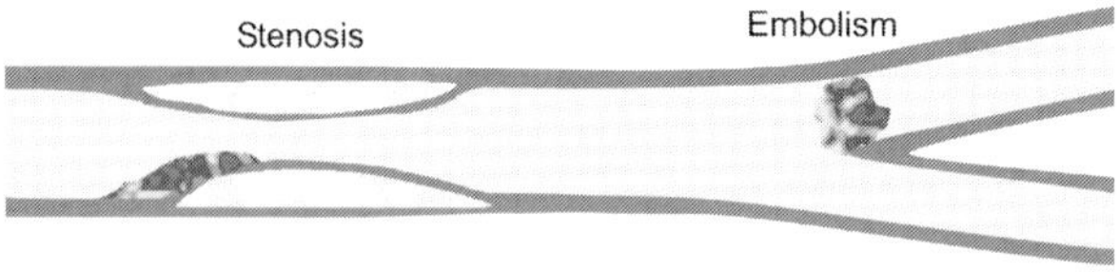

Fig. 11.2 Obstructions in the flow of blood caused by deposition of material in and on the vessel wall (stenosis, left) or 'plugging' of the vessel by a particle (embolism, right).

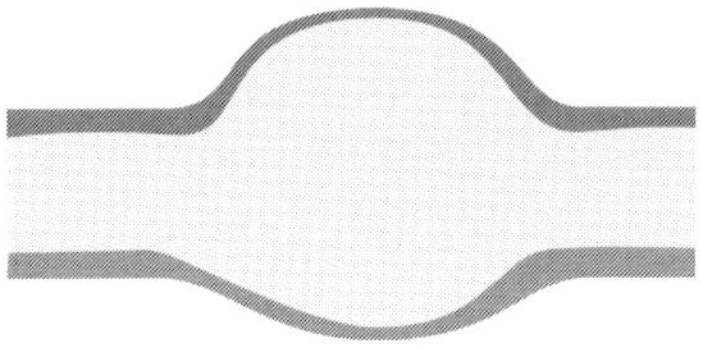

Fig. 11.3 Aneurysm: the local stretching of a blood vessel resulting in a thin vessel wall and a high risk of rupture.

in the tissue and possibly tissue death. Depending on the location, this can cause pain or result in heart attacks (myocardial infarction), strokes (ischemic stroke), etc. Obstructions are caused by the deposition of coagulated blood, fatty tissue or calcium in/on the vessel wall blocking the blood flow (stenosis, see Fig. 11.2), obstructions in blood vessels by particles such as dislodged blood clots (embolism, see Fig. 11.2) or a spasm of the blood vessel itself.

In the case of internal bleeding, problems are caused by the lack of blood beyond the rupture causing ischemia, pressure of the blood on surrounding tissue and blood loss. Internal bleeding can be the consequence of rupture of the vessel wall due to the stretching of the vessel wall to a balloon-like widening of the vessel resulting in a thin vessel wall (aneurysm, see Fig. 11.3), abnormal, thin-walled vessels as found in some arteriovenous malformation or in cases of traumata.

Problems with the circulation can occur if there are direct connections between arteries and veins (shunts). This causes a flow of oxygenated blood into the veins without reaching the organs or tissue, which can lead to ischemia or even death.

The above list of possible vascular problems is not exhaustive, but gives an idea of the kind of problems that can occur and which are sometimes diagnosed and treated using vascular catherization.

11.2.1.2 Vascular Catheterization

In traditional (cardio)vascular surgery, the physician cuts through healthy tissue to reach the affected blood vessel. In contrast, during a catherization, long, thin instruments are inserted into the blood vessels, moved to the affected area and the ailment is diagnosed and/or treated from inside the blood vessel itself. In order to appreciate the requirements and problems involved, some understanding of the procedure is required.

To insert the instruments, a short tube, called an introducer sheath (sleeve catheter), will be inserted in the blood vessel artery (Fig. 11.4). This sheath eases insertion of the instruments, keeps the access to the blood vessel open and impedes bleeding. The insertion is most often done in the groin, although the arm and neck are used in some procedures. Through this access port, a thin, flexible wire, called the guide wire, is inserted and pushed through the blood vessels. By pushing and rotating the guide wire, the physician maneuvers the guide wire to the place where the problem occurs. This is usually done under the guidance of an x-ray fluoroscope (see Section 11.3). Through the vascular system, almost every part

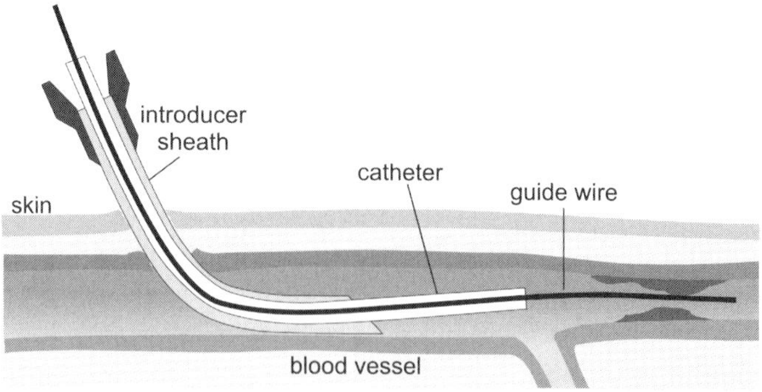

Fig. 11.4 Schematic representation of catheterization, showing the introducer sheath, catheter, and guide wire in the blood vessel.

of the body can be reached in this way. The purpose of the guide wire is to serve as a kind of railway and guide the instruments to the same location. The actual instruments are thin tubes called catheters, with a diameter varying between 1 and 5 mm depending on their use. They are pushed over the guide wire or hooked on to the guide wire.

In some narrow blood vessels, primarily the coronary arteries (the vessels supplying blood to the heart muscle), the set-up is reversed. A guiding catheter is pushed up to the problem area and a solid instrument, similar to a guide wire, is put through this for diagnosis and treatment. The further these instruments need to penetrate into the smaller blood vessels, the thinner they need to be.

11.2.1.3 **Vascular Treatment**

Once the guide wire is in place, special catheters are inserted for diagnoses and/or treatment. Treatment mainly consists of opening up obstructed blood vessels or intentionally blocking blood vessels. Catheters are also used for drainage or release/infusion of medicines.

The most common procedure for opening obstructed vessels is PTA (percutaneous transluminal angioplasty). A catheter with a deflated balloon is inserted into the blockage and inflated to compress and push aside the deposits on and in the vessel wall, thereby improving blood flow. In addition to the dilatation of the vessel, a stent is often inserted. This tiny coiled wire or mesh functions as shoring in the vessel to keep it open. Other methods to open up blocked vessels use mechanical scrapers, lasers, ultrasound heads or water jets to remove the blockage. In the case of small aneurysms and arteriovenous malformation, a procedure called embolization is performed. Here the blood vessels are blocked by small particles, a special gels or special coils. This procedure is also used to treat hypervascularized tumors.

11.2.2
Urology

Urology is another medical field where catheters are often used. A catheter is inserted in the urinary tract via the urethra into the bladder. This is done as both a diagnostic and a therapeutic measure.

Diagnostic catheterization can be carried out to sample urine for bacteriological analyses, but also for diagnosis of lower urinary tract dysfunction, e.g. urodynamics and urethrocystogram. The catheter tip in the bladder is used to measure/monitor bladder pressure and in the urethra to probe suspected strictures (calibration/measurement of pressure profile of the urethra) in cases of leakage or obstruction. It is also used as a means of measuring the residual volume of urine by draining it, if x-ray fluoroscopy or ultrasonography is not used. Additionally, catheterization can be used to differentiate between anuria (little urine production) and ischuria (reduction in urine flow due to blockage of the urethra or reduced function of bladder muscle) and to supervise the fluid balance and renal function after surgery.

Therapeutic catheterization is necessary in case of disturbed micturition (reflex to urinate, spinal cord lesion), post-operatively, or in case of obstructive prostatic adenoma (growth of the prostate) and for the elimination of a residual volume of urine caused by a neurogenic bladder or a stricture of the urethra.

11.2.3
Measurement Catheters

It is clear that there are many advantages to the use of catheters over surgery. The main disadvantage, however, is a direct result of the limited access. Whereas during open surgery the physician can see and feel what the problem is, during a catherization he or she is completely dependent on the instruments to provide the necessary information for diagnosis and monitoring of treatment. In addition to external imaging systems, such as x-ray, computed tomography (CT), magnetic resonance imaging (MRI) and ultrasound, measurement catheters are needed for a more detailed analysis of the situation.

At present, the primary tool is a catheter to measure pressure. Owing to the venturi effect, obstructions in both blood vessels and the urethra can be detected and evaluated by the changes in pressure. This, however, gives a very limited view of the ailment and more information is needed for optimal treatment. The problem is that the catheter in the flowstream, tip and shaft, seriously affects the flow profile and hence the measurements. To this end, better/smaller pressure sensors, flow sensors and intravascular ultrasound-imaging catheters are coming on to the market. Stable positioning of the actual sensor relative to flow and vessel wall is a significant issue. Using these and other new catheter-based instruments, a more precise diagnosis and treatment will be possible, opening up the possibility to treat more medical problems using catheterization instead of open surgery. In open surgery, flow can be measured by perivascular electromagnetic flow probes. Owing to their small size, silicon sensors are particularly suited for use in catheters and guide wires.

11.3 Navigation Systems

Depending on the intervention, either a guide wire is first inserted into the blood vessel followed by the catheter or the catheter is inserted directly. In both cases it is important to know the location of the tip, particularly when making a turn or coming into critical locations.

In both the medical and technical literature, a multitude of systems for instrument localization/navigation have been presented [11]. Instrument localization implies the detection of the position (orientation) of the tip of the instrument and its indication in a visual, audio or any other relevant way. When an instrument is sequentially located during the medical procedure, this is called instrument navigation or guidance. To localize the successive positions of the instruments during navigation one can use various detection methods and visualization techniques. Some of these systems use sensors located at the tip of the guide wire whereas others maintain the sensor outside the patient. In interventional radiology, the most often applied technique is fluoroscopy. This section briefly presents some existing systems.

11.3.1 Fluoroscopy

A standard fluoroscope can generate two-dimensional images of different types. For example, a radiographic image visualizes the skeletal structure and the inserted medical tool, whereas a contrast image (angiogram) shows the vasculature of the patient (Fig. 11.5). A more advanced system is three-dimensional imaging, brought on to the market with the advent of the Philips Integris 3D-RA (3D rotational angiography). A three-dimensional insight into the vascular structures is rendered by a single rotational angiographic x-ray acquisition. Owing to the multiple 2D/3D visualization possibilities offered by fluoroscopy, this technique continues to be the gold standard for peripheral vascular interventions, although the harmful effects of high radiation doses are a significant disadvantage for this application.

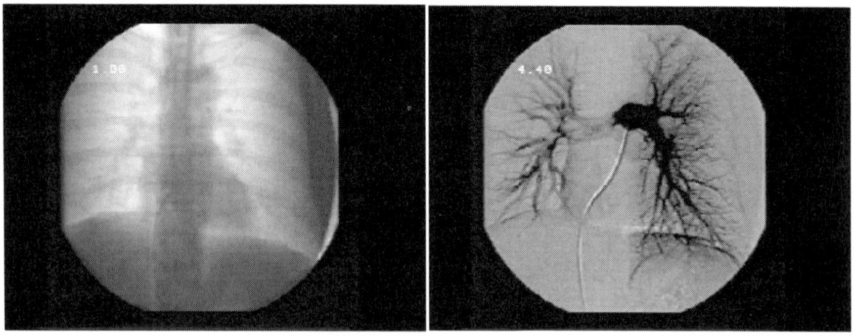

Fig. 11.5 Left: Radiographic chest image; right: angiogram of the lungs.

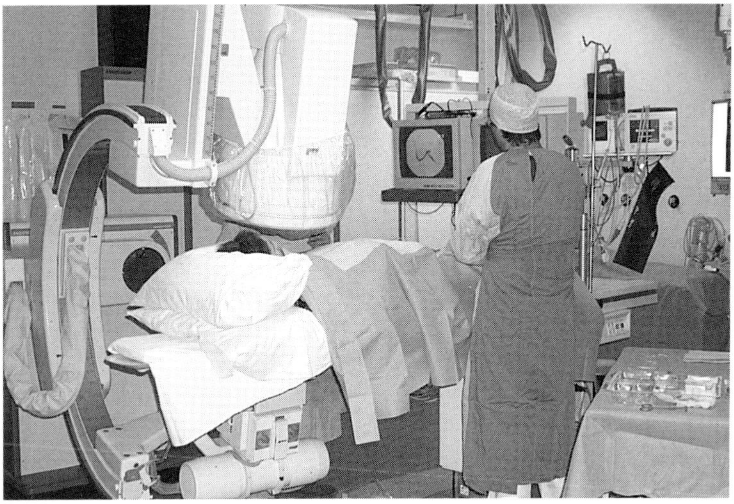

Fig. 11.6 Operating theater using an X-ray-based navigation system.

The operation therefore takes place with the x-ray source and image intensifier around the patient. This is shown in Fig. 11.6, with the x-ray source under the patient and the image intensifier above. The current x-ray image is shown on the screen where a series of x-ray images serve to form a film.

11.3.2 Ultrasound

Ultrasound is a propagated mechanical vibration of the molecules of a material [12] and has frequencies of the order of megahertz (higher than the human audible range). The vibrational energy propagates as a wave travelling from one particle to the other, through the material.

From the medical point of view, as ultrasound travels through the body, it is attenuated owing to beam divergence (the intensity along the beam axis is reduced because of the spreading of acoustic energy over a larger beam area), absorption (the energy is transferred from the beam to the tissue and eventually degrades into heat) and deflection (energy is deflected out of the beam by reflection, refraction and scattering) of acoustic energy out of the beam. It has been shown that when a sound beam encounters a boundary between two materials, some of the energy is reflected and the remainder is transmitted through the boundary. The direction of the echo (the reflected wave) depends on the orientation of the sound wave with respect to the boundary surface.

Unlike x-rays, ultrasound is non-ionizing, which means that it does not contain sufficient energy to dislodge orbiting electrons from atoms. The most common physical effect that may be generated by ultrasound is heat. Ultrasound has been used in different medical branches and for a variety of applications [13–22] and no confirmed reports of adverse effects due to the use of ultrasound in patients have been presented.

The imaging modality based on the propagation of ultrasound in tissue is called ultrasonography [23]. This technique takes various forms in practice, eg, pulse-echo ultrasound, Doppler ultrasound and intravascular ultrasound, which are presented in this section.

11.3.2.1 Pulse-Echo Ultrasound for Medical-Tool Localization/Navigation

The ultrasonic imaging system presented by Vilkomerson and Lyons [24] uses a display to vizualize the location of a medical instrument, an ultrasound transducer placed at the tip of the medical tool, an ultrasound scanner on the patient's skin and signal processing circuitry (Fig. 11.7).

The working principle of the system is based on the pulse-echo technique, by means of which internal anatomical images are generated by transmitting ultrasound waves via an ultrasound scanner to the body and processing their reflections (echoes). When an ultrasound transducer is placed at the tip of the medical tool, it will provide an electric signal each time it is struck by an ultrasound wave. In ultrasound systems, the returning echoes from the transmitted pulses provide one line in an ultrasonic image, which corresponds to the sequence of interfaces encountered by the ultrasound pulse as it propagates downwards into the body of the patient.

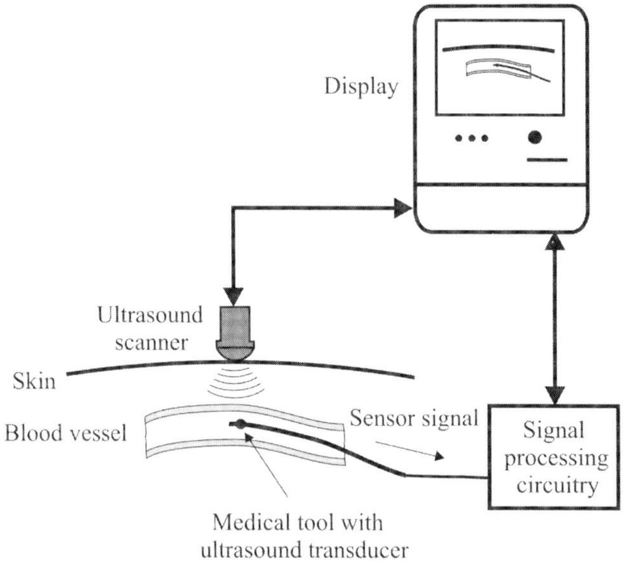

Fig. 11.7 Pulse-echo system for medical tool navigation [24].

11.3.2.2 Ultrasonic Beacon Guidance of Catheters

Vilkomerson et al. [25, 26] presented another ultrasound system, similar to that presented above. This system uses an ultrasound imager to vizualize the detected catheter location, a catheter system interface (CSI), a scan head and a sensor located at the tip of a catheter (Fig. 11.8). Ultrasound pulses are sent via the scan head to the patient's body and when a pulse is detected by the sensor, a signal is generated, which is further processed in order to produce an electronic arrow at the corresponding position in the ultrasound image. Different acoustic pulses may strike the sensor and produce electrical signals, which are sent to the CSI. In this way, the time between the transmission of a pulse and the reception of an electrical signal is known. The time delay is doubled because the time to travel from the scan head to the sensor and back is the same. Since this electrical signal is injected into the scan head at twice the one-way travel time, a signal is present at the scan head at exactly the same time as a signal produced by a reflection from the sensor would be. In this way, the CSI inserts electric signals into the scan head, which are further transmitted to the ultrasound imager. The difference between the two presented systems is the way in which the information about the location of the tip of the medical instrument is collected. The ultrasound image itself, used here as a background image, is generated in both cases by means of the ultrasound transducer, using the pulse-echo technique.

The method of ultrasonic beacon guidance has been used clinically for guiding balloon angioplasties in the leg [27, 28]. Reports from St. Mary's Hospital in London [29] indicate that 55 patients had been successfully treated and x-rays were used to determine whether the procedure was successful.

Although this system was used for interventions in the leg, it did not offer the visual information obtained with fluoroscopy. The vasculature was hardly visible and background information (e.g., skeletal structure) that should render a good anatomical overview was not available at all. Similarly to the previous system, the generated images in this case are also two-dimensional, in the plane of the hand-held

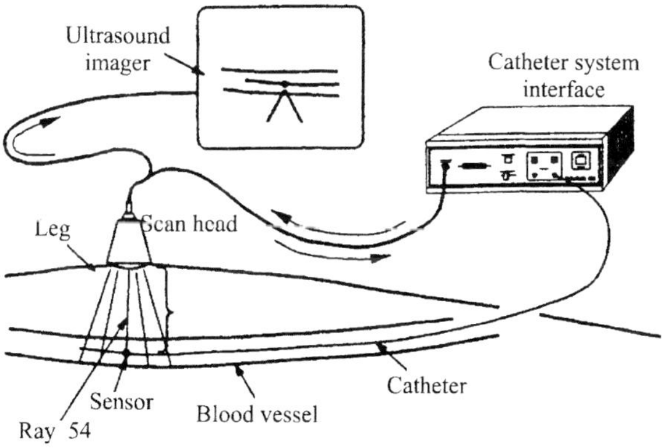

Fig. 11.8 Ultrasonic beacon guidance of catheters [25, 26].

scan head. Despite the fact that ultrasound is the most widely used medical imaging modality [30], catheters are most frequently guided by fluoroscopy.

11.3.2.3 Doppler Ultrasound for Catheter Position Monitoring

Beside the pulse-echo technique used by the previously presented systems, Doppler ultrasound may be used for instrument localization. In an approximate mathematical form, the Doppler principle is stated as follows [31]:

$$f_D = \frac{2 f_0 v \cos\theta}{c}$$

where f_D is the Doppler frequency shift, f_0 is the incident frequency, is the blood flow velocity, c is the speed of sound in tissue and θ is the angle between the ultrasound beam and the flow direction.

Generally, the speed of sound in tissue is ~1500 m/s and the blood velocity ~1.5 m/s. Since the insonation frequency is in the megahertz range and the relation between the blood velocity v and the sound velocity c is ~1:1000, the Doppler frequency shift will be in the kilohertz range. Therefore, the human ear may act as a sophisticated Doppler signal processor in a large number of clinical applications [32].

The Doppler effect is applied in interventional radiology and cardiology as a noninvasive method for catheter tip localization and for the detection of faulty tip placement [33, 34]. Precise placement of central venous catheters is necessary to prevent complications. To evaluate the use of Doppler sonography for this application, 200 patients were tested [35]. The catheters were inserted via standard routes such as the internal or external jugular vein, the basilar or the cephalic vein. A Doppler sonographic device with a probe having an insonating frequency of 2 MHz was positioned to the right sternal border such that the characteristic venous sound was most distinct. When a small quantity of saline solution was injected into the catheter, flow turbulences occurred at the catheter's tip, which were audible even without special training, indicating that the tip was under the Doppler device. It is worth mentioning that motion artefacts and muscle movement did not influence the correctness of the signals obtained.

11.3.2.4 Intravascular Ultrasound

Whereas pulse-echo ultrasound is used to visualize internal anatomical structures from outside the patient, intravascular ultrasound (IVUS) images the internal vascular area and the luminal surface of a blood vessel from inside the vessel, by means of a technique called angioscopy. A small ultrasound transducer placed at the tip of a catheter is moved longitudinally and radially with respect to the vessel, so that real-time tomographic images of both the lumen and the arterial wall are obtained (Fig. 11.9a).

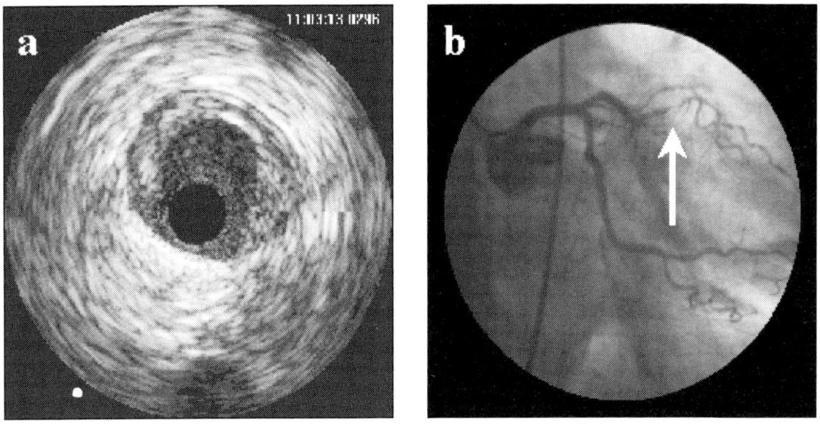

Fig. 11.9 Intravascular ultrasound image. (a) showing the lumen and the arterial wall; (b) an angiographic image showing on the same vessel segment (courtesy of Leiden University Medical Centre, Radiology-LKEB).

IVUS is useful in determining the efficacy of balloon angioplasties, in guiding atherectomy devices and in assuring appropriate placement of intravascular stents [36]. In PTA procedures, IVUS is used to provide images before and after the interventions are performed, indicating the difference between calcified and fibrous lesions. Furthermore, it allows the detection of plaque fractures and ruptures of the intima [37]. When used for atherectomy, intravascular ultrasound provides important information on the effects of different atherectomy devices on the vessel wall.

Even though IVUS provides high-resolution anatomical images, it is not used for monitoring the medical tools during the procedures. Usually, global instrument positioning is achieved with fluoroscopy, as shown in Fig. 11.9b. This is mainly because an IVUS image does not provide instrument-position information with respect to the vascular system. However, to avoid the use of fluoroscopy, other navigation systems can be used in conjunction with the IVUS device. Such a system is the Magellan coronary atherectomy device used by Gregory et al. [38] for *in vivo* coronary experiments.

The Magellan system combines ultrasound imaging and electromagnetic guidance to control laser therapy in coronary arteries. It consists of an ultrasound transducer for axial and lateral imaging located at the tip of a catheter, an electromagnetic system for angle and position detection and laser delivery optics for photo-ablative processes. The navigation of the ultrasound transducer is performed with the electromagnetic system, which uses an extra-corporeal antenna and a magnetic sensor located at the tip of the catheter, close to the ultrasound transducer. The rotation and position resolutions of the navigation system are 1.5° and 0.5 mm, respectively.

The endovascular ultrasound transducer renders 2D angioscopic images for ablative procedures, which are obtained by manually moving the catheter longitudinally and radially. Although the combination of 2D longitudinal and 2D radial images visualizes the lesions in length and width, global localization of the catheter tip is not achieved. Therefore, the magnetic navigation system is important as it can guide the ultrasound catheter to the intervention site.

The main advantage of this system is that it does not use ionizing radiation. Furthermore, it allows safe and effective treatment of vascular diseases. It should, however, be noted that the system uses magnetic instrument navigation and cannot guide the tools only by means of the ultrasound images, i.e., intravascular ultrasound does not render global positioning of the medical tools with respect to the vascular system. It has been shown [39, 40] that pulse-echo ultrasound scanners may also be combined with electromagnetic instrument-tracking systems, which use an extra-corporeal magnetic source and 3D magnetic sensors located on the ultrasound scanner. While the scanner is moved along the patient's body generating 2D ultrasound images, its position is determined with respect to the magnetic source. In this way, 2D ultrasound images can be reconstructed in 3D space.

11.3.2.5 Sonomicrometry

Sonomicrometry is another medical application of ultrasound, used to measure distances. Generally, transducers are made of piezoelectric materials that operate at frequencies of 1 MHz and higher. In order to perform a single distance measurement, one crystal transmits a burst of ultrasound and another crystal receives this ultrasound signal. The elapsed time from transmission to reception is a direct and linear representation of the physical distance between the two crystals (time-of-flight principle) [41].

This concept was used by the Sonometrics Corporation for heart-catheter navigation. The system uses piezoelectric crystals on the patient's chest and at the tip of the catheter and computes the location of the medical tool by processing the data from the ultrasound transducers (Fig. 11.10). The use of the system for heart

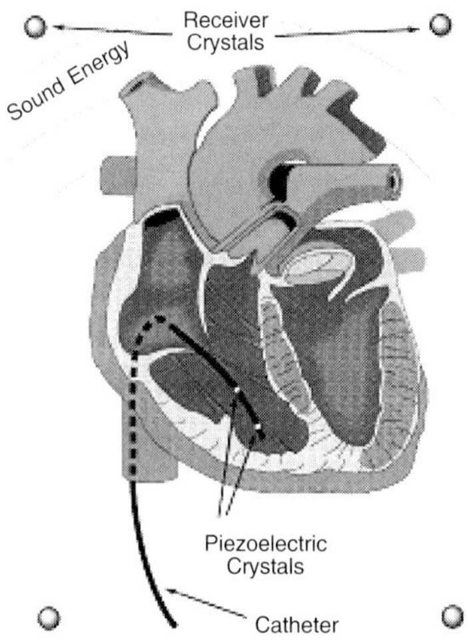

Fig. 11.10 The time-of-flight principle used for heart-catheter navigation [41].

interventions is possible because the intervention site is restricted to the heart volume and thus the medium between the transmitter and the receiver is considered to be homogeneous.

11.3.3
Magnetic Resonance Imaging (MRI)

MRI is a modern way to visualize internal anatomical structures. The technique makes use of one of the properties characteristic to the human body, namely that 75% of it is made up of water.

The nuclei of the hydrogen atoms in water molecules consist of single protons, which under normal conditions are constantly spinning. Rapidly spinning nuclei possessing a magnetic moment can be thought of as tiny bar magnets. The intrinsic magnetic field of all nuclei is randomly oriented provided that no external magnetic field is applied. However, in a static magnetic field, these magnetic dipoles tend to assume discrete orientations either parallel or antiparallel to the direction of the applied field. This is what actually happens when a person is placed inside an MRI scanner. The high magnetic field of the scanner, 0.01–4 T, will make all protons line up. In order to take an image, short bursts of radio waves are directed at the area being examined through a special antenna called a gradient. This causes the protons to flip their orientation. When the pulse is turned off, the protons return to their original position, emitting weak radio signals of a particular frequency (MR signals). These signals are analyzed by a computer and processed to create a series of cross sectional images (scans), in any direction [42]. The use of MRI in interventional radiology is rather limited, because it was not considered suitable for guiding medical tools owing to the possible effects of high magnetic fields on standard instrumentation. However, research studies are in progress to develop new interventional tools [43] and to make catheters and guide wires visible on MRI images.

When it comes to the localization of a device with MRI, one can basically think of two different approaches [44]: either to visualize the device passively by its magnetic susceptibility artefact (local image distortions) or to use a special device with a small built-in coil at the tip to identify its location actively [45].

In the passive approach, the device itself needs to produce an artefact that is large enough to allow precise localization, but small enough not to obliterate the surrounding anatomy and therefore to compromise the ability to perform the intervention. In this context, ferromagnetic materials were investigated, because their large magnetic moment results in large local magnetic susceptibility variations when they are imaged [46]. These effects may be well explained by the susceptibility theory [47].

It seems, however, that ferromagnetic materials cause artefacts that are too large, whereas plastic catheters cause no artefacts at all and cannot be seen with ultrafast MRI sequences [44]. A number of different guide wires and catheters were examined using different pulse sequences and it was shown that there are still different aspects that need further investigation, e.g., the pulse sequence parameters, the materials, the field strength and the device orientation and safety issues to be considered.

For the active approach, first suggested by Ackerman et al. [48], a small rf coil is built into the invasive device itself, for spatial localization. A method to provide active tracking of MR devices, based on the resonance equation, was developed [49]. The location of a coil incorporated into a catheter may be frequency encoded and eventually represented in the MRI image. A great advantage of MRI is that the position of the coil can be determined in any desired imaging plane [50].

Wildermuth et al. [51] showed that MR tracking is sensitive to patient motion. If motion occurs, the images need to be updated to avoid localization errors. Although MRI shows some important advantages over fluoroscopy, such as good soft-tissue contrast, no radiation exposure, no contrast agent and multi-planar imaging capabilities, it remains a rather expensive procedure. Moreover, it should be considered that both the spatial and the temporal resolution of real-time MRI are still lower than in x-ray fluoroscopy. Further development and optimization are required, as challenges pertaining to catheter and guide wire design and pulse sequence optimization still lie ahead [52].

11.3.4 Electric and Magnetic Fields

11.3.4.1 LocaLisa

For the real-time, three-dimensional localization of intracardiac electrodes, a technique known as LocaLisa, which uses an externally applied electrical field, that is detected by standard catheter electrodes, has been developed [53, 54]. An electric current externally applied through the thorax gives rise to a voltage gradient across internal organs, such as the heart. The voltage gradient can be measured by means of intracardiac electrodes placed on the catheter, indicating their position in this way.

The external electric field is generated in three directions, using electrodes positioned on the patient's skin, from left to right at the elevation of the fourth intercostal space (x-axis), from the neck to the leg (y-axis) and from the front side to the back (z-axis). Measurements are performed between a catheter electrode and a reference electrode.

With this system, a catheter is accurately localized inside the heart with a position error of 1–2 mm. Unfortunately, x-rays are still used to guide the catheter towards the heart. The measured catheter locations are indicated on a computer screen and not on a medical image with correlation to the internal anatomical structure of the patient. This implies the presence of a specialized operator, who reads and interprets the data and communicates continuously with the interventional cardiologist. Even though the interface is not user-friendly, the system was tested *in vivo* and is effective for heart ablation [55].

11.3.4.2 CARTO EP Navigation

CARTO uses magnetic fields to determine accurately the three-dimensional location of a mapping and ablation catheter [56]. The system consists of an intracardiac electrophysiological catheter with miniature coils at its tip, a magnetic-field generator located underneath the patient table, a unit that processes the output of the three-dimensional sensor and a post-processing graphical display unit.

Recorded data of the catheter location and an intracardiac electrogram at that location are used to reconstruct in real time a representation of the three-dimensional geometry of the heart chamber, color-coded with relevant electrophysiological information.

The miniature coils located at the tip of the catheter sense the ultra-low magnetic field (5–50 μT) generated by the magnetic source under the operating table [57]. When the catheter is moved in the magnetic field, the sensor output data are processed and converted in real time into the location and orientation of the catheter tip with respect to the magnetic source.

The catheter is a standard 7- or 8-French deflectable catheter, with a 4 mm tip. The location sensors are totally embedded within the catheter. The magnetic source consists of three electromagnets, which are manually placed under the operating table once the catheter is safely in the heart.

The system is in use at different hospitals and has been tested for cardiac radiofrequency ablation procedures [58, 59]. It was shown that nonfluoroscopic magnetic guidance for radiofrequency ablation is better than fluoroscopy, although x-rays are still needed to navigate the catheter to the heart. The resolution of the system is better than 1 mm for both *in vivo* and *in vitro* studies [60].

11.3.4.3 NOGA Navigation System

A similar system to that presented above is the NOGA navigation system, which acquires and records intracardiac electrical activation and endocardial motion in real time, for each acquired point. The mapping and navigation system consists of the same components as described for CARTO [61]. Since it collects endocardial electrical data, myocardial viability (the presence of normal or reduced voltage) can be accurately determined. Furthermore, the acquired map can provide global and regional contractility data of the left ventricle [62].

Compared with standard fluoroscopy techniques, the two Biosense-Webster systems provide a reduction in radiation exposure time, more accurate detection of arrhythmic foci and subsequent guidance for ablation procedures. Although the three-dimensional computer-generated images of the heart result from the anatomy of the patient, no direct visual correlation with the x-ray images is present. In these conditions, a cardiologist always works together with a well-trained operator, who sets up the system and interprets the data obtained during the intervention. As with CARTO, x-rays are needed to navigate the catheter to the heart.

11.3.4.4 TELSTAR

The Stereotaxis TELSTAR system has evolved into a computer-integrated surgical platform, which is designed to guide catheters remotely and programmatically anywhere in the body. The system was developed mainly for cardiac electrophysiology, interventional cardiology, interventional neuroradiology and minimally invasive neurosurgery, but a wider application range is expected. It integrates the magnetic control of surgical tools with x-ray and MRI scans.

The fundamental methodology employed by this system is called magnetic stereotaxis [63] and it implies the use of large electromagnetic coils, which guide a magnetic seed (a small piece of implanted permanent magnetic material) along a selected trajectory through the human body. To reduce incidental damage, the trajectory is defined preoperatively, using x-ray or MRI scans, by selecting a path that avoids important anatomical structures.

The Magnetic Surgery System (MSS) generates a magnetic field determined by the strength of the current flowing through each of six superconducting magnets. Increasing or decreasing the current to one magnet alters the contours of the total field, thus changing the direction in which the small magnet moves. A computer-controlled device provides the mechanical force to advance the guide wire or catheter, which the surgeon controls remotely from the computer console. Rather than standing at the bedside, in the x-ray field, the interventionist sits in a radiation-protected room and navigates the tools with a user-friendly interface. When the magnetic seed is maneuvered into a tumor, the seed is heated, generating localized cell death. By successive movements and heating, the tumor can be destroyed with little damage to the surrounding tissue [64]. Alternatively, the magnetic seed could be used to guide the tip of a catheter, for example to deliver drugs directly to sites inside the brain [65, 66]. Other studies were oriented towards brain biopsy procedures [67]. It has been shown that actual catheter placement was within 1.5 mm of the planned path to the biopsy site [68]. The stereotaxis system is a complex surgical workstation which requires the use of fluoroscopy and/or MRI to perform difficult neurological and cardiological interventions. It has proven its importance during different clinical studies; however, its use is restricted to interventional cardiology and neurology. Owing to the use of high-tech methodology, expensive superconducting magnets and specially designed medical tools, the price of the total system is high, limiting its use to complicated medical interventions.

11.3.4.5 The NAVION BioNavigation System

The electromagnetic system fabricated by Navion makes use of magnetic fields to sense the position of a catheter tip during central venous placement. It provides the opportunity to eliminate blind catheter placement and to reduce fluoroscopy during insertion. The system works on the principle of an alternating magnetic field, inducing a voltage in a coil that is located at the tip of a catheter [69]. The hand-held instrument contains two sets of field-generating coils: a horizontal coil, which is generally parallel to the sensor, and a vertical coil, generally perpendicular to the sensor. When the flux lines are parallel to the sensor axis, the induced sensor voltage is maximum and, as the flux lines become less parallel, the voltage decreases. When the flux lines are perpendicular to the sensor coil, no voltage is induced in the sensor.

In clinical use, an operator moves the instrument over the sensor until a high sound is detected. The highest detected pitch is an indication that the sensor is directly under the horizontal coil. A light display indicates the direction of the sensor with respect to the hand-held unit.

Generally, the common technique for placing central venous catheters is to measure the distance from the venous entry site to the optimal internal position, which is approximated by an external landmark and to insert the catheter to the measured position. Starkhammar et al. [69] show that when the NAVION system was used, the difference between the position of the tip and the optimal catheter position was <2.5 cm, while in 46% of the considered cases, the catheter was positioned exactly at the optimal location. The use of this system as an alternative to fluoroscopy implies the use of a reliable external landmark [70]. The Navion system is particularly developed to aid clinicians in attaining the correct position of central venous catheters. It is an important tool for such medical interventions; however, its use is restricted to catheter localization in a limited volume of the human body.

11.3.4.6 **Flock of Birds**

The Flock of Birds system, a magnetic tracking system developed by the Ascension Technology Corporation, is well known from numerous applications in the technical world (military applications, head tracking in flight simulators, virtual-reality games, full body tracking in animations) and is also being used in medical applications, such as the biomechanical measurement of anatomical parts [71].

The complete system includes a transmitter, a transmitter driver circuit, a receiver and signal-processing electronics. The transmitter consists of three individual antennae, which generate a d.c. magnetic field that is picked up by the receiver. The receiver itself, a three-axis magnetic sensor, measures the magnetic field generated by the transmitter. By means of a mathematical algorithm, the position and orientation of the object to which the sensor is attached can be determined. To avoid the influence of the Earth's magnetic field and of magnetic distortions (due to the generation of eddy currents in nearby conductive metals) on the output signals, pulsed d.c. currents are employed.

Although this system was not developed for a particular medical application, it may be used as a potential solution for the guidance of medical instruments. The position resolution is 0.5 mm at a transmitter–receiver distance of 30 cm and an orientation resolution 0.1°.

11.3.4.7 Pulsed Magnetic Fields

A system under development in Delft (Netherlands) uses simple coils to generate pulsed magnetic fields and a 3D sensor at the tip of the catheter [72, 73]. Practice has shown that magnets used in different medical applications had rather large dimensions and that the magnetic fields which they produced were low at large distances and difficult to measure. For example, the cylindrical magnet used by Schlageter et al. [74] is Φ 6 × 7 mm and the generated field may be detected at distances as large as 14 cm, using highly sensitive Hall sensors with integrated flux concentrators. Weitschies et al. [75] used a cylindrical capsule of Φ 5.7 × 16.1 mm, which was detected by means of a superconducting quantum interference device (SQUID) at distances as large as 20 cm. Another bar magnet having dimensions Φ 7.8 × 25 mm was used by Prakesh and Spelman [76], while Ilmoniemi et al. [77] employed a magnetic object of Φ 0.2 × 15 mm that was detected by means of a SQUID.

The most frequently used guide wires have diameters of 0.9 mm and, therefore, it is evident that magnetic capsules with dimensions such as those presented above cannot be used at a tip of a guide wire. Owing to this limitation of magnetic capsules, the sensor is placed at the tip of the medical tool while the magnetic field is generated extracorporeally (Fig. 11.11).

From the three components of magnetic field from each pulse, the position of the tip can be calculated. Since the coils are fixed to the image intensifier, the calculated position can be directly related to an x-ray image. This image is necessary to create a reference; however, the total radiation dose is significantly reduced. The set-up of the system is given in Fig. 11.12. The tip of the catheter/guide wire contains the 3D magnetic sensor (Hall plates or magneto-resistors), multiplexing and amplification. This minimizes the number of wires to the outside and also converts the signal into a format less prone to noise. Before each set of three pulses, a measurement is made without an applied field to obtain the offset and background magnetic field.

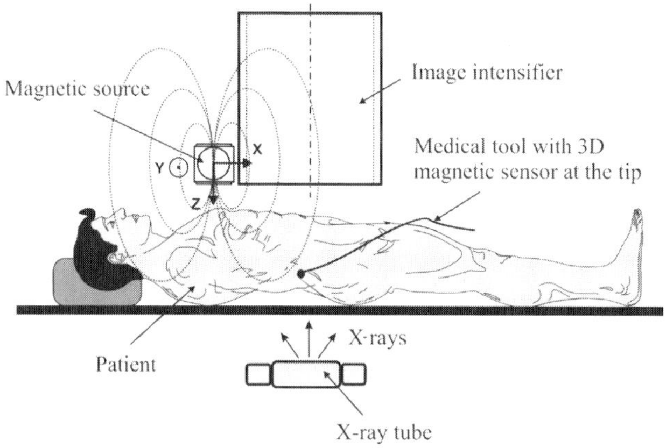

Fig. 11.11 Pulsed magnetic guidance system.

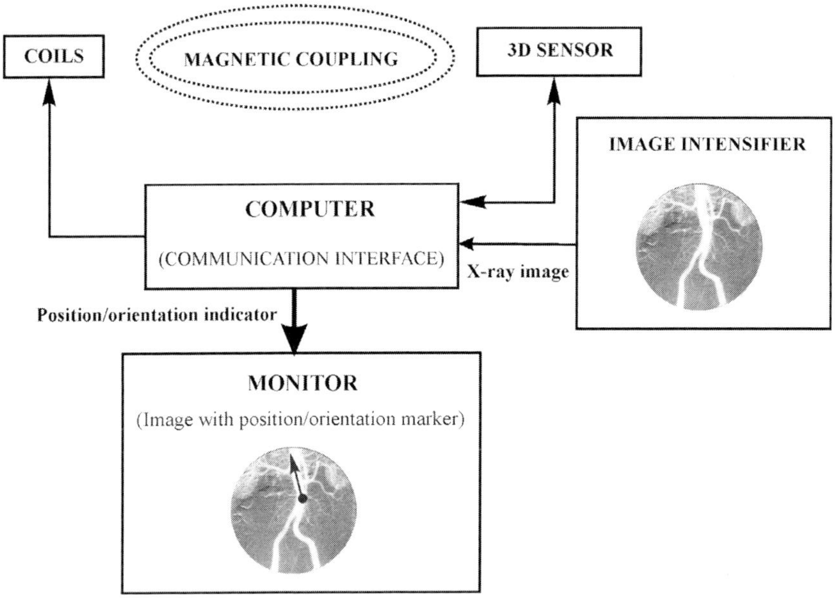

Fig. 11.12 Schematic of the pulsed magnetic field technique.

11.3.4.8 Other Electromagnetic Systems

Another system employing electromagnetic fields is the ARTMA (Advanced Real-Time Motion Analysis) Virtual Patient system, which finds the location of a surgical instrument by permanently measuring magnetic fields. When applied in endonasal sinus surgery [78], an electromagnet close to the head of the patient establishes a magnetic field, defining the reference coordinate system. Orientation in space is achieved by use of a 3D digitizer, which measures magnetic fields with Hall sensors to determine spatial position and orientation. The sensors are attached to the head of the patient and the surgical instruments, tracing every change of the magnetic field. The system handles images provided by other imaging modalities, such as MRI, CT or ultrasound, on which it virtually superimposes the detected locations of the instruments with respect to the patient's anatomical structure.

The most sophisticated feature is the creation of virtual trajectories of access to the surgical field, which lead the surgeon to the desired region. The accuracy of this system was assessed by visually judging the quality of the match between the actual location of the instrument and the calculated position. With a good calibration, an error of 1–1.25 mm was obtained [78].

The principle of magnetic field position sensing has also been applied to colonoscopy, where difficulties appear when inserting the colonoscope into the caecum with reasonable speed, comfort and safety [79]. Inappropriate maneuvers and unnecessary pain are very common during colonoscopy [80]. A magnetic system was developed to image the entire length of the endoscope, because of the need to reduce loops and to know exactly where to apply abdominal pressure [81].

Three generator assemblies, each comprising three orthogonal coils, are mounted beneath the patient's bed and are driven sequentially to produce nine different field patterns. Sixteen magnetic sensors, mounted on the endoscope, convert the field at discrete points into electrical signals, which are further processed to determine the location of each sensor and to display the path of the endoscope.

Clinical results confirmed the importance of the imaging system: it was possible to assess the large variety of loops and consequently to take the necessary steps to straighten the colonoscope and advance the tip [82]. The system was successful not only for colonoscopy, but also for percutaneous endoscopic gastrostomy [83]. A method for noninvasive examination of gastrointestinal motility uses magnetic detection by employing a small rare-earth cylindrical magnet (∅ 6 × 7 mm) and a 4 × 4 array of 16 Hall sensors with integrated flux concentrators [74].

The technique is based on recording the position of the small magnetic marker, which is ingested by the patient during the measurements. The array of sensors is placed on the abdominal skin of the patient and detects the signals emitted by the permanent magnet. The sensors' output is sent to a computer, which calculates and displays in real time the trajectory of the marker. A detection distance of 14 cm was achieved, while the position error was found to be within a few millimeters. A similar device, implemented for the same medical application, makes use of a SQUID magnetometer to establish the trajectory of the magnetic marker [75, 83], with a spatial resolution within a range of millimeters.

11.3.5
Comparison Between Navigation Systems

The previous section illustrated a number of devices employed to solve a range of medical problems that appear in different medical areas. It is the objective of this section to point out the advantages of existing systems and to present their shortcomings.

Of course, each of these systems was developed to solve a very specific medical problem and on average they have all been successfully implemented and tested, although the pulsed magnetic field system is still under development.

At present, in interventional radiology, the most frequently used method for instrument navigation is fluoroscopy. It offers high-contrast images of the vasculature and aids in assessing the location of vascular diseases. Unfortunately, it is disadvantageous owing to the harmful effects that may be produced by high radiation doses. A brief comparison between fluoroscopy and other techniques is given in Tab. 11.1.

Most of the existing navigation systems were developed for use in interventional cardiology, whereas the best navigation and visualization method for interventional radiology continues to be fluoroscopy. Research continues worldwide to improve existing systems and to develop new devices that will assist the surgeon throughout the medical procedures.

Tab. 11.1 Comparison of various navigation systems

Devices and techniques for localization and navigation	*Characteristics taken into consideration* [a]						
	TI	*L/N*	*TIm*	*IQ*	*HE*	*R*	*E*
Fluoroscope	All	L/N	2D (3D)	Very good	Yes	N/A	N/A
Pulse-echo ultrasound	IR	L/N	2D	Poor	No	N/I	N/I
Doppler ultrasound	IC	L	N/A	N/A	No	N/A	N/A
Magellan	IC	L/N	2D (3D)	Good	No	1.5°, 0.5 mm	N/I
MRI	All	L/N	2D (3D)	Good	No	N/I	1 mm
LocaLisa+ fluoroscopy	IC	L (N)	CG	Poor	No	N/I	1–2 mm
Carto + fluoroscopy	IC	L (N)	CG	Poor	No	1 mm	2 mm
Noga + fluoroscopy	IC	L (N)	CG	Poor	No	1 mm	2 mm
Telstar	All	L/N	2D (3D)	Very good	No (yes)	N/I	1.5 mm
NAVION	IC	L/N	N/A	N/A	No	N/I	<2.5 mm
Flock of Birds	N/A	L/N	N/A	N/A	No	1.5°, 0.5 mm	1–2 mm
Pulsed magnetic	IR	L/N	2D (3D)	Very good	Yes (no)	1–2 mm	6.2 mm

a) Abbreviations: TI, type of intervention for which the system (method) was developed; IC, interventional cardiology; IN, interventional neurology; IR, interventional radiology; L/N, purpose of the system, localization or navigation; TIm, types of images, 2D, 3D or CG, computer-generated; IQ, image quality, which refers to the system's capability to visualize the vascular system and to obtain good contrast between different tissues, so that internal anatomical structures are easily recognizable; HE, harmful effects, indicating whether the system may create any harmful effects in the patient/medical staff; R, resolution of the system; E, errors of the system; N/A, not applicable; N/I, not indicated in the available documentation.

11.4 Sensor Overview

Sensors for all six signal domains have been applied to catheters. The great advantage of silicon-based sensors is miniaturization.

11.4.1 Mechanical Domain

11.4.1.1 Pressure Sensors

The main mechanical sensor used in catheters is the pressure sensor for measuring local blood pressure in the blood vessels and also the pressure within balloon catheters. For the fabrication of such sensors, a wide range of technologies have been applied. These include bulk micromachining, surface micromachining and wafer-bonding. Most of these use a thin membrane and a reference chamber. The time during which these devices can operate correctly in the body can be limited by deposits on the membrane.

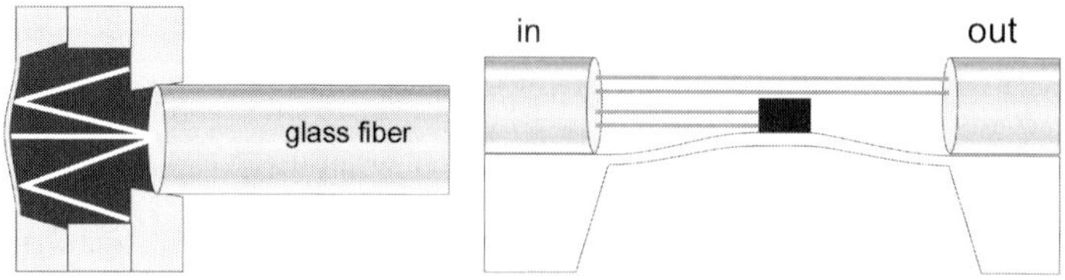

Fig. 11.13 Left: Optical pressure sensor through reflection; right: through intensity change.

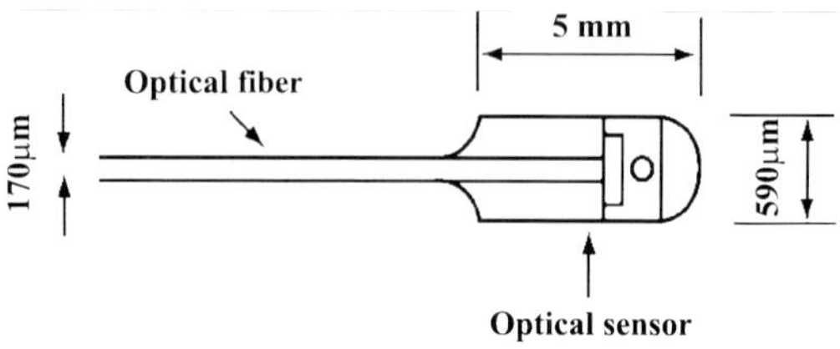

Fig. 11.14 Optical catheter pressure sensor [92].

The read-out of the devices can be piezoresistive [84, 85], capacitive [86, 87], optically, through changes in intensity of the reflected light [88, 89], or directly interferometric [90, 91]. Examples are given in Fig. 11.13.

A commercial sensor from World Commercial Instruments is shown in Fig. 11.14 [92]. The sensor tip has a diameter of 0.59 mm (1.8 French) and a length of 5 mm. Blood pressure is measured by observing the cavity's changes in length by means of an optical signal conditioner using a measurement scheme based on white light interferometry. Sensing light is transmitted to and from the sensor tip via a multimode fiber with an outer diameter of 170 mm (0.5 French). The measurement tip is virtually insensitive to any pulling or manipulation of the fiber.

In a recent device based on a Fabry-Perot interferometer, the diameter has been reduced to 360 μm, designed for a pressure range from 0 to 250 mmHg [93].

An example of the piezoresistive read-out is given in [85]. This device used an epi-poly membrane and polysilicon resistors. A further example of a piezoresistive pressure sensor is given in Fig. 11.16 [94]. In this case the diaphragm was made of

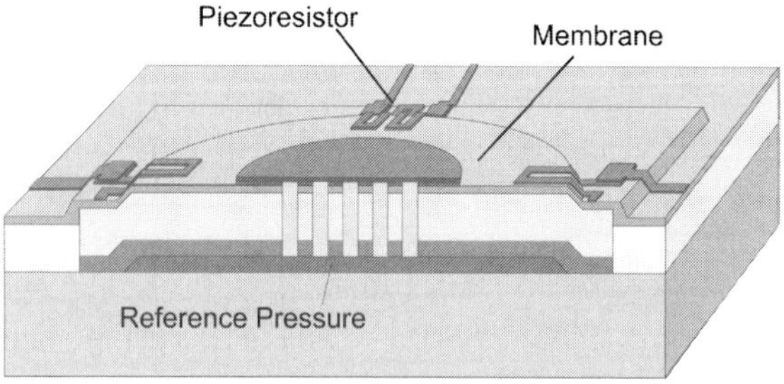

Fig. 11.15 Catheter pressure sensor fabricated using epi-poly.

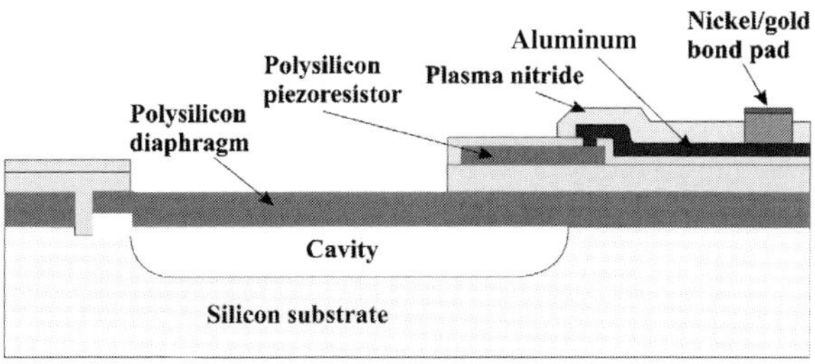

Fig. 11.16 Surface micromachined pressure sensor with piezoresistive read-outy

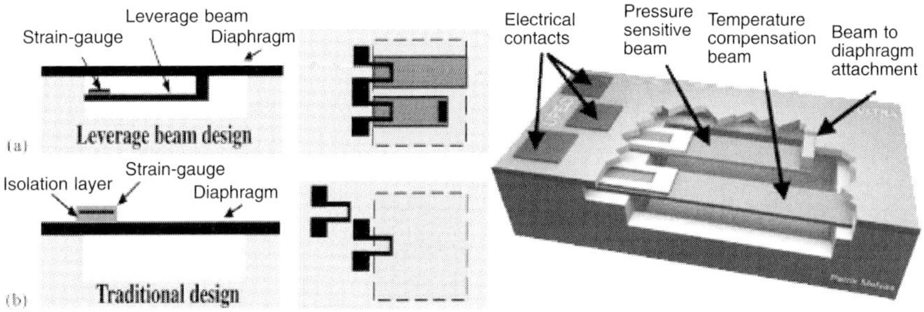

Fig. 11.17 Cross sectional view of the pressure sensor (left) and artist's impression (right) [95].

LPCVD polysilicon and had a side length of 103 μm. The device was tested in the range up to 300 mmHg having a typical sensitivity of 2 μV/V/mmHg.

Work in Stockholm has led to a temperature-compensated pressure sensor, where the piezoresistive device is placed inside the cavity. A cross section of the device, compared with the conventional device and a schematic are given in Fig. 11.17 [95].

In addition to pressure sensors designed for blood measurements, urology makes use of microsensors in catheters for bladder pressure measurements [96] (Fig. 11.18). The issues of safety and biocompatibility are different from those for blood systems. The catheter can be inserted into the bladder and drawn out to measure urethral pressure profiles.

11.4.1.2 **Blood Flow Sensors**

Blood flow is also an important parameter and a number of techniques have been proposed. Some of these measure mass flow and others flow velocity. A simple example is a plate or bar that is brought into the flow, where it will experience a force due to that flow [98, 99]. Once again the force on this obstacle can be measured in a number of ways. The disadvantage of this method is the fragility of the beam, which can create potential health hazards. An alternative technique is to use the obstacle to create a change in pressure. Thus a measure of the pressure

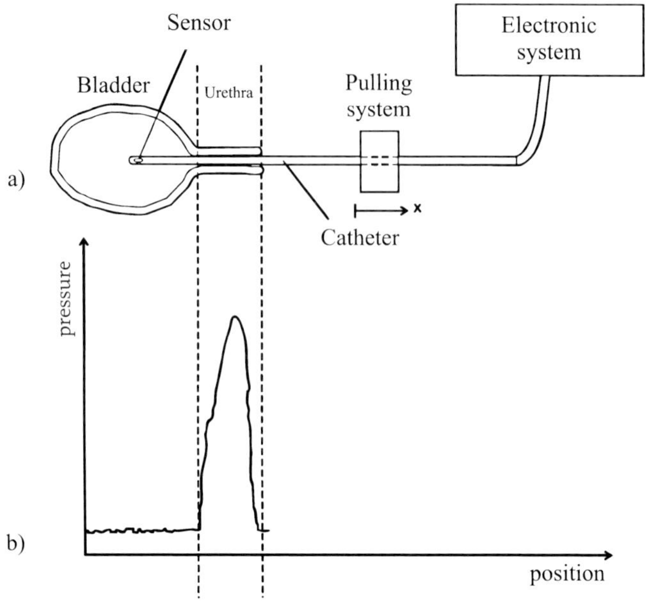

Fig. 11.18 Pressure profile measurement with a micro-tip transducer catheter (a) and pressure profile (b) [97].

before and after the obstacle will give the flow [100–102]. This method is more robust but care should always be taken when obstructing the blood flow.

There has been an attempt to use rotors to measure flow, as sometimes used in larger structures. Surface micromachining allowed sufficient miniaturization [103], but the presence of red and white blood cells meant that the device quickly seized up.

Many of the integrated flow sensors use the hot-wire technique. In this technique, the center of the chip is heated and the flow will take the heat to the down-wind side of the chip [85, 104, 105]. This is an effective method to measure flow. However, the heating of the chip should be kept below 5 K above ambient to ensure no cell damage. An example of such a structure is given in Fig. 11.19.

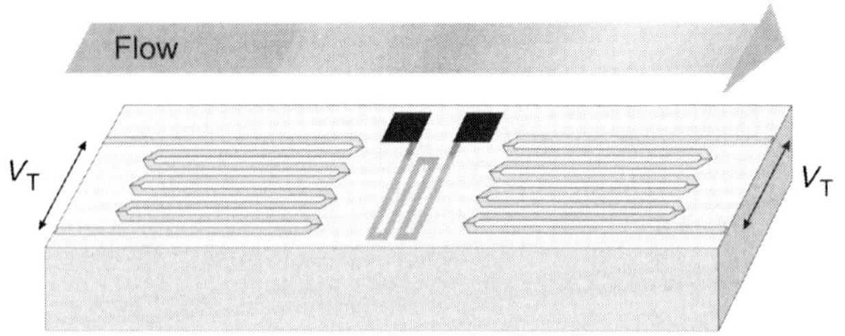

Fig. 11.19 Flow sensor based on polysilicon heater and thermopiles.

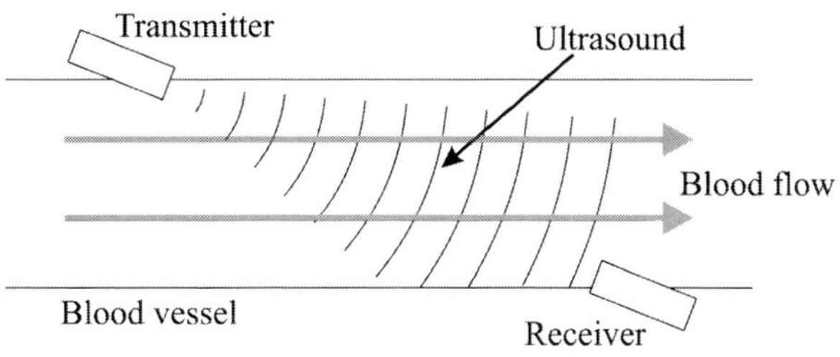

Fig. 11.20 Doppler flow measurement.

Flow measurements can also be achieved using the Doppler effect. An acoustic signal is emitted in the flow and the frequency shift when the signal reaches the receiver is a measure of the flow [106] (Fig. 11.20). This is an example of a flow velocity measurement.

11.4.1.3 Tactile Sensors

A tactile sensor can be similar to some of the pressure sensors. A bar can be used which experiences a force due to contact with an object (Fig. 11.21) [107]. This can be extended into an array to yield a pattern of pressure [108, 109]. Other sensors react to a change in the oscillation frequency when they come into contact with an object [110]. This comes through the amplification of a vibrating mass or the change in length of a vibrating beam. Read-out systems used include piezoresistive [111], piezoelectric [112] and ultrasound [113].

11.4.1.4 Movement Sensors

The term movement has a broad meaning and for medical applications it has generally been used for movements such as the heart beating, breathing and body

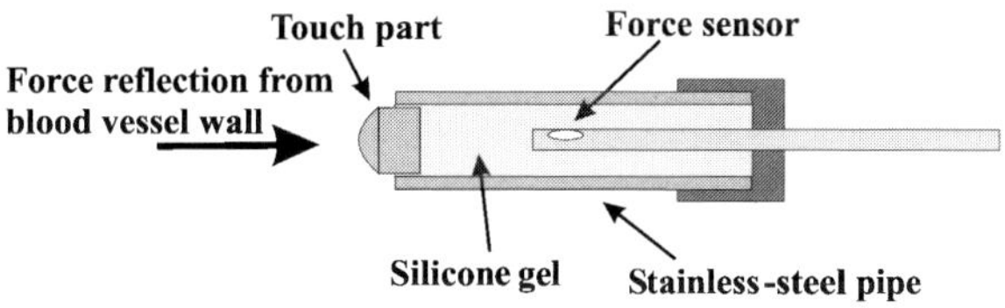

Fig. 11.21 Force sensor [106].

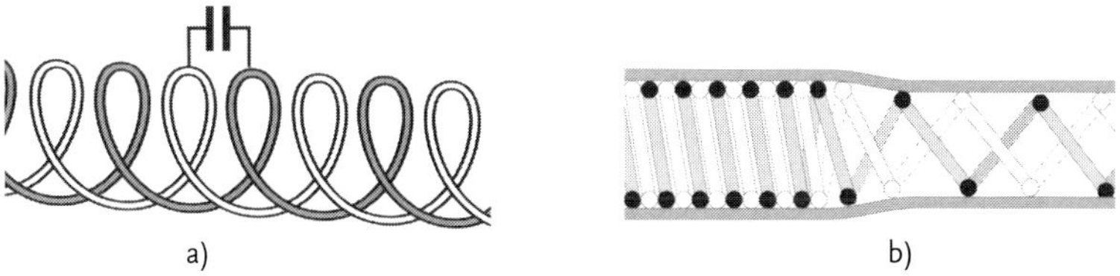

Fig. 11.22 Capacitance between coils through: a) bending; b) stretching.

movement. For catheters, movement detection has been used to measure the bending of the catheter [114]. The technique used in this device is relatively simple. It has two intertwined spirals made of metal. The readout is the capacitance between these two spirals, which will change if the catheter bends, or alternatively if they are pulled further apart. These two structures are illustrated in Fig. 11.22.

11.4.1.5 **Ultrasound**

Work at the Erasmus University in Rotterdam has led to a rotating mirror in the tip of the medical tool yielding an image not only of the thickness of the blood vessel wall but also of the material parameter elasticity (Fig. 11.23). Thus the surgeon can distinguish between fatty tissue plaques and calcified plaques.

11.4.2 Chemical

11.4.2.1 **pH Sensors**

An effective way to miniaturize the existent pH sensor is to use an ISFET (ion-sensitive field effect transistor). The ISFET can be made in a standard integrated circuit (IC) process [116], although the gate oxide preferred for this device is usually Al_2O_3 since it maintains a linear response over a wide range of pH. The sensor works on the basis of the surface charge buildup on the gate oxide due to the presence of hydrogen ions [117]. These devices are relatively simple to make, although, as with many chemical sensors, the long-term stability remains a problem. There has also been a miniaturized version using an Ag/AgCl reference elec-

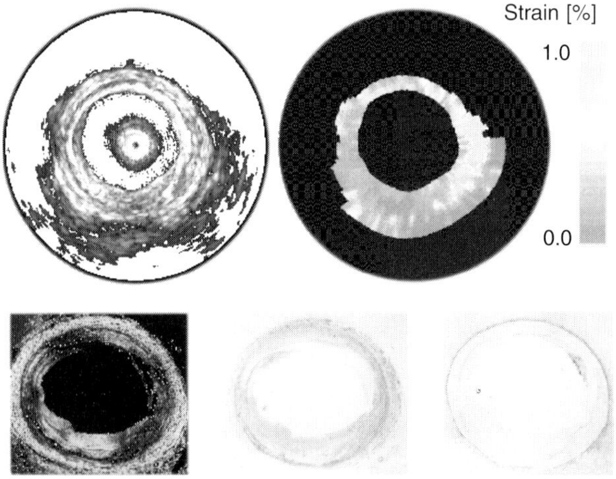

Fig. 11.23 Images of the blood vessel made from ultrasound devices incorporated in catheters [114] (reproduced with kind permission of Professor Ton van der Steen, Erasmus University Rotterdam).

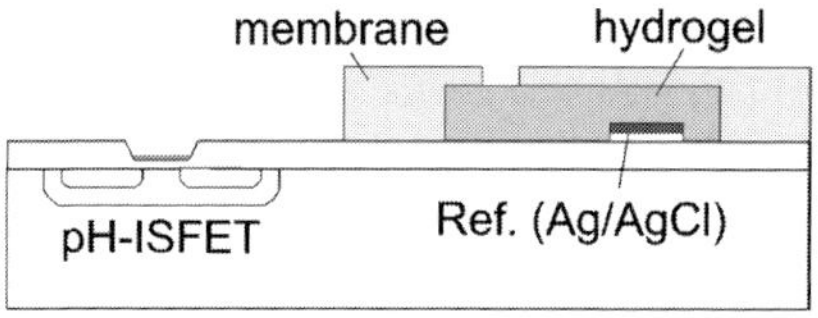

Fig. 11.24 pH sensor based on an ISFET with reference electrode.

trode [118], but this suffered from stability and reproducibility problems (Fig. 11.24). Although the accuracy of pH measurement with an intravascular ISFET sensor is less than usually claimed for off-line analysis on blood samples, it is within the normal physiological variation due to respiration. The measurement of pH can also be achieved using optical methods [119, 120]. In this case the light is reflected from a surface treated with, usually, an enzyme that reacts optically to the acidity. This is a robust method, which can easily be miniaturized. Unfortunately, the response time is rather long (9–100 s) and to date this method has not been applied to catheters. However, if some of the problems can be solved, this would be an interesting technique for pH measurement.

11.4.2.2 pO_2/pCO_2 Sensors

The standard method for off-line measurement of pO_2 in blood samples is using the Clark cell, which, just like the pH device, can be miniaturized [119] (Fig. 11.25). A version of this can also be made based on the ISFET [117]. The basic principle of the Clark cell is a reaction at the silver electrode as follows:

$$4e^- + O_2 + 2H_2O \rightarrow 4OH^-$$
$$Ag \rightarrow Ag^+ + e^-$$
$$Ag^+ + Cl^- \rightarrow AgCl$$

As can be seen in this simple reaction, the silver electrode is consumed, thus limiting the lifetime. However, for single, short-term use applications this may not be

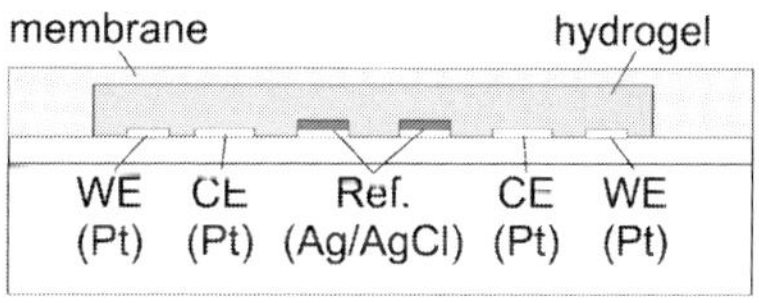

Fig. 11.25 Miniaturized Clark-cell pO_2 sensor.

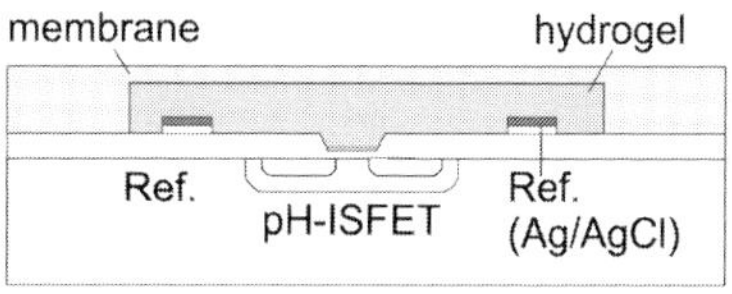

Fig. 11.26 Cross section of the CO_2 sensor.

a problem. More recently, modifications have been made to the electrode structure to increase the device lifetime significantly [121–123].

A miniaturized version of the Severinghaus electrode for pCO_2 measurement is possible based on the pH electrode described above, as shown in Fig. 11.26.

A miniature version has been made with the help of micromachining techniques [119, 124]. This device consists of a pH-ISFET with an SiN gate, an Ag/AgCl reference electrode, an internal electrolyte and a gas-permeable polysiloxane membrane. In a similar fashion, optical pH sensors for CO_2 can also be achieved [125]. It should be noted that temperature is an important factor in these measurements and should be measured simultaneously. The optical detection technique will be discussed further in the section on optical sensors.

11.4.2.3 Selective Chemical Measurement

For *in vivo* chemical measurements there are two problems which often have to be addressed: selectivity and biocompatibility. Some sensors use a special layer to trap the cells to be measured, but this has the problem that other cells may also be trapped. The second problem is finding layers which both trap the cells for the measurement and are also biocompatible enzyme-combined ISFET sensors. However, there are a number of chemical sensors with selective layers which are suitable for catheters, e.g., glucose [126, 127], lactate [128], sodium [129], ammonium and nitrates [130]. All of these sensors used electrical readout, but these can also be achieved using optical methods [131].

11.4.3 Magnetic

At first sight, it may seem that there are no medical applications for magnetic sensors in catheters. One application which uses magnetic measurements is the navigation system described in Section 11.3.4.7 [72, 73].

11.4.4 Thermal

The human body has a temperature-regulating system, but an attack from a foreign body may result in an increase in temperature, and the body tries to defeat the inflammation. This is on a global scale. It is also possible to have local increases in temperature in a blood vessel. The measurement of temperature in silicon is relatively simple since many components are temperature sensitive. A simple p–n junction will yield a linear temperature sensor covering a wide range of temperatures. However, if high accuracy is required, additional electronics will be necessary [132]. By means of localized heating (high-frequency electromagnetic fields) tumors can be treated.

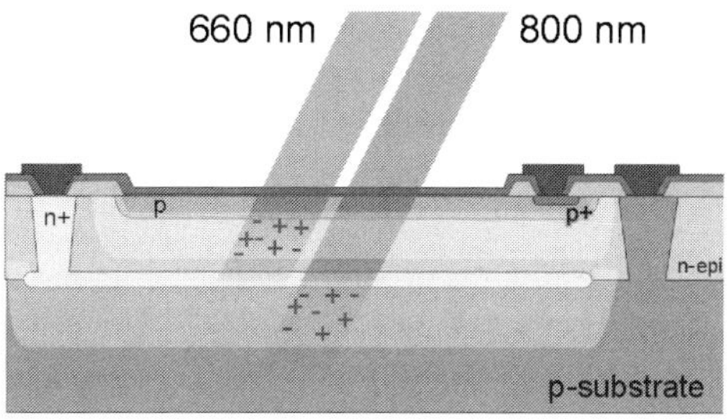

Fig. 11.27 Oxygen saturation sensor using optical techniques.

11.4.5 Radiation Sensors

Radiation sensors have been used in medicine over a wide range of frequencies. The x-ray was an early example of radiation generation and detection for medical applications. Although very effective in producing images of, for example, bones, we now know the health risks if used too frequently. Closer to the visible spectrum we have fluorescence measurements for analyses of tissue, absorption measurements for determining gases in liquids and colour measurements for estimating the condition of tissue, e.g., translumination of the skulls of babies.

The oxygen saturation level in blood is an important parameter since this is the oxygen attached to the hemoglobin and thus being transported to where it is needed. Oxygen-rich blood has a rich red color whereas when the oxygen saturation level is low the colour is darker. The oxygen saturation level can be determined by measuring the absorption at two wavelengths, 800 and 660 nm. The absorption at

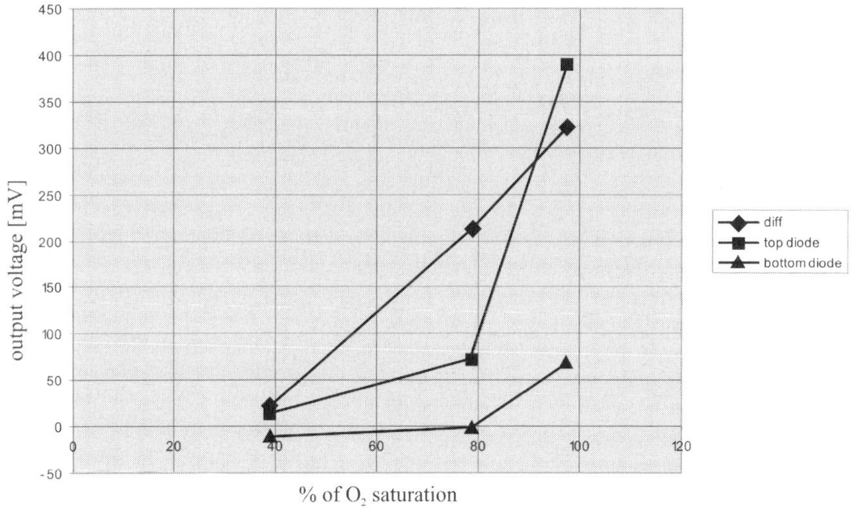

Fig. 11.28 Left: On-line measurements of oxygen saturation as a function of time; right: output as a function of oxygen saturation.

660 nm is relatively insensitive to oxy-hemoglobin and is used as a reference and at 800 nm it is highly sensitive. Fortunately, these wavelengths can be measured easily in silicon using two p–n junctions stacked vertically as shown in Fig. 11.27 [85]. The measurements of the two p–n junctions individually does not initially appear to give a clear result (Fig. 11.28), but when the difference is given as a function of oxygen saturation a linear response can be seen (also given in Fig. 11.28). To judge the oxygen transport of the blood, it is not sufficient to measure saturation of oxygen, and therefore the hematocrit is also needed (this can be done by impedance measurements).

The levels of CO_2 can also be measured in blood but this is more complicated than measuring the oxy-hemoglobin levels. In this case the CO_2 has to be extracted from the blood using a fine grating or membrane through which gases will pass but not blood. To detect the CO_2 levels we have to move further away from the visible spectrum to a wavelength of 4.23 μm. At this wavelength the absorption level is highly dependent upon CO_2 levels. Selectivity could also be a problem here if unknown gases have entered the blood system.

11.4.6 Electrical

Electrical measurements commonly used in medicine include nerve and muscle potentials, ECG (electrocardiography), EMG (electromyography), EEG (electroencephalography), EGG (electrogastrography) and EOG (electrooculography). Some of these involve simple passive electrodes. For applications such as neural probes, miniaturization, with multi-electrodes, which can be achieved with silicon, is an important tool for the physician (Fig. 11.29). A further step is the microelectrode for determining the internal potential of nerves. Holes in silicon through which axons grow contain electrodes to register signals from individual axons [133, 134]. Electrical transducers are also used for stimulation.

11.5 Multi-Sensing

The above section shows examples of sensors which are, or potentially could be, used in catheters. In many applications it is useful to be able to perform different measurements at a single location. The catheter can, of course, be replaced with another catheter having a different sensor. However, this approach does not yield simultaneous measurements. It is therefore often more useful to include more sensors on a single catheter. One combination could be temperature and plasma bicarbonate, pCO_2. An alternative is pressure-flow-oxygen saturation (Fig. 11.30). These

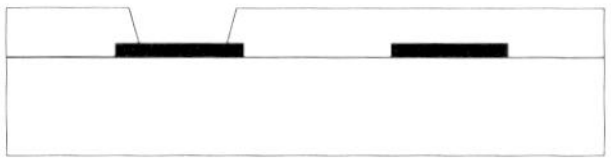

Fig. 11.29 Electrodes: left) direct contact; right) protected as reference.

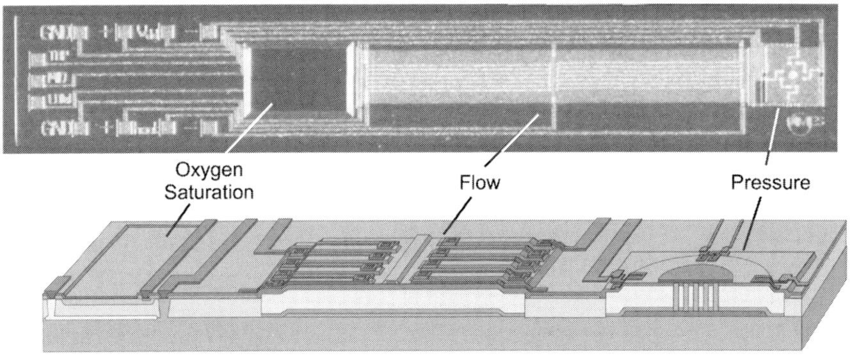

Fig. 11.30 Multi-parameter containing pressure, flow, and oxygen saturation.

individual sensors have been described above [72]. With these multi-sensors it is important to ensure that the added complexity does not lead to excessive costs.

11.6
Packaging Issues

The main issues in sensor packaging for catheters are biocompatibility, size and patient safety, and these are addressed in the following sections.

11.6.1
Size

If a sensor is to be incorporated in a small catheter or guide wire, miniaturization of the sensor chip is an important issue. Not only must the sensor(s) be small enough to fit in the instrument, but also the whole chip must be fitted in. The thickness of standard silicon wafers, which is ~520 μm, makes it impossible to fit the sensor chip, including read-out electronics to a guide wire which is smaller than, say, 700 μm (see Fig. 11.31a). This is three times the size of the smallest guide wires in use today, but no problem for more commonly used guide wires. For catheters the situation is more complicated. A catheter is a hollow tube that must fit over a guide wire in order to get to the area of interest (see Fig. 11.31 b). A sensor will have to be incorporated in the curved wall, which is in the order of 50–250 μm thick, leaving enough space for the guide wire. As the chip is not curved but flat, it will encroach on the internal opening of the catheter (see Fig. 11.31 c).

Another issue concerning the size of the sensor is its length. Although this is not as critical as the width or thickness, the length of the sensor will cause a portion of the instrument to be stiff. As the tip is usually very flexible in order to be able to maneuver it through the twists and turns of the blood vessels, this can hinder the maneuvering and cause a problem in reaching certain blood vessels. For this reason, sensor length is also limited. What that limit is, is determined by the applica-

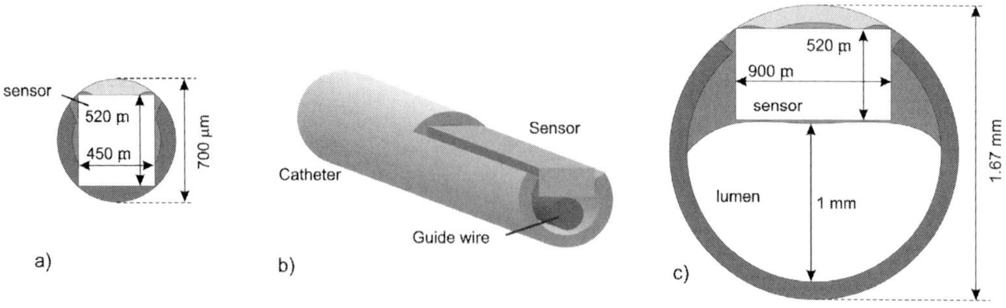

Fig. 11.31 a) Sensor chip size limitations due to guide wire diameter; b) fit of sensor chip in a catheter; c) resulting size limitations.

tion of the instrument, but four to five times the diameter of the instrument will not present problems in most cases. For esophageal and gastroenterological use the size is less critical. Even telemetric circuitry can be integrated with the sensor.

11.6.2
Working Environment and Patient Safety

Sensors on catheters and guide wires are in direct contact with the blood in order to be able to measure the desired parameters. Blood is an electricity-conducting medium. Therefore, the use of electricity for the sensors implies great risks for the heart. It seems unnecessary to mention that the functioning of the circulatory system is of vital importance and care must be taken with all procedures and instrument in this sensitive environment. The blood also flows through the whole body and any chemicals or materials introduced into the blood could go anywhere and cause damage at other locations.

Blood is a very aggressive environment where foreign objects, such as the sensor, are actively attacked. This physiological response must be controlled in order to minimize risks to the patient and keep the sensor in working order. Two responses must be considered. First, the immune system could perceive the object as an immediate biological threat and respond with an inflammation. This is a considerable risk to the patient using the device as a long-term implantation and should be avoided at all cost. The second response is the isolation of the foreign object from the body by encapsulating it, forming connective tissue around it. This leads to encapsulation and eventually the driving out of the device (when possible). Both of these reactions have consequences for the sensor design with regard to both the functioning of the sensor and the safety of the patient.

Another consideration for sensors in catheters and guide wires is that they are used in an environment with ionizing radiation. X-ray machines are used to track the progress of the catheter and to monitor the positioning of a guide wire, to image the blood vessels and to aid in diagnosis. The sensors must be able to survive this radiation, while in use. For this reason, delicate CMOS (complementary metal oxide

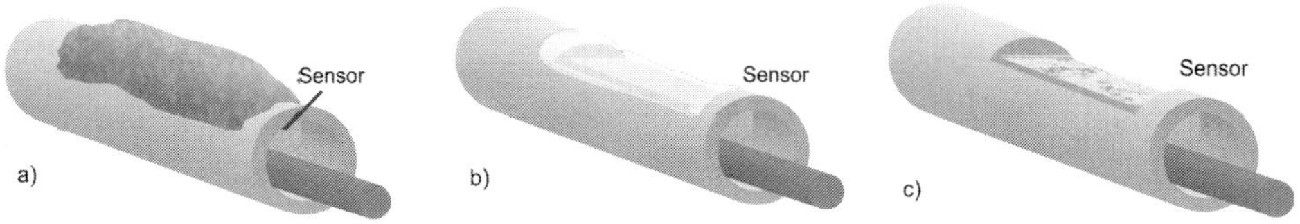

Fig. 11.32 Effects of the body detrimental to the sensor: a) immune response; b) insulation response; c) attack by chemicals.

semiconductor) processes are not always the first choice. Even if they make it possible to fabricate very small sensing devices, the electronic devices might not survive the radiation without shielding. However, not everything about this environment is bad news for the designer. The body has an excellent temperature-regulation system and the blood can be considered a stable thermal environment, compared with other environments that a sensor designer has to consider.

An infection response of the body must be avoided, not only from a health and safety point of view, as will be discussed later, but because it will also seriously affect the sensor. It will cover the sensor with material within minutes if not seconds and seriously disturb its surroundings, both chemically and mechanically, by changes in the chemical composition and mechanical changes of the blood vessel wall (attack by white blood cells) (see Fig. 11.32 a). All this makes it impossible for the sensor to function reliably. To make sure that this does not happen, the sensor and the instrument must be covered by appropriate biocompatible materials and be sterile.

Even if there is no immune response, there will always be the encapsulation to deal with (Fig. 11.32 b). It is impossible to prevent the deposition of materials completely, but the process can be slowed down. As catheters and guide wires are only used for a few hours maximum, a large buildup of material can be avoided by the use of proper materials and shapes. This will affect the sensor design, the choice of fabrication technology and the packaging. Furthermore, the sensor must be able to function when covered by a thin layer of material, as this will happen when it is introduced into the bloodstream. This will influence the fundamental measurement principles that can be used and the sensor optimization. Aspects to take into consideration are that the blood is a conductive fluid, which can cause short-circuits, and that the electrical connections on the sensor and sensor connections should be properly isolated. To this end, protective films should be free of pinholes, cracks, etc., that could cause problems.

Finally, the chemicals in the blood can attack certain materials, which could damage the working of the sensor or limit its lifetime (Fig. 11.32 c). The films used in silicon sensors are very thin and although the time in the blood is fairly short, this degradation could be enough to change sensitivities or aggravate problems with pinholes and micro-cracks.

Not only is the body a problem for the sensor, but also the sensor can present dangers for the patient. Examples of these problems are shown in Fig. 11.33, which include fragile parts which can come loose, local heating and infection.

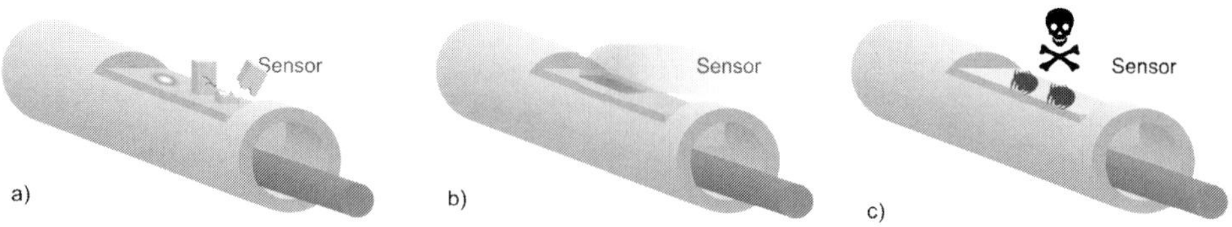

Fig. 11.33 Dangers to the body due to the sensor: a) hot or fragile parts; b) toxic materials; c) infected matter.

11.6.3
Packaging

The packaging of the sensor presents a challenge, as with all sensor packaging. The packaging must protect the sensor and the connecting wires from their environment, but still provide adequate access to this environment in order to be able to do the measurements. In addition, all the requirements for the sensor mentioned above apply also to the packaging, which should be added to a catheter or guide wire without interfering with its functioning.

The sensor chip needs to be placed on a stiff strip or in a partial tube that replaces part of the flexible wire. This not only supports the chip, it must also take up most of the mechanical stress put on the instruments when bent, pushed, etc. and may pose a problem due to the limited available space.

As the chip needs to be in contact with the blood, an opening in the side or end of the catheter or guide wire is required. To keep the sides of the instrument as smooth as possible to prevent damage to the vessel wall and facilitate cleaning and sterilization, the sensor must be placed right at the edge of this opening, as far outside the catheter as possible. However, this presents a problem with the connecting wires (see Fig. 11.34). As the sensitive side of the sensor will often be on the same side as the connections, normal bond wires would stick outside the profile of the instrument and be damaged or cause a very uneven surface for the packaging. Therefore, normal bond wires cannot be used, neither for outside connections nor for chip-to-chip connections in hybrid solutions. Other techniques, such as flexible thin films or through-wafer interconnections can be used to move the connections away from the surface. Other solutions have been used, such as soldering wires to recessed connection pads. Some attractive electrical materials like silver, copper, mercury and lead are, however, poisons for the body.

In most cases the chip surface will be covered by a protective layer, other than the standard protective layer deposited during manufacture, such as silicon nitride or silicon oxide. This can cover just the areas around the sensor and in some cases even the sensor itself. If only the area around the sensor is covered, it usually consists of a resin that is also used to seal the gaps between the chip and the frame. Otherwise it can be a silicone rubber, which is flexible, or certain plastics and resins. In some situations, intermediate layers, such as liquids or gels, can be used to separate the sensor from the hostile environment of the blood. Covering the sensor itself can only be done if the parameter to be measured is transmitted

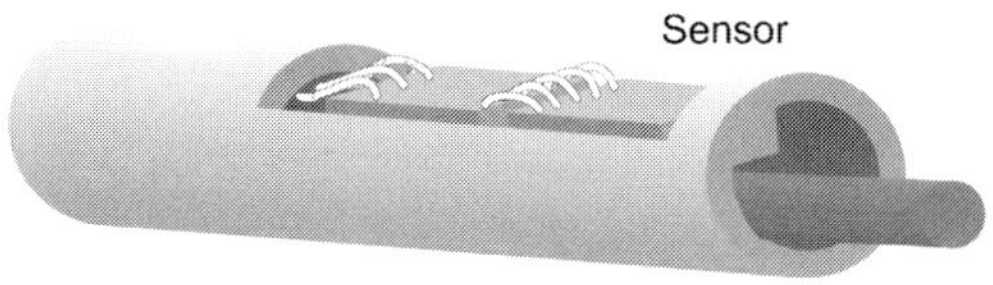

Fig. 11.34 Problems with packaging when bonding wires are on the outside of the chip.

through this layer. Pressure can be transmitted through a flexible layer and/or a liquid and optical information such as colour could be transmitted through glass or other transparent materials.

Both the covering material and the sealing material need to be compatible with the body, as discussed in Section 11.6.2 and are used not only to protect and fix the sensor chip and its connections, but also to smooth the surface and provide transitions between the flat sensor surface and the round instrument cross section.

A good packaging strategy is often as difficult as or even more difficult than the sensor design itself. Adhesion between different materials can be a problem and also shrinkage during the curing of glues and resins. In addition to having to be biocompatible, the materials have to survive the sterilization process, which can be a chemical, radioactive and thermal treatment. The high temperatures used for sterilization can introduce stresses that cause seals to fail, connections to break, etc. These stresses can also cause drift in the sensors, necessitating calibration before use. Therefore, all of these issues need to be solved before the instrument can be approved for clinical use. Two examples of packaging for catheters are given in Fig. 11.35.

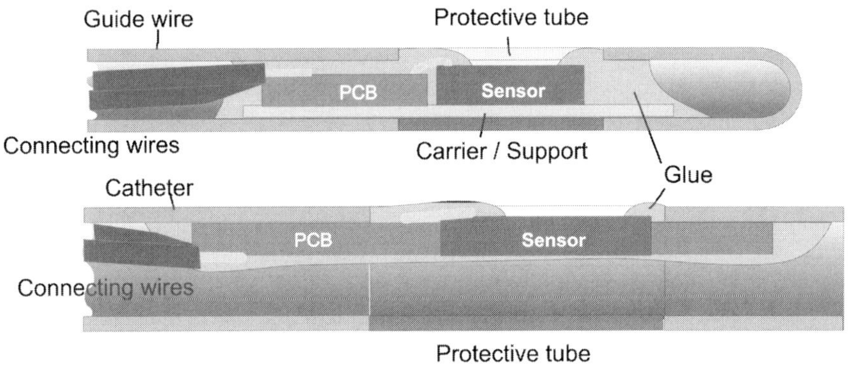

Fig. 11.35 Two examples of packaged chips in catheters.

11.7 Conclusions

This chapter has considered a number of aspects regarding the navigation of medical tools and has presented different sensors for catheters. The continuous development of silicon sensors is playing an important role in modern medicine, where more operations are being performed through minimally invasive techniques. Sensors that are able to meet these demands on biocompatibility, sterility, reliability, packaging and patient safety will open many new opportunities for medical techniques.

11.8 References

1 Mueller, R. L., Sanborn, T. A., *Am. Heart J.* **129** (1995) 146–172.
2 Forssmann, W., *Klin. Wochenschr.* **8** (1929) 2085.
3 Seldinger, S. J., *Acta Radiol.* **39** (1953) 368.
4 Dotter, C. T., Judkins, M. P., *Circulation* **30** (1964) 654–670.
5 Payne, M. M., *Tex. Heart Inst. J.* **28** (2001) 28–38.
6 King, S. B., *Circulation* **93** (1997) 1621–1629.
7 Myler, R. K., Stertzer, S. H., *Coronary and Peripheral Angioplasty: Historic Perspective, Textbook of Interventional Cardiology*, 2nd edn., Vol. 1. Topol, E. (ed.); WB Saunders, Philadelphia 1993.
8 Valji, K., Bralow, L., *Vascular and Interventional Radiology*, WB Saunders, Philadelphia, 1999.
9 Baim, D. S., Grossman, W. M. D., *Cardiac Catheterization, Angiography and Intervention*; Lippincott, Williams & Wilkins, Philadelphia 2000.
10 Crawford, M. H., DiMarco, J. P. (eds.) *Cardiology*, Mosby Publishing 2001.
11 Youngblut, C., Johnson, R. E., Nash, S. H., Wienclaw, R. A., Will, C. A., *Review of Virtual Environment Interface Technology*, Institute for Defense Analyses (IDA), IDA Paper P-3186, 1996.
12 Ziskin, M. C., *RadioGraphics* **13** (1993) 705–709.
13 Holm, H. H., *Ultrasound Med. Biol.* **24** (1998) 779–791.
14 Machi, J., Sigel, B., Kurohiji, T., Zaren, H. A., Sariego, J., *Ultrasound Med. Biol.* **16** (1990) 37–42.
15 Gunn, J., Cumberland, D. C., Siegel, R. J., *Echocardiography* **18** (2001) 213–217.
16 Abreu, J. M. M., Ceres, R., Calderon, L., Jimenez, M. A., Gonzalez-de-Santos, P., *Sens. Actuators A* **75** (1999) 131–138.
17 van der Steen, A. F. W., de Korte, C. L., Cespedes, E. I., *Ultraschall Med.* **19** (1998) 196–201.
18 Roelandt, J. R., di Mario, C., Pandian, N. G., Wenguang, L., Keane, D., Slager, C. J., de Feyter, F. J., Serruys, P. W., *Circulation* **90** (1994) 1044–1055.
19 Woodward, B., Sharp, H., *Ultrasonics* **29** (1991) 381–388.
20 Erbel, R., Roth, T., Koch, L., Ge, J., Görge, G., Serruys, P. W., Bom, N., *Minim. Invasive Ther.* **6** (1997) 195–198.
21 Salimuzzaman, M. D., Matani, A., Oshiro, O., Chihara, K., Asao, M., *Jpn. J. Appl. Phys.* **36** (1997) 3250–3254.
22 Laban, M., Oomen, J. A. F., Slager, C. J., Wentzel, J. J., Krams, R., Schuurbiers, J. C. H., den Boer, A., Birgelen von, C., Serruys, P. W., Feijter de, P. J., *Comput. Cardiol.* (1995) 325–328.
23 Goldstein, A., *RadioGraphics* **13** (1993) 701–704.
24 Vilkomerson, D., Lyons, D., *US Pat.* 5161536, 1992.
25 Vilkomerson, D., Lyons, D., *IEEE Trans. Ultrason. Ferroelec. Frequency Control* **44** (1997) 496–504.

26 Vilkomerson, D., Mezrich, R. S., Rubin, C. S., *US Pat.* 4249539, 1981.
27 Cluley, S. R., Brener, B. J., Hollier, L. H., Eisenbud, D. E., Ferrara-Ryans, M., Villanueva, A., Parsonnet, V., Vilkomerson, D., *Am. J. Surg.* **162** (1991) 117–121.
28 Ramaswami, G., al Kutoubi, A., Nicolaides, A. N., Dhanjil, S., Griffin, M., Ryan, M. F., *J. Endovasc. Surg.* **2** (1995) 27–35.
29 Ramaswami, G., al Kutoubi, A., Nicolaides, A. in: *Proceedings of the World Congress on Angiology*, London, 1995, p. 16.
30 *NSF/ERC for Emerging Cardiovascular Technologies, 7th Annual Report*; National Science Foundation, Washington, DC, 1995.
31 Taylor, K. J. W., Holland, S., *Radiology* **174** (1990) 297–307.
32 Scoutt, L. M., Zawin, M. L., Taylor, K. J. W., *Radiology* **174** (1990) 309–319.
33 Lefrant, J. Y., Cuvillon, P., Benezet, J. F., Dauzat, M., Peray, P., Saissi, G., de la Coussaye, J. E., Eledjam, J. J., *Anesthesiology* **88** (1998) 1195–1201.
34 Frazin, L. J., Vonesh, M. J., Khasho, F., Lanza, G., Chandran, K. B., Talano, J. V., McPherson, D. D., *Catheter. Cardiovasc. Diagn.* **26** (1992) 41–50.
35 Radke, J., Lange, L. P., Bartels, A., Brunsch, Ch., *Anaesthesist* **39** (1990) 283–287.
36 Scoccianti, M., Verbin, C. S., Kopchok, G. E., Back, M. R., Donayre, C. E., Sinow, R. M., White, R. A., *J. Endovasc. Surg.* **1** (1994) 71–80.
37 Gussenhoven, W. J., Zhong, Y., Li, W., van Egmond, F., Pieterman, H., van Urk, H., Gerritsen, G. P., Borst, C., Wilson, R. A., *Circulation* **86** (1992) 483–493.
38 Gregory, K. W., Aretz, H. T., Martinelli, M. A., Ledet, E. G., Hatch, G. F., Gregg, R. E., Sedlacek, T., Haase, W. C., Anderson, R. R., *Proc. SPIE* **1425** (1991) 217–225.
39 Leotta, D. F., Detmer, P. R., Gilja, O. H., Jong, J.-M., *IEEE Ultrason. Symp.* (1995) 1415–1418.
40 Leotta, D. F., Detmer P., R., Martin, R. W., *Ultrasound Med. Biol.* **23** (1997) 597–609.
41 Meyer, S. A., Wolf, P. D., *IEEE Trans. Biomed. Eng.* **44** (1997) 1061–1067.
42 *Basic Principles of MR Imaging*, 3rd edn, Philips Medical Systems.
43 Mueller, P. R., Stark, D. D., Simeone, J. F., Saini, S., Butch, R. J., Edelman, R. R., Wittenberg, J., Ferrucci, J. T., *Radiology* **161** (1986) 605–609.
44 Kochli, V. D., McKinnon, G. C., Hofmann, E., von Schulthess, G. K., *Magn. Reson. Med.* **31** (1994) 309–314.
45 Dumoulin, C. L., Souza, S. P., Darrow, R. D., *Magn. Reson. Med.* **29** (1993) 411–415.
46 Rubin, D. L., Ratner, A. V., Young, S. W., *Invest. Radiol.* **25** (1990) 1325–1332.
47 Ludeke, K. M., Roschmann, P., Tischler, R., *Magn. Reson. Imaging* **3** (1985) 329–343.
48 Ackerman, J. L., Offut, M. C., Buxton, R. B., Brady, T. J., in: *Rapid 3D tracking of small RF coils; Fifth Annual Meeting of the Society of Magnetic Resonance in Medicine,* Montreal 1986, p. 1131.
49 Dumoulin, C. L., Souza, S. P., Darrow, R. D., *Magn. Reson. Med.* **29** (1993) 411–415.
50 Leung, D. A., Debatin, J. F., Wildermuth, S., McKinnon, G. C., Holtz, D., Dumoulin, C. L., Darrow, R. D., Hofmann, E., von Schulthess, G. K., *Am. J. Radiol.* **164** (1995) 1265–1270.
51 Wildermuth, S., Debatin, J. F., Leung, D. A., Dumoulin, C. L., Darrow, R. D., Uhlschmid, G., Hofmann, E., Thyregod, J., von Schulthess, G. K., *Radiology* **202** (1997) 578–583.
52 Rasche, V., Holz, D., Kohler, J., Proksa, R., *Magn. Reson. Med.* **37** (1997) 963–968.
53 Wittkampf, F. H. M., Wever, E. F. D., Derksen, R., Wilde, A. A. M., Ramanna, H., Hauer, R. N. W., de Medina, E. O. R., *Circulation* **99** (1999) 1312–1317.
54 Wittkampf, F. H. M., *US Pat.* 5697377, 1997.
55 Molenschot, M., Ramanna, H., Hoorntje, T., Wittkampf, F., Hauer, R., Derksen, R., Sreeram, N., *Pacing Clin. Electrophysiol.* **24** (2001) 1616–1622.

56 Gepstein, L., Evans, S. J., *PACE* **21** (1998) 1268–1278.

57 Ben-Haim, S. A., Osadchy, D., Schuster, I., Gepstein, L., Hayam, G., Josephson, M. E., *Nat. Med.* **2** (1996) 1393–1395.

58 Worley, S., *PACE* **21** (1998) 1636–1645.

59 Shpun, S., Gepstein, L., Hayam, G., Ben-Haim, S. A., *Circulation* **96** (1997) 2016–2021.

60 Gepstein, L., Hayam, G., Ben-Haim, S. A., *Circulation* **95** (1997) 1611–1622.

61 Kornowski, R., Hong, M. K., Gepstein, L., Goldstein, S., Ellahham, S., Beh-Haim, S. A., Leon, M. B., *Circulation* **98** (1998) 1116–1124.

62 Bhargava, B., Kornowski, R., *Asian Cardiovasc. Thorac. Ann.* **7** (1999) 345–348.

63 Meeker, D. C., Maslen, E. H., Ritter, R. C., Creighton, F. M., *IEEE Trans. Magn.* **32** (1996) 320–328.

64 Molloy, J. A., Ritter, R. C., Broaddus, W. C., Grady, M. S., Howard III, M. A., *Med. Phys.* **18** (1991) 794–803.

65 Gillies, G. T., Ritter, R. C., Broaddus, W. C., Grady, M. S., Howard III, M. A., McNeil, R. G., *Rev. Sci. Instrum.* **65** (1994) 533–562.

66 Meeker, D. C., Maslen, E. H., Ritter, R. C., Creighton, F. M., *IEEE Trans. on Magnetics* **32(2)** (1996) 320–328.

67 Hall, W. A., Liu, H., Martin, A. J., Maxwell, R. E., Truwit, C. L., *J. Neurosurg.* **94** (2001) 67–71.

68 Grady, M. S., Howard, M. A., Dacey, R. G., Blume, W., Lawson, M., Werp, P., Ritter, R., *J. Neurosurg.* **93** (2000) 282–288.

69 Starkhammar, H., Bengtsson, M., Kay, D. A., Shapiro, A. R., *Biomed. Instrum. Technol.* **30** (1996) 164–170.

70 Starkhammar, H., Bengtsson, M., Kay, D. A., *Acta Anaesthesiol. Scand.* **34** (1990) 296–300.

71 *Flock of Birds™, Six-Degrees-of-Freedom Measurement Device, Technical Description*; Ascension Technology, Burlington, VT.

72 Tanase, D., Goosen, J. F. L., Trimp, P. J., French, P. J., *Sens. Actuators A* **97/98** (2002) 116–124.

73 Tanase, D., Bakker, N. H., van Loon, D., Goosen, J. F. L., Trimp, P. J., Reekers, J. A., French, P. J., in: *Proceedings IEEE–EMBS*; Madison, WI, 2–4 May 2002, pp. 305–308.

74 Schlageter, V., Besse, P. A., Popovic, R. S., Kucera, P., in: *Eurosensors XIV*; The Hague, The Netherlands, 1999, pp. 679–682.

75 Weitschies, W., Kotitz, R., Cordini, D., Trahms, L., *J. Pharm. Sci.* **86** (1997) 1218–1222.

76 Prakash, N. M., Spelman, F. A., in: *Proceedings of the IEEE–EMBS Conference*; Chicago, IL, 1997, pp. 2394–2397.

77 Ilmoniemi, R. J., Williamson, S. J., Boyd, A. D., *IEEE Trans. Biomed. Eng.* **35** (1988) 561–564.

78 Gunkel, A. R., Freysinger, W., Thumfart, W. F., Truppe, M. J., presented at: *Computer Assisted Radiology, 9th International Symposium and Exhibition*; Berlin, June 21–24, 1995.

79 Bladen, J. S., Anderson, A. P., Bell, G. D., Heatley, D. J., *Lancet* **341** (1993) 719–722.

80 Williams, C. B., Saunders, B. P., Bell, G. D., Bladen, J. S., Gillies, D. F., Guy, C. M., *Gastrointest. Endosc. Clin. N. Am.* **7** (1997) 469–475.

81 Bladen, J. S., Anderson, A. P., Bell, G. D., Heatley, D. J., in: *IEEE Nuclear Science Symposium and Medical Imaging Conference*; 1993, pp. 1891–1894.

82 Saunders, B. P., Bell, G. D., Williams, C. B., Bladen, J. S., Anderson, A. P., *Gut* **36** (1995) 913–917.

83 Weitschies, W., Wedemeyer, J., Stehr, R., Trahms, L., *IEEE Trans. Biomed. Eng.* **40** (1994) 192–195.

84 Sugiyama, S., Shimaoka, K., Tabata, O., in: *Proceedings of the 1991 International Conference on Solid-State Sensors and Actuators (Transducers 91)*; San Francisco, 24–28 June 1991, pp. 188–191.

85 Goosen, J. F. L., French, P. J., Sarro, P. M., in: *Proceedings of IEEE–MEMS 2000*; Miyazaki, January 2000, pp. 537–540.

86 Schöneberg, U., Schnatz, F. V., Brockherde, W., Koystynski, P., Mehlhorn, T., Obermeier, E., Benzel, H., in: *Proceedings of the 1991 International Conference on Solid-State Sensors and Actuators*

(Transducers 91)*; San Francisco, 24–28 June 1991, pp. 304–307.

87 Ji, J., Cho, S.T., Zhang, Y., Najafi, K., Wise, K.D., in: *Proceedings of the 1991 International Conference on Solid-State Sensors and Actuators (Transducers 91)*; San Francisco, 24–28 June 1991, pp. 1018–1020.

88 Tenerz, L., in: *Proceedings of the 1991 International Conference on Solid-State Sensors and Actuators (Transducers 91)*; San Francisco, 24–28 June 1991, pp. 1021–1023.

89 Tohyama, O., Kohashi, M., Yamamoto, K., Itoh, H., in: *Proceedings of the 8th International Conference on Solid-State Sensors and Actuators (Transducers 95/Eurosensors IX)*; Stockholm, 25–29 June 1995, Vol. 2, pp. 596–599.

90 Chan, M.A., in: *Proceedings of the 7th International Conference on Solid-State Sensors and Actuators (Transducers 93)*; Yokohama, 7–10 June 1993, pp. 580–583.

91 Benaissa, K., Nathan, A., in: *Proceedings of the 8th International Conference on Solid-State Sensors and Actuators (Transducers 95/Eurosensors X)*; Stockholm, 25–29 June 1995, News, pp. 77–78.

92 World Precision Instruments, *http://www.wpiinc.com/*.

93 Kaufman, K.R., Wavering, T., Morrow, D., Davies, J., Lieber, R.L., *J. Biomech.* **36** (2003) 283–287.

94 Kälvesten, E., Smith, L., Tenerz, L., Stemme, G., "The first surface micromachined pressure sensor for cardiovascular pressure measurements", *Proceedings IEEE MEMS 1998*, Heidelberg, Germany, January 1998, pp. 574–579.

95 Melvs, P., Kälvesten, E., Stemme, G., *Sens. Actuators A* **100** (2002) 46–53.

96 Poll, T., Fröhlich, G., *Urodynamik-Leitfaden*; Springer, Berlin, 1995.

97 Knoll, M., Schwartenbeck, G., Eshold, J., Ross, B., *Sens. Actuators A* **93** (2001) 52–56.

98 Schmidt, H.-J., Holtkamp, R., Benecke, W., in: *Proceedings of Micromechanics Europe (MME96)*; Barcelona, 21–22 October 1996, pp. 156–159.

99 Nishimoto, T., Shoji, S., Esashi, M., in: *Proceedings of the 7th International Conference on Solid-State Sensors and Actuators (Transducers 93)*; Yokohama, 7–10 June 1993, pp. 796–799.

100 Boillat, M.A., van der Wiel, A.J., Hogerwerf, A.C., de Rooij, N.F., in: *Proceedings of the IEEE Micro Electro Mechanical Systems Workshop (MEMS95)*; 1995, pp. 350–352.

101 Oosterbroek, R.E., Lammering, T.S.J., Berenschot, J.W., van den Berg, A., Elwenspoek, M.C., in: *Proceedings of the 1997 International Conference on Solid-State Sensors and Actuators (Transducers 97)*; Chicago, IL, 16–19 June 1997, pp. 151–154.

102 Berberig, O., Nottmeyer, K., Mizuno, J., Kanai, Y., Kobayashi, T., in: *Proceedings of the 1997 International Conference on Solid-State Sensors and Actuators (Transducers 97)*; Chicago, IL, 16–19 June 1997, pp. 155–158.

103 Rapoport, S.D., Reed, M.L., Weiss, L.E., *J. Micromech. Microeng.* **1** (1991) 60–65.

104 van der Wiel, A.J., Linder, C., de Rooij, N.F., *Sens. Actuators A* **37/38** (1993) 693–697.

105 Kersjes, P., Liebscher, F., Spiegel, E., Manoli, Y., Mokwa, W., *Sens. Actuators A* **54** (1996) 563–567.

106 Mágori, V., in: *Proceedings of the Ultrasonics Symposium 1985*; San Francisco, 16–18 October 1985, pp. 525–530.

107 *http://www.mein.nagoya-u.ac.jp/activity/1995/Micro_Force_Sensor-e.html*. University of Nagoya, Japan.

108 De Souza, R.J., Wise, K.D., in: *Proceedings of the 1997 International Conference on Solid-State Sensors and Actuators (Transducers 97)*; Chicago, IL, 16–19 June 1997, pp. 1473–1476.

109 Wolffenbuttel, M.R., Regtien, P.P.L., *Sens. Actuators A* **25–27** (1991) 257–264.

110 Weimann, M., Radius, R., Assmus, F., *Sens. Actuators A* **37/38** (1993) 715–722.

111 Hillis, W.D., *Int. J. Robot. Res.* **1**, No. 2 (1982) 33–44.

112 Nakamura, Y., Hanafusa, H., Ueno, N., *Robot Sensors, Vol. 2, Tactile and Non-Vision*; IFS Publications and Springer, Berlin, 1986.

113 Shinoda, H., Matsumoto, K., Ando, S., in: *Proceedings of the 1997 International Conference on Solid-State Sensors and Ac-*

tuators (Transducers 97); Chicago, IL, 16–19 June 1997, pp. 129–132.

114 CIMMINO, A., KLEIN, A. G., OPAT, G. I., in: *Proceedings of the 8th International Conference on Solid-State Sensors and Actuators (Transducers 95/Eurosensors IX)*; Stockholm, 25–29 June 1995, News, pp. 84–85.

115 DE KORTE, C. L., PASTERKAMP, G., VAN DER STEEN, A. F., WOUTMAN, H. A., BOM, N., "Characterization of plaque components with intravascular ultrasound elastography in human femoral and coronary arteries in vitro", *Circulation* **102** (2000) 617–623.

116 BOUSSE, L., BERGVELD, P., *Sens. Actuators* **6** (1984) 65–78.

117 KIM, C.-S., KWON, D.-H., SOHN, B.-K., in: *Proceedings of the 8th International Conference on Solid-State Sensors and Actuators (Transducers 95/Eurosensors IX)*; Stockholm, 25–29 June 1995, Vol. 2, pp. 913–916.

118 ARQUINT, PH., VAN DER BERG, A., VAN DER SCHOOT, B. H., DE ROOIJ, N. F., BÜHLER, H., MORF, W. E., DÜRSELEN, L. F. J., *Sens. Actuators B* **13** (1993) 340–343.

119 ALABBAS, S. H., ASHWORTH, D. C., BENZZAA, B., MOMIN, S. A., NARAYANASWAMY, R., *Sens. Actuators A* **51** (1996) 129–134.

120 AGAYN, V. I., WALT, D. R., *Bio/Technology* **11** (1993) 726–729.

121 SUZUKI, H., HIRATSUKA, A., SASAKI, S., KARUBE, I., *Sens. Actuators B* **46** (1998) 104–113.

122 SUZUKI, H., HIRAKAWA, T., WATANABE, I., KIKUCHI, Y., *Anal. Chim. Acta* **431** (2001) 249–259.

123 SUZUKI, H., HIRATSUKA, T., HOSHI, T., TOYOOKA, H., *Sens. Actuators B* **76** (2001) 565–572.

124 GUMBRECHT, W., STANZEL, M. J., PORRO, G., in: *Proceedings of the 8th International Conference on Solid-State Sensors and Actuators (Transducers 95/Eurosensors IX)*; Stockholm, 25–29 June 1995, Vol. 2, pp. 921–924.

125 PARKER, J. W., LAKSIN, O., YU, C., LAU, J. L., KLIMA, S., FISCHER, R., SCOTT, I., ATWATER, B. A., *Anal. Chem.* **65** (1993) 2329–2334.

126 MURAKAMI, T., NAKAMOTO, S., KUMURA, J., KURIYAMA, T., KARUBE, I., *Anal. Lett.* **19** (1986) 1973–1986.

127 KURIYAMA, T., in: *Proceedings of the 8th International Conference on Solid-State Sensors and Actuators (Transducers 95/Eurosensors IX)*; Stockholm, 25–29 June 1995, Vol. 1, pp. 447–450.

128 HINTSCHE, R., KRUSE, CH., UHLIG, A., PAESCHKE, M., LISEC, T., SCHNAKENBERK, U., WAGNER, B., *Sens. Actuators B* **26/27** (1995) 471–473.

129 UHLIG, A., SCHNAKENBERG, U., LINDNER, E., DIETRICH, F., HINTSCHE, R., in: *Proceedings of the 8th International Conference on Solid-State Sensors and Actuators (Transducers 95/Eurosensors IX)*; Stockholm, 25–29 June 1995, Vol. 2, pp. 469–472.

130 KNOLL, M., CAMMANN, K., DUMSCHAT, C., SUNDERMEIER, C., ESHOLD, J., *Sens. Actuators B* **18/19** (1994) 51–55.

131 ABDEL-LATIF, M. S., SULEIMAN, A. A., GUILBAULT, G. G., *Anal. Lett.* **21** (1988) 943–951.

132 BAKKER, A., HUIJSING, J. H., *High-Accuracy CMOS Smart Temperature Sensors*; Kluwer, Boston, 2000.

133 KEWLWY, D. T., HILLS, M. D., BORKHOLDER, D. A., OPRIS, I. E., MALUF, N. I., STROMENT, C. W., BOWER, J. M., KOVACS, G. T. A., *Sens. Actuators A* **58** (1997) 27–35.

134 VALDERRAMA, E., GARRIDO, P., CABRUJA, E., HEIDUSCHKA, P., HARSCH, A., GÖPEL, W., in: *Proceedings of the 8th International Conference on Solid-State Sensors and Actuators (Transducers 95/Eurosensors IX)*; Stockholm, 25–29 June 1995, Vol. 1, pp. 63–66.

12
Home Health Care and Telecare

T. Togawa

12.1
Introduction

At present, sensor applications in home health care and telecare are fairly limited. However, their potential importance has been recognized and stressed recently, for many reasons including to cope with the increases in the elderly population, to obtain more information for early diagnosis and disease prevention, to allow patient care at a distance, to reduce the period of admission in hospital, to reduce costs, and to provide home care for the many patients who prefer it. When a sensor or an instrument is used at home, it must be simple to use, safe, reliable enough, inexpensive, and must not disturb normal home life. Most sensors and instruments used in hospitals and medical facilities are inconvenient or inadequate for use at home. Therefore, special consideration must be paid in sensor and instrument design. Moreover, measurement principles or even measurement quantities must be modified to fit the requirements of home care.

Requirements for sensors depend on the purpose and situation of monitoring. For patient care after discharge from hospital or for high-risk patients, home monitoring is performed to confirm patient condition and reduce risks. In such cases, monitored parameters are similar to those monitored in hospitals, and physicians or other medical staff watch the real-time data or past trends through information transmission systems. Sensors used for such purposes are not essentially different from those used in hospitals, even though the instruments should be safe and simple enough for patients to use. For example, the pulse oximeter is commonly used at home when a patient is treated by oxygen therapy or a respirator.

For the elderly and for patients with chronic diseases, monitoring vital signs such as heartbeat and respiration is expected to detect emergencies automatically rather than requiring the patient to push a button in an emergency. For such purposes, sensors must be attached to the subject all the time. Although such sensors for general use are not yet available, there are some situations where monitoring is possible, such as monitoring sleep apnea. There are also some attempts to develop wearable monitors such as the ring-type pulse monitor wearable on a finger. In principle, an implantable sensor device for monitoring can be con-

Sensors in Medicine and Health Care
Edited by P. Å. Öberg, T. Togawa, and F. A. Spelman

ISBN: 3-527-29556-9

ceived, as actually realized in some cardiac pacemakers, which have a sensor, memory, and a system for data readout by telemetry.

Automatic health monitoring has been considered for health management and early diagnosis of diseases. The onset of many adult diseases such as arteriosclerosis, hypertension, and diabetes is highly correlated with life habits such as smoking, nutrition, physical exercise, and sleep. Removing risk factors of diseases will contribute to prevention, and thus lifestyle monitoring will be important to achieve healthy living. Although monitoring daily life habits is not yet well developed, there have been many attempts at installing sensors and monitoring daily life at home.

Monitoring health parameters will also contribute to finding correlations with the onset of many diseases. Identifying risk factors for diseases is a major subject of epidemiology. However, it has been done by large studies recruiting many volunteers who are asked to record their own history of daily living and to accept regular performance of medical examinations. If automatic monitoring systems can be introduced, more reliable and objective data can be obtained.

Recently, there have been many attempts to develop sensors and instruments for home monitoring based on entirely new principles and design concepts. Further, there are many possibilities of applying high technologies for this purpose. In this chapter, recent proposals for various sensors, instruments, and measurement principles, as well as expected technologies, are reviewed.

12.2 Blood Pressure

Blood pressure is an important parameter for monitoring in patients. It can vary largely in a day, and even an occasional higher blood pressure can be a risk factor for serious problems such as brain hemorrhage or rupture of an aneurysm. Monitoring normal variation of blood pressure is demanded especially when it is controlled by drugs. Frequent measurement of blood pressure requires home measurements for most subjects, even those visiting a clinic regularly. Blood pressure measurement at home is also recommended because unusually high blood pressure is often observed because of psychological stresses at the clinic. Such cases are called 'white coat' hypertension.

Medical personnel usually measure blood pressure by the auscultatory method, in which a pressure cuff is attached to the upper arm and inflated to compress the brachial artery above the systolic pressure. The cuff is then deflated gradually while listening to the Korotkoff sounds through a stethoscope placed on the brachial artery distal to the cuff. The systolic and diastolic pressures are determined by reading the manometer at the beginning and end points of the sound. However, this technique requires skill and it is not appropriate for ordinary people. It is also difficult to perform on oneself.

For blood pressure measurement at home, convenient automatic instruments have been developed, and are commercially available. Most of them use an ordinary cuff attached to the upper arm, and systolic and diastolic pressures can be determined automatically. In many instruments, inflation and deflation of the cuff can be performed automatically. To determine systolic and diastolic pressures, two different methods are commonly used in commercial instruments. In the first method, the Korotkoff sounds are detected by a microphone installed beneath the cuff, and systolic and diastolic pressures are identified from the obtained sounds and the cuff pressure measured by a pressure sensor. An advantage of this method is that the measurement principle is the same as the standard auscultatory method, and thus higher correlation with the standard method is expected as long as the system works perfectly. However, there are many sources of error because of wrong positioning of microphone, external noise, and inadequate signal processing. Despite such difficulties, fairly reliable readings can be obtained when the cuff is attached correctly.

In the second method, which is called the oscillometric method, pulsatile components in the cuff pressure detect pulsation of the artery. When the cuff pressure is reduced slowly, pulsation appears in the systolic pressure, and the amplitude of pulsation increases and then decreases again. The amplitude of pulsation is always maximum when the cuff pressure is equal to the mean arterial pressure [1]. However, it is difficult to determine diastolic pressure directly from the signals superposed on the cuff pressure. Thus, in the commercial systems, diastolic pressure is determined indirectly. A simple way is to use a rule that mean arterial pressure remains at a definite ratio that is around 40% from diastolic to systolic pressures [2]. If this rule is correct, diastolic pressure can be estimated from systolic and mean arterial pressure. While the rule is not very reliable generally, it gives a good estimate as long as the arterial waveform is normal. The oscillometric method has advantages: the attachment of the cuff is not so critical as in the case of using a microphone for detecting Korotkoff sounds, and only a sensor measuring the cuff pressure is required to detect signals. This method is used in many commercial blood pressure monitors for home use.

Blood pressure measurement is not restricted to the upper arm. It is possible at other sites of the body as long as the artery can be occluded by the applied pressure, and the arterial pulsation can be detected. Blood pressure monitors to be used at the wrist or on a finger as shown in Fig. 12.1 are available (Wrist type: Wrist Sphygmomanometer, Heine Memotronic. Finger type: Finger Blood Pressure Monitor, OMRON). However, the error due to the gravitational force, which is proportional to the relative height from the heart level, becomes larger than that of the measurement at the upper arm, and the arterial pressure waveform changes when the measurement site is changed from the upper arm to the periphery of the extremity, so that the mean pressure reduces whereas the peak pressure sometimes increases.

Some studies have shown that blood pressure monitoring can be achieved at other sites such as the temple, using a pressure pad to apply pressure [3]. If a similar method were applicable to the buttock, blood pressure could be measured automatically every time a subject sits on a toilet seat. While this technique is quite difficult, a recent study showed that it is possible (see Chapter 5).

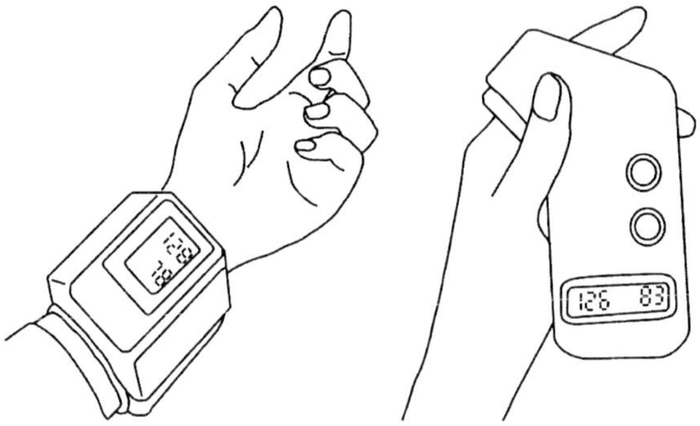

Fig. 12.1 Blood pressure monitors: left) wrist type; right) finger type.

A blood pressure monitor using pulse wave velocity has been studied. It is based on the principle that the arterial pulse wave velocity depends on the elasticity of the arterial vessel wall and, because of the nonlinearity of the elastic property of the vessel wall, elasticity depends on the arterial pressure. Thus, arterial pressure reflects on the pulse wave velocity. However, as vascular elasticity is also affected by vasomotor activities, this method has been regarded as unreliable. Even so, it has been shown that instantaneous variations in blood pressure are well reflected in the changes in pulse wave velocity, and it can be a reliable monitor if it is calibrated intermittently [4]. Pulse wave velocity can be measured non-invasively by measuring time delay of the arrival of an arterial pulse from the beginning of cardiac contraction, which is determined by the QRS component in the electrocardiogram, and thus this method will be applicable for patient care at home.

Instantaneous blood pressure monitoring systems such as the vascular unloading technique and tonometry have been used in clinical monitoring in hospitals. Because of their non-invasive nature, they will be suitable for home use if appropriate instruments can be designed. More detail about this method is given in Chapter 5.

12.3 Respiration

Respiration is the function of gas exchange between the air and the blood, and it consists of ventilation of the lung and gas transfer between the alveolar air and the blood in the pulmonary circulation. Lung ventilation can be monitored either by air flow of the ventilated air, or by volume change of the lung, which can be measured by the volume change of the thorax. Gas transfer is usually estimated by arterial blood oxygenation, which can be monitored non-invasively as described in the next section.

Fundamental parameters for respiratory monitoring are respiratory rate, respiratory amplitude, and respiratory resistance. Respiratory amplitude can be monitored either by airflow or by lung movement. A symptom for which respiratory monitoring is demanded is sleep apnea, the cessation of breathing during sleep. It occurs commonly in all ages, and may cause many problems such as high blood pressure, cardiovascular diseases, memory problems, weight gain, impotence, and headaches. Respiratory monitors are also recommended when airway obstruction is suspected or during respiratory support. Respiratory monitors are also used for training in rehabilitation after thoraco-abdominal surgery, in paraplegic or quadriplegic patients, and for weaning from mechanical ventilation.

In respiratory function tests in hospitals, spirometers are used in which either ventilatory volume or flow is measured, and many parameters relating the capacity and strength of ventilation are estimated. Although measurements usually require special skills and should be performed by physicians or technicians, some instruments can be operated easily and can be used at home. For example, the simple unit shown in Fig. 12.2 for measuring only peak flow at a mouthpiece is available at a very low price (Spirolite RPFM, Medical Systems International Corporation, Miami FL).

For continuous monitoring of respiration, it is inconvenient to use a mask or mouthpiece in home monitoring. Without measuring airflow or volume directly, lung ventilation can be estimated by thoracic and abdominal displacements. Inductance plethysmography has been used for this purpose [5, 6].

The method involves two elastic bands positioned at the rib cage and abdomen as shown in Fig. 12.3. Each band contains a coil in zigzag fashion, and the coil's inductance changes according to changes in the cross-sectional area. This system provides volume changes of rib cage and abdomen, tidal volume, and breathing rate. A respiratory monitoring system of this type is commercially available (Respitrace, Non-Invasive Monitoring Systems, Inc., Miami, FL).

Sleep apnea can be monitored by detecting respiratory motion with appropriate sensors. A simple way is to place a sensor sheet under the mattress of the patient's bed [7]. One system (BioMatt® Monitoring System, VTT) uses a thin flexible electret film that generates a voltage between electrodes laminated on both

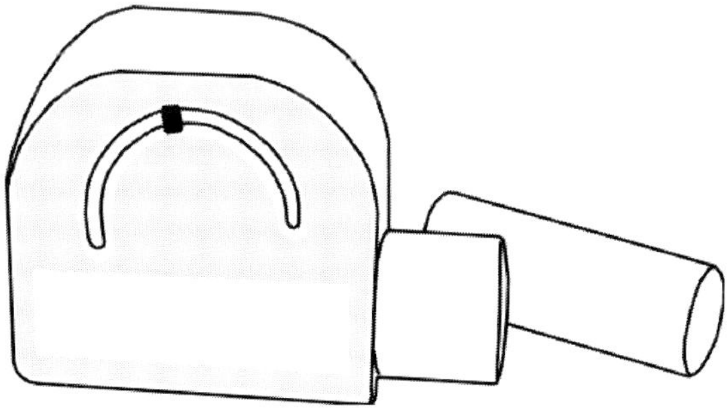

Fig. 12.2 An example of simple unit for measuring peak flow at a mouthpiece (Spirolite RPFM, Medical Systems International Corporation, Miami, FL).

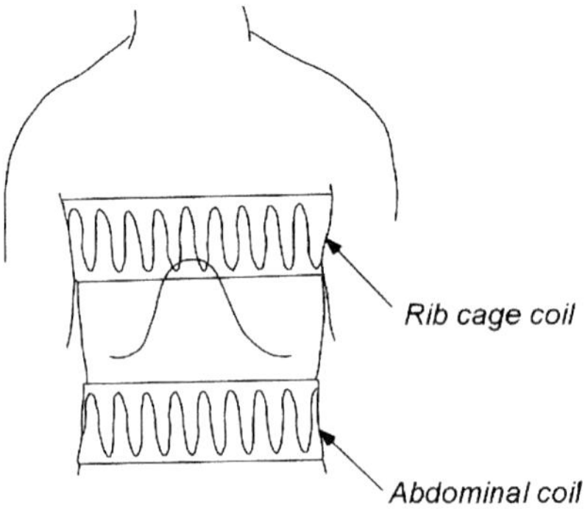

Fig. 12.3 An inductance plethysmography for detecting respiration by inductance changes of two coils contained in two bands at the rib cage and the abdomen (Respitrace, Non-invasive Monitoring Systems, Inc., Miami, FL).

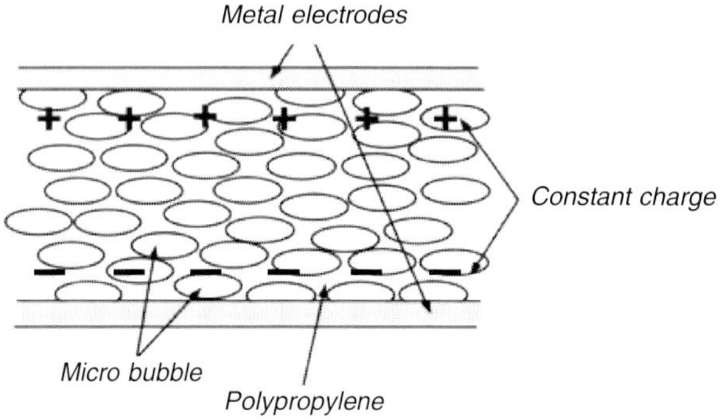

Fig. 12.4 A sensor sheet consisting of a thin flexible electret film (BioMatt® Monitoring System, VTT, Tampere, Finland).

surfaces as shown in Fig. 12.4. The system measures respiratory frequency and amplitudes as well as heart rate.

Respiration during sleep can also be detected by a change in impedance of the chest. Respiratory movement can be detected using two strips of electro-conductive cloth attached to the bed sheet under the chest [8]. It is also possible to detect respiratory movement from video images taking subtractions between successive images. Using an infrared camera and illumination, monitoring can be achieved even in the dark.

12.4 Blood Oxygenation

For patients having respiratory problems, oxygen therapy, intermittent positive pressure breathing (IPPB) therapy, or respiratory assist by a respirator are performed at home. In such situations, patient conditions must be monitored to make such treatments safe. Monitoring arterial blood oxygenation is the most effective and reliable way to achieve this, because confirmation of normal oxygen saturation of arterial blood provides assurance that respiratory function is effective enough. Blood oxygenation is commonly monitored by a pulse oximeter, in which the oxygen saturation of arterial blood can be measured non-invasively from the light transmitted through a finger.

The pulse oximeter is based on the principle that the pulsatile component in the transmitted light intensity is caused by the change in absorption of the arterial blood in the light path while the absorptions of the venous blood and the tissue remain unchanged. Because the absorption spectrum of the blood changes with oxygen saturation, oxygen saturation of the arterial blood can be determined from the time-varying spectral components in the transmitted light. In the ordinary pulse oximeter, a sensor schematically shown in Fig. 12.5 is used. This contains two LEDs and a photodiode to detect absorption changes at two different wavelengths.

The fingertip is an ideal site for pulse oximetry because the tissue blood flow is rich, and measurement of light transmission through the tissue can be achieved easily by using a clip-on probe. However, a probe with cable attached to a finger may constrain ordinary activities at home to some extent. A reflection-type pulse oximeter probe that could be attached to any part of the body may be more convenient. However, a reliable reflection-type pulse oximeter comparable with the transmission-type has not been achieved easily, although it has been shown that the performance can be improved significantly by increasing the active area of the

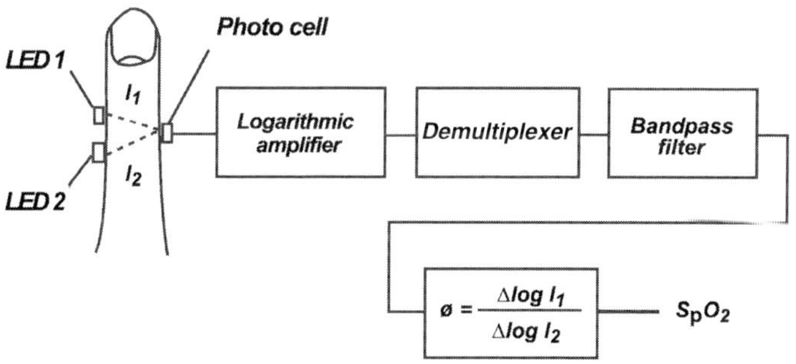

Fig. 12.5 Schematic diagram of the pulse oximeter. The finger is illuminated by two LEDs alternately at different wavelength, transmitted light intensities, I_1 and I_2, are detected by a photocell, the signal is amplified by a logarithmic amplifier, demultiplexed into two components, pulsatile components are extracted, and then taking their ratio and the arterial oxygen saturation, S_pO_2, is derived.

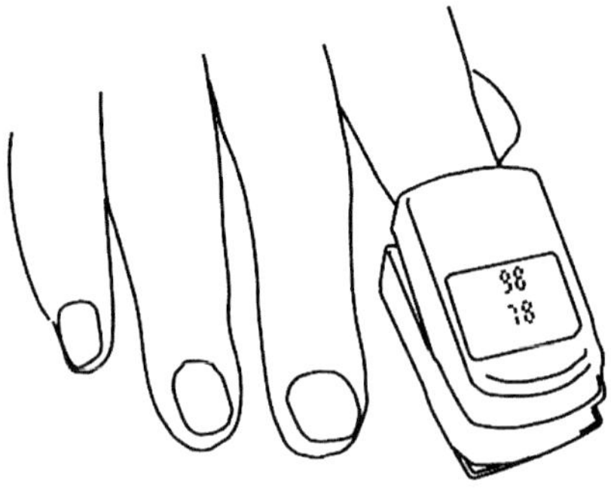

Fig. 12.6 An example of the finger-clip pulse oximeter (Onyx, Nonin Medical, Plymouth, MN).

photodetector and optimizing the separation distance between the light source and photodetector [9].

The finger-clip probe can be more convenient to use if the cable can be eliminated. This has been achieved by squeezing the pulse oximeter electronics into the finger-clip probe itself. Such a probe, shown in Fig. 12.6, is commercially supplied (Onyx, Nonin Medical). It weighs only about 60 g, is operated by two AAA alkaline batteries, and oxygen saturation and pulse rate are displayed on the face of the probe.

Ring-type sensors for monitoring oxygen saturation have been studied [10]. In this system, two-wavelength LED-photodiode pairs, amplifiers, a CPU, a transmitter, and a battery are assembled in a finger ring with a small circuit board, and the device is capable of monitoring oxygen saturation and pulse rate continuously for 24 h.

12.5 Body Temperature

Body temperature has commonly been measured at home. Detection of fever induced by infectious diseases is the commonest purpose for body temperature measurement. Frequent measurements are also required in home care of many chronic diseases. Basal body temperature measurement is also required for monitoring the menstrual cycle.

Because body temperature is always controlled in a narrow range, ordinarily within 1 K, fairly high absolute accuracy is required for clinical thermometers, typically ±0.1 K for ordinary use and ±0.05 K for basal body temperature measurement. The stand-alone mercury-in-glass clinical thermometer was used both in hospital and at home from the early 20th century until recently. It is accurate enough, stable enough, easy to use, and inexpensive. However, because mercury contamination can occur if it is broken, it has been replaced by electronic thermometers. The ordinary clinical electronic thermometer has a thermistor as the sensor, and the body temperature is displayed as a digital value. Completion of a measurement is notified by a beep. While the response time of the sensor itself is faster than that of a mercury-in-glass thermometer, the time to reach a stable temperature in the mouth is not reduced much because of the heat capacity of the tissue near the thermometer tip. Some models include an algorithm to predict the final temperature, and this has reduced the measurement time from about 3 min to 1 min or less.

The ear thermometer, which uses infrared radiation measurement, is becoming popular. As shown in Fig. 12.7, an infrared detector, usually a thermopile, is installed near the probe tip, which is inserted into the auditory canal, and radiation from the tympanum or surrounding tissue is led to the detector through a light pipe. Because the response of the detector is very fast, and the tympanic temperature is close to the deep body temperature, a measurement can be made within a few seconds. Many studies have shown that the accuracies of most ear thermometers are acceptable for ordinary clinical use [11, 12].

Continuous monitoring of body temperature has not usually been performed at home. One reason is that there is no convenient device for such purpose. While rectal or bladder thermometers have been used in hospital wards, such devices are unsuitable for use at home. The deep body thermometer, which uses the zero-heat-flow method, could be used without difficulty, but it is not popular. As shown schematically in Fig. 12.8, a deep body thermometer probe has two temperature sensors placed on both sides of an insulating layer to detect the outward heat flow from the body, and the heater current is servo-controlled so that the outward heat flow is compensated by electric heating. When the servo-control operates precisely, no heat can penetrate the probe, and thus the probe can be regarded as an ideal heat insulator. If a region of the skin surface is perfectly insulated in this way, the temperature of the skin surface beneath the probe equilibrates to the deep tissue temperature. It has been shown that temperatures obtained by attaching the probe at the chest, abdomen, or forehead are close to the deep body temperature [13].

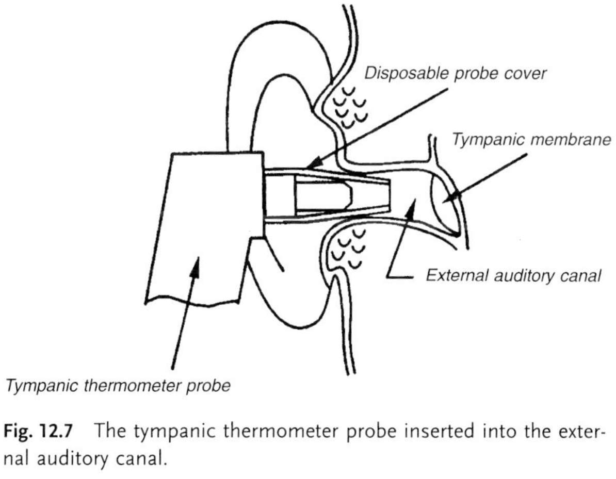

Fig. 12.7 The tympanic thermometer probe inserted into the external auditory canal.

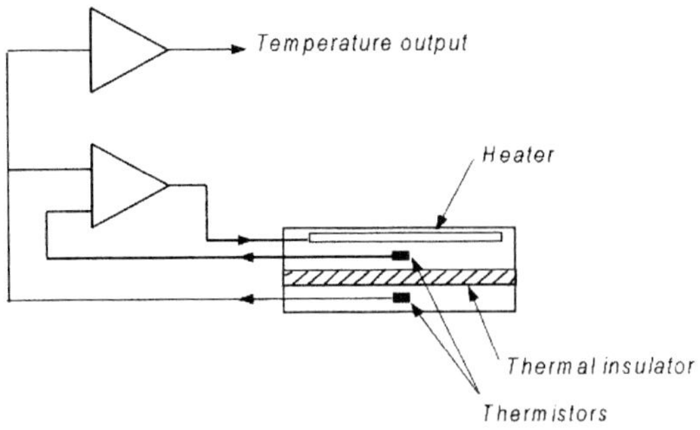

Fig. 12.8 Schematic diagram of the deep body thermometer.

An ingestible telemetering capsule with a temperature sensor can also be used for body temperature monitoring. Once a capsule is swallowed, it remains in the alimentary tract for about 24 h. Because the temperature in the stomach or intestine can be regarded as the deep body temperature, fairly accurate body temperature can be monitored for a whole day. A disposable temperature capsule is commercially available (Human Technologies, St. Petersburg, FL). In the model, a quartz resonator is used as the temperature sensor [14]. A portable receiver with a data logger is also supplied for continuous recording of body temperature, thus body temperature can be monitored even in unconstrained subjects.

12.6 Electrocardiogram

The electrocardiogram (ECG) gives cardiologists important diagnostic information. Although an expert physician's skill is required to interpret ECG waveforms, recording ECGs at home can be a great help for diagnosis by a physician for some health problems. By home monitoring, long-term records can be obtained, and thus rarely occurring abnormalities can be detected.

The Holter ECG recorder has been widely used. It is a portable recorder in which ECGs of two or more channels are recorded for 24 or 48 h, either on an ordinary audio cassette tape or in a digital memory device such as a hard disk or solid-state flash memory. Most Holter recorders are lightweight, typically 300 g or less including the battery, and they cause almost no obstruction in ordinary daily activities or even during sleep. However, they require chest electrodes, which are ordinarily attached by clinical staff in the office, and thus monitoring is performed according to a physician's instructions and it is difficult in practice to continue for many days.

There are some special situations in the home where ECGs can be recorded automatically. ECGs during sleep can be derived from a pillow and the bed sheet beneath the leg using conductive textiles, as shown in Fig. 12.9 [15]. Because the contact of the textile electrodes to the skin is not always secure, large artifacts can

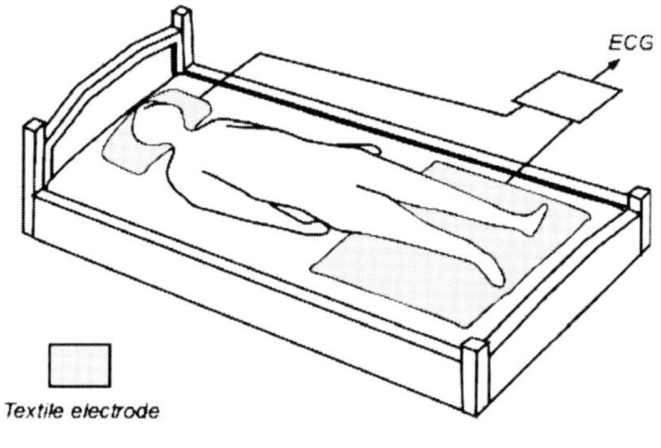

Fig. 12.9 Monitoring ECGs during sleep using conductive textiles.

arise with body movements. Nevertheless, a stable ECG can be recorded, usually more than 80% of the time, during sleep.

ECGs can also be derived during bathing. If electrodes are installed on the inside wall of a bathtub, as shown in Fig. 12.10, an ECG can be recorded through the water [16]. The ECG signal amplitude depends on the conductivity of the water. If the water conductivity is high, the water makes a short circuit to the body as a voltage source, and consequently the amplitude is reduced. In contrast, if the water conductivity is low, signal amplitude remains at levels similar to the skin surface measurement. However, because of the higher resistance of the water, a high-impedance amplifier should be used, and thus the recording system becomes more vulnerable to electromagnetic interference. Fortunately, the electric conductivity of ordinary tap water, which is of the order of 10^{-2} S/m, is within an acceptable range of measurement using a conventional ECG amplifier. When a subject enters the tub, large artifacts appear because of the movement of water near the electrodes and the body surface, but the base line is quickly stabilized, and a stable ECG can then be obtained throughout the stay in the tub.

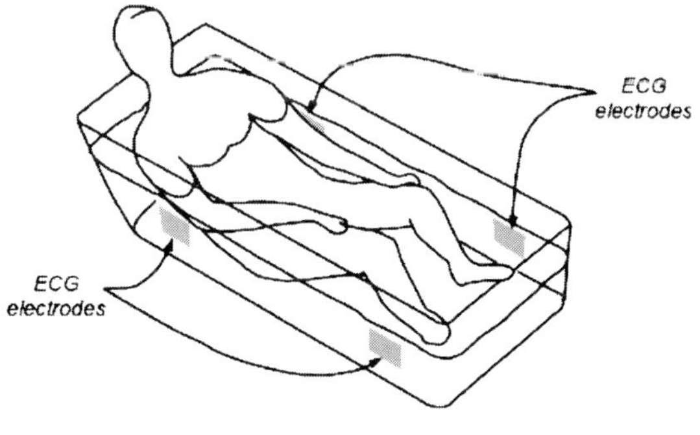

Fig. 12.10 ECG monitoring during bathing.

12.7
Heart Rate and Pulse Rate

Heart rate is the number of contractions of the heart per minute, and pulse rate is the number of arterial pulses per minute. When the cardiac function is normal, pulse rate is the same as heart rate. However, sometimes, especially when there is an arrhythmia, some contractions of the heart do not cause effective ejection of blood to the arteries, and the pulse rate becomes lower than the heart rate. Thus, heart rate and pulse rate should be distinguished in such cases. When such ineffective contractions are rare, heart rate can be substituted by pulse rate.

Physiological conditions quite often affect the heart rate. When the heart rate remains in the normal range with small fluctuations, the physiological condition may be stable. However, if the heart rate increases even in the physically and mentally relaxed state, some physiological problems such as hypoxia should be suspected. Therefore, heart rate monitoring has been widely used for high-risk patients.

Heart rate can be determined by counting the QRS complexes in an ECG or measuring R–R intervals when the cardiac rhythm is regular. In patients having arrhythmia, ECG waveforms are very different from normal, thus, to determine the heart rate, appropriate filters, or detection algorithms must be used.

Pulse rate can be determined by detecting arterial pulses by many different means such as photoplethysmograph, detecting mechanical force, detecting vibration, or impedance plethysmograph. Photoplethysmography is a method of detecting the change in light absorption in the tissue because of the pulsatile change in arterial volume. To perform pulse rate monitoring by photoplethysmography, the most convenient site is the finger. For the light source and detector, a near infrared LED and a photodiode are commonly used. Wavelengths around 800 nm are adequate for this purpose because tissue absorbance is low and absorbance of hemoglobin does not change with oxygen saturation.

In the pulse oximeter, the pulsatile component of light absorption is detected, thus pulse rate can be determined from the detected signal itself. Ordinary pulse oximeters display pulse rate together with the oxygen saturation. The ring-type pulse oximeter described before can also be used as a pulse rate monitor.

Wrist-watch-type pulse rate meters are supplied commercially. Fig. 12.11 shows an example of this kind. It has a sensor at the face of the watch, which consists

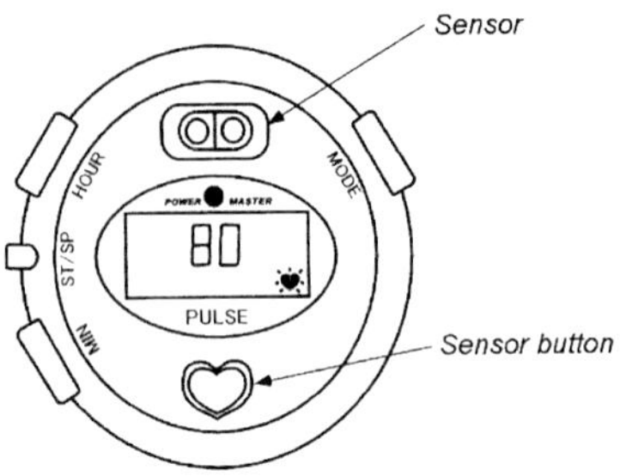

Fig. 12.11 An example of the wrist-watch-type pulse rate meter (Pulse watch, Walnut, CA). To measure pulse rate, the sensor button is pushed and a fingertip placed on the sensor.

of a reflection-type photoplethysmograph. To measure pulse rate, the sensor button is pushed and a fingertip placed on the sensor. Flashing of an icon in the display indicates detected pulses. The pulse rate is then displayed within about 5 s.

Pulse rate can be detected by a small vibration of the body, especially during sleep. As described before, the BioMatt consists of a thin flexible electret film to be placed beneath the bed sheet, and it can detect pulse rate. Arterial pulse can also be detected on the toilet seat by a vibration of the body, as described later in the section on body weight.

12.8 Blood Components

In ordinary clinical examinations, many items of blood components are analyzed. Blood analysis is commonly performed by automatic analyzers in professional facilities, but is not commonly performed at home except for blood glucose monitoring in people with diabetes. An invasive method is usually required to obtain a blood sample, thus, at least at present, blood analysis at home is limited to occasions when frequent measurement is absolutely necessary. If non-invasive or even less-invasive measurements of blood components could be achieved, monitoring of blood components at home would be accepted more widely.

In many people with diabetes, self-management of blood glucose level is introduced with frequent self-measurements of blood glucose, and insulin injections controlled according to the glucose level. Blood glucose is commonly measured at home using a glucose meter and a drop of blood taken from the finger. To take a drop of blood, a simple tool, called a lancing device, is used. When it is used, a pointed steel needle, called a lancet, is pushed into the skin for about 1.8–3.0 mm by a spring, and a small amount of blood is drawn into the lumen of the needle. The needle diameter is from 0.3 mm (30 G) to 0.8 mm (21 G). To perform blood sampling with the least pain, pricking the side of a finger is recommended because the sides have a good blood supply and fewer nerve endings than fingertips. When the patient is reluctant to prick a finger, as might be the case for a professional musician, the side of the palm or the forearm, the abdomen, or the outer thigh can also be used. There are also laser lancing devices, which use a laser beam to produce a small hole by vaporizing the skin tissue.

In taking a drop of blood, a mosquito is far smarter than any available lancing devices or puncturing systems. A mosquito penetrates a needle of about 30 μm in diameter and 2 mm long into a small vessel and sucks a few microlitres of blood in a few minutes with no pain except a slight tingling sensation. If an artificial device mimicking a mosquito could be developed by micromachining technology, and installed in a place to which the skin will make frequent contact, blood samples could be obtained automatically. A preliminary experiment was attempted in which a needle of 37–125 μm in diameter was inserted into the auricular vein of a rabbit using a computer-controlled introducer and obtained up to 200 μl of blood.

Achieving a fully automatic blood sampling system will be a valuable challenge; extreme difficulties are expected.

Once a small amount of blood can be obtained, blood components, typically blood glucose, can be analyzed by either a test strip or a glucose meter. To use a glucose test strip, blood is placed on the end of it, and after about a minute the color of the strip changes. A rough estimate of the blood glucose level can be made by matching the color of the strip to a color chart. A more accurate value can be given by using a photometric meter that detects the color change. In electrochemical glucose meters for home care, a drop of blood of about 10 μL or less is placed on the sensor chip. The blood glucose is then measured by an enzyme-based biosensor, and the result is displayed in about 30 s. Some glucose meters require only 2–3 μL of blood. In a glucose meter, lancing, blood collection, and glucose testing are automatically performed with a single press of a button (MediSense Sof-Tact).

In some instruments, two or more chemical components can be analyzed simultaneously using biosensor arrays. In one instrument (I-STAT, I STAT Co. NJ), single-use disposable cartridges are provided that allow ions, blood gases, pH, urea, glucose, and hematocrit to be analyzed by an array of biosensors using amperometric, potentiometric, and conductimetric principles.

While it is difficult to measure blood components without sampling blood, some small molecules, as well as blood gases and ions, pass the capillary wall freely, and such chemical species can be detected in the interstitial fluid at the same concentrations as those in the blood. Thus, it is expected that concentrations in the blood of such species, typically glucose, may be estimated less- or non-invasively. In a product called GlucoWatch (Cygnus, CA), a small amount of interstitial fluid is pulled out through the skin by iontophoresis, and the glucose content is detected by a biosensor. Although it must be calibrated occasionally by an ordinary glucose meter, changes of glucose level with time can be monitored by an instrument worn at the wrist like a wristwatch.

Spectrophotometric measurements are ideal because of their non-invasive nature when they can be applied for chemical analysis. However, the acceptable wavelength range for such measurements is limited from 600 to 1200 nm because of the high absorption by tissues outside this range, and absorptions of most chemical species in this range are small except for hemoglobin and bilirubin. Despite this difficulty, there have been some attempts to measure tissue glucose level from a slight change in the tissue absorption spectra because of the glucose concentration [17, 18]. Many companies are at present trying to develop spectrophotometric non-invasive glucose meters.

12.9 Urine Components

In many diseases, abnormal components appear in urine, and thus urine analysis provides diagnostic information. For example, urine glucose and ketones indicate diabetes, urine protein indicates kidney disease, urine bilirubin indicates liver or bile duct disease, and urine nitrites and white blood cells indicate urinary tract infection. In contrast with the blood test, urine tests require only normal urination, and there is no discomfort. This feature is a great advantage in home health care.

For ordinary urine analysis in hospitals and clinics, a midstream urine sample of about 50 mL is taken, and then analyzed in the laboratory. However, for home care, urine test strips have been used most commonly. Urine test strips, also called urine dipsticks, contain little pads of reagents that change color when they come in contact with the specific substances. To use a test strip, a midstream urine sample can be taken in a clean container and the test strip placed in the sample, or the strip can be passed through the urine stream. Excess urine is shaken off the strip, and the strip pad compared to the color chart on the test strip container. Urine components that can be detected by ordinary urine test strips are blood, protein, nitrite, glucose, ketones, pH, leukocytes, urobilinogen, and bilirubin. Many different urine test strips are available; they always have four or more reagent pads on a strip according to the purposes of monitoring such as renal, hepatic, pancreatic, gastro-intestinal and cardiovascular functions.

A urine test can be done using a test strip without pain and discomfort, but it still requires a little time and a procedure must be performed. To ordinary people, such a procedure is not acceptable if it disturbs daily activities. To obtain test results every day for a long period of time, an automatic test system would be preferable. No fully automatic urine test system is currently available, but there have been some attempts to realize urine analysis at home with minimal procedures. The instrument shown in Fig. 12.12 was developed. It is installed in the toilet and measures urine glucose with only one operation of a start button (TOTO Co., Tokyo). When using the toilet for urination, the subject pushes the start button, then a urine collector protrudes, collects urine automatically from the urine stream, and analyses urine glucose by an enzymatic glucose sensor. Although the system has no capability for personal identification, each individual in a family

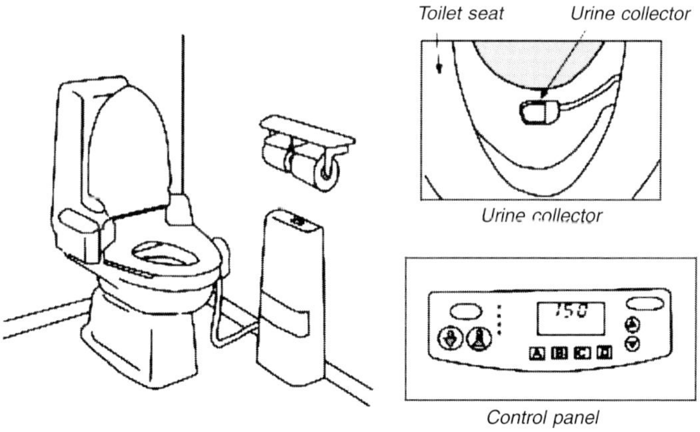

Fig. 12.12 A system which can measure urine glucose with only one operation of a start button (TOTO Co., Tokyo).

can be identified by the use of a specific start button from a set of four. Data for each individual in a family can be stored separately. The sensor must be replaced every 4 months, and the calibration solution must be replenished every 3 months or more often, depending on the frequency of use.

12.10
Body Weight

Body weight is an essential parameter for health management. When body weight is maintained in a narrow range for a long period of time without effort, overall physiological regulation will be working normally, and the person will be less vulnerable to disease. However, few health conscious people measure their own weight very often, and even when they do measure it, they seldom record the values. To use body weight for health management, data must be taken regularly and stored. In a system called the wellness monitoring system, a digital scale is connected to a laptop computer together with monitors for blood pressure, body temperature, and a bed sensor system [19].

The body-mass index, defined as the weight (in kilograms) divided by the square of the height (in meters), is sometimes used for health management. A detailed study showed that greater body-mass index was associated with higher mortality [20]. It is commonly said that people with body-mass indexes between 19 and 22 live longest, and death rates are notably higher for people with indexes above 25. A scale is available in which four family members' heights can be memorized and body-mass indexes can be displayed by selecting a button (Salter 992, TANITA Co.).

To obtain body weight data automatically for health monitoring, measurement on the toilet seat was attempted [21]. It is quite simple in principle. As shown in Fig. 12.13, a precision load cell system was installed in the floor of the toilet, and

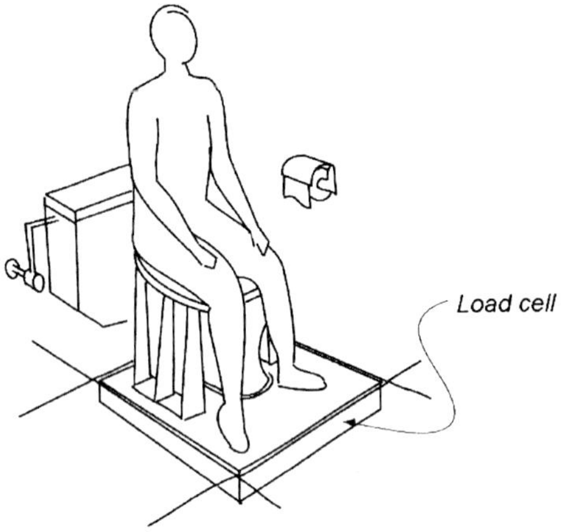

Fig. 12.13 A load-cell system installed in the floor of the toilet for automatic body weight monitoring.

the seat was supported so that the weight on the seat was transferred to the load cell. With this system, body weight data can be collected many times a day automatically. As well, this system allows measurement of the amount of urine and feces, the urine flow rate, together with habits of urination and evacuation. The study also showed that cardiac function could be monitored from a precise analysis of the load-cell signals.

12.11 Body Fat

Body composition, especially the amount of body fat, has been proposed for prediction of risks of chronic diseases. Body composition is a better determinant of health than body-mass index. To estimate percentage body fat, a derivation from body density has been commonly used. According to the Siri equation [22]

$$\text{percentage body fat} = (4.95/(\text{specific gravity of the body}) - 4.5) \times 100 \, .$$

This method requires measurement of the density or specific gravity. To measure body density, underwater weighing based on Archimedes' principle has been a standard method, but it is not convenient to perform at home. The body volume can be determined from the air volume in an airtight chamber with the body inside by measuring the compliance of the air in the chamber. A chamber for this purpose, called the Bod Pod, is commercially available (Life Measurements Instruments, Concord, CA).

Body fat can also be estimated from the thickness of subcutaneous fat, which can be measured directly using a skin-fold caliper. Chest, abdomen, and thigh for men, and triceps, suprailiac, and thigh for women are recommended as skin-fold sites. While the method is simple, it requires some skill, and the information obtained is limited to the subcutaneous fat. The waist-to-hip ratio, defined as waist circumference divided by hip circumference, is also used as an index of body fat. The measurement is made by a tape scale. This index mainly reflects the fat in the abdominal area.

As a convenient method of estimating body composition, whole-body impedance measurement has been used. Its principle is based on the fact that electrical impedance is different in different tissues. For example, resistivity in the low frequency region (less than 1 MHz) is about 0.63 Ωm in plasma, 1.5 Ωm in blood, 16 Ωm in skeletal muscle in the transverse direction, and 25 Ωm in fat [23]. Although the impedance method was first introduced for the estimation of body water content, it has also been applied to the estimation of the amount of fat.

Body impedance is usually measured by the four-electrode technique, in which a constant alternating current of 50–100 kHz and 0.1–1 mA is applied between the outer electrode pair, and the alternating voltage developed between the inner electrode pair is detected. In a commercial instrument (Tanita Co.), four electrodes are placed on a scale as shown in Fig. 12.14. Alternating current is applied

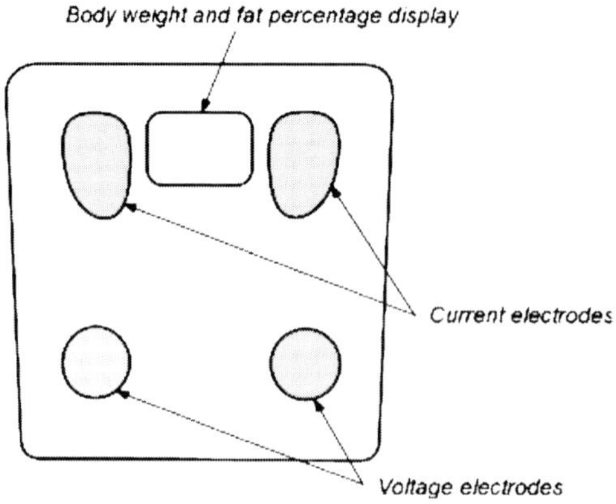

Fig. 12.14 A body scale with fat percentage estimated from the body weight and electrical impedance between heels (Tanita Co., Tokyo).

between the toes of both feet, and the voltage developed between electrodes at both heels is detected. Because the impedance between feet and the body weight can be measured simultaneously, the approximate amount of fat can be estimated from a comparison between the measured body weight and the weight estimated from impedance assuming a lean body. It was shown that the impedance technique is highly reproducible for estimating lean body mass [24].

The use of near-infrared interactance to determine body composition has been studied [25]. This method is based on the difference in absorption spectra of fat and lean tissues. In the commercial instruments, the specific wavelengths 938 and 948 nm are used, and the intensity of back-scattered light is measured (FUTREX-5000, Home model FUTREX-1100). While the measurement can be made at any site of the body, measurement at the biceps site is most representative of the whole body fat.

12.12 Daily Activity

Changes in health condition, both physical and mental, will be reflected in daily physical activities. The amount of daily physical activity can be roughly estimated as the number of steps in a day, and measured by a step counter attached to the belt or waistband. The step count for ordinary people is 3000–5000 steps per day, while 10000 steps per day or more is sometimes recommended for improving physical fitness and lowering body fat. Using precision accelerometers, behavioral patterns such as the changes in posture and walking or running can be classified to some extent. By analyzing body acceleration patterns, metabolic rate can be roughly estimated [26]. However, such a measurement requires attachment of the device to the body, and will require continual motivation to take data for keeping health.

There have been some attempts to monitor daily activities without attaching any device to the body. If some infrared sensors are installed in a house as shown in Fig. 12.15, they detect infrared radiation from a body so that the presence or absence

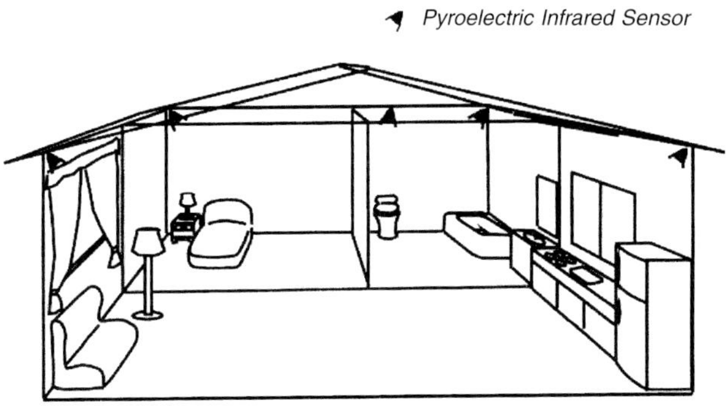

Fig. 12.15 Monitoring of the presence and absence of a subject living alone in the house using many infrared sensors.

of a subject can be monitored, and thus daily activity in the house can be roughly identified, at least while the subject is living alone in the house. Pyroelectric sensors are sensitive enough for this purpose. Although a pyroelectric sensor can detect only a change in incident radiation, entering or leaving each room can be identified by processing signals from all sensors.

Other simple sensors such as photointerrupters, electric touch sensors, and magnetic switches, as shown in Fig. 12.16, can also be used to detect activities in daily living. The use of room lights, air conditions, water taps, electric appliances such as a refrigerator, TV, or microwave oven, can be detected and used as information relating to daily living. By installing many sensors, even though each one is simple, some studies have shown that habits of daily living in a house can be recorded without any constraints on the subject [27, 28].

The sensors described above installed in a house could be used to record daily activities of an individual who is living alone in the house. However, when two or more people live in a house, it is difficult to allocate data to individuals. Even in such a case, the data obtained will reflect the life habits of the family, and thus whether or not the daily activities of a family are regular will be determined objectively.

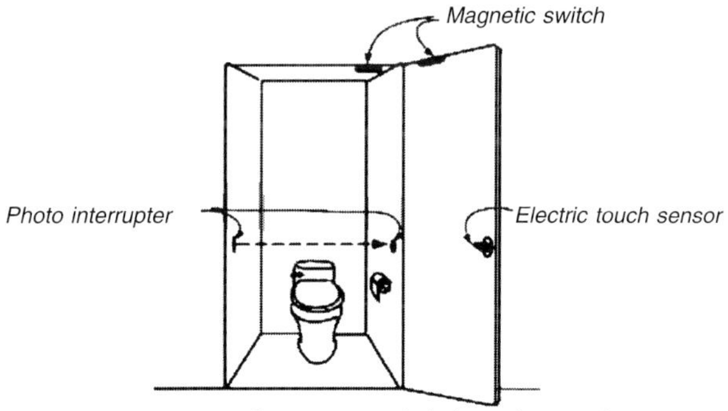

Fig. 12.16 Detection of activities in daily living by simple sensors.

12.13 Sleep

Sleep is essential to maintain health, but it can be disturbed by many factors such as psychological stresses, noises, extreme temperature, surrounding environments and medical side effects. Insufficient or poor sleep is called insomnia. Insomnia is usually identified by subjective observations such as difficulty in falling asleep, waking up frequently during the night with difficulty returning to sleep, and waking up too early in the morning. In sleep studies, electroencephalography (EEG), electrooculography (EOG), and electromyography (EMG) are required, but such measurements are unacceptable at home, and thus simple instruments are required if sleep is to be evaluated objectively.

A device called an actigraph is an instrument that detects body motion. It is wearable like a wristwatch and detects the acceleration of the wrist by a piezoelectric sensor. It has been shown that the wrist acceleration recorded by an actigraph distinguishes from wakefulness in considerable accuracy [29]. A commercial model contains a non-volatile memory in which continuous data for over 60 days can be stored (SleepWatch, AMI, New York). It has been used to estimate sleep quality and quantity, as well as for assessing pathological daytime activities such as fatigue, generic inactivity, neurological conditions, and drug effects.

Body movements during sleep can also be monitored without attaching any device to the body. For example, a pressure-sensitive sheet can be used for this purpose. A system called BioMatt consists of a pressure-sensitive film about 50 μm thick, which can be inserted under the mattress (VTT, Tampere). It is made from a biaxially oriented polypropylene film with many small bubbles in the material, and is permanently polarized, and thin metal electrodes are laminated to both surfaces. The film is quite sensitive and detects not only body motions but also respiration and heart rate. Therefore, it can be used as a sleep monitor for detecting insomnia and sleep cycle disorders, and as a patient monitor for detecting symp-

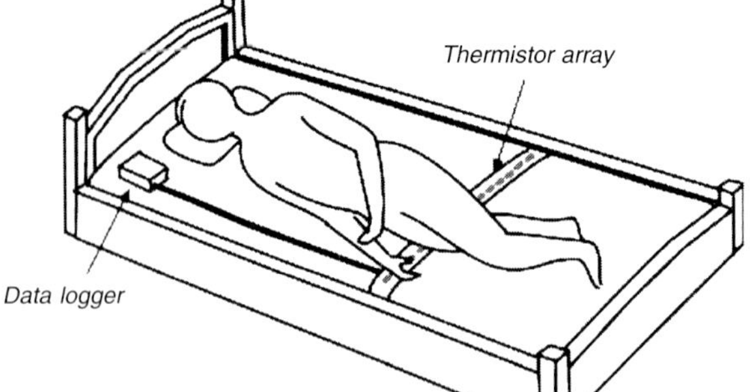

Fig. 12.17 Monitoring of body motion during sleep by a thermistor array.

toms such as sleep apnea, nocturnal asthma, nocturnal heart dysfunctions, and even coughing and grinding teeth [30–32].

Body motion during sleep can also be monitored by a thermistor array inserted in the mattress at the waist or thigh level, as shown in Fig. 12.17. It has been shown that each body movement can be detected as a change between successive temperature profiles [33, 34].

12.14 Nutrition

A healthy diet is fundamental to the maintenance of good health, and health problems such as heart disease, diabetes, and some cancers are linked to poor nutrition. Therefore, nutrition monitoring in a large population has been conducted in many countries, and risks and preventive effects of foods for the occurrence of diseases have been analyzed. However, such studies are expensive and require large manpower. In the ordinary dietary data collection, random samples in different age groups in a population are selected. Then, interviewers visit each home, ask the resident to describe everything eaten and drunk in a specific day, and enter the data in a questionnaire sheet. Sometimes, data collection from the same subject is repeated either by visiting again or asking by telephone. Although such nutrition monitoring is sometimes continued for many years for epidemiological studies, in practice, monitoring can only be conducted for a small part of the population, and the data obtained are not fed back directly to the subject to be used for health management.

Patients with diseases such as diabetes must control nutritional intake. For such patients, every meal must be prepared according to the instructions of a physician or dietician. The nutritional control is always performed by the patients or their families. However, it is difficult to confirm objectively the extent to which nutritional intake is controlled.

It may be possible in principle to monitor nutritional intake of every meal automatically. For example, if a digital camera is installed above the dining table and a photograph of the meal can be taken automatically and stored in a computer, the foods of every meal can be recorded. Then, the amount of each nutritional intake can be estimated by a dietician. Of course, it is desirable for the analysis of nutritional components from a photograph of each meal to be automated. Even if automatic analysis cannot be attained, the stored data can be used at least for occasional confirmation of diet control. Furthermore, if the data are stored for a long period of time for many subjects, occurrence of some diseases will be seen among them. If the data of many cases can be collected, by analyzing nutritional intakes retrospectively, it may be that some risk factors will be found, and thus such a technique can be an effective tool for epidemiological studies.

12.15
Environmental Parameters

Environmental parameters may influence physiological conditions. For example, extreme heat or cold may trigger some disorders such as stroke. Even when the condition is not extreme, prolonged exposure to physical stress may cause illness. For example, noise may cause sleep disorders even when the level is subliminal. House dust and pollen may cause allergic effects. On the other hand, human activities may influence environmental parameters. The temperature and humidity of a room may be changed by opening a window or operating an air conditioner, and thus daily activities may be deduced indirectly from the trends of environmental parameters.

Monitoring of environmental parameters for health care has not been attempted extensively, but it would not be difficult to perform using conventional sensors. The temperature and humidity of each room can be monitored easily. If the house is air-conditioned, it may have some sensors that will provide environmental information. Sound and vibration can be monitored in different frequency ranges. More sophisticated waveform analysis will provide information about the sources of the sound and vibration. Illumination in each room can also be monitored. Habits of daily living will be reflected by opening and closing a curtain or shutter in the daytime, and electric illumination at night. Smoke detectors are installed in buildings and houses for detecting fire, and false alarms may occur when smoking if the sensitivity of the smoke detector is too high; such a sensor can be a smoking monitor. House dust can be collected by filtering the room air for a prescribed time. A dust counter can also be used for this purpose if simple sensors become available. It may be possible to combine an electrostatic air cleaner and a dust counter. Carbon dioxide content can be monitored by a simple infrared-absorption type carbon dioxide sensor. Preliminary studies have shown that the presence of a human in a room can be detected easily by an increase in carbon dioxide content.

Many animals detect environmental information by odor. Such animals have extremely sensitive chemical sensors in their olfactory system and they can detect and discriminate different chemical species. If an odor sensor comparable to an olfactory receptor were available, it would be a great help for health and environmental monitoring. Unfortunately, sensitivity and selectivity of available gas sensors are far from those of animal olfactory receptors. While a specially designed mass spectrometer can discriminate many chemical species in exhaled air [35], such an instrument is impractical for home monitoring. Combining presently available odor sensors, some smells can be distinguished; such a system has been called an artificial nose [36]. A preliminary study showed that different odor signals can be recorded when such sensors are installed in a house [37]. Sensitive odor sensors are therefore very promising for use in home monitoring to detect odorous appearing in daily living activities as well as from the human body.

12.16 Conclusions

Up to date, very many monitoring instruments for home care and telecare have been proposed and developed, and some of them are becoming commercially available. Some instruments do not require sensor attachment to the body and are fully automatic so that health parameters can be monitored without performing any operation. Utilizing such instruments, and by connecting to the telecommunication network system, physical condition of patients of chronic diseases as well as daily life of elderly people living alone can be monitored from distant places, and monitoring data of many subjects can be analyzed at a central station.

For home monitoring of health parameters, there are many possibilities by applying existing sensor technologies. Even simple and primitive sensors can be effectively used in some purpose such as in daily activity monitoring. Applications of advanced sensor technologies, such as spectro-photometric non-invasive glucose measurement or mass-spectrometric inhaled gas analysis, have been attempted. Such instruments will be accepted if the obtained information is essential for caring the subject. Furthermore, future technologies, such as extremely sensitive smell sensors, which are comparable to olfactory receptors of animals, will be expected in home care and telecare.

Technologies of monitoring health parameters at home are now becoming available, and long-term monitoring data will be accumulated, which are hardly obtained by conventional examinations in hospitals. Such data will be utilized not only for medical and health care of the monitoring subject alone, but also for basic medical studies in a large population, such as in epidemiological studies where risk factors of diseases are identified by statistical analyses.

12.17 References

1 Ramsey, M., Noninvasive automatic determination of mean arterial pressure. *Med. Biol. Eng. Comput.* **1979**, *17*, 11–18.

2 Sapinski, A., Hetmanska, St., Standard algorithm of blood-pressure measurement by the oscillometric method. *Med. Biol. Eng. Comput.* **1992**, *30*, 671.

3 Tanaka, S., Yamakoshi, K., Ambulatory instrument for monitoring indirect beat-to-beat blood pressure in superficial temporal artery using volume-compensation method. *Med. Biol. Eng. Comput.* **1996**, *34*, 441–447.

4 Chen, W., Kobayashi, T., Ichikawa, S., Takeuchi, Y., Togawa, T., Continuous estimation of systolic blood pressure using the pulse arrival time and intermittent calibration. *Med. Biol. Eng. Comput.* **2000**, *38*, 569–574.

5 Milledge, J. S., Stott, F. D., Inductance plethysmography – a new respiratory transducer. *J. Physiol.* **1977**, *267*, 4P–5P.

6 Sackner, J. D., Nixon, A. J., Davis, B., Atkins, N., Sackner, M. A., Noninvasive measurement of ventilation during exercise using a respiratory inductance plethysmography. *Am. Rev. Respir. Dis.* **1980**, *122*, 867–871.

7 Alihanka, J., Vaatoranta, K., Saarikivi, I., A new method for long-term monitoring of the ballistocardiogram, heart

rate, and respiration. *Am. J. Physiol.* **1981**, *240*, R384–R392.

8 Ishijima, M., Long-term cardiopulmonary monitoring in bed without subject awareness. *Proc. 18th Ann. Int. Conf. IEEE Eng. Med. Biol.* **1996**, 503.

9 Menderson, Y., Ochs, B. D., Noninvasive pulse oximetry using skin reflectance photoplethysmography. *IEEE Trans. Biomed. Eng.* **1988**, *35*, 798–805.

10 Rhee, S., Yang, B. H., Chang, K., Asada, H. H., The ring type sensor: a new ambulatory wearable sensor for twenty-four-hour patient monitoring system. *Proc. 20th Ann. Int. Conf. IEEE Eng. Med. Biol.* **1998**, 1906–1909.

11 Shinozaki, T., Dean, R., Perkins, F. M., Infrared tympanic thermometer: evaluation of a new clinical thermometer. *Crit. Care Med.* **1988**, *16*, 148–150.

12 Weiss, M. E., Pue, A. F., Smith III, J., Laboratory and hospital testing of new infrared tympanic thermometers. *J. Clin. Eng.* **1991**, *16*, 137–144.

13 Togawa, T., Non-invasive deep body temperature measurement. In Rolfe, P. (ed.) Non-invasive physiological measurement. Academic Press, London **1979**, 261–277.

14 Cutchis, P. P., Hogrefe, A. F., Lesho, J. C., The ingestible thermal monitoring system. *Johns Hopkins APL Tech. Digest* **1988**, 9(7), 16–21.

15 Ishijima, M., Monitoring of electrocardiograms in bed without utilizing body surface electrodes. *IEEE Trans. Biomed. Eng.* **1993**, *BME-40*, 593–594.

16 Ishijima, M., Togawa, T., Observation of electrocardiogram through tap water. *Clin. Phys. Physiol. Meas.* **1989**, *10*, 171–175.

17 Danzer, K., Fischbacher, Ch., Jagemann, K. U., Reichelt, K. J., Near-infrared diffuse reflection spectroscopy for non-invasive blood-glucose monitoring. *LEOS Newslett.* **1998**, *12*, 9–11.

18 Malin, S. F., Ruchiti, T. L., Blank, T. B., Thennadil, S. N., Monfre, S.L., Noninvasive prediction of glucose by near-infrared diffuse reflectance spectroscopy. *Clin. Chem.* **1999**, *45*, 1651–1658.

19 Korhonen, I., Lappalainen, R., Tuomisto, T., Koobi, T., Pentikainen, V., Tuomisto, M., Turjanmaa, V., TERVA: wellness monitoring system. *Proc. 20th Ann. Int. Conf. IEEE Eng. Med. Biol.* **1998**, 1988–1991.

20 Stevens, J., Cai, J., Pamuk, E. R., Williamson, D. F., The effect of age on the association between body-mass index and mortality. *New Engl. J. Med.* **1998**, *338*, 1–7.

21 Yamakoshi, K., Kuroda, M., Tanaka, S., Yamaguchi, I., Kawarada, A., A non-conscious and automatic acquisition of body and excreta weight together with ballistocardiogram in a lavatory. *Proc. 18th Ann. Int. Conf. IEEE Eng. Med. Biol.* **1996**, 504.

22 Siri, W. E., The gross composition of the body. *Adv. Biol. Med. Phys.* **1956**, *4*, 239–280.

23 Geddes, L. A., Baker, L. E., Principles of Applied Biomedical Instrumentation, 2nd edn. Wiley, New York **1975**.

24 Segal, K. R., Gutin, B., Presta, E., Wang, J., Van Itallie, T. B., Estimation of human body composition by electrical impedance methods: a comparative study. *J. Appl. Physiol.* **1985**, *58*, 1565–1571.

25 Conway, J. M., Norris, K. H., Bodwell, C. E., A new approach for the estimation of body composition: infrared interactance. *Am. J. Clin. Nutr.* **1984**, *40*, 1123–1130.

26 Servais, S. B., Webster, J. G., Estimating human energy expenditure using an accelerometer device. *J. Clin. Eng.* **1984**, 9(2), 159–171.

27 Celler, B. G., Ilsar, E. D., Earnshaw, W., Preliminary results of a pilot project on remote monitoring of functional health status in the home. *Proc. 18th Ann. Int. Conf. IEEE Eng. Med. Biol.* **1996**, 797.

28 Ohta, S., Shinagawa, Y., Tanikawa, T., Nakamoto, H., A health-monitoring system for the elderly living alone. *J. Telemed. Telecare* **2002**, *8*, 151–156.

29 Cole, R. J., Kripke, D. F., Gruen, W., Mullaney, D. J., Gillin, J. C., Automatic sleep/wake identification from wrist activity. *Sleep* **1992**, *15*, 461–469.

30 Salmi, T., Partinen, M., Hyyppa, M. T., Kronholm, E., Automatic analysis of static charge sensitive bed (SCSB) recordings in the evaluation of sleep-related apneas. *Acta Neurol. Scand.* **1986**, *74*, 360–364.

31 SALMI, T., SOVIJARVI, A. R., BRANDER, P., PIRILA, P., Long-term recording and automatic analysis of cough using filtered acoustic signals and movements on static charge sensitive bed. *Chest* **1988**, *94*, 970–975.

32 SJOHOLM, T., POLO, O., ALIHANKA, J., Sleep movements in teethgrinders. *J. Craniomandib. Disord. Facial Oral Pain* **1992**, *6*, 184–191.

33 TOGAWA, T., TAMURA, T., MIZUKAMI, H., ISHIJIMA, T., Physiological monitoring systems attached to the bed and sanitary equipments. *Proc. 11th Ann. Int. Conf. IEEE Eng. Med. Biol.* **1989**, 1461–1463.

34 LU, L., TAMURA, T., TOGAWA, T., Detection of body movements during sleep by monitoring bed temperature. *Physiol. Meas.* **1999**, *20*, 137–148.

35 GARDNER, J. W., BARTLET, P. N., Sensors and sensory systems for an electronic nose. NATO ASI Series E, vol. 212. Kluwer, Dordrecht **1992**.

36 SPANEL, P., ROLFE, P., RAJAN, B., SMITH, D., The selected ion flow tube (SIFT) – A novel technique for biological monitoring. *Ann. Occup. Hyg.* **1996**, *40*, 615–626.

37 OYABU, T., NANTO, H., KASAHARA, H., ONODERA, T., NAKATA, N., Odor sensing characteristics in residential space using metal-oxide sensors. *Sensor and Materials* **1999**, *11*, 457–468.

Index